Offshore Structures

Design, Construction and Maintenance

Offshore Structures

Design, Construction and Maintenance

Mohamed A. El-Reedy, Ph.D.
Consultant Engineer

ELSEVIER

AMSTERDAM • BOSTON • HEIDELBERG • LONDON
NEW YORK • OXFORD • PARIS • SAN DIEGO
SAN FRANCISCO • SINGAPORE • SYDNEY • TOKYO

Gulf Professional Publishing is an imprint of Elsevier

G|P
P|

Gulf Professional Publishing is an imprint of Elsevier
225 Wyman Street, Waltham, MA 02451, USA
The Boulevard, Langford Lane, Kidlington, Oxford, OX5 1GB, UK

Library of Congress Cataloging-in-Publication Data
El-Reedy, Mohamed A. (Mohamed Abdallah)
Offshore structures : design, construction and maintenance / Mohamed A. El-Reedy.
p. cm.
ISBN 978-0-12-385475-9
1. Offshore structures. I. Title.
TC1665.E517 2012
627'.98–dc23
2011038667

British Library Cataloguing-in-Publication Data
A catalogue record for this book is available from the British Library.

For information on all Gulf Professional Publishing publications visit our Web site at *http://store.elsevier.com*

Typeset by: diacriTech, Chennai, India

This book is dedicated to the spirits of my mother and my father, and to my wife and my children Maey, Hisham and Mayar.

Contents

Preface

When a structural engineer starts work on offshore structure design, construction or maintenance, the offshore structure may appear to be a black box to him. Most engineering faculties, especially those in structural or civil engineering, focus on the design of residential, administrative, hospital and other domestic buildings from concrete or steel, while other faculty focus on harbor design.

The design of offshore structure platforms is a combination of steel structure design methods and loads applied in harbors, such as waves, current and other parameters. On the other hand, offshore platform design depends on technical practice, which depends on the experience of the engineering company itself.

While the construction of steel structures is familiar to the structural engineer, as anyone can observe construction of a new steel building, the construction and installation of an offshore structure platform are very rarely seen unless one has a direct role in the project, especially because the installation will be in the sea or ocean. There are far fewer offshore structures world-wide than there are steel structures for normal buildings on land, and the major design guidance for offshore structures lies in research and development, which is growing very fast to keep pace with development in the global oil and gas business. Therefore, all the major oil and gas exploration and production companies support and sponsor research to enhance the design and reliability of offshore structures, in order to improve revenues from their petroleum projects and their assets.

This book aims to cover the design, construction and maintenance of offshore platforms in detail, with comprehensive focus on the critical issues in design that the designer usually faces. The book also provides the simplest design tools, based on the most popular codes (such as API and ISO) and the other technical standards and practices that are usually used in offshore structure design. In addition, it is important to focus on methods for controlling and reviewing the design that most engineers will face in the review cycle, so this book covers the whole range of the offshore structure engineer's activities.

Corrosion of offshore structure platforms costs a lot of money to control and to maintain within the allowance limit so that it will not affect the structure's integrity, so methods for designing and selecting a suitable cathodic protection system and the advanced methods of protecting the structure from corrosion are very important to the structural engineer, and they are considered in depth in this text.

An offshore structure platform is a considerable asset in the oil and gas industry, so another goal of this book is to assist the structural engineer in making decisions in design that take into consideration the factors, parameters and constraints faced by the owner that control the options and alternatives in the engineering studies phase.

Furthermore, it is very important to the owner, engineering firm and contractor that the offshore project's lifespan be identified. In other words, the structural engineer should have an overview of the relation between the structure's system and its configuration from both an economic and an engineering point of view.

Most offshore structure platforms were constructed world-wide in the period of growing oil investment between 1970 and 1980, so these platforms now are over 40 years old. Consequently, a lot of mature offshore structures are going through rehabilitation designed to increase and maintain their structural reliability. Development of the integrity management system with up-to-date and advanced techniques for qualitative and quantitative risk assessment has been essential to the risk-based inspection and maintenance planning that enhance the reliability of platforms during their lifespan. Accordingly, this book provides advanced techniques for topside and underwater inspection and assessment of offshore structure platforms, as well as ways to implement the maintenance and rehabilitation plan for the platform that match business requirements.

It is also important to present case studies of repair and strengthening of platforms and the methods of decommissioning platforms when required.

This book is intended to be a guidebook for junior and senior engineers who work in design, construction, repair and maintenance of fixed offshore structure platforms.

The text serves as an overview of, and a practical guide to, traditional and advanced techniques in design, construction, installation, inspection and rehabilitation of fixed offshore structure platforms, along with the principles of repairing and strengthening the structures and the methodology for delivering a maintenance plan for the fleet of platforms.

Mohamed Abdallah El-Reedy
elreedyma@gmail.com
Cairo, Egypt

The Author

Mohamed A. El-Reedy's background is in structural engineering. His main area of research is the reliability of concrete and steel structures. He has provided consulting to different engineering companies and oil and gas industries in Egypt and to international companies, such as the International Egyptian Oil Company (IEOC) and British Petroleum (BP). Moreover, he provides different concrete and steel structure design packages for residential buildings, warehouses, and telecommunication towers and electrical projects with WorleyParsons Egypt. He has participated in Liquefied Natural Gas (LNG) and Natural Gas Liquid (NGL) projects with international engineering firms. Currently, Dr. El-Reedy is responsible for reliability, inspection, and maintenance strategy for onshore concrete structures and offshore steel structure platforms. He has performed these tasks for hundreds of structures in the Gulf of Suez in the Red Sea.

Dr. El-Reedy has consulted with and trained executives at many organizations, including the Arabian American Oil Company (ARAMCO), BP, Apachi, Abu Dhabi Marine Operating Company (ADMA), the Abu Dhabi National Oil Company, King Saud's Interior Ministry, Qatar Telecom, the Egyptian General Petroleum Corporation, Saudi Arabia Basic Industries Corporation (SABIC), the Kuwait Petroleum Corporation, and Qatar Petrochemical Company (QAPCO). He has taught technical courses about repair and maintenance for reinforced concrete structures and about advanced materials in the concrete industry world-wide, especially in the Middle East.

Dr. El-Reedy has written numerous publications and has presented many papers at local and international conferences sponsored by the American Society of Civil Engineers, the American Society of Mechanical Engineers, the American Concrete Institute, the American Society for Testing and Materials, and the American Petroleum Institute. He has published many research papers in international technical journals and has authored four books about total quality management, quality management and quality assurance, economic management for engineering projects, and repair and protection of reinforced concrete structures. He received his bachelor's degree from Cairo University in 1990, his master's degree in 1995, and his Ph.D. from Cairo University in 2000.

Chapter 1

Introduction to Offshore Structures

1.1 INTRODUCTION

Offshore structures have special economic and technical characteristics. Economically, offshore structures are dependent on oil and gas production, which is directly related to global investment, which is in turn affected by the price of oil. For example, in 2008 oil prices increased worldwide, and as a result many offshore structure projects were started during that time period.

Technically, offshore structure platform design and construction are a hybrid of steel structure design and harbor design and construction.

Only a limited number of faculty of engineering focus on offshore structural engineering, including the design of fixed offshore platforms, floating or other types, and, perhaps due to the limited number of offshore structural projects in comparison to the number of normal steel structural projects, such as residential facilities and factories. In addition, offshore steel structure construction depends on continuous research and study drawn from around the world.

All the major multinational companies that work in the oil and gas business are interested in offshore structures. These companies provide continuous support for research and development that will enhance the ability of their engineering firms and construction contractors to support their business needs.

1.2 HISTORY OF OFFSHORE STRUCTURES

As early as 1909–1910, wells were being drilled in Louisiana. Wooden derricks were erected on hastily built wooden platforms that had been constructed on top of timber piles.

Over the past 40 years, two major types of fixed platforms have been developed: the steel template, which was pioneered in the Gulf of Mexico (GoM), and the concrete gravity type, first developed in the North Sea. Recently, a third type, the tension-leg platform, has been used to drill wells and develop gas projects in deep water. In 1976, Exxon installed a platform in the Santa Barbara, CA, channel at a water depth of 259 m (850 ft). Approximately two decades earlier, around 1950, while the developments were taking place in the GoM and Santa Barbara channel, the BP (British Petroleum) company was engaged in a similar exploration off the coast of Abu Dhabi in the Persian Gulf. The water depth there is less than 30 m (100 ft) and the operation has grown steadily over the years.

The three basic design requirements for a fixed offshore platform are:

1. The ability to withstand all loads expected during fabrication, transportation, and installation.
2. The ability to withstand loads resulting from severe storms and earthquakes.
3. The ability to function safely as a combined drilling, production, and housing facility.

The importance of the second requirement, and the need to reevaluate platform design criteria, was highlighted in the 1960s, when hurricanes caused serious damage to platforms in the GoM. In 1964, hurricane Hilda, with wave heights of 13 m and wind gusts up to 89 m/s, destroyed 13 platforms. The next year, hurricane Betsy destroyed three platforms and damaged many others. Because Hilda and Betsy were "100-year hurricanes," designers abandoned the use of "25- and 50-year storms" and began designing for the more destructive 100-year storms.

1.3 OVERVIEW OF FIELD DEVELOPMENT

Estimates of global oil reserves, based on geological and geophysical studies and oil and gas discoveries as of January 1996, indicate that about 53% of the reserves are in the Middle East, a politically troubled region. Overall, 60% of reserves are controlled by the Organization of Petroleum Exporting Countries (OPEC). Obviously, OPEC and the Middle East are very important for the world's current energy needs.

Most researchers believe that the major land-based hydrocarbon reserves have already been discovered and that most significant future discoveries will be in offshore areas, the Arctic and other difficult-to-reach areas of the world.

Geological research indicates why North America, northwest Europe and the coastal areas of West Africa and eastern South America appear to have similar potential for deepwater production. During very early geological history, sediments were deposited in basins with restricted circulation and were later converted to the supersource rocks found in the coastal regions of these areas. The presence of these geological formations is the initial indication of the presence of hydrocarbons, but, before feasible alternatives for producing oil and gas from a field are identified and the most desirable production scheme is selected, exploratory work defining the reservoir characteristics has to be completed. First, geologists and geophysicists assess the location's geological formations to determine if it has potential hydrocarbon reserves.

After the geologists and geophysicists decide that a field could be economically viable, further exploratory activities are undertaken to prepare cost, schedule and financial return estimates for selected exploration and production schemes. After that, the various alternative schemes are compared and the most beneficial one is identified.

During this phase, due to the absence of detailed information about reservoir characteristics, future market conditions and field-development alternatives, experts make judgments based on their past experience and on cost and schedule estimates based on data available from previous history. The success of oil and gas companies depends on this expertise, so most companies keep experts on hand and compete with each other to recruit them. Sometimes, the experiential data are not enough, so decisions are made as a result of brain-storming sessions attended by experts and management, and these are greatly affected by a company's culture and past experiences.

The reservoir management plan is affected by the characteristics of the fluid the reservoir produces, the reservoir's size and topography, regional politics, company and partner culture and the economics of the entire field-development scheme. Well system and completion design are affected by the same factors that affect the reservoir management plan, except perhaps the political factors. Platforms, facilities for processing and production, storage systems and export systems are affected by all these factors as well.

The field-development scheme has to take into account:

- Reservoir characteristics
- Production composition (e.g., oil, gas, water, H_2S)
- Reservoir uncertainty
- Environment (e.g., water depth)
- Regional development status
- Technologies available locally
- Politics
- Partners
- Company culture
- Schedule
- Equipment
- Construction facilities
- Market
- Economics

If the preliminary economic indicators in the feasibility study phase are positive, seismic data generation and evaluation, done by geophysicists, follow. These data comprise reasonable information about the reservoir's characteristics, such as its depth, spread, faults, domes and other factors, and an approximate estimate of the recoverable reserves of hydrocarbons.

If the seismic indications are positive and the decision is to explore further, exploratory drilling commences. Depending on water depth, the environment and what's available, an appropriate exploration scheme is selected. A jack-up exploratory unit is suitable for shallow water depths. In water depths exceeding 120 m (400 ft), ships or semisubmersible drilling units are utilized.

At depths of 300 m (1000 ft), floating drilling units require special mooring arrangements or a dynamic positioning system. A floating semisubmersible drilling rig is capable of operating in water as deep as 900–1200 m (3000–4000 ft).

Exploratory drilling work follows the discovery well. This generally requires three to six wells drilled at selected points of a reservoir. These activities and production testing of the wells where oil and gas are encountered give reasonably detailed information about the size, depth, extent and topography of a reservoir, such as the fault lines, impermeable layers, etc., and its recoverable reserves, viscosity (API grade), liquid properties (e.g., the oil/water ratio), and impurities, such as sulfur or another critical component.

Reservoir information enables geologists and geophysicists to estimate the location and number of wells that will be required to produce a field and the volumes of oil, gas and water production. This information is used to determine the type of production equipment, facilities and the transport system needed to produce the field.

Obviously, the accuracy of reservoir data has a major effect on the selection of a field-development concept. In marginal or complex reservoirs, reliable reservoir data and the flexibility of the production system in accommodating changes from the reservoir appraisal are very desirable.

1.3.1 Field-Development Cost

Field development for a new project or for extending existing facilities is a multistep process. The first step is gathering input parameters, such as the reservoir and environmental data; the selection and design of major system components, such as the production drilling and the wells, facilities and offtake system; and the decision criteria, such as the economics. The next step is evaluating the different field-development options that satisfy the input requirements and establishing their relative merits with respect to the decision criteria. In this design process loop, not only alternatives for field-development systems, but also alternatives for each system, need to be taken into account.

At the next stage, a preliminary design for the selected system is started. In this phase, the selection activity is focused on the system components and detail elements. During this phase, design iterations are generated until all the members of all engineering and operation disciplines are satisfied from a technical point of view. All the system components and construction activities must be well defined. Once the design is complete, few changes to the system and its components can be made without suffering delays and cost overruns.

The operation phase includes maintenance, production, repair and reassessment and transportation activities. Viable field-development options are identified and developed and selection of the most suitable option occurs in parallel with the acquire, explore and appraise cycle. All project activities that precede the start of the basic design phase are called the FEED (front-end engineering design) phase. FEED is the most important phase of a field-development timeline. An ideal field-development schedule should allow sufficient lead-time for performance of all FEED work before basic design starts.

In the FEED phase, after the facilities required on the topside of the deck have been defined, the geometrical shape of the platform is defined. A preliminary structure analysis is run through structure analysis software to identify the section that matches the loads and the geometrical shape with appropriate deck dimensions that serve the required facilities on the deck. The structure system of the platform subsea structure, which is called a jacket, and a construction method based on the water depth are selected in this phase. In management of a new project, this is called the select phase.

Experience shows that the FEED phase will identify viable options; develop, evaluate and select concepts; and provide a conceptual design. The FEED phase usually consumes only about 2–3% of the total installed cost (TIC) of field development and has the highest impact on cost, schedule, quality and success. It is not uncommon to observe major cost overruns when a full FEED phase is not performed. Reanalysis of projects that did not have a satisfactory FEED phase because of political factors indicates that a 50% TIC reduction could have been achieved if a satisfactory FEED phase had been performed.

Figure 1.1 shows the variation in the accuracy of the TIC estimates in different phases of a project. In general, understanding of the economics and other features of a field-development system improve as we move along the field-development timeline. At the start of the FEED phase, a number of options are available and identifying the right field-development concept will profoundly influence a project's success. During the conceptual design phase of FEED, each system component, such as well systems, platform(s), topside facilities, transportation and their subcomponents, such as hull, the mooring system, tethers, living quarters, processes, utility systems, pipelines, storage and risers, are defined and a cost and schedule estimate is prepared. Selection and definition of the system components and subcomponents also have a significant impact on costs and/or the schedule. At this phase, the accuracy of the TIC estimates is approximately ±15% to ±25%.

Experience shows that a standard project with routine components and almost no innovations will experience very low operability problems. A recent survey by Karsan (2000) of offshore projects indicated that 90% of projects with substantial innovations had major operability problems; therefore, care must be taken not to introduce low value when adding new ideas and components into a field-development system.

The preliminary or basic design phase includes a definition of the process through process flow diagrams (PFDs) and preparation of the field and equipment layouts: piping and instrumentation diagrams (P&IDs), general platform drawings, materials and equipment lists, data sheets, specifications and a final engineering, procurement and construction (EPC) cost and schedule

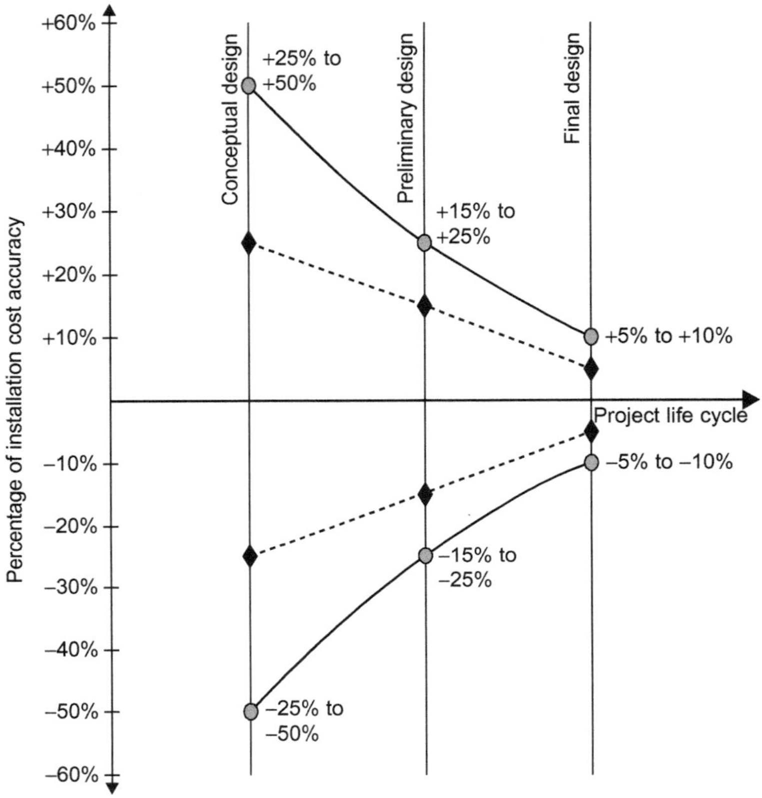

FIGURE 1.1 Accuracy of the total installation cost (TIC) estimates at each project stage.

estimate. The basic design phase allows some system optimization but only at a subcomponent and specification level. At this phase, the accuracy of TIC estimates is approximately ±15% to ±25%.

The ability to influence cost and savings decreases with progress along the field-development timeline. At the concept-development stage, the selection and development of the right concept have a major impact on the TIC. Savings in the detailed design and construction phases generally stem from good project control and execution, which result in TIC optimization. This phase provides detailed engineering analysis and design, approved-for-construction (AFC) drawings and fabrication, transportation, installation, precommissioning and the hook-up and commissioning of the field by an EPC contractor. Efficient project management in the execution of the plan; cost, schedule and quality control; verification; quality and safety assurance; and purchasing and documentation have some effect on the TIC, but not as profoundly as the FEED phase. Past experience indicates that at the start of the construction phase, the accuracy of the TIC estimates is approximately ±5% to ±10%.

In most cases, the owner is mainly focused on the early production phase. As an example, in one case, after the drilling and completion of 18 wells from a satellite well protector, a production rate of 100,000 bpd (barrels per day) is reached immediately after the installation of the production platform. After one year from the production platform installation, all 36 wells are completed and a production rate of 200,000 bpd is reached.

In another example, in the North Sea, after installation of a single self-contained tower, a period of about three months has to elapse before a production rate of 6000 bpd may be achieved, and about two and a half years after platform installation are required to drill and complete all 36 wells, reaching a production rate of 200,000 bpd.

For the self-contained tower, the highest negative cash flow occurs within the second quarter of the third year, when some oil production starts. From this point on, positive cash flow from the produced oil will start offsetting the negative cash flow from early investments and operating costs. The zero cumulative cash flow position is reached within the fourth quarter of the fourth year, when all the field investments to date have been paid off.

The maximum negative cash flow for the multiplatform concept is reached six months ahead of that for the self-contained platform, sometime within the first quarter of the third year, when oil flow from the production platform starts. Due to the heavy upfront investment on a satellite platform and the early drilling program, the maximum cash invested will be about 30% more than that for the self-contained platform. However, rapid cash recovery from early drilled production wells rapidly offsets the negative cash flow. Within the first quarter of the fourth year, the zero cumulative cash flow point will be reached, about six months ahead of the self-contained platform concept. From this point on, the multiplatform concept results in higher cumulative cash flow.

For increased gas production in later years, the addition of a gas treatment and compression platform may also be planned. In this case, satellite drilling platforms may be constructed and installed within a six-month timeframe to allow early drilling to commence. Within three years, production starts and the additional drilling platform is installed. Once a significant number of the satellite platform wells are drilled and completed, a large volume of production can start.

A two- to three-year drilling cycle for the field-development program generally follows a self-contained platform installation. In some cases, some of the wells may be drilled before the tower is set in location or a successful exploratory well may be utilized, resulting in some early production. But, in general, about three to four years have to elapse after platform installation before the total field production rate can be reached. This concept generally allows reaching top production rates about a year or so earlier. The upfront cash available for investment may vary from one oil company to another. If plenty of cash is available from other operations, and some tax hedging is desired, an early high cash investment option may be preferred. However, if the company is cash starved or a higher corporate cash discount rate must be imposed due to many other competing investment options or interest rates, a lower upfront cash investment option may be preferred.

Of course, this is a somewhat simplistic presentation of the economic factors affecting the platform concept selection. Many other economic factors other than the cash flow, including the tax and discount rates, inflation and the time value of money, also need to be considered, resulting in complex net-present-value calculations. It is worth mentioning that investment analysis specialists generally perform these calculations.

1.3.2 Multicriteria Concept Selection

At the FEED stage, external factors, such as country requirements and characteristics, technology transfer and environmental pollution potential, as well as the culture, politics, economics and infrastructure of the host nation and the operating oil company and its partners, may have major influences on the concept selection. Not easily comparable criteria, such as the economics, design completeness and maturity and external factors, have to be weighed against each other and used for concept ranking and selection. In a multicriteria process, first the goal of the exercise is defined. Then the viable field-development options are identified. This is followed by the identification of a multitude of selection criteria that are grouped and ordered in a hierarchy, after which experts determine the importance of each criterion by comparing it to the others in a pairwise manner. The comparisons are then passed through an analytical process to obtain rankings for each comparison and the alternatives. Currently, there are several such processes in use.

The basic design defines the platform, production facility and structural configurations and dimensions in enough detail to allow the detailed design to start. Basic design results enable reliable cost and schedule estimates and the ordering of long-lead major equipment and structural components. This also allows the contractor to provide a reliable lump-sum price bid for a detailed EPC contract for the platform.

The basic design phase includes the following tasks:

- A well-defined field-development plan.
- A conceptual design based on field characteristics, operational and environmental parameters, foundation conditions, platform configurations, global materials selection and other information and assumptions used for concept development.
- Conceptual drawings showing major component configurations for platforms, topside facilities layouts, well locations and well systems, reservoir maps and production profiles, storage (if needed) and offloading systems, pipelines to shore and preliminary sizes.

- A general platform structure configuration, as defined by conceptual drawings that show side elevations and plans for legs and major bracing. Preliminary PFDs and major equipment lists may also be available.
- A concept cost estimate (±40%) and schedule for the entire development plan (including the capital and operational expenses, cash flow diagrams and net present value (NPV) of the total investment).

A conceptual design package that includes this information is prepared by the owner or specialized engineering team (this called the appraise phase) and is given to the design contractor as input to the preliminary design phase to start the FEED stage.

1.4 FEED REQUIREMENTS

The FEED process ends with the completion of the conceptual design. At this point, the FEED phase has provided the following deliverables:

- Basic design drawings for all major platform and deck structures and components (i.e., jacket, deck, piles and conductors). These should contain enough detail to enable reliable field-development cost and schedule estimates. This information is particularly important if owners wish to solicit bids and to enter into a lump-sum EPC agreement with a contractor.
- A basis of detailed design (BOD) document for the detailed design phase. The BOD defines the detailed design requirements, including the platform configuration and environmental parameters (e.g., metocean, seismic, ice, etc.).
- Site-specific information (e.g., water depth, temperature, soil characteristics, mudslides, shallow gas pockets, etc.).
- Definition of nongenerated loads (e.g., equipment and wet/dry supplies and operating loads, such as dynamic vibrations from rotating machinery, mud pumps and operations, etc.). Design life for the structure should be defined considering the operation requirement and fatigue effect along its lifetime.
- Definition of accidental loads (e.g., collisions with boats, dropped objects, fire and explosions).
- Load combinations (extreme environmental, operational, serviceability, transportation, lift and launch). Damage stability and/or redundancy requirements (e.g., missing member or flooded leg or compartment).
- Preferred material classes.
- Design regulations, codes and recommended practices.
- Definition of appurtenances and their locations (e.g., escape and evacuation equipment, escape routes, stairs, boat landings, barge bumpers, conductors, mud-mats, etc.) with different discipline engineers.
- Corrosion protection requirements (e.g., sacrificial anode or impressed current, and types of anodes).
- A narrative of the construction methods and procedures that will affect the platform configuration and size (e.g., skidding and load-out procedures,

pulling points, lifting eyes, launch skids, etc.) and the verification and certification requirements.
- Any other owner requirements that will affect the detailed design (e.g., jack-up drilling unit clearances, tender rig sizes and weights).

If a contractor has not already been selected, a detailed engineering bid document is prepared. This document, which contains most of the information just listed, is issued to all qualified bidders. It is very important to note that most contractors will not be interested in bidding a lump-sum EPC contract if a basic design package is not available.

The following tasks should also be completed in this phase:

- Preparation of detailed and final process flow diagrams (PFDs). If a detailed design is attempted without final PFDs, process changes during the final design phase may cause significant rework and schedule and cost overruns.
- Preparation of P&IDs.
- Preparation of final deck and facilities layouts, providing adequate space and clearance for all equipment and operations.
- Preparation of equipment and material lists, data sheets and specifications.
- Preparation of detailed engineering, transportation, precommissioning and commissioning scopes of work.
- Preparation of detailed design schedules and cost estimates.
- Performance of global in-place analyses to confirm that major structure members and equipment will fit.

1.5 TYPES OF OFFSHORE PLATFORMS

The types of fixed offshore platforms are:

- Drilling/well-protector platforms
- Tender platforms
- Self-contained platforms (template and tower)
- Production platforms
- Quarters platforms
- Flare jacket and flare tower platforms
- Auxiliary platforms
- Bridges
- Heliports

Each of these platform types has its own unique characteristics from a functionality point of view.

Drilling/Well-protector Platforms

Oil and gas wells are drilled from this platform, so the rig will approach this platform to drill new wells or to perform any work over the life of the platform. Platforms built to protect the risers on producing wells in shallow water are called well protectors or well jackets. Usually a well jacket serves from one to four wells.

Tender Platforms

Tender platforms are not used as commonly now as they were 40 years ago. Tender platforms function as the drilling platform, but the drilling equipment rests on the the platform topside. Nowadays it is common to use a jack-up drilling rig, which does not rest on the platform deck.

With tender platforms, the derrick and substructure, drilling mud, primary power supply and mud pumps are placed on the platform.

As mentioned, since tender platforms are not used much anymore, they will not be encountered in new designs or new projects, but they may be encountered in existing structures, so you must be familiar with them in order to perform assessments for an existing drilling platform.

Figure 1.2 shows a tender platform. It is shown with two beams along the platform deck length; the beams are used as a railway to the tender tower above the deck for drilling activity.

Self-contained Platforms

The self-contained platform is large, usually with multiple decks that have adequate strength and space to support the entire drilling rig with its auxiliary equipment and crew quarters, and enough supplies and materials to last through the longest anticipated period of bad weather when supplies cannot be brought in. There are two types of self-contained platform: the template type and the tower type.

FIGURE 1.2 Tender platform.

Production Platforms

Production platforms support control rooms, compressors, storage tanks, treating equipment and other facilities. Figure 1.3 shows a production platform carrying the separators and other facilities for production purposes.

Quarters Platforms

The living accommodations platform for offshore workmen is commonly called a quarters platform.

Flare Jacket and Flare Tower Platforms

A flare jacket is a tubular steel truss structure that extends from the mud line to approximately 3–4.2 m (10–13 ft) above the mean water line (MWL). It is secured to the bottom by driving tubular piles through its three legs.

Auxiliary Platforms

Sometimes small platforms are built adjacent to larger platforms to increase available space or to permit the carrying of heavier equipment loads on the principal platforms. Such auxiliary platforms have been used for pumping or compressor stations, oil storage, quarters platforms or production platforms.

FIGURE 1.3 Production platform.

Sometimes they are free standing, and other times they are connected by being braced to the older structure.

Bridges

A bridge 30–49 m (100–160 ft) in length that connects two neighboring offshore structures is called a catwalk. The catwalk supports pipelines, pedestrian movement or materials handling. The different geometries of bridges are shown in Figure 1.4, and Figure 1.5 is a photo of a bridge between two platforms.

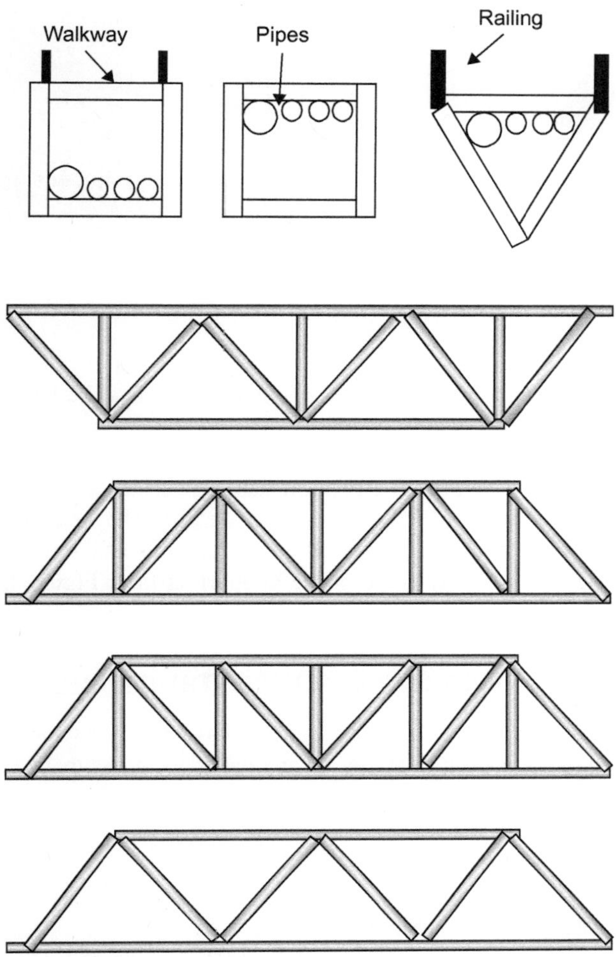

FIGURE 1.4 Different bridge geometries.

FIGURE 1.5 Bridges connecting a helideck platform.

Heliport

The heliport is the landing area for a helicopter, so it must be large enough to handle loading and unloading operations.

A square heliport has a side length of one and a half to two times the largest helicopter's expected length. The heliport landing surface should be designed for a concentrated load of 75% of the gross weight. The impact load is two times the gross weight for the largest helicopter, and this load must be sustained in an area of 24″ × 24″ anywhere in the heliport surface. Figure 1.5 shows a heliport platform.

1.6 DIFFERENT TYPES OF OFFSHORE STRUCTURES

Different types of offshore structural systems have been developed over time due to the requirements for obtaining oil and gas in locations that have a greater water depth. These types of platforms are as follows.

Concrete Gravity Platform

The concrete gravity platform shown in Figure 1.6 is a concrete platform that was constructed in 1997 for Shell.

Figure 1.7 illustrates a complex platform that consists of a production and drilling platform connected by bridges.

FIGURE 1.6 Concrete gravity platform.

In areas where there is a low oil reserve, only one well will be drilled. Many alternatives were devised to address this situation and to obtain the business target. One solution is to have a subsea well that is connected to the nearest platform by a pipeline. This solution is costly, but it is now used widely in deep water.

Another solution is to use a minimal offshore structure, as is shown in elevation and plan views in Figure 1.8(a) and (b). The concept for this platform is to use the conductor itself as the main support for the small deck. There are also two diagonal pipes that are connected to the soil by two piles, as shown in the elevation view (Figure 1.8(a)). Figure 1.9 shows the shape of the topside of this three-legged platform.

Floating Production, Storage and Offloading

The first floating production, storage and offloading (FPSO) platform was the Shell Castellon, built in Spain in 1977. The first conversion of an LNG (liquefied natural gas) carrier (Golar LNG owned the Moss-type LNG carrier) into an LNG floating storage and regasification unit was carried out in 2007 by the Keppel Shipyard in Singapore. In the last few years, concepts for LNG FPSOs have also been launched. An LNG FPSO works under the same principles as an oil FPSO, but it produces only natural gas, condensate and/or liquefied petroleum gas (LPG), which is stored and offloaded.

FIGURE 1.7 Complex platform consisting of production and drilling platforms connected by bridges.

FPSO vessels are particularly effective in remote or deepwater locations where seabed pipelines are not cost effective. FPSOs eliminate the need to lay expensive long-distance pipelines from the oil well to an onshore terminal. They can also be used economically in smaller oil fields that can be exhausted in a few years and do not justify the expense of installing a fixed oil platform. Once the field is depleted, the FPSO can be moved to a new location. In areas of the world subject to cyclones (such as northwest Australia) or icebergs (Canada), some FPSOs are able to release their mooring/riser turret and steam away to safety in an emergency. The turret sinks beneath the waves and can be reconnected later.

The FPSO operating at the deepest water depth is the FPSO Espirito Santo of Shell America; it is operated offshore by SBM Offshore N.V. The FPSO is moored in water 1800 m deep in the Campos Basin in Brazil and is rated for 100,000 bpd. The EPC contract was awarded in November 2006 and was

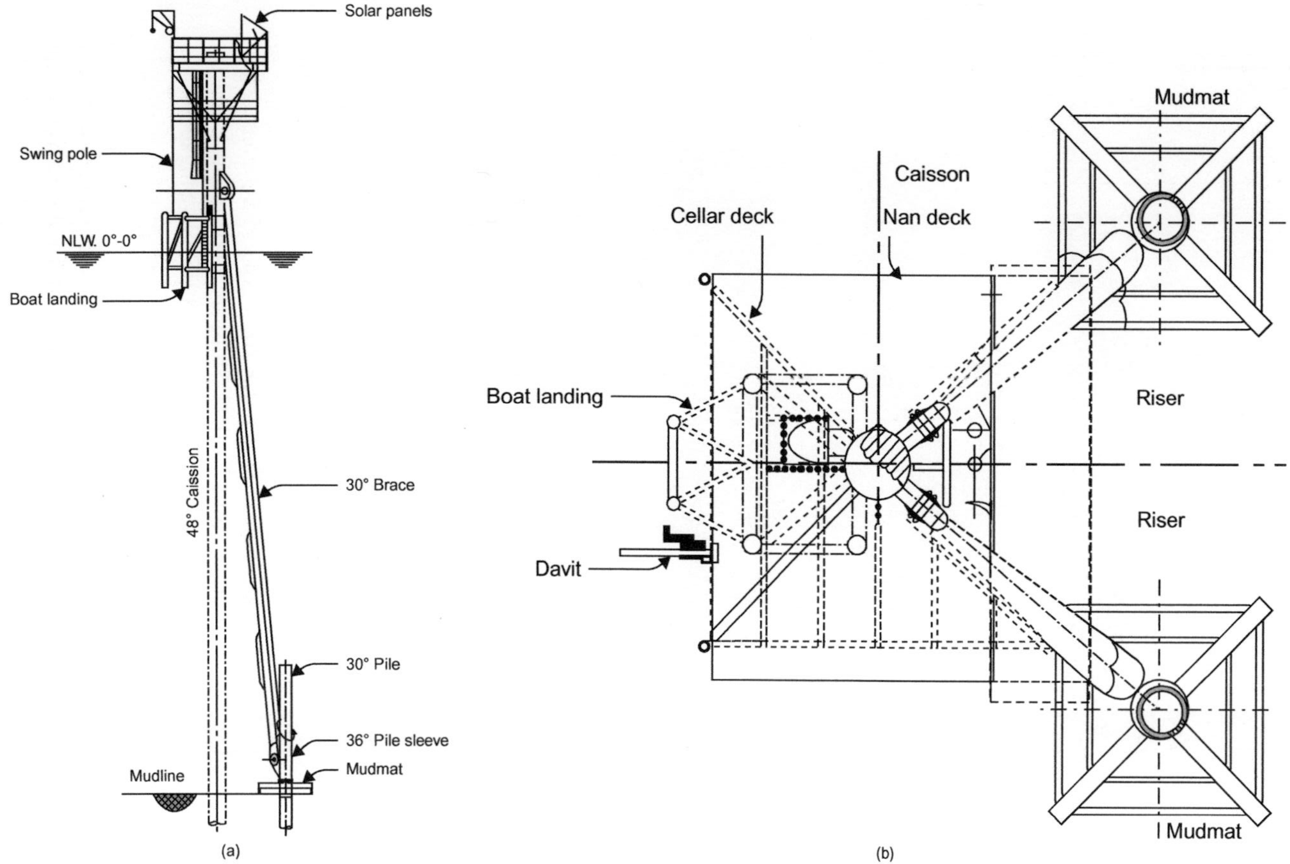

FIGURE 1.8 Three-legged platform elevation view (a) and plan view (b).

FIGURE 1.9 Photograph of three-legged platform.

scheduled for first oil production in December 2008. The FPSO conversions and internal turret were done at the Keppel Shipyard in Singapore and the topsides were fabricated in modules at Dyna-Mac and BTE in Singapore.

Tension-Leg Platform

Nowadays there is a trend to use gas that in the previous 40 years would have been burned off in the air, so there are many projects aimed at discovering gas for production, and gas exploration has been extended into deep water. The conventional fixed offshore structure cannot be used, so researchers and engineering firms have used the legs of tension wire platforms in deep water.

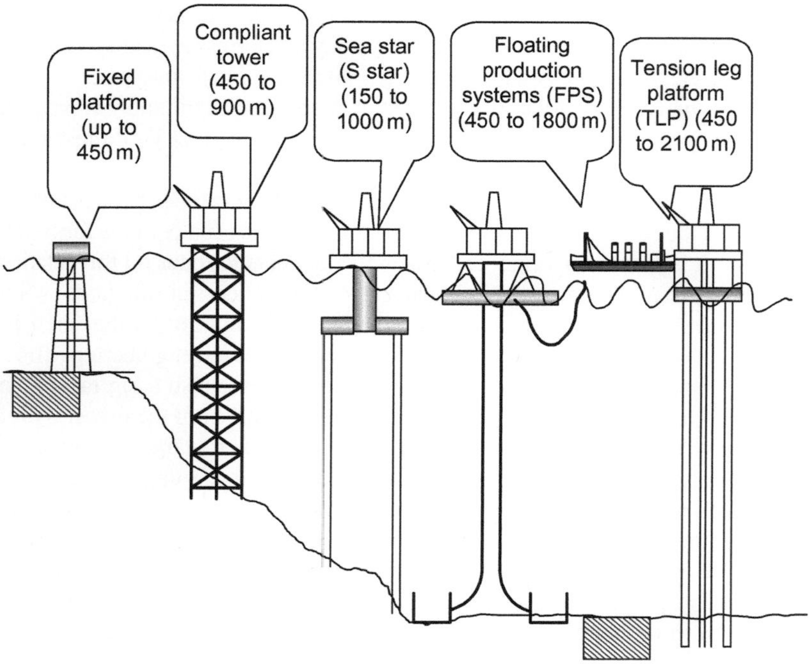

FIGURE 1.10 Different types of offshore structures.

A tension-leg platform is a vertically moored floating structure normally used for the offshore production of oil or gas, and it is particularly suited for water depths greater than 300 m (about 1000 ft).

The first tension-leg platform was built for Conoco's Hutton field in the North Sea in the early 1980s. The hull was built in the dry-dock at Highland Fabricator's Nigg yard in the north of Scotland, with the deck section built nearby at McDermott's yard at Ardersier. The two parts were mated in the Moray Firth in 1984. The Magnolia extended tension-leg platform was constructed for a water depth of over 1425 m (~4700 ft).

Figure 1.10 summarizes the different types of platform structures, their water-depth range and their functions. Note that water-depth ranges change over time, as new research and development allow construction in deeper water.

1.7 MINIMAL OFFSHORE STRUCTURE

The minimal offshore structure was created about 25 years ago, in response to demand for production companies to investigate low-volume reservoirs. The minimal structure consists of one conductor for oil production that is also used as a pile and on the other hand, a lot of platforms constructed over the years are minimal offshore structures, formed by one pile and using the conductor itself as a support, two additional inclined members as a support for the pile. Furthermore, in the last 20 years many projects have used concrete gravity platforms fabricated from reinforced concrete, and there is a lot of research into these platform types.

Besides the types of offshore structures that are used in oil and gas projects, we need to focus on the economics and policy decisions that guide and direct the projects.

Worldwide, the relationship between multinational companies in the petroleum industry and the countries that have oil and gas reserves is usually in the form of an agreement. It is just as important for structural engineers to focus on this relationship as on stresses, strain, structural analysis, codes and design standards, which are the main elements of the structural engineer's job, but not all that their job entails. Creation of a structural configuration that matches the engineering office's knowledge and capability with the owner's expectations is also a big factor. The contractors, the engineering firm and the engineering staff from the owner's organization have to be on the same page to achieve the owner's target and goals.

The organization's target and goals are based on business targets and profits, the expected oil and gas reserve, expected oil and gas prices and the last important factor, the country that owns the land and the reserves. Therefore, the terms and conditions of, and the political situation in, the country will directly affect investment in the project. Consequently, any engineer working on the project should keep in mind an overview of all the constraints on the project, because these constraints affect the engineering solutions, options and alternative designs. In fact, this overview is very important knowledge for all staff, from senior management down to the junior level.

1.8 PREVIEW OF THIS BOOK

Load calculation is the first step in a structural analysis. Loads on platforms are usually defined based on the owner's specifications. The studies that govern the loads and other technical practices are discussed in detail in Chapter 2.

In order to design a steel structural platform, one must understand the principles of steel structure design, the effects of environmental and other loads on offshore structures and how designs are governed by API and ISO standards. These concepts are discussed in Chapter 3.

The foundation is very critical to the safety of the structure, but the foundation is subject to many variations and uncertainties. Therefore, Chapter 4 presents methods for assessing the pile capacity for offshore structures on sand and clay soil. All the tools and data required to estimate pile capacity are illustrated.

The construction of an offshore structure is very costly. So the fabrication, erection, launch and installation of the platform are discussed in detail in Chapter 5.

Corrosion protection for an offshore structure is very critical, and an appropriate system needs to be chosen for both existing structures and retrofitting projects. There are different types of cathodic protection, and design of the sacrificial anode and cathodic protection (CP) system is presented in Chapter 6. The structural engineers who are responsible for a design usually don't have much knowledge of the corrosion process and its effects and ways of protecting an offshore structure's structural integrity and maintaining the structure's reliability. Yet, structural engineers need to understand corrosion protection in order to select the optimal corrosion protection system. Knowledge of up-to-date corrosion protection is also very important to the corrosion engineer, who is responsible for designing the protection system, and to the construction group that will do the system installation. The maintenance engineer should know the pitfalls of each system and, from reading the CP system during inspection, should have a sense of the structure's condition from a corrosion point of view.

Oil and gas platforms are facing the problem of aging, as many platforms were constructed over 40 years ago. API standards for fixed offshore structure platform design were started in 1969, so all the old platforms were designed according to the engineering office's experience.

It is worth noting that, in years past, companies used paper copies of reports and drawings, a lot of which have been destroyed over time. In addition, in most cases there wasn't a change policy, so extant drawings may not match actual conditions. All these factors affect many rehabilitation projects worldwide, so Chapter 7 discusses the assessment and evaluation of offshore structures. Recently, risk-based inspection has been used to establish maintenance plans for existing platforms, so qualitative risk assessment is discussed in detail in Chapter 8.

BIBLIOGRAPHY

Earthly Issues. Gulf of Mexico Deep Water Horizon Oil Spill. http://www.earthlyissues.com/gulfspill.htm.

International Organization for Standardization, Petroleum and Natural Gas Industries, Offshore Structures, Part 1. Fixed Steel Offshore Structures, ISO/DIS 19902:2004.

Demire I. Karasan, 2000. General Design of Fixed Offshore Structure, University of Texas at Austin, Short Course.

Shell International Ltd. Oil rig photos. http://www.oilrig-photos.com/picture/number202.asp.

Chapter 2

Offshore Structure Loads and Strength

2.1 INTRODUCTION

Fixed offshore platforms are unique structures since they extend to the ocean floor and their main function is to hold industrial equipment that services oil and gas production and drilling.

Robust design of fixed offshore structures depends on accurate specification of the applied load and the strength of the construction materials used. Most loads that laterally affect the platform, such as wind and waves, are variable, so the location of the platform determines the metocean data.

In general, the loads that act on the platform are:

- Gravity loads
- Wind loads
- Wave loads
- Current loads
- Earthquake loads
- Installation loads
- Other loads such as impact load from boats

2.2 GRAVITY LOADS

Gravity loads consist of the dead load and the live (imposed) load.

2.2.1 Dead Load

The dead load is the platform's own overall weight and, in addition, the weight of the equipment, such as piping, pumps, compressors, separators, and other mechanical equipment, used during operation of the platform. The overall weight of platform structure upper decks (topside) includes the piling, superstructure, jacket, stiffeners, piping and conductors, corrosion anodes, decking, railing, grout, and other appurtenances. Sealed tubular members are considered either buoyant or flooded, whichever produces the maximum stress in the structure analysis.

The main function of the topside components is to carry the load from the facilities and drilling equipment. The weights of the topside components and their percentage of the total are shown for a sample structure in Table 2.1. Table 2.2 gives the overall weight of the topside for an eight-legged platform in 91 m of water. In calculating the overall weight of a platform, a contingency allowance of 5% should be used to cover variations in the loads.

TABLE 2.1 Weights and Weight Percentages for an Eight-legged Drilling/Production Platform

		Weight (Tons)	Percentage of Total
1.	Deck		
	Drilling deck		
	Plate	72	11
	Production deck		
	Plate	52	7.8
	Grating	1.0	0.16
	Subtotal	125	18.8
2.	Deck beams		
	Drilling deck	174	26.3
	Production deck	56	8.5
	Subtotal	230	34.8
3.	Tubular trusses	146	22.1
4.	Legs	105	15.9
5.	Appurtenances		
	Vent stack	6	0.9
	Stairs	12	1.8
	Handrails	4	0.6
	Lifting eyes	2	0.3
	Drains	6	0.9
	Fire wall	4	1.7
	Stiffeners	14	2.2
		661	100

TABLE 2.2 Jacket Weight for an Eight-legged Drilling/Production Platform in 91 m of Water

ID	Component Description	Weight, Ton	% of Total Weight	Sub-system % of Total Weight
1.	Legs			
	Joint can	177	14.6	
	In between tubular and others	309	25.4	40
2.	Braces			
	Diagonal in vertical plan	232	19.1	
	Horizontal	163	13.4	
	Diagonal in horizontal plan	100	8.2	40.7
3.	Other framing			
	Conductor framing	35	2.9	
	Launch trusses and runners	82	6.7	
	Miscellaneous framing	2	0.2	9.8
4.	Appurtenances			
	Boat landing	28	2.3	
	Barge bumpers	29	2.4	
	Corrosion anodes	22	1.8	
	Walkways	16	1.3	
	Mud mats	5	0.4	
	Lifting eyes	2	0.2	
	Closure plates	2	0.2	
	Flooding system	7	0.6	
	Miscellaneous	4	0.3	9.5
				100%

Virtually all the decisions about the design of a platform depend on the number of jacket legs. Soil conditions and foundation requirements often control the leg size. The function of the jacket is to surround the piles and to hold the pile extensions in position all the way from the mud line to the deck substructure. Moreover, the jacket legs provide support for boat landings, mooring bits, barge bumpers, corrosion protection systems and many other platform components. The golden rule in design is to minimize the projected area of the structure member near the water surface in high wave zones in order to minimize the load on the structure and to reduce the foundation requirements.

2.2.2 Live Load

Live load is the load imposed on the platform during its use; live loads change from one mode of operation to another. They include:

- The weight of drilling and production equipment.
- The weight of living quarters, heliport and other life-support equipment.
- The weight of liquid in storage tanks.
- The forces due to deck crane usage.

The live load depends on the owner's requirements, and normally it is included in the statement of requirements (SOR) or basis of design (BOD) documents. See Table 2.3 for guidelines on live loads.

For general deck area loading, the topside deck structure should be designed for the specified imposed loads (outlined in Table 2.4) applied to open areas of the deck, where the equipment load intensity is less than the values shown.

DNV (2008) states that the variable functional loads on deck areas of the topside structure should be based on the values in Table 2.5. These values are considered guidelines, so they should be defined in the design criteria, which will be approved by the owner. If the owner needs to increase the load more than is noted in the code, it should be stated in the BOD and the detailed drawings should include the load on the deck. In Table 2.5, the loads that are identified for the local design are used for the design of the plates, stiffeners,

TABLE 2.3 Guidelines for Live Loads

	Uniform Load on Beams and Decking, kN/m^2 (lbs/ft^2)	Concentrated Live Load on Decking, kN/m^2 (lbs/ft)	Concentrated Load on Beams, kN (kips)
Walkways and stairs	4.79 (100)	4.378 (300)	4.44 (1)
Areas over 400 ft^2	3.11 (65)		
Areas of unspecified light use	11.97 (250)	10.95 (750)	267 (60)
Areas where specified loads are supported directly by beams		7.3 (500)	

TABLE 2.4 Live Loads Values Based on Structure Member

	Loading, kN/m²				
	Member Category				
Area	Deck Plate, Grating and Stringers	Deck Beams	Main Framing	Jacket and Foundation	Point Load, kN
Laydown areas	12	10	*a*		30
Open deck areas and access hatches	12	10	*a*	*c*	15
Mechanical handling routes	10	5	*a*	*c*	30
Stairs and landing	2.5	2.5	*b*	—	1.5
Walkways and access platforms	5	2.5	*a*	*c*	5 *d*

[a]*For the design of the main framing, two cases should be considered:*

- *Maximum operating condition: all equipment, including future items and helicopter, together with 2.5 kN/m² on the laydown area.*
- *Live load condition: all equipment loads but no future equipment, together with 2.5 kN/m² on the laydown areas, and a total additional live load of 50 tons. This live load should be applied as a constant uniformly distributed load over the open areas of the deck.*

[b]*Loading for deck plate, grating and stringers should be combined with structural dead loads and designed for the most onerous of the following:*

- *Loading over entire contributory deck area.*
- *A point load (applied over a 300 mm × 300 mm footprint).*
- *Functional loads plus design load on clear areas.*

[c]*For substructure design, deck loading on clear areas in extreme storm conditions may be reduced to zero, in view of the fact that the platform is not normally manned during storm conditions. A total live load of 200 kN at the topside center of gravity should be assumed for the design of the jacket and foundations.*

[d]*Point load for access platform beam design is to be 10 kN and 5 kN for deck grating and stringers, respectively.*

beams and brackets. The loads for the primary design should be used in the design of girders and columns. The loads for the global design should be used for the design of the deck main structure and substructure.

From Table 2.5, the wheel loads should be added to distributed loads where relevant. (Wheel loads can be considered as acting on an area of 300 × 300 mm.) Point loads, which should be applied on an area 100 × 100 mm, and at the most severe position, should not be added to wheel loads or distributed loads.

The value of q in Table 2.5 is to be evaluated for each case. Laydown areas should not be designed for less than 15 kN/m². The value of f in Table 2.5 is obtained from:

$$f = \min\{1.0, (0.5 + 3/\sqrt{A})\} \tag{2.1}$$

where A is the loaded area in m².

TABLE 2.5 Variable Functional Loads on Deck Areas

	Load Design		Primary Design	Global Design
	Distributed Load, kN/m^2	Point Load, kN	Apply Factor for Distributed Load	Apply Factor to Primary Design Load
Storage areas	q	$1.5q$	1.0	1.0
Laydown areas	q	$1.5q$	f	f
Lifeboat platforms	9.0	9.0	1.0	May be ignored
Area between equipment	5.0	5.0	f	May be ignored
Walkways, staircases and platform crew spaces	4.0	4.0	f	May be ignored
Walkways and staircases for inspection only	3.0	3.0	f	May be ignored
Areas not exposed to other functional loads	2.5	2.5	1.0	May be ignored

Where f is calculated from equation (2.1).

Global loads should be established based on the worst-case-scenario characteristic load combinations complying with the limited global criteria for the structure. For buoyant structures, these criteria are established by requirements for the floating position in still water, and intact and damage stability requirements, as documented in the operational manual, are considered for variable loads on the deck and in tanks.

In calculating the dry weight of piping, valves and other structure supports, there should be a 20% increase as a contingency for all estimates of piping weight, because in most cases there are changes in piping dimensions and location during the lifetime of the structure. In addition, all the piping and fittings are calculated in the operating condition, assuming they are full of water with specific gravity equal to 1, with a 20% contingency.

When calculating the dry weight of all equipment, equipment skid, storage and helicopter, a contingency allowance of 10% should be included.

From a practical point of view, Table 2.6 presents the minimum uniform live load values from industrial practices.

TABLE 2.6 Minimum Uniform Loads from Industrial Practices

Platform Deck	Uniform Load, kN/m^2 (lb/ft^2)
Helideck	
Without helicopter	14 (350)
With Bell 212	2 (40)
Mezzanine deck	12 (250)
Production deck	17 (350)
Access platforms	12 (250)
Stairs/walkways	4.7 (100)
Open area used in conjunction with the equipment operating and piping loads for operating and storm conditions	2.4 (50)

TABLE 2.7 Impact Load Factor

	Load Direction	
Structural Item	Vertical	Horizontal
Rated load of cranes	100%	100%
Support of light machinery	20%	0%
Support of reciprocating machinery	50%	50%
Boat landings	200 kips (890 kN)	200 kips (890 kN)

2.2.3 Impact Load

For structural components carrying live loads that could face impact, the live load must be increased to account for the impact effect, as shown in Table 2.7.

2.2.4 Design for Serviceability Limit State

The serviceability of the topside structures can be affected by excessive relative displacement or vibration in vertical or horizontal directions. Limits for either can be dictated by the following:

- Discomfort to personnel.
- Integrity and operability of equipment or connected pipework.
- Limits to control deflection of supported structures, such as flare structures.

- Damage to architectural finishes.
- Operational requirements for drainage (free surface or piped fluids).

Unless stricter limits are established by the platform owner's company or regulator, the limits of deflection (presented in "Deflections" below) should apply.

Vibrations

All sources of vibration should be considered in the design of the structure. At a minimum, the following should be reviewed for their effect on the structure:

- Operating mechanical equipment, including that used in drilling operations.
- Vibrations from variations of fluid flow in piping systems, in particular slugging.
- Oscillations from vortex shedding on slender tubular structures.
- Global motions from the effect of environmental actions on the total platform structure.
- Vibrations due to earthquake and accidental events.

Design limits for vibration should be established from operational limits set by equipment suppliers and from the requirements for personnel comfort, health and safety.

It is important to note that large cantilevers, whether they are simple beams or trusses forming an integral part of the topside platforms, but excluding masts or booms, normally should be proportioned to have a natural period of less than 1 second in the operating condition.

Deflections

The final deflected shape, Δ_{max}, of any element or structure has three components:

$$\Delta_{max} = \Delta_1 + \Delta_2 - \Delta_0 \tag{2.2}$$

where Δ_0 is any precamber of a beam or element in the unloaded state if it exists, Δ_1 is the deflection from the permanent loads (actions) immediately after loading and Δ_2 is the deflection from the variable loading and any time-dependent deformations from permanent loads.

The maximum values for vertical deflections, based on ISO 9001, are given in Table 2.8.

The limiting values for vertical deflection based on a load and resistance factor design (LRFD) are given in Table 2.9.

Lower limits may be necessary to limit ponding of surface fluids and to ensure that drainage systems function correctly.

Horizontal deflections generally should be limited to 0.3% of the height between floors. For multifloor structures, the total horizontal deflection should

TABLE 2.8 Maximum Vertical Deflection Based on ISO 9001

Structural Element	Δ_{max}	Δ_2
Floor beams	$L/200$	$L/350$
Cantilever beams	$L/100$	$L/150$
Deck plate		$2t$ or $b/150$

Note: L is the span, t is the deck thickness and b is the stiffener spacing.

TABLE 2.9 Limiting Values for Vertical Deflection Based on LRFD

Structure Member	Δ_{max}	Δ_2
Deck beams	$L/200$	$L/300$
Deck beams supporting plaster or other brittle finish or nonflexible partitions	$L/250$	$L/350$

Note: L is the beam span. For cantilevers, L is twice the projecting length of the cantilever.

not exceed 0.2% of the total height of the topside structure. Limits can be defined to limit pipe stresses and to avoid riser or conductor overstress or failure. Some designers allow higher deflections for structural elements where serviceability is not compromised by deflection.

2.2.5 Helicopter Landing Loads

The maximum dynamic local actions from an emergency landing may be determined from the collapse load of the landing gear. This should be obtained from the helicopter's manufacturer.

Alternatively, default values may be used for design by considering an appropriate distribution of the total impact load of 2.5 times the maximum take-off weight (MTOW).

The local loads used in design should correspond to the configuration of the landing gear. A single main rotor helicopter may be assumed to land simultaneously on its two main undercarriages or skids. A tandem main rotor helicopter may be assumed to land on the wheels of all main undercarriages simultaneously. For a single main rotor helicopter, the total loads imposed on the structure should be taken as concentrated loads on the undercarriage centers of the specific helicopter divided equally between the two main undercarriages. For tandem main rotor helicopters, the total loads imposed on the structure should be taken as concentrated loads on the undercarriage centers of the specific helicopter and distributed between the main undercarriages in the proportion in which they carry the maximum static loads.

The concentrated undercarriage loads should normally be treated as point loads but, where it is advantageous, a tire contact area may be assumed in accordance with the manufacturer's specification. The MTOW and undercarriage centers for which the platform has been designed and the maximum size and weight of helicopters for which the deck is suitable should be recorded.

Information on the dimensions and MTOW of specific helicopters is given in Table 2.10.

Based on CAP 437, the take-off and landing area should be designed for the heaviest and largest helicopter anticipated to use the facility, as shown in Table 2.10. Helideck structures should be designed in accordance with the International Civil Aviation Organization (ICAO) requirements in the *Heliport Manual*, International Standards Organization (ISO) codes for offshore structures and, for a floating installation, the relevant International Maritime Organization (IMO) code. The maximum size and mass of helicopter for which the helideck has been designed should be stated in the installation or vessel operations manual and verification and classification documents.

TABLE 2.10 Helicopter Weights, Dimensions and *D* Value for Different Types

Type	*D* Value[a] (m)	Perimeter *D* Marking	Rotor Height (m)	Rotor Diameter (m)	Max Weight (kg)
Augusta A109	13.05	13	3.30	11	2600
Dauphin SA365N2	13.68	14	4.01	11.93	4250
Sikorsky S76B&C	16.00	16	4.41	13.40	5307
Bell 212	17.46	17	4.80	14.63	5080
Super Puma AS 332L2	19.50	20	4.92	16.20	9300
Bell 214ST	18.95	19	4.68	15.85	7936
Sikorsky S61N	22.20	22	5.64	18.90	9298
EH101	22.80	23	6.65	18.60	14,600

[a]*The D value is the largest overall dimension of the helicopter when rotors are turning. This dimension will normally be measured from the most forward position of the main rotor tip path plane to the most rearward position of the tail rotor tip path plane or the most rearward extension of the fuselage in the case of Fenestron or Notar tails. The D circle is a circle, usually imaginary unless the helideck itself is circular, the diameter of which is the D value of the largest helicopter the helideck is intended to serve.*

Loads for Helicopter Landings

The helideck should be designed to withstand all the forces likely to act when a helicopter lands, including:

a. Dynamic load due to impact landing. This should cover both a heavy normal landing and an emergency landing. For the former, an impact load of 1.5 × maximum take-off mass (MTOM) of the helicopter should be used. This should be treated as an imposed load, applied together with the combined effects of (b) to (f) below in any position on the safe landing area so as to produce the most severe landing condition for each element concerned. For an emergency landing, an impact load of 2.5 × MTOM should be applied in any position on the landing area together with the combined effects of (b) to (f) inclusive. Normally, the emergency landing case governs the design of the structure.

b. Sympathetic response of landing platform. The above dynamic load should be increased by a structural response factor, depending on the natural frequency of the helideck structure, after considering the design of its supporting beams and columns and the characteristics of the designated helicopter. It is recommended that a structure response factor of 1.3 be used unless further information is available to allow a lower factor to be calculated. Information required to do this will include the natural periods of vibration of the helideck and dynamic characteristics of the designated helicopter and its landing gear.

c. Overall superimposed load on the landing platform. To allow for snow, personnel and other live loads in addition to wheel loads, an allowance of 0.5 kN/m^2 should be added over the whole area of the helideck.

d. Lateral load on landing platform supports. The landing platform and its supports should be designed to resist concentrated horizontal imposed loads equivalent to 0.5 × MTOM of the helicopter, distributed between the undercarriages in proportion to the applied vertical loading in the direction that will produce the most severe loading on the element being considered.

e. Dead load of structural members.

f. Wind loading. Wind loading should be allowed for in the design of the platform. This should be applied in the direction that, together with the imposed lateral loading, will produce the most severe loading condition on each element.

g. Punching shear. A check should be made for the punching shear from an undercarriage wheel with a contact area of $65 \times 103\ mm^2$ acting in any probable location. Particular attention to detail should be taken at the junction of the supports and the platform deck.

Loads for Helicopters at Rest

The helideck should be designed to withstand all the applied forces that could result from a helicopter at rest. The helideck components should be designed to resist the following simultaneous actions in normal landing and at rest situations:

- Helicopter static loads (local landing gear, local patch loads)
- Area load
- Helicopter tie-down loads, including wind loads from a secured helicopter
- Dead loads
- Helideck structure and fixed appurtenances self-weight
- Wind loading
- Installation motion

Helicopter Static Loads

All parts of the helideck accessible to a helicopter should be designed to carry an imposed load equal to the MTOW of the helicopter. This should be distributed at the landing gear locations in relation to the position of the center of gravity of the helicopter, taking into account different orientations of the helicopter with respect to the installation.

Area Load

To allow for personnel, freight, refueling equipment and other traffic, snow and ice, rotor downwash and other, a general area load of 2.0 kN/m^2 should be included.

Helicopter Tie-down Loads

Each tie-down should be designed to resist the total wind load imposed by "100-year storm" winds on the helicopter.

Sufficient flush-fitting tie-down points should be provided for securing all the helicopter types for which the landing area is designed. They should be located and be of such construction so as to secure the helicopter when subjected to weather conditions of a severity unlikely to be exceeded in any one year. They should also take into account any recommendations made by the aircraft manufacturer and, where significant, the inertial forces resulting from movement of floating platforms.

Wind Loading

Wind loading on the helideck structure should be applied in the direction that, together with the horizontal imposed loading, will produce the most severe loading condition for the element considered.

Consideration should also be given to the additional wind loading from a secured helicopter, as noted above.

Installation Motion

The effect of acceleration forces and other dynamic amplification forces arising from the predicted motions of the installation in the 100-year storm condition should be considered.

Safety Net Arms and Framing

Safety nets for personnel protection should be installed around the landing area except where structural protection exists. The netting used should be flexible, with the inboard edge fastened level with, or just below, the edge of the helicopter landing deck. The net itself should extend at least 1.5 m in the horizontal plane and be arranged so that the outboard edge is slightly above the level of the landing area, but not by more than 0.25 m, so that it has an upward and outward slope of at least 10°. The supporting structure associated with the safety net should be capable of withstanding, without damage, a 75 kg weight being dropped from a height of 1 m onto an area of 0.25 m^2.

The entire helideck, including any separate parking or runoff area, should be designed to resist an imposed load equal to the MTOM of the helicopter. This load should be distributed between all the undercarriages of the helicopter. It should be applied in any position on the helicopter platform so as to produce the most severe loading condition for each element considered.

The values for these overall superimposed loads, dead loads, and wind loads should be considered to act in combination with the dynamic load impact, as discussed above. Consideration should also be given to the additional wind loading from any parked or secured helicopter.

Based on the American Petroleum Institute (API) RP2L design for heliports, the flight deck, stiffeners and supporting structure should be designed to withstand the helicopter landing load encountered during exceptionally hard landing after power failure while hovering. Examples of helicopter parameters that should be obtained are shown in Table 2.11 and it is recommended that parameters similar to those given in the table be obtained from the manufacturer of any helicopter considered in the helideck design.

The maximum contact area per landing gear used to design deck plate bending and shear should conform to the manufacturer's values given in Table 2.11. For multiwheeled landing gear, the value of the contact area is the sum of the areas for each wheel. The contact area for float or skid landing gear is the area of the float or skid around each support strut.

The load distribution per landing gear, in terms of percentage of gross weight, is given in Table 2.11.

The design landing load is the landing gear load based on a percent of a helicopter's gross weight times an impact factor of 1.5. For percentages and helicopter gross weight, see Table 2.11.

Design Load Conditions

The heliport should be designed for at least the following combinations of design loads:

- Dead load plus live load.
- Dead load plus design landing load. If icing conditions are prevalent during normal helicopter operations, superposition of an appropriate live load should be considered.
- Dead load plus live load plus wind load.

TABLE 2.11 Technical Parameters for Helicopters

Manufacturer Model	Common Name	Gross Weight (kg)	Overall Length (m)	Type	Number		Contact Area		Percentage of Gross Weight		Distance between Fore and Aft Gears (m)	Width between Gears (m)
					Fore	Aft	Fore (cm^2)	Aft (cm^2)	Fore	Aft		
Aerospatiale												
315B	Lama	2305	12.9	Skid					38	62		2.4
316B		2205	12.9	Wheel	1	2	297	594	28	72	3.1	2.6
318C	Alouette II	1656	12.1	Skid								2.3
330J	Puma	7400	18.2	Wheel	2	4	1200	2142	36	64	5.3	3.0
332L	Super Puma	8351	18.7	Wheel	2	2	465	735	40	60	4.5	3.0
341G	Gazelle	1800	12.0	Skid					51	49		2.1
350B/d	ASTAR	1950.5	13.0	Skid					51	49		2.1
360	Dauphin	2799	13.4	Wheel	2	1					7.2	2.0
360C		2994	13.4	Wheel	2	1	213	123	84	16	3.32	2.4
360C		2994	13.4	Skid					84	16	3.32	2.3
365C		3401	13.4	Wheel	2	1	213	123	84	16	3.32	2.4
365C		3401	13.4	Skid								2.3
365N	Dauphin 2	3850	13.5	Wheel	2	2	245	426	22	78	3.6	2.0
A109	Hirando	2450	13.1	Wheel	1	2	129	129			3.5	2.3

Bell Helicopter												
47G		1338	13.3	Skid			174	174			1.6	2.3
205A-1		4309	17.4	Skid			310	310			2.3	2.7
206B	Jet Ranger	1451	12.0	Skid			174	174	19	81	1.4	1.8
212	Twin	5080	17.5	Skid			310	310	22	78	2.3	2.5
214B	Big Lifter	7257	19.0	Skid								2.6
214ST	Super Transport	7938	19.0	Wheel	2	2	247	581	22	78	4.8	2.8
222B		3742	15.3	Wheel	1	2	123	413	19	81	3.7	2.8
222UT		3742	15.3	Skid			310	310	32	68	2.4	2.4
412		5262	17.1	Skid			310	310	20	80	2.4	2.5
BO105C		2300	11.8	Skid								2.6
BO105CBS		2400	11.9	Skid			181	181	36	64		2.5
CH47234		22,680	30.2	Wheel	4	2	1007	503			6.9	3.4
FH-1100		1247	12.7	Skid								2.2
Hiller												
UH-12L4		1406	12.4	Skid								2.3
UH12E/E4		1270	12.4	Skid								2.3

Most classification societies with jurisdiction in the U.K. sector of the North Sea have classification notes available that include helideck specifications. Because the main duties of some of these authorities is primarily classification of floating structures, the helideck specifications generally appear only within rules for mobile offshore units and the like. Two, however, Lloyd's Register of Shipping (LRS) and Det Norske Veritas (DNV), have adopted identical specifications for fixed platforms.

Tables 2.12 and 2.13 summarize the loading requirements for most regular specifications. Table 2.12 is concerned with the load specifications during helicopter landings, while Table 2.13 relates to helicopters in the stowed position.

From both tables, it can be seen requirements vary considerably among the specifications, with particular variation in:

- The factor on MTOW (M) for emergency landing conditions.
- Whether a deck response factor is considered.
- Whether the level of superimposed load is considered simultaneously or separately.
- Whether a lateral load is considered simultaneously with the emergency landing load.

Example of Helicopter Load

The helicopter loads presented in Table 2.12 are used here for the skid loading model. In the SACS input loading menu for a Bell-212 helicopter, the MTOW = 50 kN (5 tons).

The helicopter has two skids. Each skid's geometrical dimensions are assumed to be 3.68 (L) × 2.856 (W) meters, with the center of gravity (C.o.G.) point of application at the middle of the skid. The height of the helicopter fuselage is taken as 2 m.

1. CAP-437 imposed live load = 0.5 kN/m^2
2. At-rest condition = 1 × MTOW = 50.0 kN
3. Normal operating condition = 1.5 × 1.3 × MTOW = 97.50 kN
4. Emergency landing condition = 2.5 × 1.3 × MTOW = 162.50 kN

The landing conditions, either normal operating or emergency, are combined with lateral horizontal force = 0.5 × MTOW = 25.0 kN applied on both skids.

2.2.6 Crane Support Structures

The crane support structures comprise the crane pedestal and its connections to the topside structure's primary steelwork. They do not include the slew ring or its equivalent or the connections between the slew ring and the pedestal.

Crane support structures should, where practical, be attached at the intersection of topside primary trusses and connected at main deck elevations with minimal eccentricities.

TABLE 2.12 Helicopter Landing Loading Specifications of Various Authorities

	ISO	CAP	HSE	ABS	B.V.	DNV	GL	LRS
Heavy landing	—	1.5 M						
Emergency landing	2.5 M	2.5 M	2.5 M	1.5 M[a]	3.0 M	2.0 M	1.5 M	1.5 M[b] 2.5 M[c]
Deck response factor	1.3	1.3[d]	1.3[d]	—	—	—	—	—
Superimposed load, kN/m^2	0.5	0.5	0.5	2.0[e]	2.0[e]	As normal class	0.5	— 0.2 M[c]
Lateral load	0.5 M	0.5 M	0.5 M	—	—	0.4 M	—	0.5 M
Wind load	Max. oper.			Normal design		$V_w = 30$ m/s	$V_w = 25$ m/s	—

Note: M is the maximum take-off weight and V_w is the wind velocity. The authorities are: American Bureau of Shipping (ABS), Bureau Veritas (B.V.), Det Norske Veritas (DNV), Germanischer Lloyd (GL) and Lloyd's Register of Shipping (LRS).

[a] *Manufacturer's recommended wheel impact loads.*

[b] *For design of plating.*

[c] *For design of stiffening and supporting structure.*

[d] *Additional frequency-dependent values given for Chinook helicopter.*

[e] *Considered independently.*

TABLE 2.13 Loading Specifications for a Helicopter at Rest from Various Authorities

	ISO	CAP	HSE	ABS	B.V.	DNV	GL[a]	LRS
Self-weight	M	M	M	M	M	M	1.5 M	M
Super-imposed load, kN/m^2	2.0	0.5	0.5	0.49	0.5	As normal class	2.0	2.0
Wind load	100-yr storm					V_w = 55 m/s	V_w = 50 m/s	—

Note: M is the maximum take-off weight and V_w is the wind velocity.
[a]*Fixed platforms; for floating platforms, also include lateral load of 0.6 (M + W) where W is the deck weight considering the platform side sway.*

The pedestal should be included in the analytical model of the primary structure as its stiffness can have a significant effect on load distribution. When they are located in accordance with this guideline, the crane support structures' performance will generally be governed by static actions with negligible dynamic amplification. The structures are, however, subject to fatigue damage and should always be checked to ensure that fatigue life is satisfactory for the required service conditions.

The maximum rotation at the top of the pedestal or in the plane of the effective point of support should not exceed the manufacturer's recommended requirements and in no case should it exceed 1° for the most onerous case of loading. Where this criterion cannot be met, the dynamic response should be checked.

Several situations are considered in the design of the crane support structures:

- Crane working in calm conditions.
- Crane working at maximum operating wind conditions. The maximum operating wind may be different for platform lifts and for sea lifts and for lifts to or from an adjacent vessel, and it may also vary depending on the weight being lifted.
- Crane collapse scenario. This situation is included to ensure that, in the event of a gross overload of the crane causing collapse of any part of the crane structure (most commonly the boom or A-frame), no damage to the crane support structure is suffered and progressive collapse is resisted.

The loading in each of these cases should be checked as follows, based on ISO 9001.

Crane Working without Wind

$$F_c = F_G + fF_L + F_H \qquad (2.3)$$

where F_c is the total crane load, F_G is the vertical load due to dead weight of components, F_L is the vertical load due to the suspended load, including sheave blocks, hooks and others, F_H is the horizontal load due to offlead and side-lead and f is a dynamic coefficient that should be taken as 2.0 for sea lifts and 1.3 for platform lifts.

Crane Working with Wind

$$F_{cw} = F_G + fF_L + F_H + F_W \tag{2.4}$$

where F_{cw} is the crane lateral load due to wind, F_G, F_L, F_H and f are as above, F_W is the operating wind action and f should be taken as 2.0 for sea lifts and 1.3 for platform lifts for a maximum crane operating wind.

Crane at Rest (Not Working), Extreme Wind

$$F_{cr} = F_G + F_{W,\max} \tag{2.5}$$

where F_{cr} is the crane load at rest, F_G is as above and $F_{W,\max}$ is the extreme wind action.

The action factors used with each of the above should be those for normal operating conditions.

For the crane working in calm conditions and the crane working at maximum operating wind conditions, F_L should be selected to check the lifted load applicable to both maximum and minimum crane radius, for sea and platform lifts.

For the crane working at maximum operating wind conditions and the crane at rest in extreme wind conditions, the most onerous wind directions should be checked.

It should be demonstrated that the crane pedestal and its components are designed to safely resist the forces and moments from the most onerous loading condition applicable to the prevailing sea state together with associated offlead and side-lead forces. These values should be obtained from the crane manufacturer, and the angles used should not be less than the following:

- Offlead angle: 6°
- Side-lead angle: 3°

The crane support structures should be designed so that their failure load exceeds the collapse capacity of the crane.

The crane manufacturer's failure curves, for all crane conditions, should be used to determine the worst loading on the pedestal. It should be assumed that the maximum lower bound failure moment of the weakest component will place an upper bound on the forces and moments to which the pedestal can be subjected.

The design moment for the crane failure condition should be taken as the lower bound failure moment described above, multiplied by a safety factor of 1.3.

It is not normal practice for the support structure of offshore cranes to be subject to a dynamic analysis. The process of fatigue design incorporates an average dynamic amplification factor (DAF) for all lifts and this has been found to provide satisfactory results in practice. A dynamic analysis is required only where the design is unconventional or the maximum rotation at the plane of support exceeds the limit of 1°, and experience or engineering judgment indicates that the performance of the crane support structure may be adversely affected by its dynamic response to actions. Where dynamic analysis is considered necessary, the model employed should be sufficiently detailed to ensure that the coupled response of the crane and its supports are realistically represented. Any mechanical damping device incorporated within the design of the crane should be taken into consideration. Where appropriate, the results of the dynamic analysis may be used to modify the fatigue design process.

2.3 WIND LOAD

The wind data are provided by the owner according to the metocean study, which defines the prevailing wind direction and the maximum wind speed every 1 year, 50 years and 100 years. Figures 2.1 and 2.2 are examples of wind data from a metocean study report.

The most important design considerations for an offshore platform are the storm wind and storm wave loadings it will be subjected to during its service life. These data are usually available to the owner and are officially submitted to the engineering firm. If these data are not available, they should be retrieved from an experienced authority in the country where the platform will be located or other specific international offices.

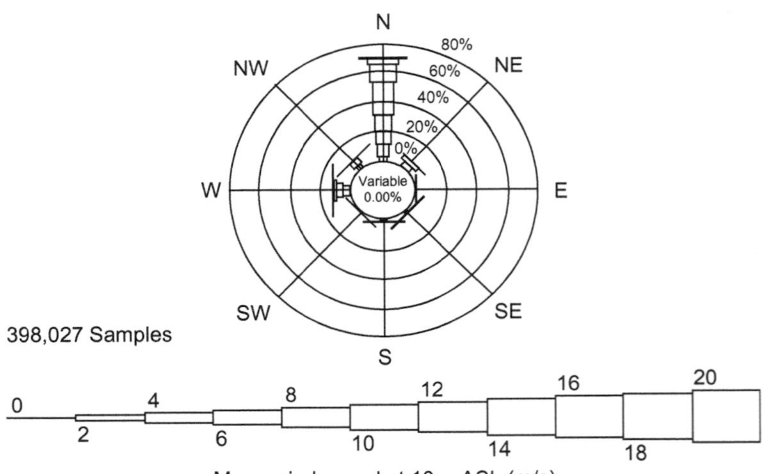

FIGURE 2.1 Example of wind at a location in the Gulf of Suez in the Red Sea (GoS).

Mean wind speed at 10 m ASL (m/s)	N	NE	E	SE	S	SW	W	NW	Total	CFD
0.0–2.0	0.34	0.28	0.24	0.24	0.19	0.22	0.39	0.43	2.3	100.0
2.0–4.0	1.9	1.2	0.61	0.60	0.32	0.34	2.0	2.9	9.8	97.7
4.0–6.0	8.2	1.9	0.16	0.47	0.22	0.31	4.8	3.6	19.8	87.9
6.0–8.0	19.3	1.7	0.01	0.35	0.18	0.25	3.6	0.93	26.2	68.1
8.0–10.0	21.6	0.74	<0.01	0.23	0.13	0.15	1.4	0.13	24.4	41.8
10.0–12.0	12.6	0.18	<0.01	0.11	0.05	0.03	0.62	0.02	13.6	17.4
12.0–14.0	3.2	0.02	<0.01	0.03	0.01	<0.01	0.18	<0.01	3.4	3.9
14.0–16.0	0.32	<0.01		<0.01	0.01	<0.01	0.04	<0.01	0.38	0.42
16.0–18.0	<0.01			<0.01	<0.01	<0.01	0.01		0.03	0.04
18.0–20.0	<0.01			<0.01			<0.01	<0.01	<0.01	<0.01
20.0–22.0							<0.01		<0.01	<0.01
Total	67.4	6.0	1.0	2.0	1.1	1.3	13.0	8.0	100.0	

FIGURE 2.2 The tabulated data for mean wind speed from different directions.

The wind speed at any elevation above a water surface is presented as:

$$V_z = V_{10}(z/10)^{1.7} \tag{2.6}$$

The wind velocity, vertical wind profile and time averaging duration in relation to the dimensions and dynamic sensitivity of the structure's components should be determined. In special cases, dynamic response to wind action can be significant and should be taken into account.

In addition, the following meteorologic data should be obtained:

- Lowest air temperature
- Highest air temperature
- Average air temperature
- Maximum relative humidity
- Minimum relative humidity
- Average relative humidity
- Annual fog days
- Annual thunderstorm days
- Total annual rainfall
- Maximum rainfall in one day
- Annual rainfall days
- Sea water surface temperature
- Maximum degree of salt
- Minimum degree of salt

- Average degree of salt
- Mud temperature at seabed at 0 m, 1.0 m, 1.5 m and 2.0 m
- Snow maximum thickness and average thickness

For all directions and wind angles of approach to the structure, wind actions on vertical cylindrical objects may be assumed to act in the direction of the wind. Actions on cylindrical objects that are not in a vertical attitude should be calculated using appropriate formulae that take into account the direction of the wind in relation to the altitude of the object. Actions on walls and other flat surfaces that are not perpendicular to the direction of the wind should also be calculated using appropriate formulae that account for the skewness between the direction of the wind and the plane of the surface. Where appropriate, loads caused by wind should be calculated with due account for increased exposure area and surface roughness due to ice. Local wind effects, such as pressure concentrations and internal pressures, should be considered by the designer where applicable.

In accordance with the above observations, attention should be given to which velocity (in magnitude and direction) and which area are to be used. Similarly, the direction of the resulting action vector should be carefully considered. The basic relationship between the wind velocity and the wind load on an object is presented by the following equation:

$$F = 1/2\,\rho_a V_w C_s A \tag{2.7}$$

where F is the wind action on the object, ρ_a is the mass density of air (at standard temperature and pressure), V_w is the wind speed, C_s is the shape coefficient and A is the area of the object.

In the absence of data indicating otherwise, the shape coefficients in Table 2.14 are recommended for perpendicular wind approach angles with respect to each projected area.

TABLE 2.14 Shape Coefficients, C_s, for Perpendicular Wind Approach Angles

Component	Shape Coefficients, C_s
Flat walls of buildings	1.5
Overall projected area of structure	1.0
Beams	1.5
Cylinder	0.5
Smooth $R_e > 5 \times 10^5$	0.65
Smooth $R_e \leq 5 \times 10^5$	1.2
Rough all R_e	1.05
Covered with ice, all R_e	1.2

Wind loads on downstream components can be reduced due to shielding by upstream components. The extreme quasistatic global action caused by wind should be calculated as the vector sum of the total above wind loads on all objects.

When wind loads are important for structural design, wind pressures and resulting local loads should be determined from wind tunnel tests on a representative model or from a computational model representing the structure and considering the range and variation of wind velocities. Computational models should be validated against wind tunnel tests or full-scale measurements of similar structures.

The code equation, which is the same for the different codes, can be used to calculate the wind force on the structure:

V_{10} = wind speed at height 10 m
10 = reference height, m
z = desired elevation, m

Table 2.15 lists some design wind pressures for a 100-year storm with a sustained wind velocity of 125 mph.

For the wind load on the topside, there are usually a series of beams or trusses that serve as a shield from the wind. Therefore, according to Table 2.16 (from API), there is a reduction (shielding) factor for the wind load. For example,

TABLE 2.15 Design Wind Pressures at 125 mph for a 100-Year Storm

Structure Member	Pressure, kN/m^2 (lb/ft^2)
Flat surfaces, such as wide flange beams, gusset plates, sides of building, etc.	2.9 (60)
Cylindrical structural members	2.3 (48)
Cylindrical deck equipment ($L = 4D$)	1.4 (30)
Tanks standing on end ($H \leq D$)	1.16 (25)

Note, L is the member length, D is the member diameter and H is the cylindrical tank height

TABLE 2.16 Shielding Factors

Component	Shielding Factor
Second in a series of trusses	0.75
Third or more in a series of trusses	0.50
Second in a series of beams	0.50
Third or more in a series of beams	0.00
Second in a series of tanks	1.00

there is a shielding factor for the second series of trusses and beams and for the third or more in a series of beams, the reduction is great enough that there is no wind load effect. Furthermore, if there is a short object behind a long object, the short object will be free from any wind load effect.

According to DNV, the wind pressure acting on the surface of helidecks may be calculated using a pressure coefficient $C_p = 2.0$ at the leading edge of the helideck, linearly reducing to $C_p = 0$ at the trailing edge, taken in the direction of the wind. The pressure may act both upward and downward.

2.4 STAIR DESIGN

The first bay of the stair that connects the main deck at level 13.525 m to the helideck is the most critical stair bay.

2.4.1 Gravity Loads

- Uniform gravity loads/one stair channel = (dead load + live load) × ½ stair width + handrail weight/meter + channel weight/meter
- Dead load = 0.5 kN/m′
- Live load = 50 psf = 2.5 kN/m′ (extracted from Table 2.17 or as per project specification)
- Handrail weight/meter = 0.4 kN/m′
- Channel weight/meter = 0.379 kN/m′
- Uniform gravity loads (ton/m)/one stair channel = (0.5 + 2.5) × 0.5 × 1.2 + 0.4 + 0.379 = 2.6 kN/m′
- Moment = (uniform load/m) × inclined length × projected length/8 = 2.6 × 6.22 × 4.4/8 = 8.9 m/kN

TABLE 2.17 Live Load on the Fixed Platform from Technical Practice

	Loadings, kN/m^2		
	Member Category		
Area	Deck Plate, Grating and Stringers	Deck Beams	Main Truss Framing, Girders, Jacket and Foundation
Cellar and main decks	14.4	9.6	9.6
Walkway, stairs and access decks	4.8	2.4	7.2
Laydown areas	19	14.4	7.2

- Actual stress = moment/section modulus = 8.9 (10^6)/371 = 24,000 kN/m^2
- Unsupported length = 6.22 m
- Allowable stress (ksi) = 12,000 C_b/(member length × depth/area of flange) = 12,000 × 1/[622 × 26/(9 × 1.4)] = 9.35 ksi = 64,466 kN/m^2 > actual stress

2.4.2 Wind Loads

- Wind loads/one stair channel = $C_d \times ½ \times \rho \times V^2 \times ½$ stair width
- C_d (drag coefficient) = 2
- ρ (air intensity) = 1.3 kg/m^3
- V (wind velocity) = 29 m/sec
- Wind loads/one stair channel = 2 × ½ × 1.3 × 29^2 × 0.6 = 655.98 N/m = 0.66 kN/m^2
- Moment = (wind load/m) × inclined length × projected length/8 = 0.66 × 6.22 × 4.4/8 = 0.23 m/kN
- Actual stress (gravity + wind) = moment/section modulus = (8.9 + 0.23) (10^6)/371 = 24,609 kN/m^2 < allowable stress × 1.33

Note that wind loads on a channel in the minor axis direction are neglected because horizontal bracing is provided.

2.5 OFFSHORE LOADS

Offshore loads that most affect platforms are waves, tide and current. For this discussion, some terminology related to sea water levels has to be introduced:

- High-high water level (HHWL)
- Mean-high water level (MHWL)
- Mean water level (MWL), equal to mean sea level (MSL) and still water level (SWL)
- Low-low water level (LLWL), equal to the chart datum (CD)
- Mean-low water level (MLWL)

The mean-low water level is less than the mean-high water level by 304 mm (1 ft).

Tides are the rise and fall of sea levels caused by the combined effects of the gravitational forces exerted by the moon and the sun and the rotation of the Earth. Most places in the ocean usually experience two high tides and two low tides each day (semidiurnal tide), but some locations experience only one high and one low tide each day (diurnal tide). The times and amplitude of the tides at the coast are influenced by the alignment of the sun and moon, by the pattern of tides in the deep ocean, as shown in Figure 2.3, and by the shape of the coastline and near-shore bathymetry.

Most coastal areas experience two high and two low tides per day. The gravitational effect of the moon on the surface of the Earth is the same when it is at its zenith as when it is at its nadir. The moon orbits the Earth in the same

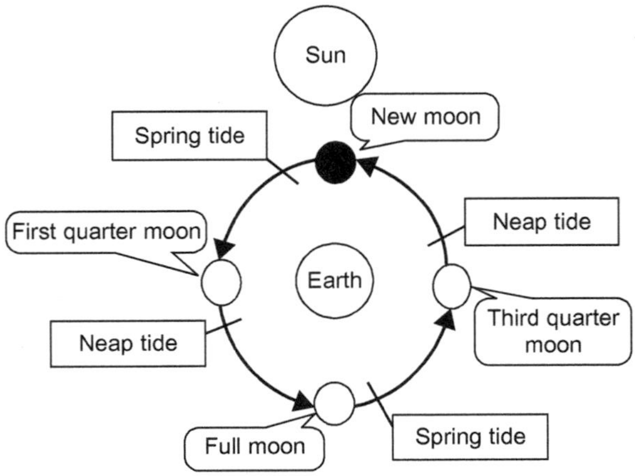

FIGURE 2.3 Times of neap and spring tides.

direction the Earth rotates on its axis, so it takes slightly more than a day—about 24 hours and 50 minutes—for the moon to return to the same location in the sky. During this time, it has passed its zenith once and its nadir once, so in many places the period of strongest tidal forces is 12 hours and 25 minutes. High tides do not necessarily occur when the moon is at its zenith or nadir, but the period of the forces still determines the time between high tides.

Storm surge is an offshore rise of water associated with a low-pressure weather system, typically a tropical cyclone. Storm surges are caused primarily by high winds pushing on the ocean's surface. The wind causes the water to pile up higher than the ordinary sea level. Low pressure at the center of a weather system also has a small secondary effect, as can the bathymetry of the body of water. It is this combined effect of low pressure and persistent wind over a shallow water body that is the most common cause of storm surge flooding problems. The term *storm surge* in casual (nonscientific) use is storm tide; that is, it refers to the rise of water associated with the storm, plus tide, wave run-up and freshwater flooding. When referencing storm surge height, it is important to clarify the usage, as well as the reference point.

2.5.1 Wave Load

Large forces result when waves strike a platform's deck and equipment. Where insufficient air gap exists, all actions resulting from waves, including buoyancy, inertia, drag and slam, should be taken into account (see ISO 19901-1 and ISO 19902).

Typically, waves in the ocean often appear as a confused and constantly changing sea of crests and troughs on the water's surface because of the irregularity of wave shape and the variability in the direction of propagation. The direction of wave propagation can be assessed as an average of the direction of individual waves. Wave direction is considered to affect the platform omnidirectionally, as per API RP2A.

In general, actual water-wave phenomena are complex and difficult to describe mathematically because of nonlinearities, three-dimensional characteristics and apparent random behavior. However, there are two classic theories that describe simple waves, one developed by Airy in 1845 and the other by Stokes in 1880. The Airy and Stokes theories generally predict wave behavior better where water depth relative to wavelength is not too small. For shallow water regions, conidial wave theory, originally developed by Korteweg and DeVries in 1895, provides a reliable prediction of the waveform and associated motions for some conditions. Recently, the work involved in using conidial wave theory has been substantially reduced by the introduction of graphical and tubular forms of function by Wiegel in 1960 and Masc and Wiegel in 1961; however, application of the theory is still complex.

In 1880, Stokes developed a finite amplitude theory that is more satisfactory. Only the second-order Stokes equation is presented, but the use of higher-order approximations is sometimes justified for the solution of practical problems.

Another widely used theory, known as the stream function theory, is a nonlinear solution similar to the Stokes fifth-order theory, as both use summations of sine and cosine waveforms to develop a solution to the original differential equation.

The theory to be used for a particular offshore design is determined by the policy under which the designing engineers are working.

The selection of the best method is defined by curves produced by Atkins (1990) and modified by the API Task Group. In general, the stream function method covers the broadest range of dimensionless wave steepness (H/gT) and dimensionless relative depth (d/gT), where d is the mean water depth, T is the wave period, H is the wave height, and g is the acceleration of gravity. The Airy theory is used for intermediate and deep water depths with relative water depth greater than 0.0025 and it can be used for relative depth greater than 0.05 with wave steepness up to 0.001. Stokes fifth-order theory should be used for relative depth greater than 0.01 and wave steepness greater than 0.001.

Metocean data serve as the basis for Figure 2.4, which shows the probability of wave height at a particular platform location. Sample metocean data are shown in Table 2.18.

The sea state (E) is calculated by dividing a waveform into small slides, noting that for each slide, the height, H_i, is squared, and then the values for all the slides are added together and averaged. Then E is calculated as:

$$E = 2\frac{\sum_{i=1}^{N} H_i}{N}$$

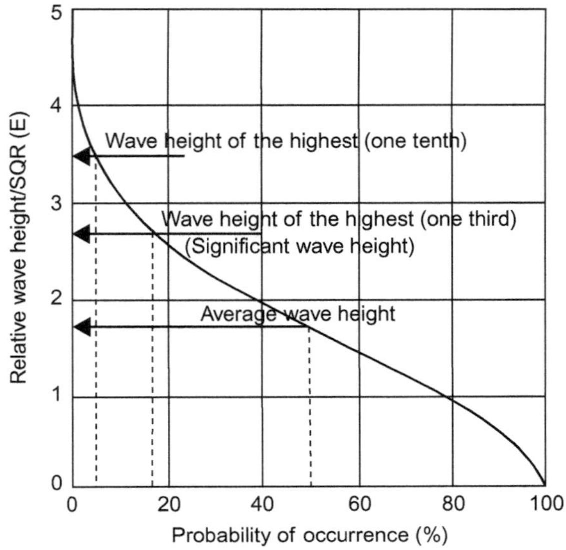

FIGURE 2.4 Probability of wave height.

TABLE 2.18 Example of Metocean Data

Return Period (years)	W_s (m/s)	H_s (m)	T_z (s)	T_p (s)	H_{max} (m)	T_{ass} (s)	H_c (m)	U (cm/s)
1	16.1	2.2	4.7	6.2	3.2	6.3	1.9	1
10	17.7	2.8	5.4	7.1	5.3	7.2	3.2	5
50	18.7	3.2	5.9	7.7	6.6	7.7	4.0	11
100	19.2	3.3	6.0	7.9	7.2	8.0	4.3	15
10,000	22.0	4.5	7.1	9.4	10.9	9.4	6.6	51

Note: W_s is the 1-hour mean wind speed at 10 m above sea level; H_s is the significant wave height, estimated from the wave energy spectrum, equivalent to the mean height of the highest one-third of the wave in a sea state; H_{max} is the maximum wave height, highest individual zero-crossing wave height in a storm of 24-hr duration; T_z is the mean zero-crossing wave period, the average period of the zero-crossing wave heights in a sea state estimated from the wave energy spectrum; T_p is the peak wave period, the period associated with the peak in the wave energy spectrum, estimated from the wave energy spectrum; T_{ass} is the wave period associated with the maximum wave height; H_c is the crest height, the highest crest to mean-level height of an individual wave in a storm of 24-hr duration; U is the horizontal wave orbital velocity at 3 m above the seabed, estimated from H_{max} and T_{ass} using stream function wave theory.

Figure 2.5 shows a two-dimensional, simple progressive wave propagation in the positive x-direction. The symbol η denotes the displacement of the water surface relative to the SWL, which is a function of x and time, t, at the wave crest, η, equal to one-half of the wave height.

Water particle displacement is presented in Figure 2.6 for deep and shallow water. Water particle displacement is an important factor in linear wave mechanics, which deal with the displacement of individual water particles within a wave. Water particles generally move in elliptical paths in shallow water or transitional water and in circular paths in deep water (see Figure 2.6). If the mean particle position is considered to be at the center of the ellipse or circle, then vertical particle displacement with respect to the mean position cannot exceed one-half the wave height. Therefore, the displacement of the fluid particle is small if the wave height is small.

Figure 2.7 presents the horizontal and vertical velocities and acceleration for various locations of the particles, which is very important in calculating the wave forces on any subsea structural member, as the drag force and inertia force are functions of the particle velocity and acceleration, respectively. The following equations are used to calculate the wave velocity and acceleration.

$$F_1 = \left(2\pi(z+d)/L\right) \tag{2.8}$$

$$F_2 = (2\pi d/L) \tag{2.9}$$

$$F_3 = (2\pi x/L) - (2\pi t/T) \tag{2.10}$$

- Velocity

$$U = [(H/2)(gT/L)\cosh F_1/\cosh F_2]\cos F_3 \tag{2.11}$$

$$W = [(H/2)(gT/L)\sinh F_1/\cosh F_2]\sin F_3 \tag{2.12}$$

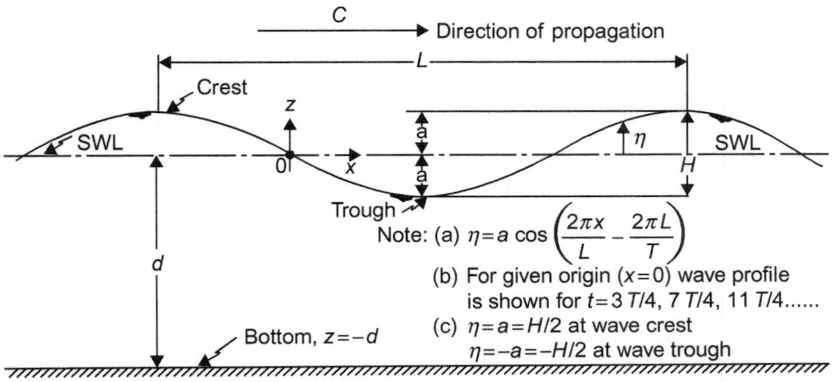

FIGURE 2.5 Simple sinusoidal progressive wave-propagation curve.

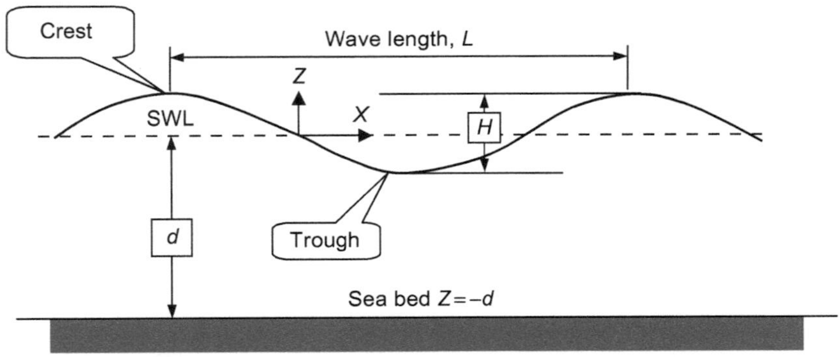

FIGURE 2.6 Water particle displacement for deep and shallow water.

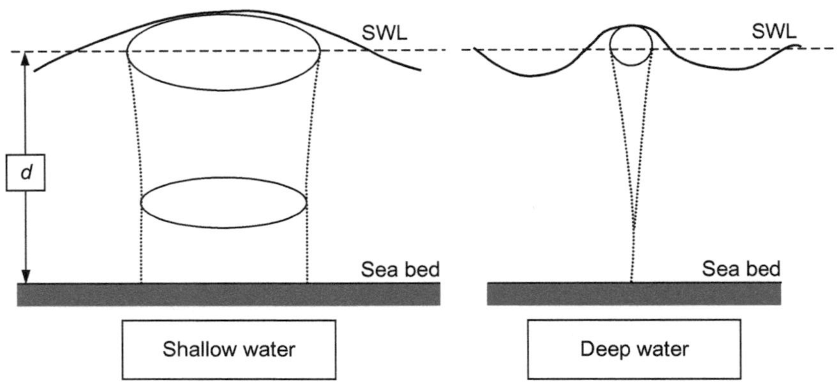

FIGURE 2.7 Local fluid velocities and acceleration.

- Acceleration

$$a_x = [+gpH/L \cosh F_1/\cosh F_2] \sin F_3 \tag{2.13}$$

$$a_z = [-gpH/L \sinh F_1/\cosh F_2] \cos F_3 \tag{2.14}$$

The theories concerning the modeling of ocean waves were developed in the nineteenth century. Practical wave force theories concerning actual offshore platforms were not developed until 1950, when the Morison equation was presented and the wave forces on a vertical pipe were shown to be as illustrated in Figure 2.8:

$$F = F_D + F_I \tag{2.15}$$

where F_D is the drag force and F_I is the inertia force.

- Drag force: The drag force due to a wave acting on an object can be found by:

$$F_D = 1/2\,\rho CdV^2A \tag{2.16}$$

  where F_D is the drag force (N), Cd is the drag coefficient (no units), V is the velocity of the object (m/s), A is the projected area (m^2) and ρ is the density of water (kg/m^3).
- Inertia force: The inertia force due to a wave acting on an object can be found by:

$$F_I = \pi\rho aCmD^2/4 \tag{2.17}$$

  where F_I is the inertia force (N), Cm is the mass coefficient (no units), a is the horizontal water particle acceleration (m^2/s), D is the diameter of the cylinder (m) and ρ is the density of water (kg/m^3).

Wave Load Calculation

- The values of *Cd* and *Cm* are dimensionless and the values most often used in the Morison equation are 0.7 and 2.0, respectively. API recommends 0.65 and 1.6, respectively, for smooth surfaces or 1.05 and 1.2, respectively, for rough surfaces, such as surfaces with marine organism growth.

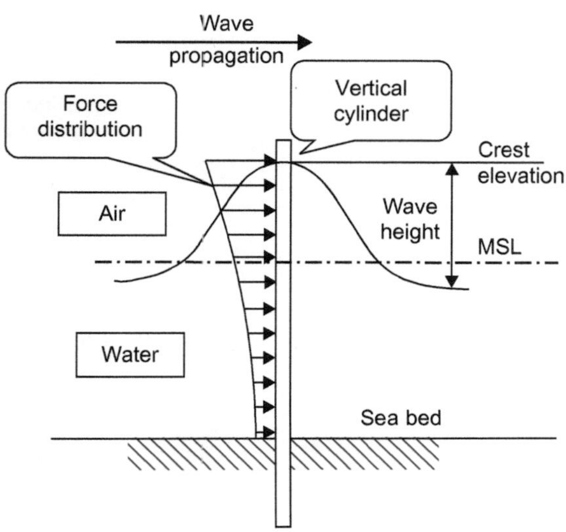

FIGURE 2.8 Wave force distribution on a vertical pipe.

- Water particle velocity and acceleration are functions of wave height, wave period, water depth, distance above the sea bottom and time.
- The most elementary wave theory was presented by Airy in 1845.
- Another widely used theory, known as the stream function theory, is a nonlinear solution similar to the Stokes fifth-order theory.

Figure 2.9 presents an example of the principal wave directions in the Gulf of Mexico (GoM) through 290° read clockwise (i.e., wave directions applied to the platform from the north, northeast, east, southeast, and so on through eight directions as wave height versus water depth). Note that the principal wave direction, with the maximum wave height value, will be defined in the metocean report for the location of the platform. Based on API-RP2A, a reduction factor should be applied on the other direction, as shown in Figure 2.9 for a water depth higher than 12 m, noting that the directions are rotated every 22.5° to define the eight directions regardless of the platform's orientation.

Comparison of Wind and Wave Calculations

Calculation of the force affecting the structure due to wind takes the drag force into consideration and neglects the inertia force, but in the case of waves, drag force and inertia force are considered in the calculation. The following example demonstrates the reason for neglecting the inertia force in wind load.

If the platform is oriented as shown in Figure 2.9, so that the direction of wind is not perpendicular to the faces of the platform, during calculation of the wind

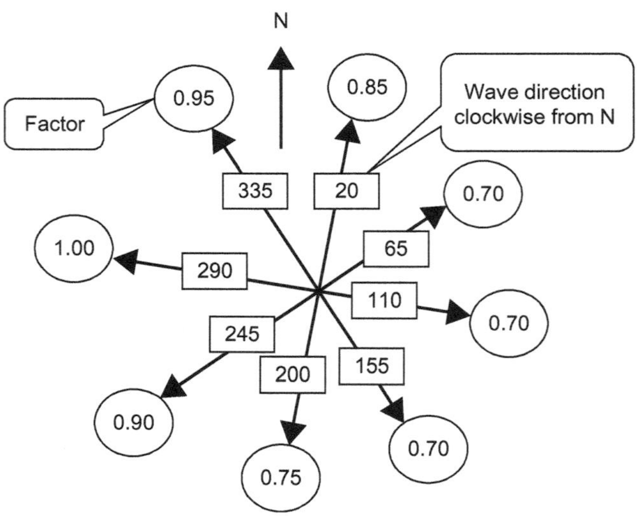

FIGURE 2.9 The design shows wave directions and factors to apply to the omnidirectional wave height in the GoM.

load, the angle of inclination of the wind direction to the platform should be considered in each direction.

Example

Pipe Dia. = 0.4 m

$$V_{air} = 25\ m/s \quad V_{water} = 1\ m/s$$

$$a_{air} = 1\ m^2/s \quad a_{water} = 1\ m^2/s$$

$$\rho_{air} = 1.3\ kg/m^3 \quad \rho_{water} = 1000\ kg/m^3$$

$$F_d = (1/2) \times Cd \times \rho \times V^2 \times A$$

$$F_m = Cm \times \rho \times \pi \times (D^2/4) \times a$$

Air

$$F_d = (1/2)(0.8)(1.3)(25)^2(0.4) = 130\ N$$

$$F_m = 2(1.3)(\pi)(0.4)^2/4\,(1) = 0.33\ N$$

Water

$$F_d = (1/2)(0.8)(1000)(1)^2(0.4) = 160\ N$$

$$F_m = 2(1000)(\pi)(0.4)^2/4\,(1) = 251\ N$$

Conductor Shielding Factor

Depending on the configuration of the structure, the number of well conductors can add a significant portion to the total wave forces. If the conductors are closely spaced, the forces on them may be reduced by hydrodynamic shielding.

However, there will be no reduction factor due to shielding when the spacing between conductors is equal to or greater than 4 times the conductor diameter. With closer spacing, the following equation can be applied:

$$S_f = 0.25(S/D_c)$$

where S_f is the shielding reduction factor, S is the spacing between conductors, and D_c is the conductor diameter.

2.5.2 Current Force

According to ISO 9002, the most common categories of ocean currents are:

- Wind-generated currents
- Tidal currents
- Circulational currents
- Loop and eddy currents

- Soliton currents
- Longshore currents

Wind-generated currents are caused by wind stress and the atmospheric pressure gradient during a storm.

Tidal currents are regular and follow the harmonic astronomical motions of the planets. The maximum tidal current precedes or follows the highest and lowest astronomical tides, HAT and LAT, respectively. Tidal currents are generally weak in deep water, but they are strengthened by shoreline configurations. Strong tidal currents exist in inlets and straits in coastal regions.

Circulational currents are steady, large-scale currents of the general oceanic circulation (i.e., the Gulf Stream in the Atlantic Ocean). Parts of circulation currents may break off from the main circulation to form large-scale eddies. Current velocities in such eddies (*loop and eddy currents*) can exceed the velocity of the main circulation current (i.e., loop current in the GoM).

Soliton currents are due to internal waves generated by density gradients. *Loop/eddy currents* and soliton currents penetrate deeply in the water column.

Longshore currents in coastal regions run parallel to the shore as a result of waves breaking at an angle on the shore; they are also referred to as *littoral* currents.

Earthquakes can cause unstable deposits to run down continental slopes and thereby set up gravity-driven flows. Such flows are called *turbidity currents*. Sediments in the flow have a higher density than the ambient water. Such currents should be accounted for in the design of pipelines crossing a continental slope with unstable sediments. Strong underwater earthquakes can also lead to the generation of *tsunamis,* which in coastal regions behave like a long, shallow water wave similar to a strong horizontal current.

The effects of currents on ships and offshore structures should be considered in their design, construction and operation. The effects of currents that should be considered in the design of offshore structures include:

- Currents can cause large, steady excursions and slow drift motions of moored platforms.
- Currents can create drag and lift forces on submerged structures.
- Currents can cause vortex-induced vibrations of slender structural elements and vortex-induced motions of large-volume structures.
- Interaction between strong currents and waves leads to change in wave height and wave periods.
- Currents can create seabed scouring around bottom-mounted structures.

Information on the statistical distribution of currents and their velocity profile is generally scarce for most areas of the world. Current measurement campaigns are recommended during the early phases of an offshore oil exploration development.

Site-specific measurements should extend over the water profile and over a period that captures several major storm events. Some general regional information on current conditions is given in ISO 19901-1 (2005), "Metocean Design and Operating Considerations."

If sufficient joint current-wave data are available, joint distributions of parameters and corresponding contour curves (or surfaces) for given exceedance probability levels can be established. Otherwise, conservative values using combined events should be applied (NORSOK N-003, DNV-OS-C101).

The presence of currents in the water produces some minor effects; the most affected variable is the current velocity, which should be added vectorially to the horizontal water particle velocity.

From a practical point of view, designers sometimes increase the maximum wave height by 5–10% to account for the current effect and the current is neglected in calculations.

Design Current Profiles

When detailed field measurements are not available, the variation in shallow water of tidal current velocity along the water depth will be modeled as a simple power law, assuming unidirectional current:

$$V_{c,tide}(Z) = V_{c,tide}(0)\left(\frac{d+z}{d}\right)^{\alpha} \quad \text{for} \quad Z \leq 0 \tag{2.18}$$

The variation of wind-generated current can be taken as either a linear profile from $z = -d_o$ to still water level (SWL),

$$V_{c,wind}(Z) = V_{c,wind}(0)\left(\frac{d_o+Z}{d_o}\right) \quad \text{for} \quad -d_o \leq Z \leq 0 \tag{2.19}$$

or a slab profile,

$$V_{c,wind}(z) = V_{c,wind}(0) \quad \text{for} \quad -d_o < Z < 0 \tag{2.20}$$

The profile giving the highest loads for the specific application should be applied.

Wind-generated currents can be assumed to vanish at a distance below the SWL:

$$V_{c,wind}(Z) = 0 \quad \text{for} \quad Z < -d_o \tag{2.21}$$

where $V_c(Z)$ is the total current velocity at level Z, z is the distance from the SWL measured positively upward, $V_{c,tide}(0)$ is the tidal current velocity at the SWL, $V_{c,wind}(0)$ is the wind-generated current velocity at the SWL, d is the water depth to the SWL measured positively upward, d_o is the reference depth for wind-generated current, $d_o = 50$ m, and α is the exponent, typically $\alpha = 1/7$.

In deep water along an open coastline, wind-generated current velocities at the SWL may, if statistical data are not available, be taken as $V_{c,wind}(0) = k\ (U_1$ hour, 10 m), where $k = 0.015 - 0.03$ and (U_1 hour, 10 m) is the 1-hour sustained wind speed at height 10 m above sea level.

The variation of current velocity over depth depends on the local oceanographic climate, the vertical density distribution and the flow of water into or out of the area. This may vary from season to season. Deepwater profiles may be complex. Current direction can change 180° with depth. A current profile for a location in the GoM is shown in API-RP2A (2007). API mentions that the minimum speed is 0.31 km/hr (see Figure 2.10).

Current Profile

The recommended current profile given in ISO 19901-1 (2005) is:

$$U_c(z) = U_{cs}\left(\frac{z+d}{d}\right)^{1/7} \tag{2.22}$$

where $U_c(z)$ is the current speed at elevation $z(z \leq 0)$, $U_{cs}(z)$ is the surface current speed at $z = 0$, d is the still-water depth and z is the vertical coordinate measured positively upward from the SWL.

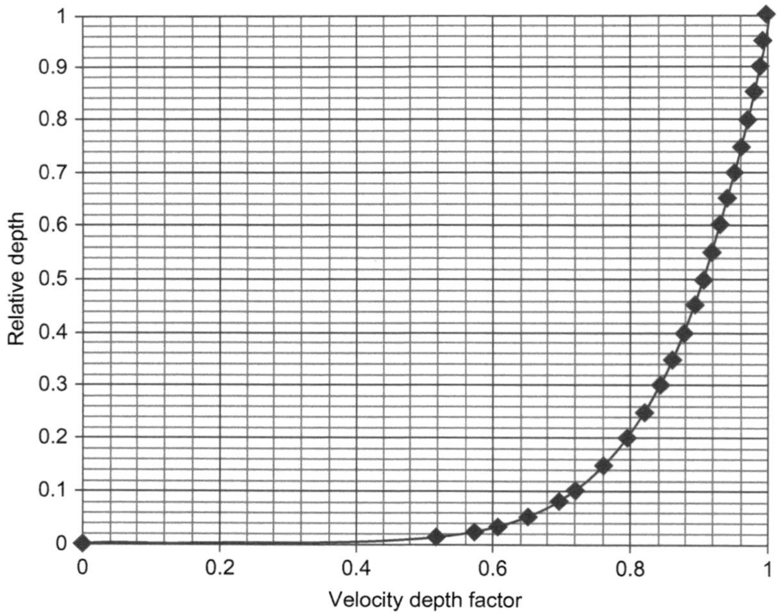

FIGURE 2.10 Current profile.

A current profile is illustrated in Figure 2.10, and Table 2.19 gives the current velocity factors along the water depth to be applied to the depth-mean current speed, to give current speeds at different depths.

TABLE 2.19 Relative Depth and Current Velocity Reduction Factor Along Depth

Relative Depth	Velocity Depth Factor
0	0
0.01	0.5180
0.02	0.5719
0.03	0.6060
0.05	0.6518
0.08	0.6971
0.1	0.7197
0.15	0.7626
0.2	0.7946
0.25	0.8203
0.3	0.8420
0.35	0.8607
0.4	0.8773
0.45	0.8922
0.5	0.9057
0.55	0.9181
0.6	0.9296
0.65	0.9403
0.7	0.9503
0.75	0.9597
0.8	0.9686
0.85	0.9771
0.9	0.9851
0.95	0.9927
1	1

The current profile presented by API RP2A (2007) focuses on the Gulf of Mexico (GoM) only. For other locations, the current profile should be obtained before the design is started, because it is different from location to another. In general, for GoM, levels below SWL from 80 to 180 m will have a nearly constant current speed, while the current speed will increase gradually from 90 m to 60 below the SWL and then will be constant from 60 m below SWL until the SWL.

2.5.3 Earthquake Load

Approximately 100 conventional steel pile-supported platforms have been installed in high-activity earthquake regions, such as offshore California, Alaska, New Zealand, Japan, China and Indonesia. Offshore structures have also been installed in new areas in high-activity earthquake regions, such as Venezuela, Trinidad and the Caspian Sea.

Ductility level (DLE) requirements are intended to ensure that the platform has sufficient reserve capacity to prevent its collapse during rare intense earthquake motion, although structural damage may occur.

In areas of low seismic activity, less than 0.05 g, the design for environmental loading from everything other than earthquakes will provide sufficient resistance against earthquakes. In seismic activity of 0.05–0.1 g, the important items are the deck appurtenances, such as piping, facilities and others. See Table 2.20 for more information about the seismic zone factor in API, which is the same as that of the American National Standard Institute (ANSI) for minimum load, and the Uniform Building Code (UBC; 1997) seismic zone factor, which is more traditionally used for seismic load.

The platform should be checked for earthquake resistance using a dynamic analysis procedure, such as spectrum analysis or time history analysis. The spectrum analysis method is for structures with a uniform shape and structure,

TABLE 2.20 Comparison Between the Seismic Zone Factor and Horizontal Ground Acceleration for API and UBC

Seismic zone (relative seismicity) factor, *Z**	1	2	3	4	5
Horizontal ground acceleration to gravitational acceleration, g in API	0.05	0.10	0.20	0.25	0.40
Horizontal ground acceleration to gravitational acceleration, g in UBC	0.075	0.15	0.20	0.30	0.40

**The zones in UBC are 1, 2A, 2B, 3, and 4, which corresponds to 1, 2, 3, 4, and 5 in API, respectively.*

and a structure height between 100 m and 150 m, or for structures where the ratio between the heights and the horizontal dimensions is more than 5% in the earthquake load direction.

The effect of seismic activity on a structure is a static lateral load on the floor level of the structure, and its value is calculated from the dynamic properties of the natural period and natural mode, which are calculated by modal analysis. The calculated lateral force should not be less than 80% of the lateral load calculated from the previous method.

Earthquake loading should be determined by response spectrum analysis in accordance with API RP2A or UBC 1997 and the way of calculating the seismic load should be stated in the basis of design document. Figure 2.11 shows the spectral acceleration, spectral velocity and spectral displacement values for three soil types, which are classified as follows:

- Soil type A: Hard rock and rock crystalline, conglomerate, or shale-like material generally having shear wave velocities in excess of 914 m/sec (3000 ft/sec).
- Soil type B: Shallow, very dense sand, silts and stiff clays with undrained shear strength in excess of about 100 kPa (2000 psf) limited to depths of

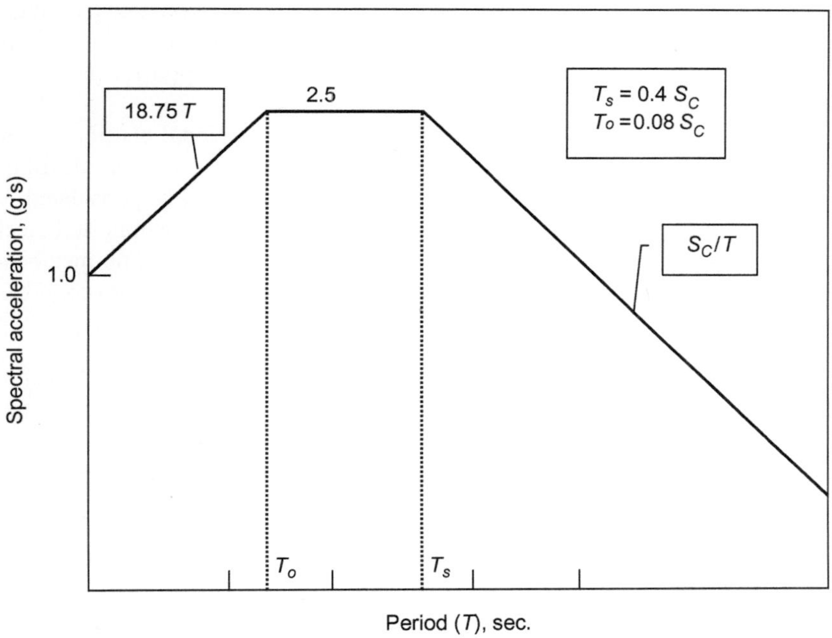

FIGURE 2.11 Normalized response spectra for design.

less than about 200 ft (61 m), and overlying rock-like materials with standard penetration test for cohesionless soil layers provide higher than 50 blows per foot.

- Soil type C: Deep, strong sands, silts and stiff clays with thicknesses in excess of about 61 m (200 ft) and overlying rock-like materials with undrained shear strength from 50 to 100 kPa with standard penetration test from 15 to 50 blows per foot.

Table 2.21 provides the values of the seismic factor, S_C, which are related to the seismic zone and the soil type. For design purposes, these values are used to obtain the response spectra for the required location.

Both the strength level (SLE) and ductility level (DLE) relative to earthquakes should be considered in the design, and a period of 200 years should be used for SLE.

Strength Requirements

Twice the SLE peak accelerations should be used in the DLE analysis. The design of the platform for an earthquake load depends on the dynamic analysis. The mass used in the dynamic analysis should consist of the mass of the platform associated with gravity loading, the mass of the fluids enclosed in the structure and the appurtenances and the added mass. The added mass may be estimated as the mass of the displaced water for motion transverse to the longitudinal axis of the individual structural framing and appurtenances. In motions along the longitudinal axis of the structural framing and appurtenances, the added mass may be neglected.

The analytical model should include the three-dimensional distribution of platform stiffness and mass. Asymmetry in platform stiffness or mass distribution may lead to significant torsional response, which should be considered.

In computing the dynamic characteristics of braced, pile-supported steel structures, uniform model damping ratios of 5% of critical structural members should be used for an elastic analysis. Where substantiating data exist, other damping ratios may be used.

TABLE 2.21 Seismic Factor S_C

	Seismic Zone				
Soil Type	1	2	3	4	5
A	1.0	1.0	1.0	1.0	1.0
B	1.44	1.39	1.33	1.36	1.40
C	1.5	1.45	1.42	1.5	1.45

According to ISO 19002, when the response spectrum method is used and one design spectrum is applied equally in both horizontal directions, the complete quadratic combination (CQC) method may be used for combining modal responses and the square root of the sum of the squares (SRSS) may be used for combining the directional responses. If other methods are used for combining modal responses, such as SRSS, care should be taken not to underestimate corner pile and leg loads. For the response spectrum method, as many modes should be considered as required for an adequate representation of the response. At least two modes having the highest overall response should be included for each of the three principal directions, plus significant torsional modes.

If the time history method is used, the design response should be calculated as the average of the maximum values for each of the time histories considered.

Earthquake loading should be combined with other simultaneous loadings, such as gravity, buoyancy and hydrostatic pressure. Gravity loading should include the platform's dead weight (comprised of the weight of the structure, equipment and appurtenances), actual live loads and only 75% of the maximum supply and storage loads.

In the calculation of member stresses, the stresses due to earthquake-induced loading should be combined with those due to gravity, hydrostatic pressure and buoyancy. For the strength requirement, the basic AISC allowable stresses may be increased by 70%. Pile soil performance and pile design requirements should be determined on the basis of special studies. The studies should consider the design loadings, installation procedures, earthquake effects on soil properties and characteristics of the soils as appropriate to the axial or lateral capacity algorithm being used. Both the stiffness and capacity of the pile foundation should be addressed in a compatible manner for calculating the axial and lateral response.

Ductility Requirements

Ductility requirements are intended to ensure that platforms located in seismically active areas have adequate reserve capacity to prevent collapse under a rare, intense earthquake. When implemented in the strength design of certain platforms, structure-foundation systems will not require an explicit analytical demonstration of adequate ductility. These systems are similar to those for which adequate ductility has already been demonstrated analytically in seismically active regions where the intensity ratio of the rare, intense earthquake ground motions to SLE earthquake ground motions is 2% or less.

No ductility analysis of conventional jacket-type structures with eight or more legs is required if the structure is to be located in an area where the intensity ratio of rare, intense earthquake ground motion to SLE earthquake ground motion is 2% or less; the piles are to be founded in soils that are stable to underground motions imposed by the rare, intense earthquake; and the following conditions are adhered to in configuring the structure and proportioning members:

- Jacket legs, including any enclosed piles, are designed using twice the SLE seismic loads.
- Diagonal bracing in the vertical frames is configured such that shear forces between horizontal frames or in vertical runs between legs are distributed approximately equally to both tension and compression diagonal braces, and "K" bracing is not used where the ability of a panel to transmit shear forces is lost if the compression brace buckles. Where these conditions are not met, including areas such as the portal frame between the jacket and the deck, the structural components should be designed using twice the SLE seismic loads.
- Horizontal members are provided between all adjacent legs at horizontal framing levels in vertical frames and these members have sufficient compression capacity to support the redistribution of loads resulting from the buckling of adjacent diagonal braces.
- The slenderness ratio (Kl/r) of primary diagonal bracing in vertical frames is limited to 80% and their ratio of diameter to thickness is limited to $1900/Fy$, where Fy is in ksi ($13{,}100/Fy$ for Fy in MPa).

All nontubular members at connections in vertical frames are designed as compact sections in accordance with the AISC specifications or using twice the SLE seismic loads.

Structure-foundation systems should be analyzed to demonstrate their ability to withstand the rare, intense earthquake without collapsing. The characteristics of the rare, intense earthquake should be developed from site-specific studies of local seismicity. Demonstration of the stability of the structure-foundation system should be by analytical procedures that are rational and reasonably representative of the expected response of the structural and soil components of the system to intense ground shaking.

Models of the structural and soil elements should include their characteristic degradation of strength and stiffness under extreme load reversals and the interaction of axial forces and bending moments, hydrostatic pressures and local inertial forces, as appropriate. The P-delta effect of loads acting through elastic and inelastic deflections of the structure and foundation should be considered.

Topside Structure, Appurtenances and Equipment

Topside structure design for earthquake loads should include the effects of the global dynamic response of the structure and, if appropriate, the effects of local dynamic response of the topside and appurtenances.

It is recommended that topside response spectra or other in-structure response spectra be obtained from time-history analyses of the complete structure. The topside response spectra are recommended to be the average values from at least four time-history analyses. Direct spectra-to-spectra generation techniques may also be used; however, such methods should be calibrated against the time-history method. The topside response spectra should be broadened to account for uncertainties in structure frequencies and soil-structure interaction.

Seismic actions on topside equipment, piping and appurtenances should be derived by dynamic analysis using either:

1. An uncoupled analysis with deck-level floor response spectra as input.
2. A coupled analysis that properly includes a simple dynamic model of the relevant part of the topside or the appurtenance in the global structural model.

Equipment, piping and other topside appurtenances should be designed and supported to resist extreme-level earthquake loads. On the other hand, displacements and deformations of the topside in extreme earthquake load should be limited or designed against to avoid damage to the equipment, piping, appurtenances and supporting structures.

The topside structure and any structural system carrying critical equipment from a safety perspective should be designed to function during and after an abnormal level of earthquake. The ISO 19902 recommendation for abnormal level of earthquake is the same as the API recommendation for the ductile requirement: the structural systems that carry equipment containing hazardous materials, such as the separators for example, should be designed to prevent catastrophic failure or rupture during an abnormal earthquake.

Therefore, in ISO the partial load factor, E, is increased from 0.9 to 1.15 for abnormal level analysis of deck-supported structures, topside equipment and equipment tie downs.

2.5.4 Ice Loads

Ice loads are particularly relevant on offshore structures in Alaska. It is also very important to know that drifting ice travels at a speed about 1–7% of the wind speed. A typical ice island in Alaska is about 1 km in diameter, with 1 km thickness, and has a travel speed of 3 knots. Generally, the ratio of the thickness of ice above water to the thickness of that under water is about 1:2, but it can range from 1:1 to 1:7. API recommends the following formula to calculate the ice impact force:

$$F_i = C_i S_{ci} A_o \tag{2.23}$$

The importance of grouping effects in ice loads depends on the spacing of individual members. Generally, the following rules are used:

1. Spacing ≥ 6 diameters. Ice will crush against the tubular members and pass through and around the platform if the tubular spacing is greater than six times the tubular diameter. For groups of tubular members of different sizes, the average tubular diameter should be used to determine the spacing.
2. Spacing ≤ 4 diameters. As the tubular spacing decreases, interference effects may occur that influence both the ice load on the tubular members and the

failure mode with ice. At a tubular spacing less than four diameters, or with closely spaced conductors between platform legs, ice blocks may wedge inside the structure and the effective contact area becomes the out-to-out dimension across the closely spaced tubular members in the direction of the ice movement. In this case, the total ice load on the structure should be calculated with D taken as the out-to-out dimension across the closely spaced tubular members.

3. Spacing of 4 to 6 diameters. Ice forces should be determined by linear interpolation between loads for spacing of four to six diameters.

Note that shielding occurs when tubular members are located in the lee of other structural members. The loads on these piles may be considerably less, as the ice may fail in another mode or may simply be cleared away under pack ice pressures. The clearing forces are estimated as the product of the pack ice pressure (estimated at 2 ton/m width based on floe area, floe profile, wind speed and current velocity) and pile diameter.

2.5.5 Other Loads

Other loads attributable to the configuration of the platform and the environmental conditions include:

- Marine growth
- Scour

Other factors affecting loads on the structure include:

- Materials selection, corrosion, stress analysis, welding, structure analysis, design for fabrication and installation
- Marine civil engineering, such as installation equipment, installation methods and navigation safety instrumentation
- Naval architecture, such as flotation/buoyancy, towing, launching and controlled flooding

Marine Growth

Marine organism growth and adherence to the surface increase the diameter of the jacket member, increasing the drag force, according to API, on the surface 1.5 inches from MHHW to −150 ft, with MHHW 300 mm (1 foot) higher than MLLW. Smaller or larger values of thickness may be obtained from a site-underwater-specific survey and from previous studies in the same site.

Structural members are considered smooth if they are above MHHW or lower than 45 m (150 ft), where marine growth is light to ignorable. In the zone between MHHW and 45 m, they are considered rough.

Scour

Seabed scour affects both lateral and axial pile performance and capacity, but scour prediction is an uncertain art. Sediment transport studies may assist in defining scour design criteria, but local experience is the best guide. Practical measures used in design assume the scour diameter is 1.5 × the pile diameter.

2.6 DESIGN FOR ULTIMATE LIMIT STATE (ULS)

An action factor should be applied to each of the nominal external actions in the combinations given in Clause 7 of ISO 19902. (Action factors depend on the national or regional building code in use.) This is to ensure that the levels of reliability for topside design achieved are similar to that implied in other ISO 19900 series standards.

The combination of factored nominal actions causes amplified internal forces, S. A resistance factor is applied to the nominal strength of each component to determine its factored strength. Each component should be proportioned to have sufficient factored strength to resist S. The appropriate strength and stability criteria should be taken from the appropriate national or international building code. These criteria are the formulae for the nominal strength of the component and the associated resistance factors.

In some conditions, particularly during construction and installation, the internal forces should be computed from unfactored nominal actions and then the action factors should be applied to the internal forces to arrive at S, as discussed in ISO 19902.

Deformation actions can arise from the effects of fabrication tolerance, foundation settlement and the indeterminate effects of transportation and lift. For the primary structure supported by a multicolumn gravity base structure (GBS), movements of the column tops can also constitute significant deformation actions. They can also occur from operational or accidental thermal effects. All such actions should be considered in combination with operating actions to ensure that serviceability and ultimate limit states are not exceeded.

2.6.1 Load Factors

The partial action factors to be used when ISO 19902, AISC-LRFD, EC3, NS 3472, or BS 5950 Part 3 is the chosen code are given in Table 2.22. The action factors cover maximum gravity and extreme environmental and operating environmental combinations. Other relevant codes or standards may be used, in which case appropriate action factors should be evaluated to achieve a similar level of reliability to that implied in the international standard. The procedure should be followed to derive appropriate sets of action factors, as necessary.

The internal force, S, resulting from the design action, F_d, should be calculated using

$$F_d = \gamma_G(G_1 + G_2) + \gamma_Q(Q_1 + Q_2) \tag{2.24}$$

TABLE 2.22 Partial Action Factors for Maximum Gravity from Different International Standards and Specifications

Code	Permanent, γ_G	Variable, γ_Q
ISO 19902	1.30	1.50
AISC-LRFD	1.25	1.40
NS 3472	1.25	1.45
EC3	1.30	1.50
BS 5950	1.45	1.65
BS 5950, Part 3	1.25	1.45

where G_1 is permanent load imposed on the structure by the self-weight of the structure with associated equipment and other objects, G_2 is permanent load imposed on the structure by self-weight of equipment and other objects that remain constant for long periods but which can change from one mode or operation to another or during a mode of operation, Q_1 is variable load imposed on the structure by the weight of consumable supplies and fluids in pipes, tanks and stores, the weight of transportable vessels and containers used for delivering supplies, and the weight of personnel and their personal effects on the structure, and Q_2 is the short-duration variable load imposed on the structure from operations, such as lifting of drill string, lifting by cranes, machine operations, vessel mooring, and helicopter loadings.

2.6.2 Extreme Environmental Situation for Fixed Offshore Platforms

The internal force, S, resulting from the design action, F_d, should be calculated using

$$F_d = \gamma_G(G_1 + G_2) + \gamma_Q Q_1 + \gamma_E(E_e + 1.25De) \tag{2.25}$$

When the internal forces due to gravity forces oppose those due to wind, wave and current forces, the internal force, S, resulting from the design action, F_d, should be calculated using reduced partial action factors

$$F_d = (1/\gamma_G)(G_1 + G_2) + (1/\gamma_Q)Q_1 + \gamma_E(E_e + 1.25De) \tag{2.26}$$

For this combination, G_2 and Q_1 should exclude any actions that cannot be ensured to be present during an extreme storm to maximize the difference between the opposing action effects.

TABLE 2.23 Partial Action Factors for Extreme Environmental Conditions from Different International Standards and Specifications

	Partial Load Factor		
Code	Permanent, γ_G	Variable, γ_Q	Environmental, γ_E
ISO 19902	1.10	1.10	1.00 γ_{ELs}
AISC-LRFD	1.05	1.05	0.96 γ_{ELs}
NS 3472	1.05	1.05	0.96 γ_{ELs}
EC3	1.10	1.10	1.00 γ_{ELs}
BS 5950	1.20	1.20	1.11 γ_{ELs}
BS 5950, Part 3	1.05	1.05	0.96 γ_{ELs}

Note: γ_{ELs} is the appropriate partial factor for the substructure.

The appropriate partial action factors for the environmental load are dependent on the location of the installation.

ISO 19902 allows a value of γ_E of 1.35 when no other information is available. The partial action factors for selected codes and standards are given in Table 2.23.

2.6.3 Operating Environmental Situations—Fixed Platforms

Platform operations are often limited by environmental conditions and differing limits may be set for different operations. Examples of operations that might be limited by environmental conditions include:

- Drilling and workover
- Crane transfer to and from supply vessels
- Crane operations on deck
- Deck and overside working
- Deck access

Each operating situation that might be restricted by environmental conditions should be assessed as shown in Equation (2.27), in which E_o and D_o represent the environmental action limiting the operations. The value of Q_2 should be that associated with the particular operating situation being considered.

The internal force, S, resulting from the design action, F_d, should be calculated using

$$F_d = (1/\gamma_G)(G_1 + G_2) + 1/\gamma_Q(Q_1 + Q_2) + \gamma_E(E_o + D_o) \qquad (2.27)$$

The action factors for operating environmental conditions in selected codes and standards are given in Table 2.24.

TABLE 2.24 Partial Action Factors for Operating Environmental Conditions in Different International Standards and Specifications

	Partial Load Factor		
Code	Permanent, γ_G	Variable, γ_Q	Environmental, γ_E
ISO 19902	1.30	1.50	1.20
AISC-LRFD	1.25	1.40	1.15
NS 3472	1.25	1.45	1.15
EC3	1.30	1.50	1.20
BS 5950	1.45	1.65	1.35
BS 5950, Part 3	1.25	1.45	1.15

2.6.4 Partial Action Factors for Platform Design

Each member, joint and foundation component should be checked for strength using the internal force due to load effect, *S*, resulting from the design action, calculated by Equations (2.28) and (2.29):

$$Q = 1.1\,G_1 + 1.1\,G_2 + 1.1\,Q_1 + 0.9\,E \qquad (2.28)$$

where *E* is the inertia action induced by the extreme level earthquake (ELE) ground-motion and determined using dynamic analysis procedures, such as response spectrum analysis or time-history analysis. G_1, G_2 and Q_1 should include loads that are likely to be present during an earthquake.

When contributions to the internal forces due to weight oppose the inertia actions due to the earthquake, the partial load factors for permanent and variable loads should be reduced so that:

$$Q = 0.9\,G_1 + 0.9\,G_2 + 0.8\,Q_1 + 0.9\,E \qquad (2.29)$$

where G_1, G_2 and Q_1 should include only loads that are reasonably certain to be present during an earthquake.

For global assessment of the offshore structure platform, Tables 2.25 and 2.26 present a matrix for load combination. This is the traditional load combination used in input for the design or assessment of fixed offshore platforms. Table 2.25 presents the load combination for 1-year storm conditions and Table 2.26 presents the load combination for 100-year storm conditions. Tables 2.27 and 2.28 show a matrix for the load combination versus the applied load in working stress design (WSD) under 1-year and 100-year storm conditions, respectively. The load combination for the maximum pile tension condition is illustrated in Table 2.29.

TABLE 2.25 Matrix for Load Combination, 1-Year Storm Conditions

Load Case	Load Condition	Combination
1	1	Dead load + buoyancy
2	2	Unmodeled dead load (jacket and deck)
3	3	Blanket live load on main deck
4	4	Blanket live load on helideck
5	11	Wind + wave + current hitting 0.0°
6	12	Wind + wave + current hitting 45°
7	13	Wind + wave + current hitting 90°
8	14	Wind + wave + current hitting 135°
9	15	Wind + wave + current hitting 180°
10	16	Wind + wave + current hitting 225°
11	17	Wind + wave + current hitting 270°
12	18	Wind + wave + current hitting 315°

TABLE 2.26 Matrix for Load Combination,100-Year Storm Conditions

Load Case	Load Condition	Combination
1	1	Dead load + buoyancy
2	2	Unmodeled dead load (jacket and deck)
3	3	Blanket live load on main deck
4	4	Blanket live load on helideck
5	21	Storm wind + wave + current hitting 0.0°
6	22	Storm wind + wave + current hitting 45°
7	23	Storm wind + wave + current hitting 90°
8	24	Storm wind + wave + current hitting 135°
9	25	Storm wind + wave + current hitting 180°
10	26	Storm wind + wave + current hitting 225°
11	27	Storm wind + wave + current hitting 270°
12	28	Storm wind + wave + current hitting 315°

TABLE 2.27 Matrix for Load Combination versus Applied Load in 1-Year Storm Conditions

Load	Load Condition											
Combination	1	2	3	4	11	12	13	14	15	16	17	18
30	1.1	1.0	1.0	1.0	1.0							
31	1.1	1.0	1.0	1.0		1.0						
32	1.1	1.0	1.0	1.0			1.0					
33	1.1	1.0	1.0	1.0				1.0				
34	1.1	1.0	1.0	1.0					1.0			
35	1.1	1.0	1.0	1.0						1.0		
36	1.1	1.0	1.0	1.0							1.0	
37	1.1	1.0	1.0	1.0								

TABLE 2.28 Matrix for Load Combination versus Applied Load in 100-Year Storm Conditions

Load Combination	Load Condition											
	1	2	3	4	21	22	23	24	25	26	27	28
40	1.1	1.0	1.0	1.0	1.0							
41	1.1	1.0	1.0	1.0		1.0						
42	1.1	1.0	1.0	1.0			1.0					
43	1.1	1.0	1.0	1.0				1.0				
44	1.1	1.0	1.0	1.0					1.0			
45	1.1	1.0	1.0	1.0						1.0		
46	1.1	1.0	1.0	1.0							1.0	
47	1.1	1.0	1.0	1.0								1.0

TABLE 2.29 Matrix for Load Combination versus Applied Load in Maximum Pile Tension Conditions

	Load Condition											
Load Combination	1	2	3	4	11	12	13	14	15	16	17	18
40	0.9	0.9	0.75		1.0							
41	0.9	0.9	0.75			1.0						
42	0.9	0.9	0.75				1.0					
43	0.9	0.9	0.75					1.0				
44	0.9	0.9	0.75						1.0			
45	0.9	0.9	0.75							1.0		
46	0.9	0.9	0.75								1.0	
47	0.9	0.9	0.75									1.0

2.7 COLLISION EVENTS

If a rigorous impact analysis requires, accidental (collision event) design should be established representing bow, stern and beam-on impacts on all exposed components.

The collision events should encompass both fairly frequent conditions, during which the structure would suffer only insignificant damage, and rare events, where emphasis is on avoiding a complete loss of integrity of the structure.

Two energy levels should be considered:

1. Low energy level, representing the frequent condition, based on the type of vessel that would routinely approach alongside the platform, such as a supply boat, with velocities representing normal maneuvering of the vessel as it approaches, leaves or stands alongside the platform.
2. High energy level, representing a rare condition, based on a vessel operating in the vicinity of the platform and drifting out of control in the worst sea state in which it is allowed to operate close to the platform.

In design for both collision situations, the first energy level represents a serviceability limit state for which the owner can set the requirements based on practical and economical considerations. The second energy level represents an ultimate limit state in which the structure is damaged but progressive collapse should not occur.

In both cases, the analysis should account for the vessel's mass, its added mass, orientation and velocity. Effective operational restrictions on vessel approach sectors can limit the collision exposure of some areas of the structure.

The vertical height of the impact zone should be established based on the dimensions and geometry of the structure and the vessel, and it should account for tidal ranges, operational sea state restrictions, vessel draft and motions of the vessel.

2.7.1 Vessel Collision

Accidental damage should be considered for all exposed elements of an installation in the collision zone. The vertical extent of the collision zone should be assessed on the basis of visiting vessel draft, maximum operational wave height and tidal elevation.

Accidental Impact Energy

Total Kinetic Energy

The total kinetic energy involved in collisions can be expressed as:

$$E = \tfrac{1}{2} a m V^2 \tag{2.30}$$

where m is the vessel displacement (kg); a is the vessel added mass coefficient, which is 1.4 for sideways collision and 1.1 for bow or stern collision; and V is the impact speed (m/s).

The total kinetic energy, E, should be taken to be at least:

- 14 MJ for sideways collision
- 11 MJ for bow or stern collisions

This corresponds to a vessel of 5000 displacement tons with an impact speed of 2 m/sec. A reduced impact energy may be acceptable in cases where the size of visiting vessels and/or their operations near the installation are restricted. In this instance, a reduced vessel size and/or reduced impact speed may be considered.

The reduced impact speed, V, in m/sec, may be estimated numerically from the empirical relation:

$$V = \tfrac{1}{2} Hs (\mathrm{m/sec}) \tag{2.31}$$

where Hs is the maximum permissible significant wave height in meters for vessel operations near the installation, per OTI (1988).

The energy-absorbing mechanisms effective during the collision should be evaluated. Typically, local member denting, elastic and plastic deflection of the impacted member, global elastic and plastic response of the whole structure and denting of the ship are the main mechanisms.

In a rigorous impact analysis, the collision actions should be evaluated based on a dynamic time simulation. The duration of the simulation should be sufficient to cover all relevant phases of the collision and the energy-dissipation process.

Dropped Objects

When evaluating the impact risk from dropped objects, the nature of all crane operations in the platform vicinity should be taken into account. If the probability of impact is not negligible, relevant accidental design situations should be defined and evaluated following the requirements. Depending on the consequences for the structural integrity, the need for a rigorous impact analysis should be determined.

Irrespective of whether a rigorous analysis is required, robustness in relation to impact loads (vessel collisions and dropped objects) should be incorporated into the design by indirect means, such as:

- Avoiding weak elements in the structure (particularly at joints).
- Selecting materials with sufficient toughness.
- Ensuring that critical components are not placed in vulnerable locations.

2.8 FIRES AND EXPLOSIONS

Hydrocarbon-pool fires on the sea surface can cause heating of, and hence degradation of the properties of, structural components. Sources of hydrocarbons include conductor or riser fracture or spillage from the topside after a process vessel rupture, while ignition sources can include radiation from oil burners and flares.

If the probability of exposure of the structure to fires is not negligible, relevant accident-related design situations should be defined and evaluated. Damage resulting from fires and explosions should be provided for in the structure design.

Generally, conventional steel-framed structures with relatively small topside structures that do not have enclosed compartments containing flammable fluids or ignition sources do not require design against explosion. However, if explosion studies of the topside indicate significant or unusual actions on the structure or indicate the need for specific support requirements to ensure that topside integrity is maintained, then such actions or support requirements should be provided for in the structure design.

2.9 MATERIAL STRENGTH

Steel should conform to a definite specification and to the minimum strength level, group and class specified by the designer. Certified mill test reports or certified reports of tests made by the fabricator or a testing laboratory in accordance with ASTM A6 or A20, as applicable to the specification listed in Table 2.30, constitute evidence of conformity with the specification. Unidentified steel should not be used.

2.9.1 Steel Groups

Steel may be grouped according to strength level and welding characteristics as follows:

- Group I are mild steels with specified minimum yield strengths of 279 MPa or less. These steels may be welded by any of the welding processes described in AWS D1.1.

TABLE 2.30 Steel Chemical Composition Based on ASTM

C	0.18% max
Mn	1.5% max
Nb	0.10% max
V	0.015% max
S*	0.025% max
CE**	0.42% max

*For plate designated on the order as Z grade, with through-thickness properties, sulfur content should be limited to 0.008% max.

**Carbon equivalent (CE) should be based on product analysis and should be calculated using the following equation:

$$\text{Carbon equivalent (CE)} = C + (Mn/6) + (Ni + Cu/15) + (Cr + Mo + V/5)$$

- Most of the platform's structural members are made from Group I steel, as shown in Table 2.31.
- Group II are intermediate-strength steels with specified minimum yield strengths of over 279 MPa through 360 MPa. These steels require the use of low-hydrogen welding processes.
- Group III are high-strength steels with specified minimum yield strengths in excess of 52 ksi (360 MPa).

Table 2.31 presents the steel groups, their yield strength ranges and their charpy toughness.

Steels may be used provided that each application is investigated with regard to:

1. Weldability and special welding procedures that may be required.
2. Fatigue problems that may result from the use of higher working stresses.
3. Notch toughness in relation to other elements of fracture control, such as fabrication, inspection procedures, service stress and environmental temperature.

The steel should be made by the basic oxygen or basic electric arc furnace process. All steel should be fully killed and made to fine grain practice. Rimming steel should not be supplied. The minimum rolling reduction ratio of material used for plates should be 4:1.

In most offshore structures using ASTM A572 Grade 50 the material should conform to the requirements of ASTM A572 and ASTM A6, except as noted below. All steel should be supplied in the normalized condition.

TABLE 2.31 Materials Category and Selection for Structure

Steel Group	Yield Strength Range, MPa	Charpy Toughness	Structural Element
I	220–275	20 J	Primary and secondary bracing Brace end stubs at node Legs Piling Conductor panels Boat landings and walkways Stiffener elsewhere
II	280–395	35 J	Joint cans Brace end stubs at node Legs Stiffener at nodes Piling with thick wall at sea floor
III	400–455	45 J	Legs Bracing in area of high collision

The chemistry should conform to ASTM A572 (see Table 2.30), with the following additional requirements for product analysis:

$$\text{Carbon equivalent (CE)} = \text{C} + (\text{Mn}/6) + (\text{Ni} + \text{Cu}/15) + (\text{Cr} + \text{Mo} + \text{V}/5)$$

The ratio of soluble aluminum to nitrogen should be a minimum of 2:1. The supplier should submit a full chemical analysis, identifying maximum and minimum levels, with their bid.

Product analysis should be determined twice per cast and should be determined on the test sample used for verification of mechanical properties.

For mechanical testing, the tensile samples should be cut with the longitudinal axis of the test samples transverse to the principal direction of rolling. Test specimens should be prepared for testing from material in the delivered condition (i.e., from a plate). A separate piece is not acceptable.

For plate designated on the order as Z Grade, through-thickness testing requirements S4 of ASTM A6 should apply.

Charpy impact testing should be carried out in accordance with the requirements of S5 of ASTM A6. All tests should be carried out at 0°C. Minimum average absorbed energy should be 50 J, with a minimum individual value of 38 J. Frequency of testing should be in accordance with ASTM A673. No impact testing is required for 6 mm thickness or less.

When steel is used for topside offshore structures, the material should conform to the requirements of ASTM A36 and ASTM A6, except as noted below. All steel should be supplied in the normalized condition.

In this grade the carbon equivalent should not exceed 0.42% and the carbon content should not exceed 0.20% by product analysis. Rimming steel should not be permitted.

The mechanical testing and the tensile testing should be in accordance with ASTM A6 and the Charpy impact tests should be carried out in accordance with the requirements of S5 of ASTM A6. All tests should be carried out as follows:

- Test temperature = 0°C.
- Absorbed energy = 27 J (average) with 20 J (minimum individual).
- Frequency of testing should be in accordance with Section 5.1 of ASTM A673.
- No impact testing is required on material of 6 mm thickness or less.

For tubular members, steel-grade API 5L tubular members should be supplied in the following conditions:

- Diameters = 18 inches (457 mm)—seamless
- Diameters >18 inches (457 mm)—double-sided submerged arc welded

Tubular members should be supplied either in the normalized or quenched and tempered condition.

Product analysis should be undertaken and the chemistry should comply with API 5L X52 (see Table 2.32).

The yield and tensile strength of the parent material and the weld metal (where applicable) should comply with Table 2.33.

TABLE 2.32 Steel Chemical Composition Based on API 5L X52

C	0.17% max
Mn	0.8–1.5% max
Cr	0.30% max
Mo	0.12% max
V	0.06% max
Nb	0.02% max
Ni	0.08% max
Cu	0.05% max
S	0.025% max
P	0.02% max
Si	0.15–0.45% max
Al (total)[a]	0.06% max
N[a]	0.014% max
CE[b]	0.42% max
Pcm[c]	0.20% max

Note: Chemistry for API 5L Grade B shall comply with API 5L.
[a] *The minimum soluble aluminum to nitrogen ratio should be 2:1; soluble aluminum content should be defined as the total content (0.005%).*
[b] $CE = C + (Mn/6) + (Cr + Mo + V/5) + (Ni + Cu)/15$
[c] *Pcm = phase change material, which is used to express weldability, and is obtained from the following equation*

$$Pcm = C + Si/30 + (Mn + Cu + Cr)/20 + Mo/15 + V/10 + 5B$$

TABLE 2.33 Steel Mechanical Properties

Steel Type	Yield Strength (N/mm^2)	Tensile Strength (N/mm^2)
API 5L X52	359	455
API 5L B	241	414

Charpy impact tests should be taken in the transverse direction at 0°C and the frequency of testing should be in accordance with Section 5.1 of ASTM A673.

Tubular members should be clean and free from visible defects. Surface marks/imperfections, such as tears, laps, slivers, gouge scabs and seams should be dressed and the remaining thickness confirmed by ultrasonic test (UT). Magnetic particle inspection (MPI) should confirm defect removal. All dressed areas should blend smoothly into the contour of the tubular element.

Tubular members should not contain any dents greater than 3 mm or 1% × OD, whichever is less. The length of the dent should not exceed 25% × OD.

Pipe sizes available commercially are shown in Table 2.34. Most pipes used are standard: schedule 40 (extra strong) or schedule 80 (double extra strong). The weight and thickness of the pipes are presented in imperial and metric units.

All tubulars should be UT inspected for laminations over the full length in accordance with API 5L, except the depth of the reference notch should be 5% of the thickness. In addition, UT inspection using compression wave techniques should be carried out for a minimum distance of 50 mm from the end of the tubular member to ensure freedom from laminations.

UT inspection of welds in welded tubular members should be subject to 100% UT and should comply with API 5L.

2.9.2 Steel Classes

In selection of steel, consideration should be given to notch toughness characteristics suitable for the conditions of service. For this purpose, steels may be classified as follows:

- Class H steels are usually used for primary structure members, piling, jacket braces and legs and deck beams. Because this class of steel has a good record of application in welded structures at service temperatures above freezing, it is usually used in the Middle East and Africa, which have hot climates.
- Class N is used where the service temperature is 10°C to 0°C.
- Class C steels are used in subfreezing service temperatures (lower than 0°C), as in the North Sea or another cold climate. Because in the Atlantic the temperature can reach –40°C, the Charpy requirement test can be achieved at −20°C to −40°C (−4° to −40°F).

Impact testing frequency for Class C steels should be in accordance with the specification under which the steel is ordered; in the absence of other requirements, heat lot testing may be used.

Unless otherwise specified by the designer, plates should conform to one of the specifications listed in Table 2.35. Structural shape and plate specifications

TABLE 2.34 Pipe Dimensions and Properties

Nominal Pipe Size, in	Nominal Pipe Size, mm	OD, in	OD, mm	Schedule Designations ANSI/ASME	Wall Thickness, in	Wall Thickness, mm	Weight, lbs/ft	Weight, kg/m
1/2	15	0.84	21.3	STD/40S	0.109	2.77	0.851	1.27
1/2	15	0.84	21.3	XS/80S	0.147	3.73	1.088	1.62
1/2	15	0.84	21.3	XX	0.294	7.47	1.714	2.55
3/4	20	1.05	26.7	STD/40S	0.113	2.87	1.131	1.68
3/4	20	1.05	26.7	XS/80S	0.154	3.91	1.474	2.19
3/4	20	1.05	26.7	XX	0.308	7.82	2.441	3.63
1	25	1.315	33.4	STD/40S	0.133	3.38	1.679	2.5
1	25	1.315	33.4	XS/80S	0.179	4.55	2.172	3.23
1	25	1.315	33.4	XX	0.358	9.09	3.659	5.45
1-1/4	32	1.66	42.2	STD/40S	0.14	3.56	2.273	3.38
1-1/4	32	1.66	42.2	XS/80S	0.191	4.85	2.997	4.46
1-1/4	32	1.66	42.2	XX	0.382	9.7	5.214	7.76
1-1/2	40	1.9	48.3	STD/40S	0.145	3.68	2.718	4.05
1-1/2	40	1.9	48.3	XS/80S	0.2	5.08	3.631	5.4
1-1/2	40	1.9	48.3	XX	0.4	10.16	6.408	9.54
2	50	2.375	60.3	STD/40S	0.154	3.91	3.653	5.44

TABLE 2.34 Pipe Dimensions and Properties—cont'd

Nominal Pipe Size, in	Nominal Pipe Size, mm	OD, in	OD, mm	Schedule Designations ANSI/ASME	Wall Thickness, in	Wall Thickness, mm	Weight, lbs/ft	Weight, kg/m
2	50	2.375	60.3	XS/80S	0.218	5.54	5.022	7.47
2	50	2.375	60.3	XX	0.436	11.07	9.029	13.44
2-1/2	65	2.875	73	STD/40S	0.203	5.16	5.793	8.62
2-1/2	65	2.875	73	XS/80S	0.276	7.01	7.661	11.4
2-1/2	65	2.875	73	XX	0.552	14.02	13.69	20.37
3	80	3.5	88.9	STD/40S	0.216	5.49	7.576	11.27
3	80	3.5	88.9	XS/80S	0.3	7.62	10.25	15.25
3	80	3.5	88.9	XX	0.6	15.24	18.58	27.65
3-1/2	90	4	101.6	STD 40S	0.226	5.74	9.109	13.56
3-1/2	90	4	101.6	XS /80S	0.318	8.08	12.5	18.6
3-1/2	90	4	101.6	XX	0.636	16.15	22.85	34.01
4	100	4.5	114.3	STD /40S	0.237	6.02	10.79	16.06
4	100	4.5	114.3	XS /80S	0.337	8.56	14.98	22.29
4	100	4.5	114.3	XX	0.674	17.12	27.54	40.99
5	125	5.563	141.3	STD /40S	0.258	6.55	14.62	21.76

(Continued)

TABLE 2.34 Pipe Dimensions and Properties—cont'd

Nominal Pipe Size, in	Nominal Pipe Size, mm	OD, in	OD, mm	Schedule Designations ANSI/ASME	Wall Thickness, in	Wall Thickness, mm	Weight, lbs/ft	Weight, kg/m
5	125	5.563	141.3	XS /80S	0.375	9.53	20.78	30.93
5	125	5.563	141.3	XX	0.75	19.05	38.55	57.37
6	150	6.625	168.3	STD/40S	0.28	7.11	18.97	28.23
6	150	6.625	168.3	XS /80S	0.432	10.97	28.57	42.52
6	150	6.625	168.3	XX	0.864	21.95	53.16	79.12
8	200	8.625	219.1	STD/40S	0.322	8.18	28.55	42.49
8	200	8.625	219.1	XS/80S	0.5	12.7	43.39	64.58
8	200	8.625	219.1	XX	0.875	22.23	72.42	107.78
10	250	10.75	273.1	STD/40S	0.365	9.27	40.48	60.24
10	250	10.75	273.1	XS/80S	0.5	12.7	54.74	81.47
10	250	10.75	273.1	XX	1	25.4	104.13	154.97
12	300	12.75	323.9	STD/40S	0.375	9.53	49.56	73.76
12	300	12.75	323.9	XS/80S	0.5	12.7	65.42	97.36
12	300	12.75	323.9	XX	1	25.4	125.49	186.76
14	350	14	355.6	40	0.438	11.13	63.44	94.41
14	350	14	355.6	XS/80S	0.5	12.7	72.09	107.29

TABLE 2.34 Pipe Dimensions and Properties—cont'd

Nominal Pipe Size, in	Nominal Pipe Size, mm	OD, in	OD, mm	Schedule Designations ANSI/ASME	Wall Thickness, in	Wall Thickness, mm	Weight, lbs/ft	Weight, kg/m
16	400	16	406.4	STD/40S	0.375	9.53	62.58	93.13
16	400	16	406.4	XS/80S	0.5	12.7	82.77	123.18
18	450	18	457.2	STD/40S	0.375	9.53	70.59	105.06
18	450	18	457.2	XS/80S	0.5	12.7	93.45	139.08
20	500	20	508	STD/40S	0.375	9.53	78.6	116.98
20	500	20	508	XS/80S	0.5	12.7	104.13	154.97
24	600	24	609.6	STD/40S	0.375	9.53	94.62	140.82
24	600	24	609.6	XS/80S	0.5	12.7	125.49	186.76
30	750	30	762	STD/40S	0.375	9.53	118.65	176.58
30	750	30	762	XS/80S	0.5	12.7	157.53	234.44
36	900	36	914.4	STD/40S	0.375	9.53	142.68	212.34
36	900	36	914.4	XS/80S	0.5	12.7	189.57	282.13

40S = schedule 40.
STD = standards.
80S = schedule 80.
XS = extra strong.
XX = double extra strong.

TABLE 2.35 Mechanical Properties of Structural Steel Plates

Group	Class	Specifications and Grade	Yield Strength, MPa	Tensile Strength, MPa
I	H	ASTM A36, to 50 mm thick	250	400–550
		ASTM A131 Grade A, to 12 mm thick	235	400–490
		ASTM A285 Grade C, to 19 mm thick	205	380–515
I	N	ASTM A131 Grades B, D	235	400–490
		ASTM A516 Grade 65	240	450–585
		ASTM A573 Grade 65	240	450–530
		ASTM A709 Grade 36T2	250	400–550
I	C	ASTM A131 Grades CS, E	235	400–490
II	H	ASTM A572 Grade 42, to 50 mm thick	290	415 min
		ASTM A572 Grade 50, to 50 mm thick (S91 required over 12 mm)	345	450 min
II	N	API Spec 2 MT1	345	483–620
		ASTM A709 Grades 50T2, 50T3	345	450 min
		ASTM A131 Grade AH32	315	470–585
		ASTM Grade AH36	350	490–620
II	C	API Spec 2H Grade 42	290	430–550
		Grade 50, to 62 mm thick	345	483–620
		Grade 50, over 62 mm thick	325	483–620
		API Spec 2W Grade 50, to 25 mm thick	345–517	448 min
		Grade 50, over 25 mm thick	345–483	448 min
		API Spec 2Y Grade 50, to 25 mm thick	345–517	448 min
		Grade 50, over 25 mm thick	345–483	448 min
		ASTM A131 Grades DH32, EH32	315	470–585
		Grades DH36, EH36	350	490–620
			345	485–620

TABLE 2.35 Mechanical Properties of Structural Steel Plates—cont'd

Group	Class	Specifications and Grade	Yield Strength, MPa	Tensile Strength, MPa
		ASTM A537 Class I, to 62 mm thick		
		ASTM A633 Grade A	290	435–570
		Grades C, D	345	485–620
		ASTM A678 Grade A	345	485–620
III	C	ASTM A537 Class II, to 62 mm thick	415	550–690
		ASTM A678 Grade B	415	550–690
		API Spec 2W Grade 60, to 25 mm thick	414–621	517 min
		Grade 60, over 25 mm thick	414–586	517 min
		API Spec 2Y Grade 60, to 25 mm thick	414–621	517 min
		Grade 60, over 25 mm thick	414–586	517 min
		ASTM A710 Grade A Class 3,		
		quenched and precipitation heat treated		
		To 50 mm	515	585
		50 mm to 100 mm	450	515
		Over 100 mm	415	485

and mechanical properties are listed in Table 2.36. Steels above the thickness limits stated may be used and will be considered by the designer.

Structural Steel Pipe

Unless otherwise specified, seamless or welded pipe should conform to one of the specifications listed in Table 2.37. Table 2.37 presents the mechanical properties of the pipe. Pipe should be prime quality unless the use of limited service, structural grade or reject pipe is specifically approved by the designer.

Structural pipe should be fabricated in accordance with API Spec. 2B, ASTM A139, ASTM A252, ASTM A381 or ASTM A671 using the grades listed in Table 2.37 except that hydrostatic testing may be omitted.

TABLE 2.36 Mechanical Properties for Structural Steel Shapes

Group	Class	Specifications and Grade	Yield Strength, MPa	Tensile Strength, MPa
I	H	ASTM A36, to 50 mm thick	250	400–550
		ASTM A131 Grade A, to 12 mm thick	235	400–550
I	N	ASTM A709 Grade 36T2	250	400–550
II	H	API Spec 2 MT2 Class C	345	450–620
		ASTM A572 Grade 42, to 50 mm thick	290	415 min
		ASTM A572 Grade 50, to 50 mm thick (S91 required over 12 mm thick)	345	450 min
		ASTM A992	345–450	450 min
II	N	API Spec 2 MT2 Class B	345	450–620
		ASTM A709 Grades 50T2, 50T3	345	450 min
		ASTM A131 Grade AH32	315	470–585
		ASTM Grade AH36	350	490–620
II	C	API Spec 2MT2 Class A	345	450–620
		ASTM A913 Grade 50	345	450 min

Selections for Conditions of Service

Consideration should be given to the selection of steels with toughness characteristics suitable for the conditions of service. For tubes cold-formed to *D/t* less than 30, and not subsequently heat-treated, due allowance should be made for possible degradation of notch toughness, for example, by specifying a higher class of steel or by specifying notch-toughness tests run at reduced temperature.

Steel for tubular joints is subject to local stress concentrations that may lead to local yielding and plastic strains at the design load. During the service life, cyclic loading may initiate fatigue cracks, making additional demands on the ductility of the steel, particularly under dynamic load. These demands are particularly severe in heavy wall joint-cans designed for punching shear.

For the underwater portion of offshore fixed platforms, the steel members or joint cans (e.g., jacket leg joint cans, chords in major X and K joints and

TABLE 2.37 Mechanical Properties of Structural Steel Pipes

Group	Class	Specifications and Grade	Yield Strength, MPa	Tensile Strength, MPa
I	H	API 5L Grade B	240	415 min
		ASTM A53 Grade B	240	415 min
		ASTM A135 Grade B	240	415 min
		ASTM A 139 Grade B	240	415 min
		ASTM A500 Grade A (Round)	230	310 min
		(Shaped)	270	310 min
		ASTM A501	250	400 min
I	N	ASTM A106 Grade B, normalized	240	415 min
		ASTM A524 Grade I, to 10 mm thick	240	415 min
		Grade II, over 10 mm thick	205	380–550
I	H	ASTM A333 Grade 6	240	415 min
		ASTM A334 Grade 6	240	415 min
II	H	API 5L Grade X42, 2% max cold expansion	290	415 min
		API 5L Grade X52, 2% max cold expansion	360	455 min
		ASTM A500 Grade B (Round)	290	400 min
		(Shaped)	320	400 min
		ASTM A618	345	485 min
II	N	API 5L Grade X52 with SR5 or SR6	360	455 min

through-members in joints designed to overlap) should meet notch-toughness criteria at the temperatures given in Table 2.38.

The toughness criteria defined as the Charpy V-notch energy should be 15 ft-lbs (20 J) for Group I steels, 25 ft-lbs (34 J) for Group II steels, and 35 ft-lbs (47 J) for Group III steels (transverse test).

For water temperatures of 40° F (4°C) or higher, these requirements may normally be met by using steel ASTM A333 Grade 6, as listed in Table 2.37.

For above-water joints exposed to lower temperatures and possible impact from boats, or for critical connections at any location in which it is desired to

TABLE 2.38 Input Testing Conditions

D/t	Test Temperature[a]	Test Condition
Over 30	36° F (20°C) LAST	Flat plate
20–30	54° F (30°C) LAST	Flat plate
Under 20	18° F (10°C) LAST	As fabricated

[a]*LAST = lowest anticipated service temperature.*

prevent all brittle fractures, the tougher Class C steels should be considered, for example, API Spec. 2H, Grade 42 or Grade 50. For 345 MPa yield and higher-strength steels, special attention should be given to welding procedures.

For critical connections involving high restraint (including adverse geometry, high yield strength and/or thick sections), through-thickness shrinkage strains and subsequent through-thickness tensile loads in service, consideration should be given to the use of steel having improved through-thickness (z-direction) properties.

Although the brace ends at tubular connections are also subject to stress concentration, the conditions of service are not quite as severe as for joint-cans. For critical braces, for which brittle fracture would be catastrophic, consideration should be given to the use of stub-ends in the braces having the same class as the joint-can, or one class lower. This provision need not apply to the body of braces between joints.

Cement Grout

If the design requires grout to be poured in the annulus (the space between the pile and the leg), laboratory tests of the grout mix should be performed before construction, in compliance with ASTM C109 (2011). The mixing and pouring procedures should follow the manufacturer's precautions and recommendations. The compressive strength of grout specimens after 28 days should meet ACI 214 criteria. The compressive strength should not be less than the specified concrete strength in the project's specifications, and, in accordance with ACI 214 (2002), the concrete strength should not be less than 17 MPa.

REFERENCES

American Concrete Institute (ACI), 2002. 214-93, Evaluation of Strength Test Results of Concrete, American Concrete Institute, Farmington Hills, MI, USA.

American Institute of Steel Construction, Dec. 1999. Load and Resistance Factor Design Specification for Structural Steel Buildings, AISC LRFD.

American National Standard, 1982. Minimum Design Loads for Buildings and Other Structures, ANSI, A58.

American Petroleum Institute, 1993. Recommended Practice for Planning, Designing, and Constructing.

American Petroleum Institute, 1996. Working Stress Design, API RP2A-WSD, 20th Ed.

ASTM C109/C109M-11, 2011. Standard Test Method for Compressive Strength of Hydraulic Cement Mortars (Using 2-in. or [50-mm] Cube Specimens), American Society of Testing Materials, USA.

Atkins Engineering Services, 1990. Fluid Loading on Fixed Offshore Structures, OTH 90 322.

British Standard (BS) 5400, Part 3, Steel, Concrete and Composite Bridges, Code of Practice for Design of Steel Bridges.

British Standard (BS) 5950, Structural Use of Steelwork in Buildings, Code of Practice for Design.

DNV (Det Norske Veritas), Offshore Standard DNV-OS-C101: Design of Offshore Steel Structures, General (LRFD Method), October 2008. http://dc140.4shared.com/doc/Pflypj7r/preview.html

Design and Plastic Design, June 1989.

ENV 1993-1-1:1992, Eurocode 3, Design of Steel Structures, General Rules and Rules for Buildings.

Norwegian Standard (NS) 3472, 2004, Steel Structures, Design Rules, Rev.4. Standard Norway, Norway.

Uniform Building Code, 1997, Volume 2, International Conference of Building Officials, Whittier, CA, USA.

Chapter 3

Offshore Structure Platform Design

3.1 INTRODUCTION

Offshore fixed platform design has three main phases, the first of which is design of the deck that carries the topside facility. The dimensions of the deck depend on the function of the platform and the facilities that will be located on it.

The second part is design of the jacket, which depends on the water depth, the wave and current loads, and the other loads described in Chapter 2. The configuration of the jacket structural system is chosen based on the water depth and the designer's experience.

The third phase of design is to assess the robustness of the deck and jacket design for lifting, pullout, transportation, launching and installation.

The three main components of a steel template platform are the topside facilities (decks), the jacket and the piles.

Topside facilities frequently have three decks: a drilling deck, a well head/ production deck and a cellar deck. The decks are supported by a gridwork of girders, trusses, and columns. Figures 3.1 and 3.2 show a plan of the main deck and two elevation views, including the three common levels on every platform (the main deck, cellar deck, spider deck or mezzanine deck in some cases).

The main function of the topside is to carry the load from the facilities and drilling equipment. The function of the jacket is to surround the piles and to hold the pile extensions in position all the way from the mud line to the deck substructure. Moreover, the jacket provides support for boat landings, mooring bits, barge bumpers, the corrosion protection system and many other platform components. Examples of jacket drawings are presented in Figure 3.3. Plan views at different levels are presented in Figure 3.4 for the highest jacket level, in Figure 3.5 for the mud-mat level and Figure 3.6 for the horizontal frames at different levels.

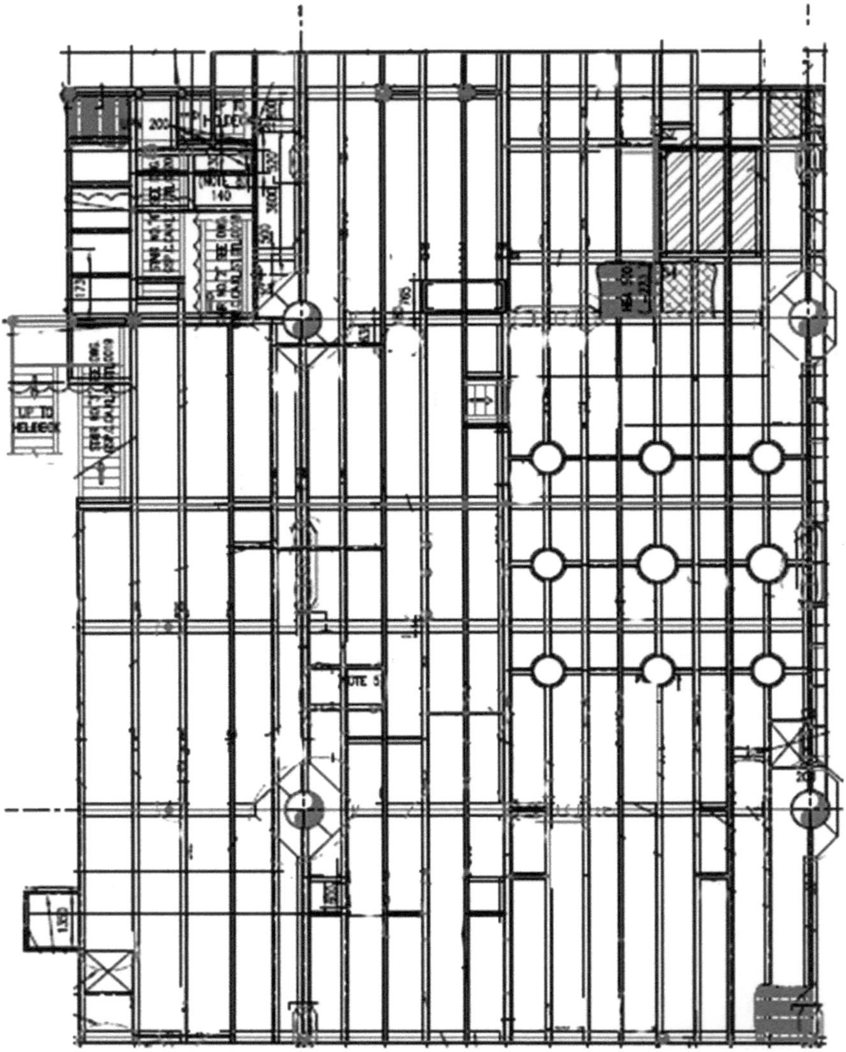

FIGURE 3.1 Plan of main deck.

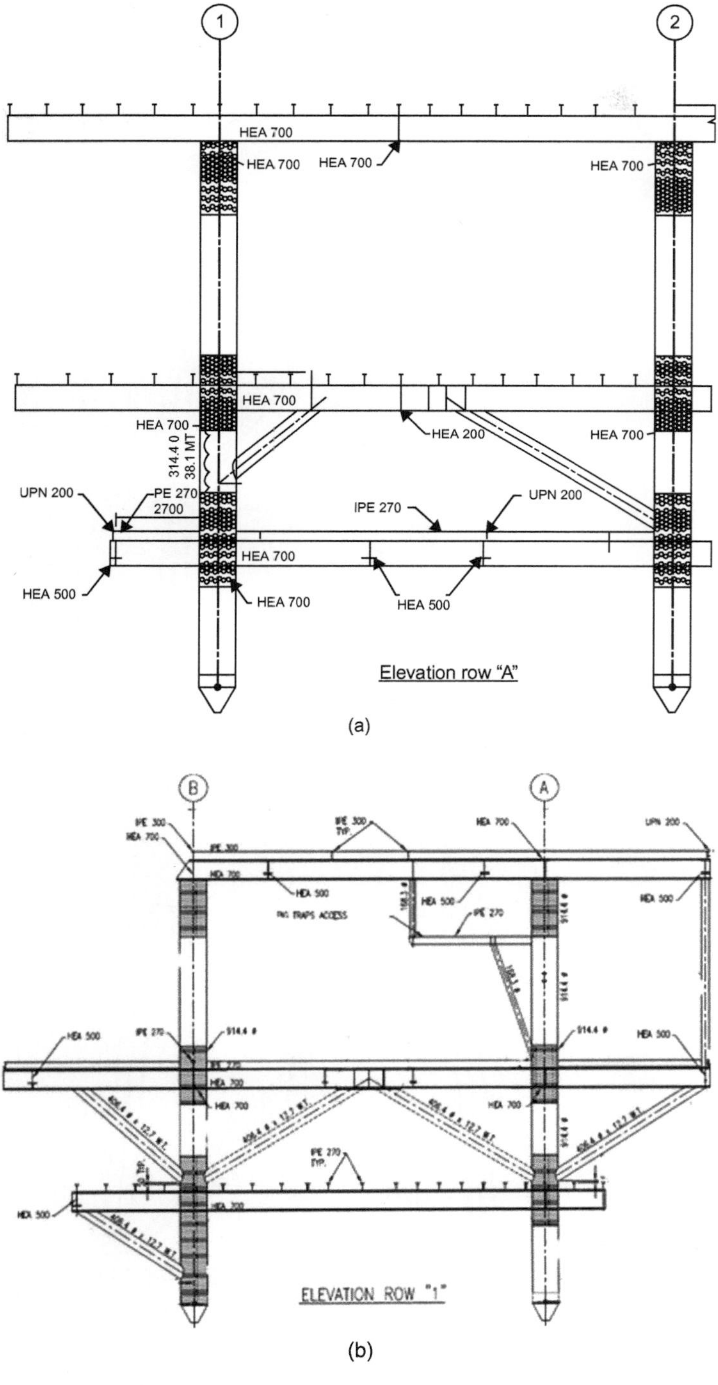

FIGURE 3.2 **(a)** Elevation at row A; **(b)** Elevation at row 1.

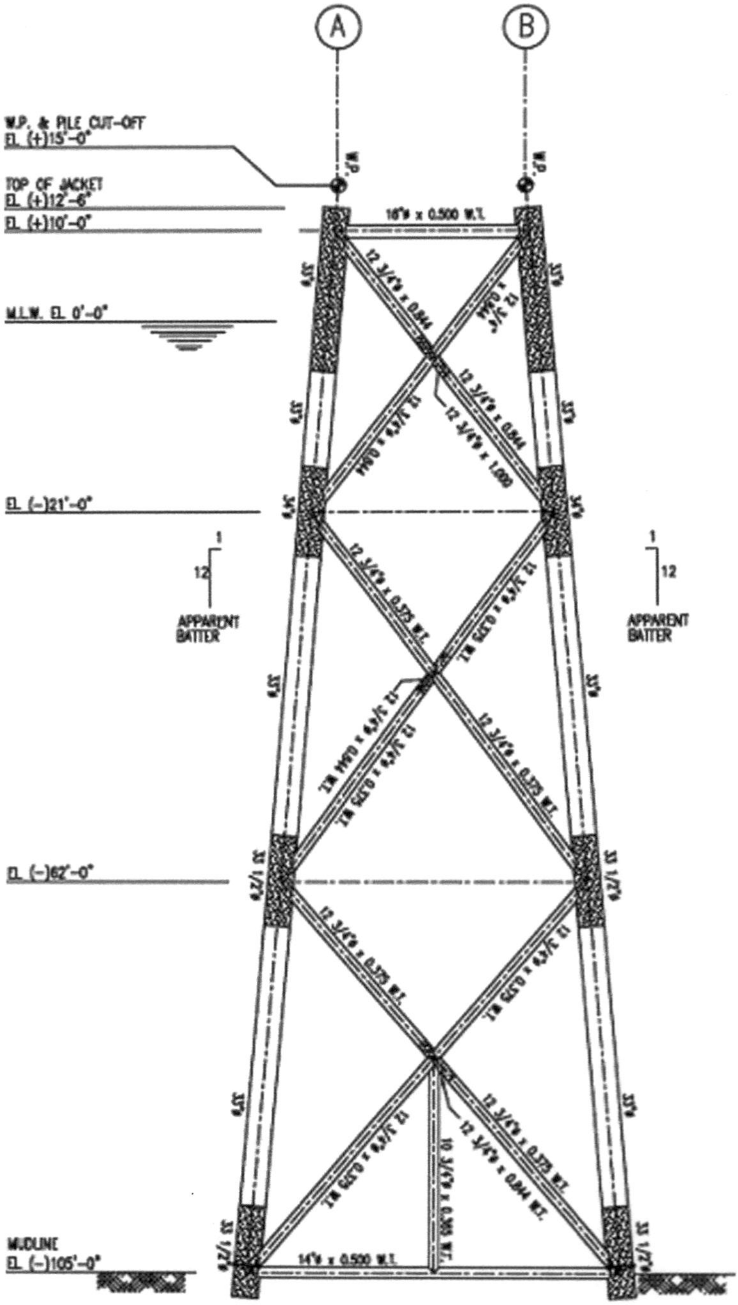

(a)

FIGURE 3.3 **(a)** Jacket view; **(b)** another view of the jacket.

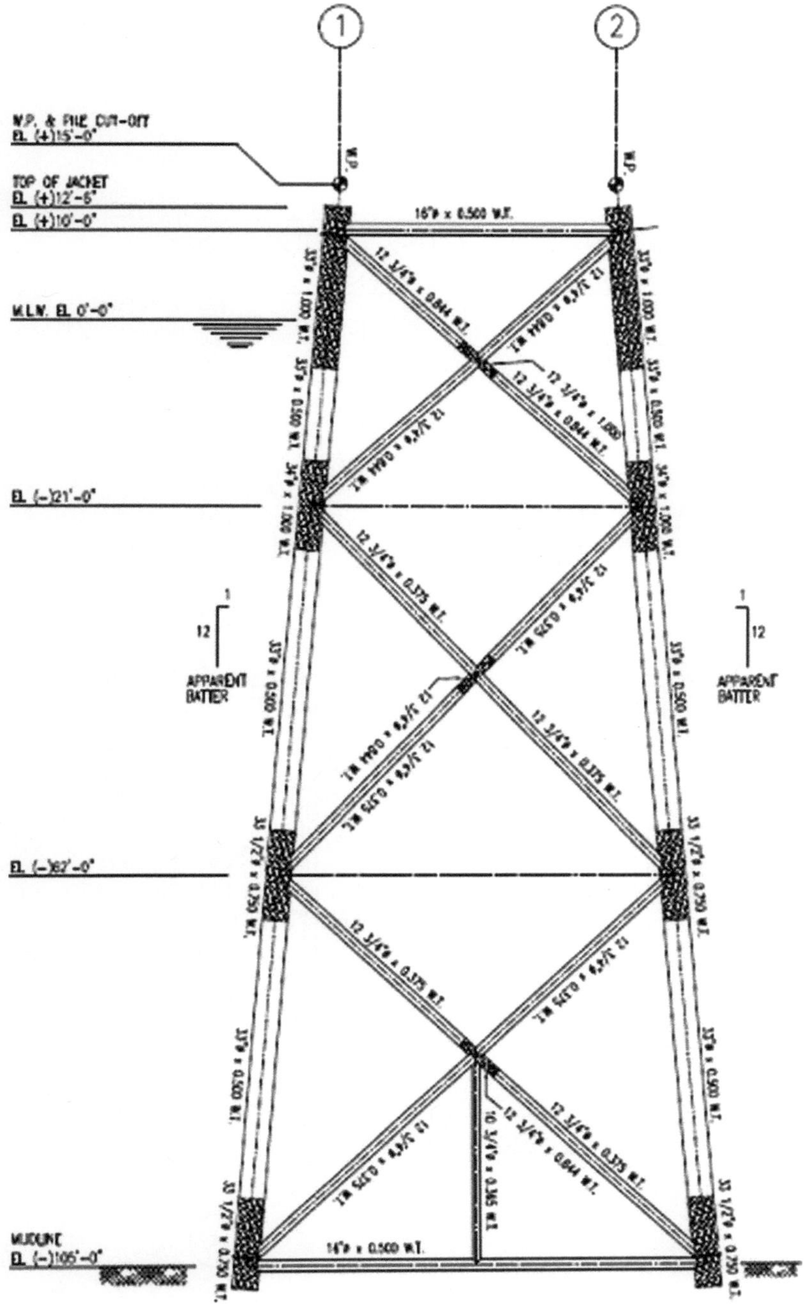

(b)

FIGURE 3.3 *(Continued)*

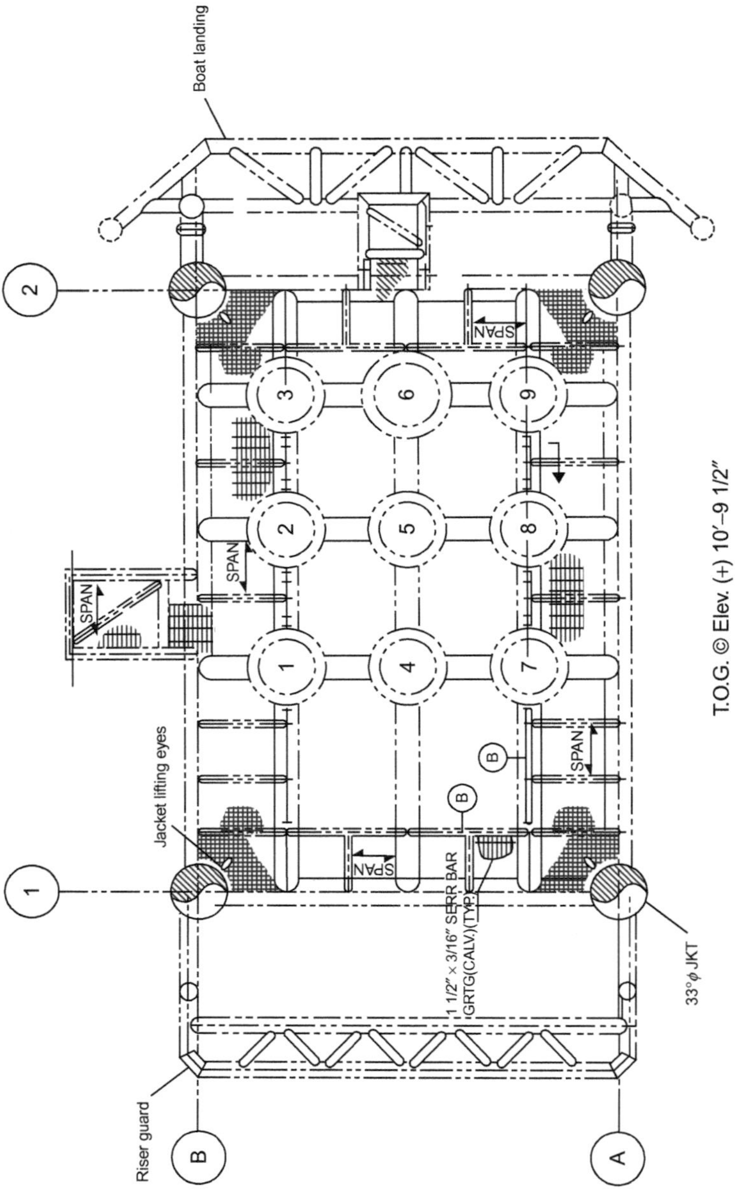

FIGURE 3.4 Plan view at the highest jacket level.

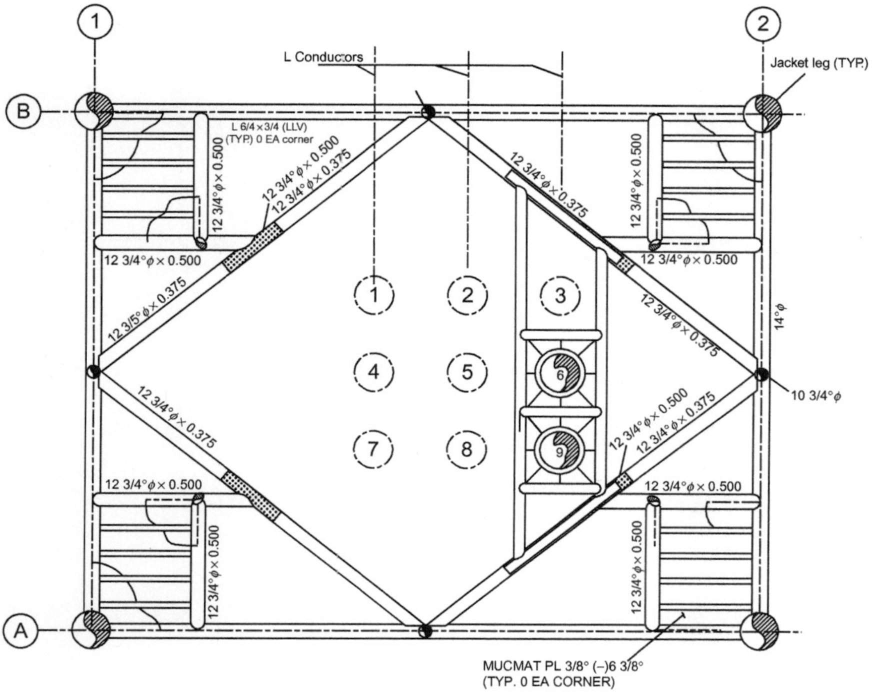

FIGURE 3.5 Plan view at the mud mat.

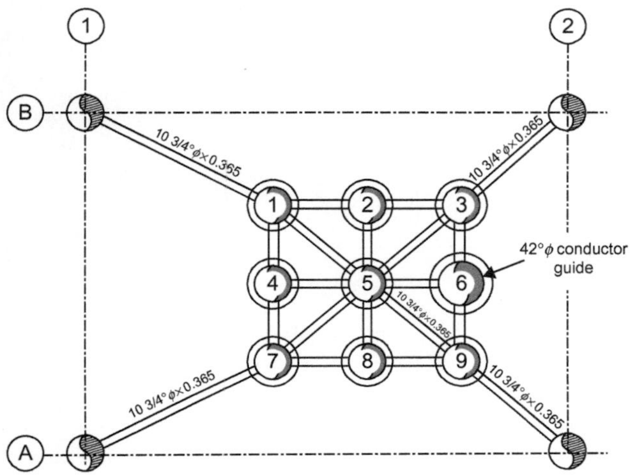

Jacket framing plan © EL.(–)21'–0"

(a)

FIGURE 3.6 **(a)** Plan view of the jacket at an elevation of 21 ft; **(b)** plan view of the jacket at an elevation of 62 ft; **(c)** plan view of the jacket at an elevation of 10 ft.

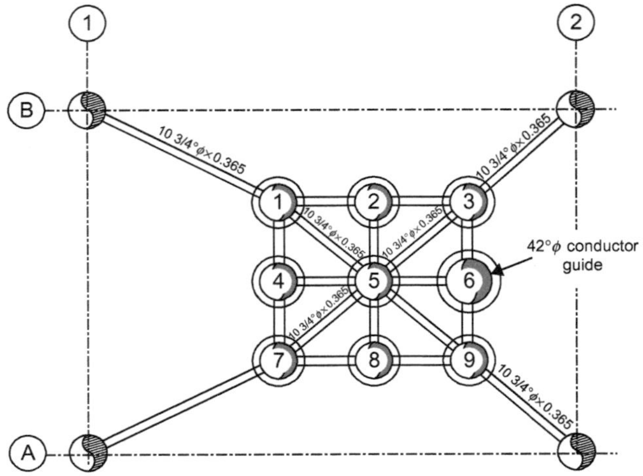

Jacket framing plan © EL. (–)62'–0"

(b)

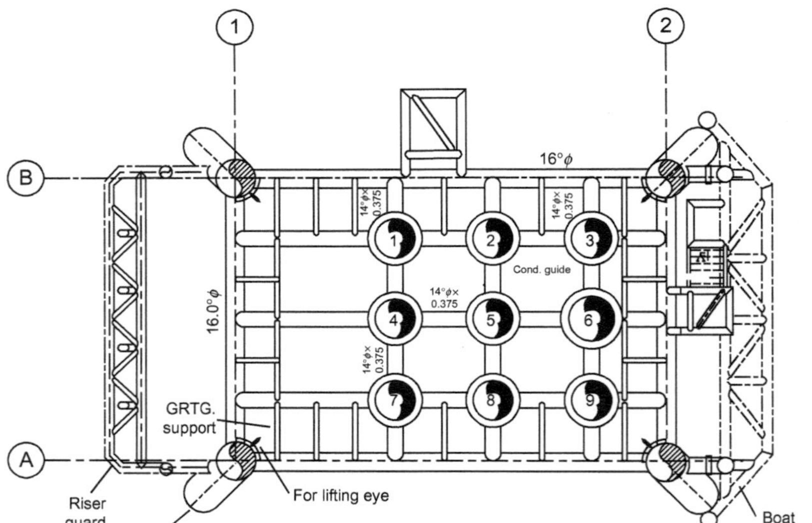

Jacket framing plan @ EL. (+)10'–0"

(c)

FIGURE 3.6 (*Continued*)

3.2 PRELIMINARY DIMENSIONS

Topside structures are designed based on requirements of the American Institute of Steel Construction (AISC), with Allowable Stress Design (ASD) or Load Resistance Factor Design (LRFD) specifications according to the project requirement. The main supporting element is the plate girder or tubular truss, and in most cases a tubular member is preferred because it incurs little wind load.

The floor is covered with steel plates, usually about 38 mm (1½ inches) thick. The thickness of the deck framing depends on the spacing between floor beams and the anticipated load on the deck.

Figure 3.7 is a platform elevation view showing the main parts of the offshore platform and affected load.

3.2.1 Approximate Dimensions

- Large forces result when waves strike a platform's deck and equipment.
- Therefore, the air gap should be at least 1.5 m (5 ft) added to the wave crest height from the omnidirectional guideline for the maximum wave height for the 100-year return period, as shown in Figure 3.7.

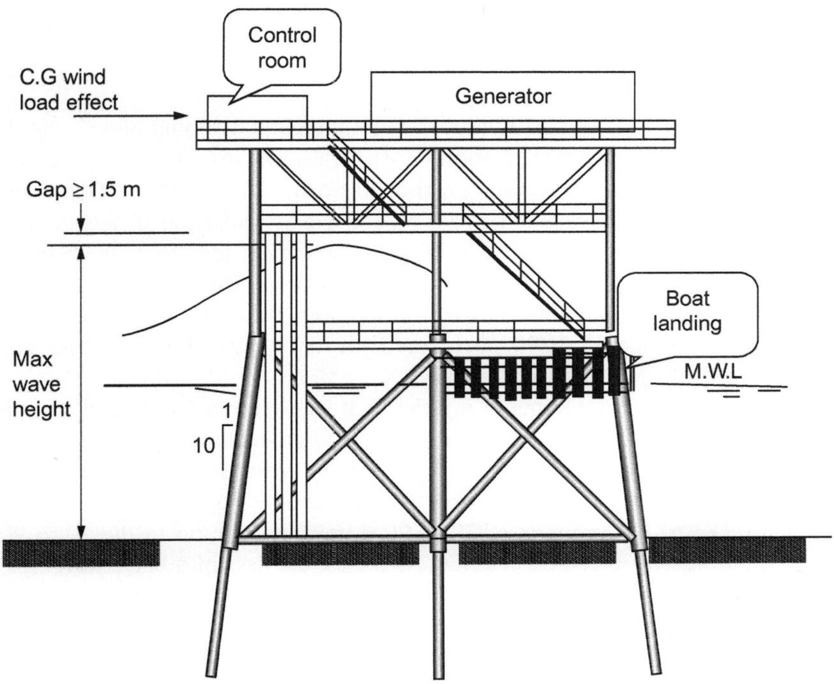

FIGURE 3.7 Platform elevation view.

- From a practical point of view, sea deck levels are usually considered to be at an elevation of 10–14 ft (3–4 m) above the mean water level (MWL).
- The jacket walkway is above the normal everyday waves that pass through the jacket.
- In the case of the 8-legged platform, the spacing between the legs is about 12–18 m (40–60 ft) and is usually set by the availability of launch barges and the spacing of launch runners on the barges.
- In the transverse direction, the leg spacing is approximately 14 m (45 ft). This dimension is always constrained by the dimension of the drilling and production packages that will be placed on the deck.
- The length of the cantilever overhang is usually about 3.5–4.5 m (12–15 ft).
- Allowing 25 mm (1 inch) annular clearance between the pile and the inside of a leg, for a pile of 60 inches and 48 inches OD, the legs will have internal diameters of 62 inches and 50 inches, respectively.
- Jacket legs are battered to provide a larger base for the jacket at the mud line and thus assist in resisting environmentally induced overturning moments. The legs' batter (slope) is 1:8 or 1:7.
- Preliminary sizes are selected based on experience.
- Conductor and riser numbers and sizes will be provided; the conductor is always 18, 20, 24 or 30 inches and the risers are always 14–20 inches.

3.3 BRACING SYSTEM

The bracing system consists of vertical, horizontal and diagonal tubular members connected to jacket legs, forming a stiff truss system. This system transfers the horizontal load acting on the platforms to the piles. There are variations of the platform-bracing pattern, and every system has its advantages and disadvantages. For example, the K brace has fewer members intersecting at joints, so it has reduced welding and assembly costs. But its disadvantage is that K bracing has less redundancy than X bracing. This is based on the results of a BOMEL study in 1999 performed jointly by different oil and gas companies.

Figure 3.8 presents a frame of a jacket that was subjected to lateral load in the test workshop. The load applied was increased gradually until complete failure. The figure shows the buckling of the bracing member.

It is recommended to choose a brace member diameter that has a slenderness ratio (KL/r) in the range of 70 to 90. Limiting the ratio to the 70 to 90 range is an industry-accepted practice. As the slenderness of a brace increases, its allowable axial stress (F_a) decreases. At KL/r equal to 80, the allowable axial stress (F_a) for A-36 steel (248 MPa) is 71% of that allowable for a nonslender member, i.e., that with a KL/r equal to zero. In the case of steel of 345 MPa (50 KSI) at a KL/r equal to 80, F_a is reduced to about 63% of that of a nonslender member.

FIGURE 3.8 Buckling of beam in the jacket.

At a high KL/r ratio, the high-yield pipe is less efficient than at lesser values. Note that lower slenderness ratios also encourage higher *D/t* ratios for tubular members that may compound local buckling problems.

For sizes up to and including 450 mm (18 inches), use the wall thickness for standard pipe to start. For sizes up to and including 700 mm (27 inches), try 12 mm. For 750 to 900 mm, start with 16 mm.

It is practical to keep the *D/t* ratio of the members between 19 and 60. For a *D/t* less than 19, pipe is difficult to buy or to make. For A-36 steel, a *D/t* higher than 60 can present local buckling problems. From the practical point of view, for a water depth of *h* (ft), begin to check the hydrostatic problems when the *D/t* is higher than $250/(h)^{0.3333}$.

In general, the legs of the jacket are interconnected and rigidly held by diagonal bracing in vertical planes and horizontal and diagonal bracing in horizontal planes. In most designs, horizontal bracing is spaced 12–16 m and near the water's surface the span is approximately 12 m.

The benefits and general functions of the bracing system are:

- Transmission of the horizontal load to the soil through the pile foundation
- Provision of structural integrity during fabrication and installation
- Resistance to the wrenching motion of the installed jacket-pile system
- Support for the corrosion anodes and well conductors

3.4 JACKET DESIGN

Virtually all of the decisions about design depend on the jacket leg. The soil conditions and foundation requirements often control the leg size.

The golden rule in jacket design is to minimize the projected area of the member near the water surface (high wave zone), to minimize the load on the structure and to reduce the foundation requirements.

From the joint industrial project study (1999), Figures 3.9 and 3.10 present the prototype in three dimension and the buckling of the K bracing member, respectively.

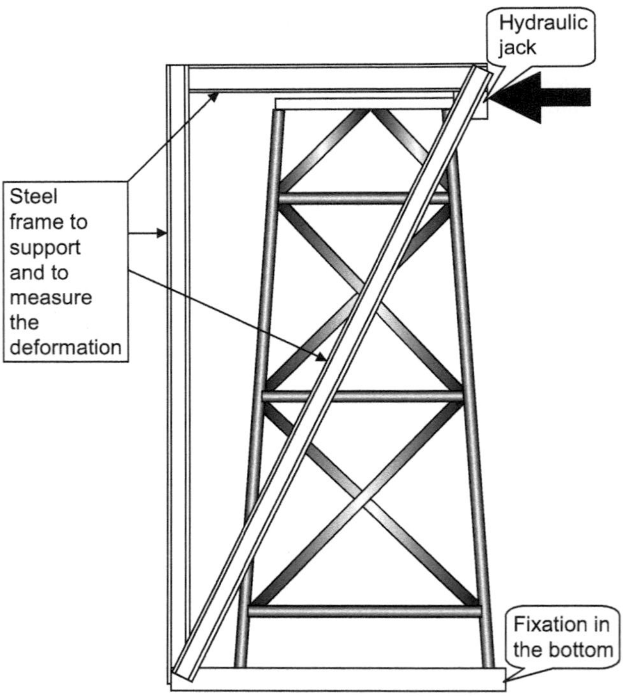

FIGURE 3.9 Shape of the jacket model in two dimensions for the load test.

FIGURE 3.10 Buckling in the member for K bracing.

The results of the study are presented in Figure 3.11. The relationship between the applied load and displacement for X bracing with horizontal bracing and without a joint-can is presented in Figure 3.11(a). Figure 3.11(b) presents the relation between load and displacement in the case of existing horizontal bracing and with a joint-can. It can be seen that, in the case of a joint-can, the jacket can carry more load than specified for the design based on API. In addition, ductility is higher with a joint-can.

If there is no horizontal bracing and without a joint-can, as in Figure 3.11(c), the jacket will carry less load than in the case with a horizontal member.

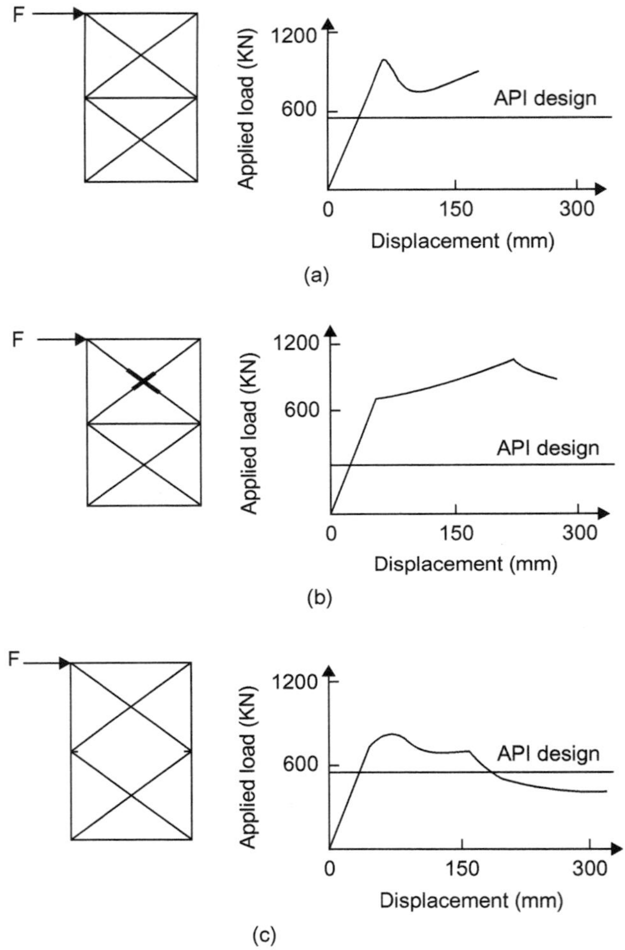

FIGURE 3.11 **(a)** X bracing with horizontal bracing and without a joint-can; **(b)** X bracing with horizontal bracing and with a joint-can; **(c)** X bracing without horizontal bracing and without a joint-can.

The test was performed for K bracing with different β, which is the relation between the bracing diameter and the chord diameter (d/D), and with a different gap between the bracing (the gap is denoted by g).

Figure 3.12 presents the relation between the applied load and the frame displacement for designs according to American Petroleum Institute (API) and Det Norske Veritas (DNV). With decreasing β values, the redundancy will increase. Also, with an increased gap between the braces, the ductility will increase, as seen in the comparison of cases (c) and (d) in the figure.

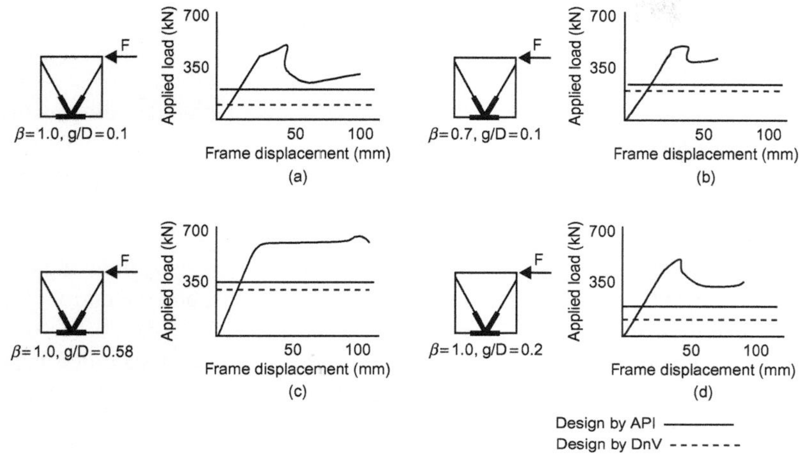

FIGURE 3.12 Relation between applied load and displacement for different K-bracing geometries.

3.5 STRUCTURE ANALYSIS

For environmental and gravitational loads, all the necessary parameters required for automatic load generation by the software program are input by the engineer. Most software programs on the market use default values for many of these parameters. Usually, the engineer must override the defaults and input project-specific values. The following should be defined in the program:

- Self-weight
- Buoyancy for flooded and unflooded members
- Wind (direction, terrain category, gust duration, drag coefficients, etc.)
- Waves (wave theory, direction, height, period, drag and inertia coefficient (C_D, C_M), wave kinematics, etc.)
- Current direction, speed variation with depth, current blockage factor and other
- Marine growth (thickness variation with depth, roughness, etc.)

In most cases, the software does not use the correct default coefficients so it should be reviewed carefully.

The structure's pile foundations should be modeled sufficiently to reflect the actual stiffness of the foundation:

- Simple (fixed, pinned, sliding, etc.)
- Linear springs
- Nonlinear stiffness

3.5.1 Global Structure Analysis

The steps of using software in design are:

1. The designer must define the structure in terms of physical dimensions, member size and materials properties.
2. The designer must input the soil conditions (as interpreted by the soil specialist); in some programs this requires a *p-y* curve.
3. All loads must be entered into the program.
4. The wave load is applied through the structure at several azimuth angles to determine the direction that produces higher reactions, and the current load is as presented in Figure 3.13.
5. The computer advances the wave through the structure at specified increments, calculating total shear and overturning moment on the structure at the mud line.
6. For each load condition, the computer analysis provides:
   A. Total base shear and overturning moment
   B. The member end forces and moments
   C. Joint rotation and deflection
   D. External support reaction
7. After calculating the stresses, the computer compares them with the allowable stresses as defined by the American Institute of Steel Construction (AISC).
8. The pile is replaced by lateral springs in two directions, an axial spring and a moment spring.

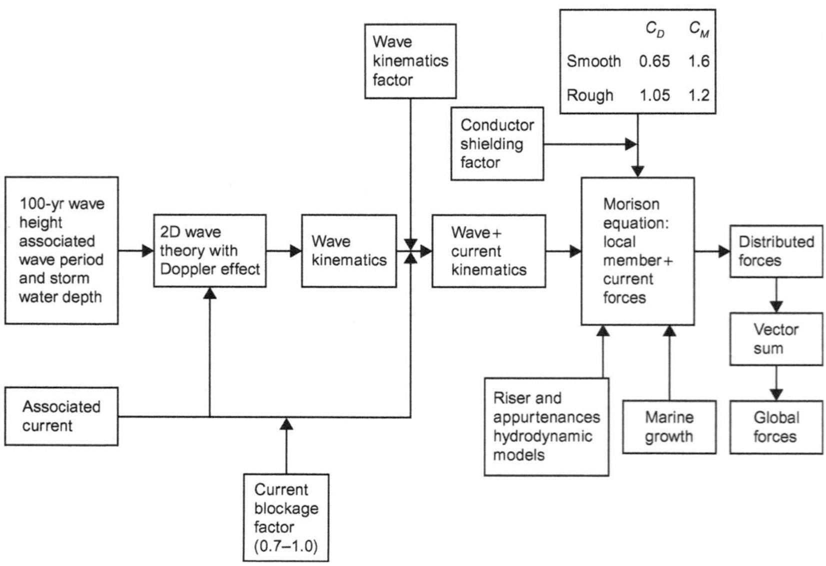

FIGURE 3.13 Applied environmental loads.

9. The piling interacts with the surrounding soil in an inelastic manner, and the computer linearizes its response to generate the equivalent elastic spring, as shown in Figure 3.14.
10. In the case of sea ice abrasion, an allowance of 0.1 mm/year is to be considered for all steel between the elevations of −1.3 m to 3.0 m and usually results in a minimum decrease in thickness of 2.5 mm over the life of the platform.

The process of applying the wave load and current to the offshore structure is illustrated in Figure 3.13, as described in API RP2A.

The horizontal members in the wave zone should be designed for wave slam forces, in accordance with API RP2A. Bending stresses due to both horizontal and vertical slam forces should be investigated. However, the current velocity components should not be included in the wave kinematics when calculating wave slam loading. For X braces, members are assumed to span the full length. Member lengths are reduced to account for the jacket leg ratio. Wave slam calculations are carried out during detailed design, not basic design.

The static structure analysis for the offshore structure is the same as in normal structures, because the software uses the stiffness matrix to calculate the deflection and then the internal forces and stresses on each member. However, for offshore structures, the problem is the interaction between the structure and the piles, as the structure will be elastic and the piles will be inelastic. So the structure analysis steps are:

1. Set-up of the geometric of the jacket with material specifications and preliminary member sections with the dimensions. The software calculates the stiffness matrix for the jacket excluding the piles.

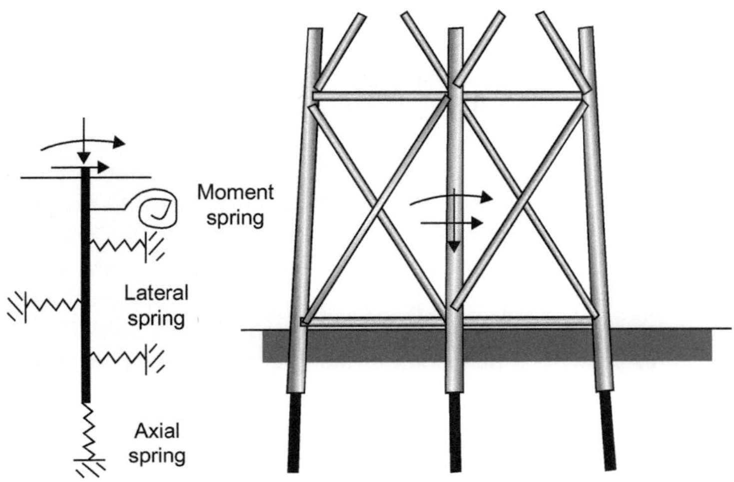

FIGURE 3.14 Foundation piling model.

2. Apply the loads on the structure jacket with different load cases, but the software cannot be run because there are no supports applied to the structure system
3. Order the nodes that reflect the degree of freedom in K, F so that the nodes p that will connect the piles are together at the end of the stiffness matrix and follow the nodes j that are slowly connected to jacket members. The stiffness matrix and force vector can be:

$$\begin{bmatrix} j_1 & j_2 & & p_1 & p_2 \\ \\ \\ k_{jj} & & & k_{jp} & \\ k_{pj} & & & k_{pp} & \end{bmatrix} \begin{Bmatrix} \delta_j \\ \\ \\ \delta_p \end{Bmatrix} = \begin{Bmatrix} F_j \\ \\ \\ F_p \end{Bmatrix} \tag{3.1}$$

4. In order to analyze foundation behavior, the stiffness of the jacket and loading on the jacket as felt by the piles are required. Detailed behavior of the jacket is not required at this stage, as an equation of form $K_s\delta_p = F_s$ is wanted, where K_s and F_s represent the stiffness and applied forces on the structure as seen at the pile connecting nodes p.
5. Form the model of the foundation by developing the stiffness matrix for the foundation at zero deflection.
6. Assume no load is applied directly to the piles and the only load is applied through the structure. The foundation forcing vector contains the element from the forces on the structure but located in the appropriate positions for the same degree of freedom in the foundation stiffness matrix.
7. Add the foundation and the jacket substructure and solve, then recalculate the pile foundation stiffness and displacement, as the first nodal displacement along the pile and at connection to the jacket is only a first estimate because it has been based on step 5, in which the stiffness of p-y and t-z curves is at zero deflection. Note that once the pile deflections have been estimated, a better estimate of p-y and t-z stiffness can be made. The model of the foundation is shown in Figure 3.14. The stiffness may be represented by either a secant or tangent stiffness. Note that the secant stiffness method is generally slower but more stable than the tangent stiffness approach.
8. Repeat the sequence from step 5 until the stiffnesses have converged; then the nodal deflections can be used to determine the forces, shears and moment in the piles.
9. The deflections at the link to the jacket are also known now. They can be applied as prescribed deflection to the pile nodes on the original jacket model for step 2.
10. Given the jacket deflections, the jacket member forces are calculated from separate member stiffness properties.

Figure 3.15 illustrates the global structure analysis procedure in a flowchart.

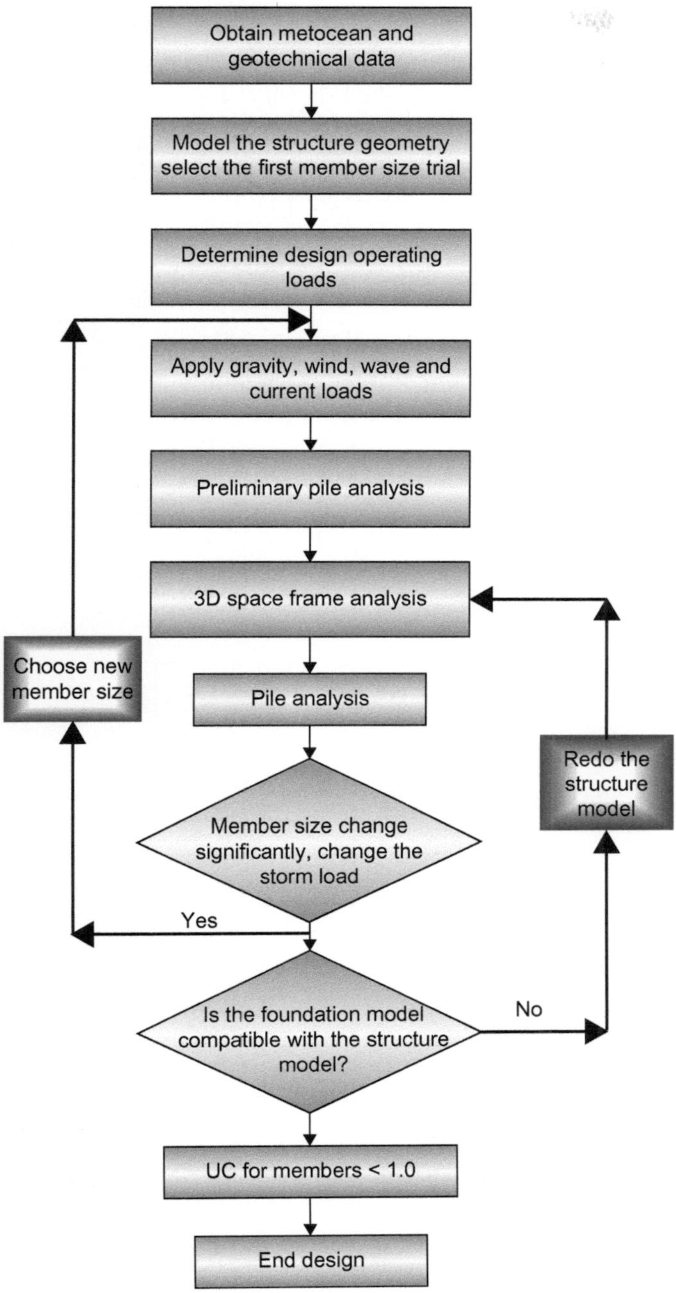

FIGURE 3.15 Fixed offshore platform design procedure.

3.5.2 The Loads on Piles

The loads on the piles are calculated by the software, but preliminary calculation can be done by hand. The method shown in Figure 3.16 presents the calculation parameters.

$$N = \frac{(M - sh)\cos\alpha}{2(h + d_f)}$$

$$A_1 = \frac{1}{2}\left(\frac{V}{\cos\alpha} - \left(s + \frac{(M - sh)\cos^2\alpha}{h + d_f}\right)\frac{1}{\sin\alpha}\right) \quad (3.2)$$

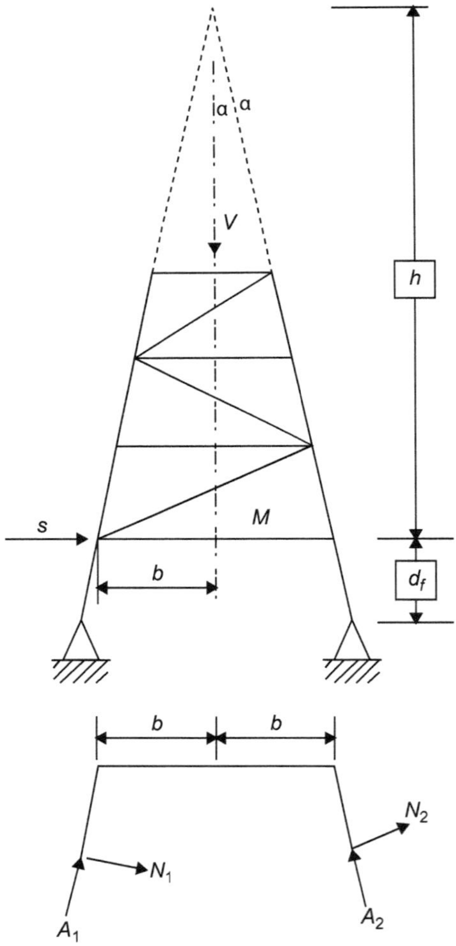

FIGURE 3.16 Calculation of loads on piles.

$$A_2 = \frac{1}{2}\left(\frac{V}{\cos\alpha} + \left(s + \frac{(M - sh)\cos^2\alpha}{h + d_f}\right)\frac{1}{\sin\alpha}\right) \tag{3.3}$$

Another way to obtain the pile load is by obtaining the overturning moment value at the mud line.

The calculation of the reaction force at each pile in the jacket plan at the mud line is as shown in Figure 3.17, where M is the total overturning moment, D_x and D_y are the distances in the x and y directions from the natural axis and θ is the axis wave angle.

It is assumed that the base is rigid. The resultant force at each pile is calculated by assuming that M is constant with different wave angle θ.

$$R = \frac{M_x d_y}{\sum A d_y^2} + \frac{M_y d_x}{\sum A d_x^2}$$

Where R = vertical pile reaction, M = total overturning moment, A = relative axial pile stiffness, D_{xy} = distance from neutral angle, and θ = axis wave angle.

3.5.3 Modeling Techniques

The following is a guideline and recommendation for using software in modeling any steel structure and specifically an offshore platform structure.

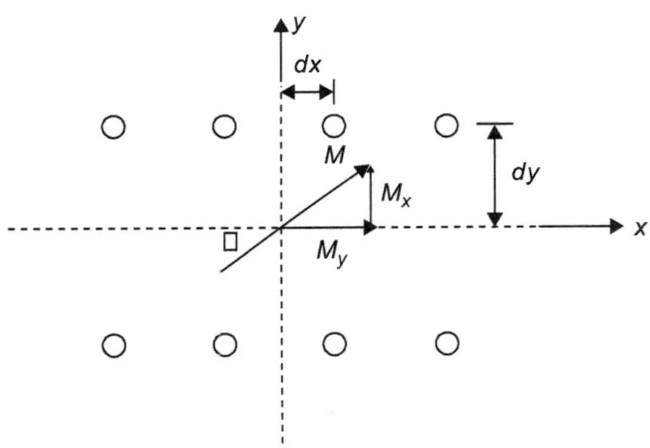

FIGURE 3.17 Jacket plan at mud line.

As a guide, the global axis system should be orientated as noted below. The origin should be at the center of the platform or structure at chart datum, MSL or mud line, as determined by the project.

- *x* axis points toward platform east
- *y* axis points toward platform north
- *z* axis points vertically upward

Note: The axes convention may differ for each project.

Joint numbers are assigned by the engineer. Allowing the program to automatically assign joint numbers should not be permitted. It is important to follow a strict numbering system when creating or editing a model. This allows easier interpretation and use of the analysis results. An example of modeling is shown in Figure 3.18.

Joint Coordinates

To facilitate the checking process when using software, the joint coordinates are always input and presented using a single set of units (i.e., m or ft). Take care to avoid using dual units (m/cm or ft/in), which is the usual mistake. Figure 3.19 presents the structure geometry with node, tubular element and pile modeling.

Offshore structure fixed platforms usually have a sliding connection between the structure elements that should be considered in the modeling, and the two most common cases are:

- Jacket piles are welded off at the top of the jacket and guided within the legs by spacers, as shown in Figure 3.20.
- Conductors are restrained horizontally but not vertically by conductor frames, as shown in Figure 3.21.

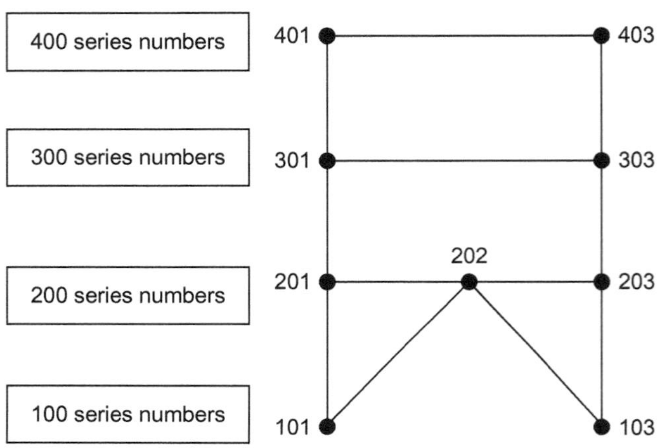

FIGURE 3.18 Node modeling to enhance quality assurance.

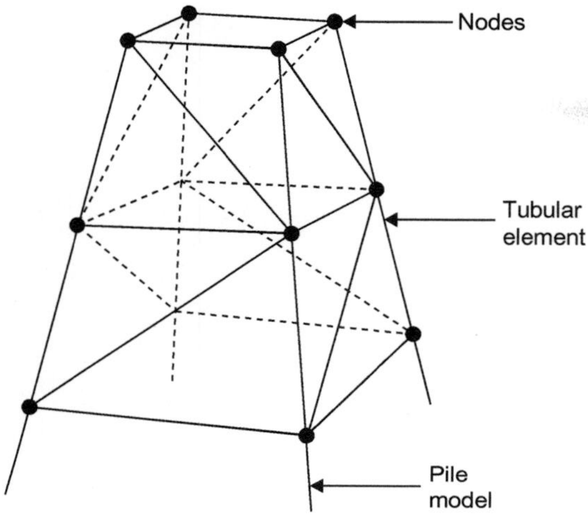

FIGURE 3.19 Modeling for the jacket structure.

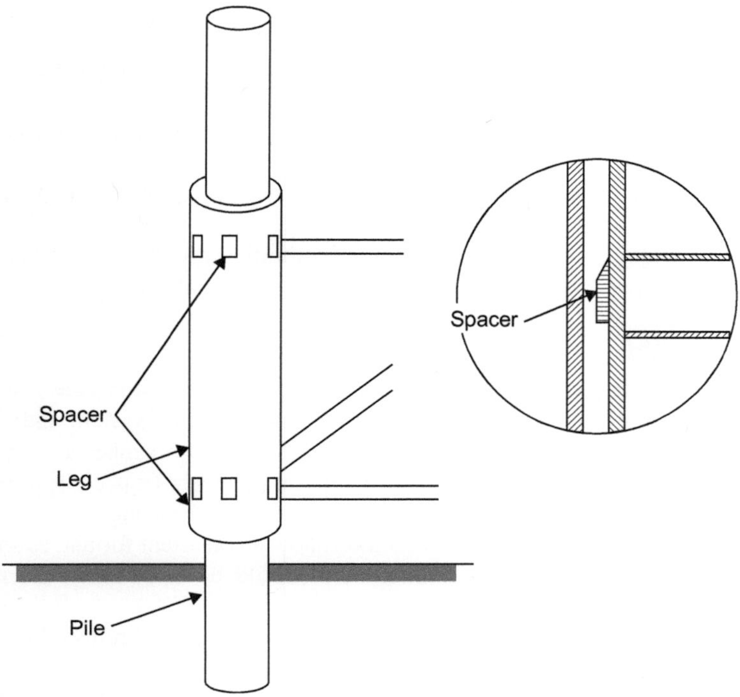

FIGURE 3.20 Spacer between legs and piles.

FIGURE 3.21 Conductor guides.

The boundary conditions should be clearly defined and should reflect the actual support conditions for the structure.

Any model should generally consist of all primary framing members. Secondary members need not be explicitly modeled unless they facilitate the input of loads or contribute to the structural action of a primary member.

Primary and secondary steelwork are classified as follows:

- Topside primary steel includes all truss members, girders and horizontal bracing.
- Jacket primary steel includes legs, diagonal bracing, horizontal bracing and piles.
- Topside major secondary steel includes deck plate, grating, deck beams, walkways, stairs and the crane pedestal.
- Jacket major secondary steel includes cathodic protection, boat landing, barge bumpers, walkways, appurtenance supports and mud mats.

Local Member Axes

When constructing the model and running the software, the engineer should review and appreciate the program's default member axis system and adopt this system where possible. In addition, the default local member axis system should be taken into consideration with vertical members (especially I-beams and channels), because it will affect the orientation of the flanges.

The orientation of the members should follow a consistent format, as shown in Figure 3.22. That is, all like members should be oriented in the same direction.

Member-end releases should be clearly defined and should reflect the actual connection constraints for the member.

In most cases, the member-end offsets may be used where there are large joint thicknesses. The offset should extend only to the face of the joint.

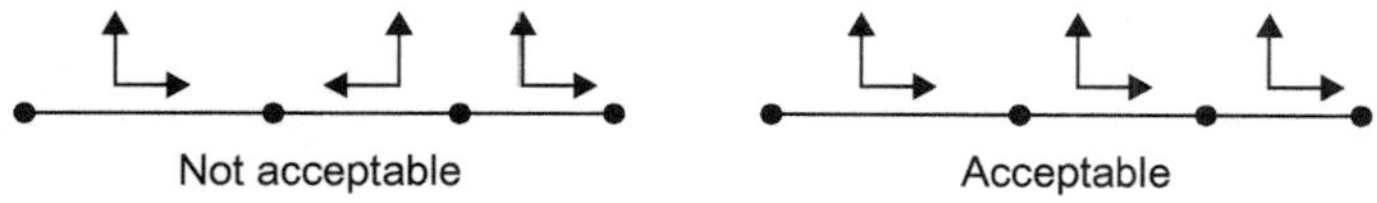

FIGURE 3.22 Local axis technique.

Member Effective Lengths

The effective length of a member under axial compression should reflect the relative joint stiffnesses at the end of the member. The appropriate effective length factor *K* should be selected from the recommended values in the design codes. Consideration should be given to the constraining effect provided by intermediate members along the length of the member; the effective length of a member buckling about its *y-y* axis is often different from the effective length about the *z-z* axis.

The compression flange (or critical flange) of a member may buckle under bending lateral torsional buckling. The effective length of a member under bending should reflect the degree of torsional restraint offered by the end connections of the member and by intermediate members along the length of the member. The bending effective length of a member should be calculated using the appropriate factors given in the design codes.

Joint Eccentricities

Eccentric joints in jacket structures should be modeled using member-end offsets. For topside type structures, joint eccentricities should be modeled using discrete elements, thereby allowing easy extraction of joint forces from the output.

When required, the deck plate should be modeled as a structural element using it as a membrane plate. Note that the plate elements need not be offset.

Alternatively, pinned-end axial brace members may be used in lieu of plate elements.

A problem the structural engineer always faces is how to model the pile inside the leg. Most software provides a wishbone member that should be modeled at all horizontal bracing levels of the jacket to account for pile-to-jacket leg interaction, as shown in Figure 3.23. If the pile-to-jacket leg annulus is to be grouted, then a rigid connection between the pile and leg should be modeled.

Generally, appurtenances do not contribute to the structural stiffness of the primary structure. Appurtenances may be modeled to facilitate automatic load generation by the program and are sometimes referred to as nonstructural members. Appurtenances may be modeled by assigning a small modulus of elasticity or small stiffness properties to these members. It is important to ensure any member-end releases, etc., accurately reflect the actual support conditions for the appurtenance and that no spurious forces enter the structure due to poor

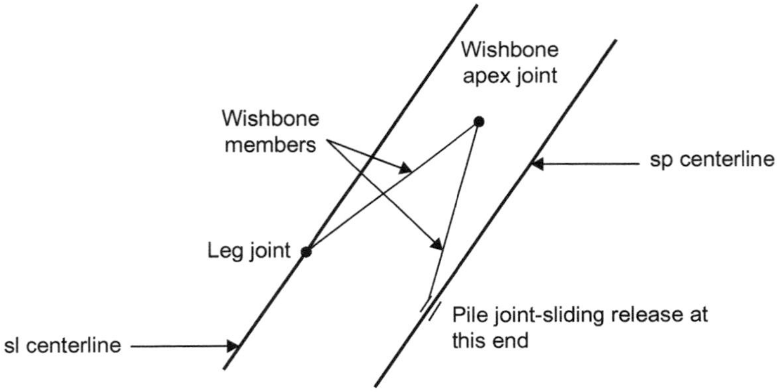

FIGURE 3.23 Pile-to-leg annulus modeling.

modeling techniques. When using this modeling method, the engineer should verify the analysis and ensure there are not compatibility problems caused by the small stiffnesses.

3.5.4 Dynamic Structure Analysis

Dynamic analysis is becoming increasingly important for: 1) larger and more costly structures, where dynamic analysis translates into huge construction cost savings; 2) complex offshore structures, where ordinary static analysis tends to result in higher risk; 3) harsh environmental conditions at many deep water sites that cannot be adequately modeled by static analysis. Software for dynamic analysis is now available so the process has become easier. The equation of motion will be:

$$M\ddot{x} + C\dot{x} + kx = P(\dot{x}, \ddot{x}) \tag{3.4}$$

where

M = diagonal matrix of virtual mass;
C = matrix for structural and viscous damping;
K = square linear structure stiffness matrix;
$P(\dot{x}, \ddot{x})$ = the load vector where $\dot{x}$ and $\ddot{x}$ are the water velocity and acceleration, respectively;
$\ddot{x}$ = structural acceleration;
$\dot{x}$ = velocity; and
x = displacement.

Natural Frequency

The first step in the dynamic analysis is to calculate the natural frequency of the structure.

In general, there are two methods for calculating the natural frequency of linear single-degree-of-freedom structural models: the Rayleigh method, which is based on the energy principle, and the direct method, which is based on the equation of motion, and the latter method is illustrated here.

The direct method of obtaining the natural frequency of a single-degree-of-freedom structure is to use its equation of motion. The structure's natural frequency, ω_o, is defined as the frequency compatible with an undamped structure of constant mass, with a restraint force that varies linearly with the displacement coordinate, and with no external excitation force. Under these conditions, the governing equation of motion, written in terms of the displacement coordinate x, is:

$$m\ddot{x} + kx = 0 \tag{3.5}$$

A special case of Equation (3.5) occurs when the applied force is zero and there is no damping; this is called a simple harmonic motion.

$$x = x_o \sin \omega_n t \tag{3.6}$$

Substituting Equation (3.2) and its second derivative into Equation (3.5) results in:

$$(-m\omega_n^2 + k)x_o \sin \omega_n t = 0$$

However, the term sin $\omega_n t$ is not zero for all time t and thus the term in brackets must be zero. This leads to the following equation for the natural frequency of the structure:

$$\omega_n = \sqrt{\frac{k}{m}} \tag{3.7}$$

where ω_n is the natural frequency in rad/s.

The natural period T and the natural frequency f are calculated from:

$$T = 2\pi/\omega = 2\pi\sqrt{\frac{M}{k}}\ s \tag{3.8}$$

$$f = 1/T = 1/2\pi\sqrt{\frac{k}{M}}\ \text{cycle/s (Hz)} \tag{3.9}$$

In multiple-degree-of-freedom systems, as in the case of the structure in general and the fixed offshore structure shown in Figure 3.23, every level of the structure has its own mass and stiffness. So the values of the mass, stiffness, displacement and acceleration will be represented in matrix form as:

$$[M]\{\ddot{x}\} + [k]\{x\} = 0 \tag{3.10}$$

$$\{x\} = (A \sin \omega t + B \cos \omega t)\{\phi\} \tag{3.11}$$

where ϕ is the vibration shape.

$$\{\ddot{x}\} = -\omega^2(A \sin \omega t + B \cos \omega t)\{\phi\} \tag{3.12}$$

$$[k]\{x\} = \omega^2[M]\{x\} \tag{3.13}$$

Equation (3.10) is the standard eigenvalue problem, which may be solved by the natural frequency and natural modes.

The output results for the dynamic analysis will be:

1. Time-history of member-end forces
2. Time-history of joint displacement
3. Maximum values of joint displacement
4. Time-history of interstory shear
5. Time-history of overturning moment
6. Time-history of base shear
7. Time-history of axial pile loads

The simple harmonic motion with mass and stiffness is shown in Figure 3.24.

$$\begin{bmatrix} m_1 & 0 & 0 & 0 \\ 0 & m_2 & 0 & 0 \\ 0 & 0 & m_3 & 0 \\ 0 & 0 & 0 & m_4 \end{bmatrix} \begin{Bmatrix} \ddot{x}_1 \\ \ddot{x}_2 \\ \ddot{x}_3 \\ \ddot{x}_4 \end{Bmatrix} + \begin{bmatrix} K_{11} & K_{12} & K_{13} & K_{14} \\ K_{21} & K_{22} & K_{23} & K_{24} \\ K_{31} & K_{32} & K_{33} & K_{34} \\ K_{41} & K_{42} & K_{43} & K_{44} \end{bmatrix} \begin{Bmatrix} x_1 \\ x_2 \\ x_3 \\ x_4 \end{Bmatrix} = \begin{Bmatrix} 0 \\ 0 \\ 0 \\ 0 \end{Bmatrix} \tag{3.14}$$

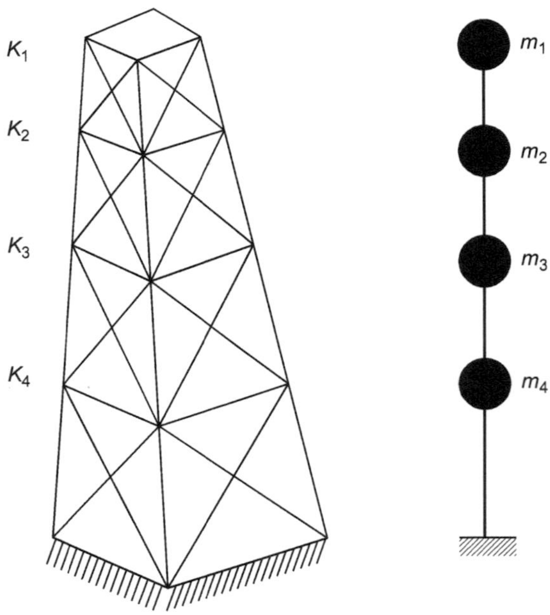

FIGURE 3.24 Mass distribution for the jacket.

Figure 3.25 presents the movement of point X_1 at applied unit load at this point and its effect at all nodes.

The modes will be as shown above, but in the space structure as in the jacket structure analysis, there are modes of direction of displacement, as shown in Figure 3.26.

The analysis can be done by mode supposition: the total response can be obtained by adding the individual modes, as presented, for the four-level structure shown in Figure 3.27.

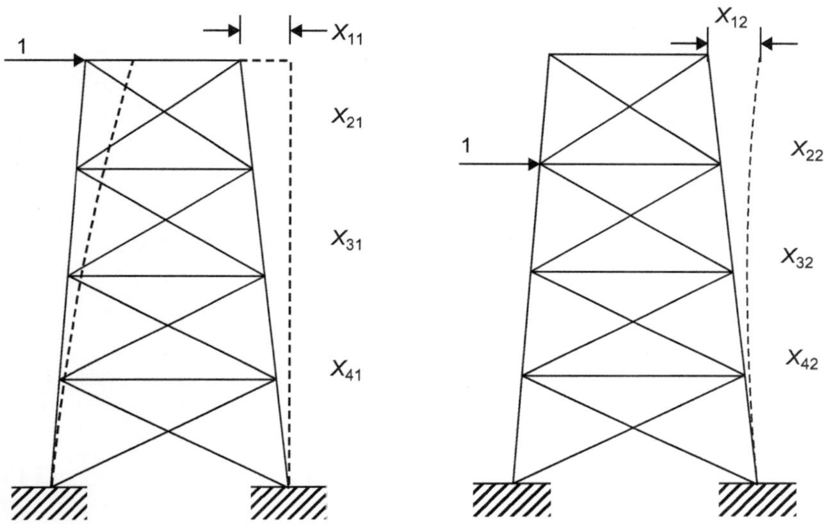

FIGURE 3.25 Relation between applied unit load at each level and deflection.

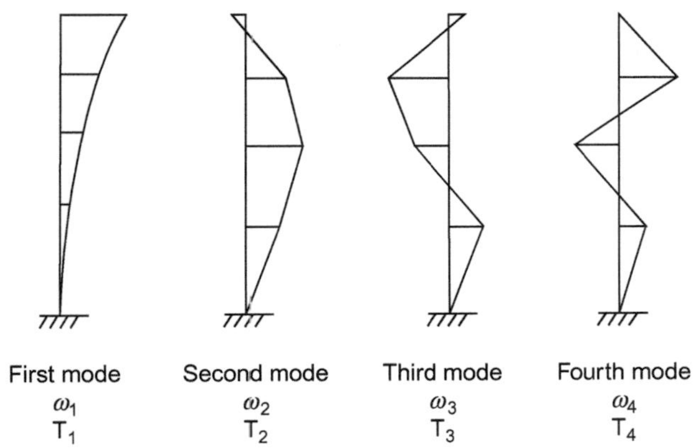

FIGURE 3.26 Modes of deformation.

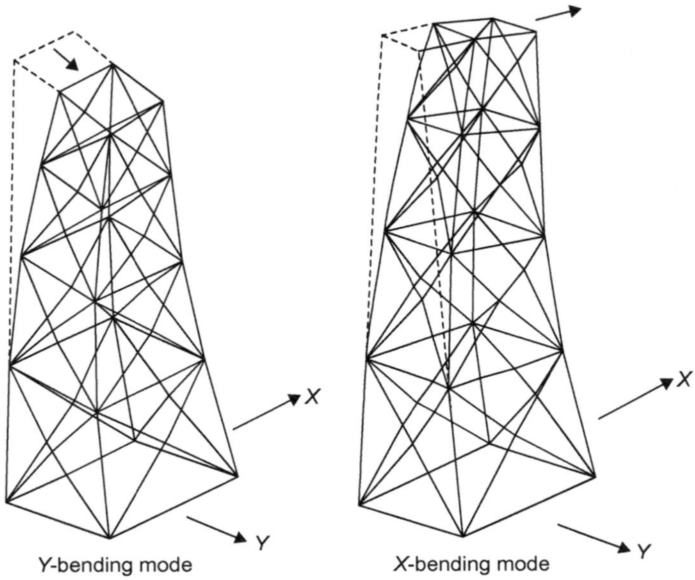

FIGURE 3.27 Modes of deflection.

$$\begin{Bmatrix} X_1 \\ X_2 \\ X_3 \\ X_4 \end{Bmatrix} = \{\phi_1\}x_1 + \{\phi_2\}x_2 + \{\phi_3\}x_3 \tag{3.15}$$

The total procedure may be summarized as:

1. Calculate the natural frequency (ω_n) and mode shapes (ϕ_n)
2. Calculate generalized masses for each mode

$$M_m = \{\phi_m\}^T[M]\{\phi_m\} \tag{3.16}$$

3. Calculate generalized stiffness for each mode

$$K_m = \{\phi_m\}^T[K]\{\phi_m\} \tag{3.17}$$

4. Calculate generalized force for each mode

$$P_m = \{\phi_m\}^T\{P(t)\} \tag{3.18}$$

5. Calculate the response for each mode from the following equation:

$$M_m\ddot{x} + K_m = P_m(t) \tag{3.19}$$

6. The equations of step 5 are:

$$\ddot{x}_1 + \omega_1^2 X_1 = P_m / M_m \quad X_1(t) = \frac{P_m}{M_m \omega^2}(1 - \cos \omega_1 t) \tag{3.20}$$

$$\ddot{x}_2 + \omega_2^2 X_2 = P_m / M_m \quad X_2(t) = \frac{P_m}{M_m \omega^2}(1 - \cos \omega_2 t) \tag{3.21}$$

$$\ddot{x}_3 + \omega_3^2 X_3 = P_m / M_m \quad X_3(t) = \frac{P_m}{M_m \omega^2}(1 - \cos \omega_3 t) \tag{3.22}$$

$$\ddot{x}_4 + \omega_4^2 X_4 = P_m / M_m \quad X_4(t) = \frac{P_m}{M_m \omega^2}(1 - \cos \omega_4 t) \tag{3.23}$$

$$\begin{Bmatrix} X_1 \\ X_2 \\ X_3 \\ X_4 \end{Bmatrix} = \{\phi_1\} X_1(t) + \{\phi_2\} X_2(t) + \{\phi_3\} X_3(t) + \{\phi_4\} X_4(t) \tag{3.24}$$

Using the above equation, we can define the values of drift with time for each floor.

3.5.5 In-place Analysis According to ISO 19902

To perform in-place analysis of the load resistance factor design according to ISO 19902, the general equation for determining the design load (action) (F_d) for in-place situations is:

$$\begin{aligned} F_d = {} & \gamma_{f,G1} G_1 + \gamma_{f,G2} G_2 + \gamma_{f,Q1} Q_1 + \gamma_{f,Q2} Q_2 + \gamma_{f,Eo}(E_o + \gamma_{f,D} D_o) \\ & + \gamma_{f,Ee}(E_e + \gamma_{f,D} D_e) \end{aligned} \tag{3.25}$$

where G_1 and G_2 are the permanent loads; Q_1 and Q_2 are the variable loads; E_o is the environmental load, which is defined by the owner as the operating wind, wave and current parameters; D_o is the equivalent quasi-static action representing dynamic response, but caused by the wave condition that corresponds with that for E_o; E_e is the extreme quasi-static action due to wind, waves and current; D_e is the equivalent quasi-static action representing dynamic response; and $\gamma_{f,G1}$, $\gamma_{f,G2}$, $\gamma_{f,Q1}$ and $\gamma_{f,Q2}$ are the partial load (action) factors for the various permanent and variable actions and for which values for different design situations are given in Table 3.1.

On the other hand, $\gamma_{f,E}$ and $\gamma_{f,D}$ are the partial action factors for the environmental actions and for which appropriate values should be defined by the owner through the statement of requirement document (SOR), $\gamma_{f,Eo}$ and $\gamma_{f,Ee}$ are partial action factors applied to the total quasi-static environmental action plus equivalent quasi-static action representing dynamic response for operating and extreme environmental conditions, respectively, and for which values for different design situations are given in Table 3.1.

TABLE 3.1 Partial Load Factors for In-Place Situation

	Partial Load Factors[a]					
Design Situation	$\gamma_{f,G1}$	$\gamma_{f,G2}$	$\gamma_{f,Q1}$	$\gamma_{f,Q2}$	$\gamma_{f,Eo}$	$\gamma_{f,Ee}$
Permanent and variable actions only	1.3	1.3	1.5	1.5	0.0	0.0
Operating situation with corresponding wind, wave and/or current conditions[b]	1.3	1.3	1.5	1.5	0.9 $\gamma_{f,E}$	0.0
Extreme conditions when the action effects due to permanent and variable actions are additive[c]	1.1	1.1	1.1	0.0	0.0	$\gamma_{f,E}$
Extreme conditions when the action effects due to permanent and variable actions are opposed[d]	0.9	0.9	0.8	0.0	0.0	$\gamma_{f,E}$

[a]*A value of 0 for a partial action factor means that the action is not applicable to the design situation.*
[b]*For this check G_2, Q_1 and Q_2 are the maximum values for each mode of operation.*
[c]*For this check G_1, G_2 and Q_1 include those parts of each mode of operation that can reasonably be present during extreme conditions.*
[d]*For this check G_2 and Q_1 exclude any parts associated with the mode of operation considered that cannot be guaranteed to be present during extreme conditions.*

3.6 CYLINDER MEMBER STRENGTH

Traditionally, the member in the jacket and in some cases for the topside is a cylinder member, so this section focuses on design of the tubular member according to ISO 19902, which is principally concerned with load resistance factor design (LRFD), as in the AISC. In addition, the strength of the cylinder member is also presented according to API RP2A, which focuses on the working stress design.

3.6.1 Cylinder Member Strength Calculation According to ISO 19902

According to ISO 19902, tubular members subjected independently to axial tension, axial compression, bending, shear or hydrostatic pressure should be designed to satisfy certain strength and stability requirements.

Axial Tension

Tubular members subjected to axial tensile forces should be designed to satisfy the condition:

$$f_t \le \frac{F_t}{\gamma_{R,t}}$$

where f_t is the axial tensile stress due to forces from factored actions; F_t is the representative axial tensile strength, $F_t = F_y$, and F_y is the representative yield strength, in stress units; and $\gamma_{R,t}$ is the partial resistance factor for axial tensile strength, $\gamma_{R,t} = 1.05$.

The member unity check U_c under axial tension is calculated from:

$$U_c = \frac{f_t}{F_t/\gamma_{R,t}} \tag{3.26}$$

Axial Compression

Tubular members subjected to axial compressive forces should be designed to satisfy the following condition:

$$f_c \leq \frac{F_c}{\gamma_{R,c}} \tag{3.27}$$

where f_c is the axial compressive stress due to forces from factored actions; F_c is the representative axial compressive strength, in stress units; and $\gamma_{R,c}$ is the partial resistance factor for axial compressive strength, $\gamma_{R,c} = 1.18$.

The member unity check U_c under axial compression should be calculated from Equation (3.28):

$$U_c = \frac{f_c}{F_c/\gamma_{R,c}} \tag{3.28}$$

Column Buckling

In the absence of hydrostatic pressure, the representative axial compressive strength for tubular members should be the smaller of the in-plane and the out-of-plane buckling strengths determined from the following equations:

$$F_c = [1.0 - 0.278\lambda^2]F_{yc} \quad \text{for} \quad \lambda \leq 1.34 \tag{3.29}$$

$$F_c = \frac{0.9}{\lambda^2} F_{yc} \quad \text{for} \quad \lambda > 1.34 \tag{3.30}$$

where

$$\lambda = \sqrt{\frac{F_{yc}}{F_e}} = \frac{KL}{\pi r}\sqrt{\frac{F_{yc}}{E}} \tag{3.31}$$

where F_c is the representative axial compressive strength, in stress units; F_{yc} is the representative local buckling strength, in stress units; λ is the column slenderness parameter; F_e is the smaller of the Euler buckling strengths in the y- and z-direction, in stress units; E is Young's modulus of elasticity; K is the effective length factor; L is the unbraced length in the y- or z-direction; r is the radius of gyration $r = \sqrt{\frac{I}{A}}$; I is the moment of inertia of the cross-section; and A is the cross-sectional area.

Local Buckling

The representative local buckling strength, F_{yc}, should be calculated from the following equations:

$$F_{yc} = F_y \quad \text{for} \quad F_y/F_{xe} \leq 0.170 \tag{3.32}$$

$$F_{yc} = \left[1.047 - 0.274\frac{F_y}{F_{xe}}\right] F_y \quad \text{for} \quad F_y/F_{xe} > 0.170 \tag{3.33}$$

$$F_{xe} = 2C_x Et/D \tag{3.34}$$

where F_y is the representative yield strength, in stress units; F_{xe} is the representative elastic local buckling strength, in stress units; C_x is the critical elastic buckling coefficient (see below); E is Young's modulus of elasticity; D is the outside diameter of the member; and t is the wall thickness of the member.

The theoretical value of C_x for an ideal tubular member is 0.6. However, a reduced value of $C_x = 0.3$ should be used in Equation (3.34) to account for the effect of initial geometric imperfections within the tolerance limits, as presented in Chapter 5. A reduced value of $C_x = 0.3$ is also implicit in the limits for F_y/F_{xe} given in Equations (3.32) and (3.33).

Bending

Tubular members subjected to bending moments should be designed to satisfy the following condition:

$$f_b = \frac{M}{Z_e} \leq \frac{F_b}{\gamma_{R,b}} \tag{3.35}$$

where f_b is the bending stress due to forces from factored actions (when $M > M_y$, f_b is to be considered an equivalent elastic bending stress, M/Z_e); F_b is the representative bending strength, in stress units (see below); $\gamma_{R,b}$ is the partial resistance factor for bending strength, $\gamma_{R,b} = 1.05$; M is the bending moment due to factored actions; M_y is the elastic yield moment; and Z_e is the elastic section modulus:

$$Z_e = \frac{\pi}{64}[D^4 - (D - 2t)^4] \Big/ \left(\frac{D}{2}\right)$$

The utilization of a member as described in ISO or unity check U_c, as described in most software, under bending moments should be calculated from:

$$U_c = \frac{f_b}{F_b/\gamma_{R,b}} = \frac{M/Z_e}{F_b/\gamma_{R,b}} \tag{3.36}$$

The representative bending strength for tubular members should be determined from:

$$F_b = \left(\frac{Z_p}{Z_e}\right) F_y \quad \text{for} \quad F_y D/Et \leq 0.0517 \tag{3.37}$$

$$F_b = \left[1.13 - 2.58\left(\frac{F_y D}{Et}\right)\right]\left(\frac{Z_p}{Z_e}\right)F_y \quad \text{for} \quad 0.0517 < F_y D/Et \le 0.1034 \qquad (3.38)$$

$$F_b = \left[0.94 - 0.76\left(\frac{F_y D}{Et}\right)\right]\left(\frac{Z_p}{Z_e}\right)F_y \quad \text{for} \quad 0.1034 < F_y D/Et \le 120F_y/E \qquad (3.39)$$

where, additionally, F_y is the representative yield strength, in stress units; D is the outside diameter of the member; t is the wall thickness of the member; Z_p is the plastic section modulus, and, calculating from the following equation,

$$Z_p = \frac{1}{6}[D^3 - (D - 2t)^3]$$

Shear

Tubular members subjected to beam shear forces should be designed to satisfy the following condition:

$$f_{v,b} = \frac{2V}{A} \le \frac{F_v}{\gamma_{R,v}} \qquad (3.40)$$

where

$$F_v = \frac{F_y}{\sqrt{3}}$$

and where $f_{v,b}$ is the maximum beam shear stress due to forces from factored actions; F_v is the representative shear strength, in stress units; $\gamma_{R,v}$ is the partial resistance factor for shear strength, $\gamma_{R,v} = 1.05$; V is the beam shear due to factored actions, in force units; and A is the cross-sectional area.

The member unity check U_c under beam shear is calculated from:

$$U_c = \frac{f_{v,b}}{F_v/\gamma_{R,v}} = \frac{2V/A}{F_v/\gamma_{R,v}} \qquad (3.41)$$

Torsional Shear

Tubular members subjected to torsional shear forces should be designed to satisfy the following condition:

$$f_{v,t} = \frac{M_{v,t}D}{2I_p} \le \frac{F_v}{\gamma_{R,v}} \qquad (3.42)$$

where $f_{v,t}$ is the torsional shear stress due to forces from factored actions; $M_{v,t}$ is the torsional moment due to factored actions; and

$$I_p = \frac{\pi}{32}[D^4 - (D - 2t)^4]$$

where I_p is the polar moment of inertia.

The partial resistance factor for shear, $\gamma_{R,v}$, is the same for both torsional shear and beam shear.

The member unity check U_c under torsional shear should be calculated from:

$$U_c = \frac{M_{v,t} D/2I_p}{F_v/\gamma_{R,v}} \tag{3.43}$$

Hydrostatic Pressure

The effective depth at the location being checked should be calculated taking into account the depth of the member below still water level (SWL) and the effect of passing waves. The factored water pressure (p) is calculated from:

$$p = \gamma_{f,G1} \rho H_z \tag{3.44}$$

where $\gamma_{f,G1}$ is the partial action factor for permanent loads, as shown in Table 3.1; ρ is the density of seawater, which may be taken as 1.025 kg/m^3; H_z is the effective hydrostatic head (m)

$$H_z = -z + \frac{H_w}{2} \frac{\cosh[k(d+z)]}{\cosh(kd)} \tag{3.45}$$

z is the depth of the member relative to SWL (measured positively upward); d is the still water depth to the sea floor; H_w is the wave height; and k is the wave number, $k = 2\ \pi/\lambda$, where λ is the wave length.

For installation conditions, z should be the maximum submergence during launch or the maximum differential head during the upending and installation sequence plus an amount to allow for deviations from the planned sequence, and $\gamma_{f,G1}$ in Equation (3.44) should be replaced by $\gamma_{f,T}$, which is equal to 1.1 when permanent and variable action predominate and equal to 1.35 when the environmental load is predominate, as in transportation and installation calculations.

Hoop Buckling

Tubular members subjected to external pressure should be designed to satisfy the following condition:

$$f_h = \frac{pD}{2t} \leq \frac{F_h}{\gamma_{R,h}} \tag{3.46}$$

where f_h is the hoop stress due to forces from factored hydrostatic pressure; p is the factored hydrostatic pressure, as calculated from Equation (3.44); D is the outside diameter of the member; t is the wall thickness of the member; F_h is the representative hoop buckling strength, in stress units; $\gamma_{R,h}$ is the partial resistance factor for hoop buckling strength and $\gamma_{R,h} = 1.25$.

For tubular members satisfying out-of-roundness tolerances, as presented in Chapter 5, F_h should be determined from:

$$F_h = F_y \quad \text{for} \quad F_{he} > 2.44F_y \tag{3.47}$$

$$F_h = 0.7(F_{he}/F_y)^{0.4}F_y \leq F_y \quad \text{for} \quad 0.55F_y < F_{he} \leq 2.44F_y \tag{3.48}$$

$$F_h = F_{he} \quad \text{for} \quad F_{he} \leq 0.55F_y \tag{3.49}$$

where F_y is the representative yield strength, in stress units, and F_{he} is the elastic hoop buckling strength, in stress units.

The elastic hoop buckling strength (F_{he}) is determined from:

$$F_{he} = 2C_h Et/D \tag{3.50}$$

where the critical elastic hoop buckling coefficient C_h is:

$$C_h = 0.44t/D \quad \text{for} \quad \mu \geq 1.6D/t$$
$$C_h = 0.44t/D + 0.21(D/t)^3/\mu^4 \quad \text{for} \quad 0.825D/t \leq \mu < 1.6D/t$$
$$C_h = 0.737/(\mu - 0.579) \quad \text{for} \quad 1.5 \leq \mu < 0.825D/t$$
$$C_h = 0.80 \quad \text{for} \quad \mu < 1.5$$

where μ is a geometric parameter,

$$\mu = \frac{L_r}{D}\sqrt{\frac{2D}{t}} \tag{3.51}$$

and where L_r is the length of tubular between stiffening rings, diaphragms or end connections.

For members that violate the allowable tolerance and have out-of-roundness greater then 1% and less than 3%, the reduced value of F_{he} will be:

$$F'_{he} = F_{he}\left(1 - 0.2\sqrt{\frac{D_{\max} - D_{\min}}{0.01D_n}}\right)\Big/0.8 \tag{3.52}$$

where $D_{\max}$ and $D_{\min}$ are the maximum and minimum values of any measured outside diameter at a cross-section and D_n is the nominal diameter.

The unity check U_c of a member under external pressure should be calculated from:

$$U_c = \frac{pD/2t}{F_h/\gamma_{R,h}} \tag{3.53}$$

Tubular Members Subjected to Combined Forces without Hydrostatic Pressure

Members subjected to combined forces, which give rise to global and local interactions between axial forces and bending moments, without hydrostatic pressure, have different requirements. Generally, the secondary moments from factored global actions and the associated bending stresses (P-Δ effects) do not need to be considered. However, when the axial member force is substantial, or when the component on which the axial force acts is very flexible, the secondary moments due to P-Δ effects from factored global actions should be taken into account.

Axial Tension and Bending

Tubular members subjected to combined axial tension and bending forces should be designed to satisfy the following condition at all cross-sections along their length:

$$\frac{\gamma_{R,t} f_t}{F_t} + \frac{\gamma_{R,b}\sqrt{f_{by}^2 + f_{bz}^2}}{F_b} \leq 1.0 \tag{3.54}$$

where f_{by} is the bending stress about the member's y-axis (in-plane) due to forces from factored actions and f_{bz} is the bending stress about the member's z-axis (out-of-plane) due to forces from factored actions.

Axial Compression and Bending

Tubular members subjected to combined axial compression and bending forces should be designed to satisfy the following conditions at all cross-sections along their length:

$$\frac{\gamma_{R,c} f_c}{F_c} + \frac{\gamma_{R,b}}{F_b}\sqrt{\left[\left(\frac{C_{m,y} f_{by}}{1 - f_c/F_{ey}}\right)^2 + \left(\frac{C_{m,z} f_{bz}}{1 - f_c/F_{ez}}\right)^2\right]} \leq 1.0 \tag{3.55}$$

and

$$\frac{\gamma_{R,c} f_c}{F_c} + \frac{\gamma_{R,b}\sqrt{f_{by}^2 + f_{bz}^2}}{F_b} \leq 1.0 \tag{3.56}$$

where $C_{m,y}$ and $C_{m,z}$ are the moment reduction factors corresponding to the y- and z-axes, respectively; F_{ey} and F_{ez} are the Euler buckling strengths corresponding to the y- and z-axes respectively, in stress units, such that:

$$F_{ey} = \frac{\pi^2 E}{(K_y L_y / r_y)^2} \tag{3.57}$$

$$F_{ez} = \frac{\pi^2 E}{(K_z L_z / r_z)^2} \tag{3.58}$$

where K_y and K_z are the effective length factors for the y- and z-directions, respectively; and L_y and L_z are the unbraced lengths in the y- and z-directions, respectively.

Tubular Members Subjected to Combined Forces with Hydrostatic Pressure

A tubular member below the water line is subjected to hydrostatic pressure unless it has been flooded due to installation procedure requirements. Platform legs are normally flooded in order to assist in their upending and placement and for pile installation. Even where members are flooded in the in-place condition, they can be subjected to hydrostatic pressures during launch and installation. The analysis of the structure can take the axial components of hydrostatic pressure on each member (capped-end actions) into account, or these effects can be included subsequently.

The requirements are presented in terms of axial stresses, which include capped-end forces f_{ac}.

For analyses using factored actions that include capped-end actions, f_{ac} is the axial stress resulting from the analysis. For analyses using factored actions that do not include the capped-end actions:

$$f_{ac} = |f_a \pm \gamma_{f,G1} f_q| \tag{3.59}$$

where f_a is the axial stress resulting from the analysis without capped-end actions and f_q is the compressive axial stress due to the capped-end hydrostatic actions; f_q should be added to f_a if f_a is compressive and subtracted from f_a if f_a is tensile. Note that the condition for which f_a is tensile and $f_a < \gamma_{f,G1} f_q$ is one of axial compression.

The capped-end stresses (f_q) may be approximated as half the hoop stress due to forces from factored hydrostatic pressure:

$$f_q = |0.5 f_h| \tag{3.60}$$

In reality, the magnitude of these stresses depends on the restraint on the member provided by the rest of the structure and its value can be more or less than the value in Equation (3.60). The approximation $|0.5 f_h|$ may be replaced by a stress computed from a more rigorous analysis.

In all cases, Equation (3.46) should be satisfied in addition to the requirements below.

Axial Tension, Bending and Hydrostatic Pressure

Tubular members subjected to combined axial tension, bending and hydrostatic pressure should be designed to satisfy the following requirements at all cross-sections along their length.

$$\frac{\gamma_{R,t} f_{ac}}{F_{th}} + \frac{\gamma_{R,b}\sqrt{f_{by}^2 + f_{bz}^2}}{F_{b,h}} \leq 1.0 \tag{3.61}$$

where f_{ac} is the tensile axial stress due to forces from factored actions that include capped-end actions ($f_{ac} > 0$) and F_{th} is the representative axial tensile strength in the presence of external hydrostatic pressure, in stress units, such that:

$$F_{th} = F_y\left[\sqrt{1 + 0.09B^2 - B^{2\eta}} - 0.3B\right] \quad (3.62)$$

$F_{b,h}$ is the representative bending strength in the presence of external hydrostatic pressure, in stress units, and

$$B = \frac{\gamma_{R,h} f_h}{F_h} \quad B \leq 1.0$$

$$\eta = 5 - 4\frac{F_h}{F_y}$$

Axial Compression, Bending and Hydrostatic Pressure

Tubular members subjected to combined axial compression, bending and hydrostatic pressure should be designed to satisfy the following requirements at all cross-sections along their length.

$$\frac{\gamma_{R,c} f_{ac}}{F_{yc}} + \frac{\gamma_{R,b}\sqrt{f_{by}^2 + f_{bz}^2}}{F_{bh}} \leq 1.0 \quad (3.63)$$

If $f_a < 0$, i.e. the member is in compression regardless of the capped-end stresses, Equation (3.64) should also be satisfied.

$$\frac{\gamma_{R,c} f_c}{F_{c,h}} + \frac{\gamma_{R,b}}{F_{bh}}\sqrt{\left[\left(\frac{C_{m,y} f_{by}}{1 - f_c/F_{ey}}\right)^2 + \left(\frac{C_{m,z} f_{bz}}{1 - f_c/F_{ez}}\right)^2\right]} \leq 1.0 \quad (3.64)$$

where, additionally, $F_{c,h}$ is the representative axial compressive strength in the presence of external hydrostatic pressure, in stress units.

$$F_{c,h} = 0.5F_{yc}\left[(1.0 - 0.278\lambda^2) - \frac{2f_q}{F_{yc}} + \sqrt{(1.0 - 0.278\lambda^2)^2 + 1.12\lambda^2\frac{f_q}{F_{yc}}}\right]$$

$$\text{for } \lambda \leq 1.34\sqrt{\left(1 - \frac{2f_q}{F_{yc}}\right)^{-1}} \quad (3.65)$$

$$F_{c,h} = \frac{0.9}{\lambda^2}F_{yc} \quad \text{for } \lambda > 1.34\sqrt{\left(1 - \frac{2f_q}{F_{yc}}\right)^{-1}} \quad (3.66)$$

If the maximum combined compressive stress $f_x = f_b + f_{ac}$ (if $f_{ac} \leq 0$) or $f_x = f_b - f_{ac}$ (if $f_{ac} < 0$) and the elastic local buckling strength F_{xe} exceeds the limits given below, then Equation (3.68) should also be satisfied.

$$f_x > 0.5\frac{F_{he}}{\gamma_{R,h}} \text{ and } \frac{F_{xe}}{\gamma_{R,c}} > 0.5\frac{F_{he}}{\gamma_{R,h}} \tag{3.67}$$

$$\frac{f_x - 0.5\dfrac{F_{he}}{\gamma_{R,h}}}{\dfrac{F_{xe}}{\gamma_{R,c}} - 0.5\dfrac{F_{he}}{\gamma_{R,h}}} + \left[\frac{\gamma_{R,h} f_h}{F_{he}}\right]^2 \le 1.0 \tag{3.68}$$

where F_{he} is the elastic hoop buckling strength and F_{xe} is the representative elastic local buckling strength from Equation (3.34).

Effective Lengths and Moment Reduction Factors

The effective lengths and moment reduction factors may be determined using a rational analysis that includes joint flexibility and side-sway. In lieu of such a rational analysis, values of effective length factors (K) and moment reduction factors (Cm) may be taken from Table 3.2. Table 3.2 does not apply to cantilever members, and it is assumed that both member ends are rotationally restrained in both planes of bending.

Lengths to which the effective length factors K are applied are normally measured from centerline to centerline of the end joints. However, for members framing into legs, the following modified lengths may be used, provided that no interaction between the buckling of members and legs affects the utilization of the legs.

Face-of-leg to face-of-leg for main diagonal braces.

Face-of-leg to centerline of end joint for K braces.

K factors lower than those of Table 3.2 may be used if they are supported by a more rigorous analysis. In case of connecting by more than one member, the buckling factor K can be obtained from the chart alignment in Figure 3.28.

The subscripts A and R in Figure 3.28 refer to the joints at the two ends of the column section being considered. G is defined as

$$G = \frac{\Sigma \dfrac{I_c}{L_c}}{\Sigma \dfrac{I_G}{L_G}} \tag{3.69}$$

in which Σ indicates a summation of all members rigidly connected to that joint and lying in the plane in which buckling of the column is being considered; I_c is the moment of inertia and L_c the unsupported length of the column section; and I_G is the moment of inertia and L_G the unsupported length of a girder or other restraining member. I_c and L_G are taken about axes perpendicular to the plane of buckling being considered.

TABLE 3.2 Effective Length and Moment Reduction Factors for Member Strength Checking

Structural Component	K	Cm
Topside legs		
Braced	1.0	0.85
Portal (unbraced)	K^b	0.85
Structure legs and piling		
Grouted composite section	1.0	Cm = 1.0 − 0.4 • (f_c/F_e), or 0.85, whichever is less
Ungrouted jacket legs	1.0	Cm = 1.0 − 0.4 • (f_c/F_e), or 0.85, whichever is less
Ungrouted piling between shim points	1.0	Cm = 0.6 − 0.4 • $M1/M2$
Structure brace members		Cm = 0.6 − 0.4 • $M1/M2$[a] or Cm = 1.0 − 0.4 • (f_c/F_e), or 0.85, whichever is less
Primary diagonals and horizontals	0.7	
K braces[c]	0.7	
X braces	0.7	Cm = 1.0 − 0.4 • (f_c/F_e), or 0.85, whichever is less
Longer segment length[c]	0.8	
Full length[d]	0.7	
Secondary horizontals	0.7	

[a] *$M1/M2$ is the ratio of smaller to larger moments at the ends of the unfaced portion of the member in the plane of bending under consideration. $M1/M2$ is positive when the member is bent in reverse curvature, negative when bent in single curvature. $F_e = F_{ey}$ or F_{ez} as appropriate.*
[b] *Use effective length alignment chart as in Figure 3.28.*
[c] *For either in-plane or out-of-plane effective lengths, at least one pair of members framing into a K or X joint should be in tension, if the joint is not braced out of plane.*
[d] *When all members are in compression and the joint is not braced out of plane.*

3.6.2 Cylinder Member Strength Calculation

The design of the cylinder member in accordance with API is based on AISC allowable stress design.

Axial Tension

The allowable tensile stress, F_t, for cylindrical members subjected to axial tensile loads should be determined from:

$$F_t = 0.6F_y \tag{3.70}$$

where F_y = yield strength, in ksi (MPa).

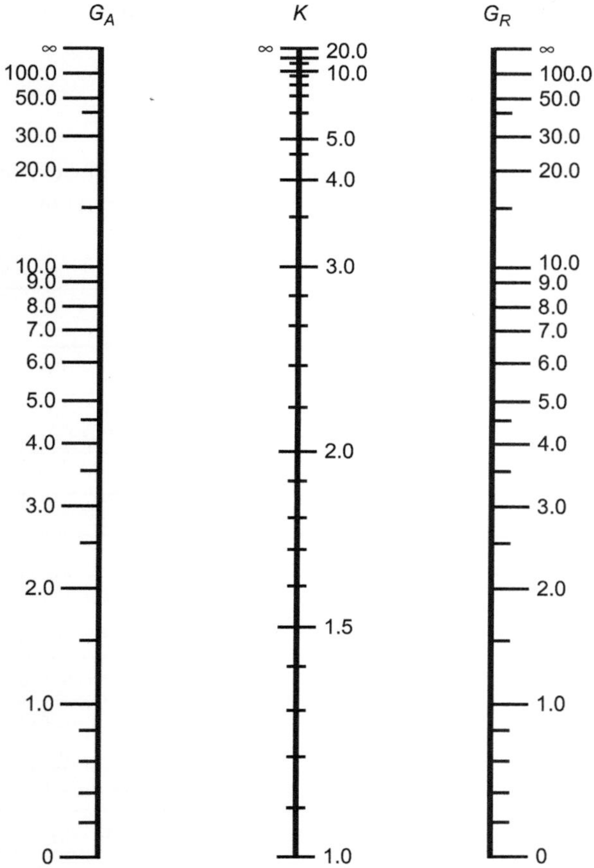

FIGURE 3.28 Chart alignment.

Axial Compression

The allowable axial compressive stress, F_a, should be determined from the following AISC formulas for members with a *D/t* ratio equal to or less than 60:

$$F_a = \frac{\left[1 - \frac{(kl/r)^2}{2C_c^2}\right] F_y}{5/3 + \frac{3(kl/r)}{8C_c} - \frac{(kl/r)^3}{8C_c^3}} \quad \text{for } kl/r < C_c \tag{3.71}$$

$$F_a = \frac{12\pi^2 E}{23(kl/r)^2} \quad \text{for } kl/r \geq C_c \tag{3.72}$$

K and Cm values can be obtained from Table (3.3) for different structural members, where

$$C_c = \sqrt{\frac{2\pi^2 E}{F_y}}$$

E is Young's modulus of elasticity, in ksi (MPa); K is the effective length factor; l is the unbraced length, in m (in); and r is the radius of gyration, in m (in).

TABLE 3.3 ***K*** **and** ***Cm*** **Values Based on API RP2A**

Structure element	Effective Length Factor K	Reduction Factor Cm
Superstructure legs		
Braced	1.0	0.85
Portal (unbraced)	K^a	0.85
Jacket legs and piling		
Grouted composite section	1.0	Min of {1 − 0.4(f_a/F_e)} or 0.85
Ungrouted jacket legs	1.0	Min of {1 − 0.4(f_a/F_e)} or 0.85
Ungrouted piling between shim points	1.0	0.6 − 0.4($M1/M2$), 0.4 < Cm < 0.85
Deck truss web members		0.6 − 0.4($M1/M2$), 0.4 < Cm < 0.85
In-plane action	0.8	0.6 − 0.4($M1/M2$), 0.4 < Cm < 0.85 or 0.85
Out-of-plane action	1.0	0.85
Jacket braces		0.6 − 0.4($M1/M2$), 0.4 < Cm < 0.85 or Min of {1 − 0.4(f_a/F_e)} or 0.85
Face-to-face length of main diagonals	0.8	
Face-of-leg to centerline of joint length of K braces[b]	0.8	Min of {1 − 0.4(f_a/F_e)} or 0.85
Longer segment length of X braces[b]	0.9	Min of {1 − 0.4(f_a/F_e)} or 0.85
Secondary horizontals	0.7	Min of {1 − 0.4(f_a/F_e)} or 0.85
Deck truss chord members	1.0	0.85 or 6 − 0.4($M1/M2$), 0.4 < Cm < 0.85 or Min of {1 − 0.4(f_a/F_e)} or 0.85

[a]*As in Figure 3.28.*
[b]*At least one pair of members framing into a joint must be in tension if the joint is not braced out of plane.*

For members with a *D/t* ratio greater than 60, substitute the critical local buckling stress (F_{xe} or F_{xc}, whichever is smaller) for F_y in determining C_c and F_a.

Local Buckling

Unstiffened cylindrical members fabricated from structural steels should be investigated for local buckling due to axial compression when the *D/t* ratio is greater than 60. When the *D/t* ratio is greater than 60 and less than 300, with wall thickness $t > 0.25$ inch (6 mm), both the elastic (F_{xe}) and inelastic local buckling stress (F_{xc}) due to axial compression should be determined from Equation (3.83). Overall column buckling should be determined by substituting the critical local buckling stress (F_{xe} or F_{xc}, whichever is smaller) for F_y in Equation (3.81) and in the equation for C_c.

Elastic Local Buckling Stress

The elastic local buckling stress, F_{xe}, should be determined from Equation (3.34).

Inelastic Local Buckling Stress

The inelastic local buckling stress, F_{xc}, should be determined from:

$$\begin{aligned} F_{xc} &= F_y[1.64 - 0.23(D/t)^{0.25}] \le F_{xe} \\ F_{xc} &= F_y \quad \text{for} \quad (D/t) \le 60 \end{aligned} \tag{3.73}$$

Bending

The allowable bending stress, F_b, should be determined from the following equations:

$$F_b = 0.75F_y \quad \text{for} \quad D/t \le 10{,}340/F_y (\text{SI units})$$

$$F_b = \left[0.84 - 1.74\frac{F_y D}{Et}\right]F_y \quad \text{for} \ 10{,}340/F_y < D/t \le 20{,}680/F_y \ \text{in SI units} \tag{3.74}$$

$$F_b = \left[0.72 - 0.58\frac{F_y D}{Et}\right]F_y \quad \text{for} \ 20{,}680/F_y < D/t \le 300 \ \text{in SI units} \tag{3.75}$$

For *D/t* ratios greater than 300, refer to API Bulletin 2U.

Shear

The maximum beam shear stress, f_v, for cylindrical members is:

$$f_v = \frac{V}{0.5A} \tag{3.76}$$

where f_v = the maximum shear stress, in MPa (ksi); V = the transverse shear force, MN (kips); and A = the cross-sectional area, m^2 (inches^2).

The allowable beam shear stress, F_v, should be determined from:

$$F_v = 0.4F_y \tag{3.77}$$

Torsional Shear

The maximum torsional shear stress, f_{vt}, for cylindrical members is:

$$f_{vt} = \frac{M_t(D/2)}{I_p} \tag{3.78}$$

where f_{vt} = maximum torsional shear stress (in MPa); M_t = torsional moment, MN-m; and I_p = polar moment of inertia, in m^4.

The allowable torsional shear stress, F_{vt}, should be determined from Equation (3.78).

Pressure on (Stiffened and Unstiffened Cylinders)

For tubular platform members satisfying API Spec 2B out-of-roundness tolerances, the acting membrane stress, f_h, in ksi (MPa), should not exceed the critical hoop buckling stress, F_{hc}, divided by the appropriate safety factor:

$$f_h \leq F_{hc}/SF_h \tag{3.79}$$

$$f_h = pD/2t \tag{3.80}$$

where f_h = hoop stress due to hydrostatic pressure, in ksi (MPa); p = hydrostatic pressure, in ksi (MPa); and SF_h = safety factor against hydrostatic collapse.

Design Hydrostatic Head

The hydrostatic pressure calculation in API is the same as presented in Equation (3.55) and the depth below still water surface including tide, in ft (m), z, is positive measured downward from the still water surface. For installation, z should be the maximum submergence during the launch or differential head during the upending sequence, plus a reasonable increase in head to account for structural weight tolerances and for deviations from the planned installation sequence. Seawater density is equal to 0.01005 MN/m^3 (64 lbs/ft^3).

Hoop Buckling Stress

The elastic hoop buckling stress, F_{he}, and the critical hoop buckling stress, F_{hc}, are determined from the following formulas.

Elastic Hoop Buckling Stress

The elastic hoop buckling stress determination is based on a linear stress-strain relationship from:

$$F_{he} = 2C_h Et/D \tag{3.81}$$

where the critical hoop buckling coefficient C_h includes the effect of initial geometric imperfections within API Spec 2B tolerance limits.

$$C_h = 0.44t/D \quad \text{for} \quad M \geq 1.6D/t$$
$$C_h = 0.44t/D + 0.21(D/t)^3/M^4 0.825D/t \leq M < 1.6D/t$$
$$C_h = 0.736/(M - 0.636) 3.5 \leq M < 0.825D/t$$
$$C_h = 0.736/(M - 0.559) 1.5 \leq M < 3.5$$
$$C_h = 0.8\, M < 1.5$$

The geometric parameter M is defined as:

$$M = \frac{L}{D}\sqrt{\frac{2D}{t}} \tag{3.82}$$

where L = length of cylinder between stiffening rings, diaphragms, or end connections, in inches (m).

Note: For $M > 1.6D/t$, the elastic buckling stress is approximately equal to that of a long unstiffened cylinder. Thus, stiffening rings, if required, should be spaced such that $M < 1.6D/t$ in order to be beneficial.

Critical Hoop Buckling Stress

The material yield strength relative to the elastic hoop buckling stress determines whether elastic or inelastic hoop buckling occurs and the critical hoop buckling stress, F_{hc}, in ksi (MPa), is defined by the appropriate formula.

Elastic buckling:

$$F_{hc} = F_{he} \quad \text{for} \quad F_{he} \leq 0.55F_y$$

Inelastic buckling:

$$F_{hc} = 0.45F_y + 0.18 + F_{he} \quad \text{for} \quad 0.55F_y < F_{he} \leq 1.6F_y$$
$$F_{hc} = \frac{1.31F_y}{1.15 + (F_y/F_{he})} \quad \text{for} \quad 1.6F_y < F_{he} < 6.2F_y \tag{3.83}$$
$$F_{hc} = F_y \quad \text{for} \quad F_{he} > 6.2F_y$$

Combined Stresses for Cylindrical Members

The method of calculating the applied combined stress between bending with compression and tensile stress in addition to the hydrostatic stress is discussed in the following section based on AISC.

Combined Axial Compression and Bending

Cylindrical members subjected to combined compression and flexure should be proportioned to satisfy both the following requirements at all points along their length.

$$\frac{f_a}{F_a} + \frac{C_m\sqrt{f_{bx}^2 + f_{by}^2}}{\left(1 - \frac{f_a}{F'_e}\right)F_b} \leq 1.0 \tag{3.84}$$

$$\frac{f_a}{0.6F_y} + \frac{\sqrt{f_{bx}^2 + f_{by}^2}}{F_b} \leq 1.0 \quad (3.85)$$

where the undefined terms used are as defined by the AISC *Specification for the Design, Fabrication, and Erection of Structural Steel for Buildings.*

When $f_a/F_a \leq 0.15$, the following formula may be used in lieu of the foregoing two formulas.

$$\frac{f_a}{F_y} + \frac{\sqrt{f_{bx}^2 + f_{by}^2}}{F_b} \leq 1.0 \quad (3.86)$$

Equation (3.84) assumes that the same values of C_m and F'_e are appropriate for f_{bx} and f_{by}. If different values are applicable, the following general formula should be used instead of Equation (3.84):

$$\frac{f_a}{F_a} + \frac{\sqrt{\left(\frac{C_{mx} f_{bx}}{1 - \frac{f_a}{F'_{ex}}}\right)^2 + \left(\frac{C_{my} f_{by}}{1 - \frac{f_a}{F'_{ey}}}\right)^2}}{F_b} \leq 1.0 \quad (3.87)$$

Member Slenderness

The slenderness of the steel section is defined by (Kl/r), where l is the member length, k is the buckling factor and r is the radius of gyration, which is also applied for cylindrical compression. Members should be in accordance with the America Institute of Steel Construction (AISC). To obtain the buckling factor, the joint end should be defined by its fixity and the joint movement. Moreover, a rational definition of the reduction factor should consider the character of the cross-section and the loads acting on the member. In lieu of such an analysis, the following values may be used.

Combined Axial Tension and Bending

Cylindrical members subjected to combined tension and bending should be proportioned to satisfy Equation (3.85) at all points along their length, where f_{bx} and f_{by} are the computed bending tensile stresses.

Axial Tension and Hydrostatic Pressure

The hydrostatic pressure is usually applied to the member at the sea mean water level. So the following interaction equation applies if the member is under longitudinal tensile stresses and hoop compressive stresses at the same time:

$$A^2 + B^2 + 2\nu |A| B \leq 1.0 \quad (3.88)$$

where

$$A = \frac{f_a + f_b - (0.5f_h)}{F_y}(SF_x) \tag{3.89}$$

The term A should reflect the maximum tensile stress combination

$$B = f_h / F_{hc}(SF_h) \tag{3.90}$$

where ν is Poisson's ratio = 0.3; F_y = yield strength, in ksi (MPa); f_a = absolute value of acting axial stress, in ksi (MPa); f_b = absolute value of acting resultant bending stress, in ksi (MPa); f_h = absolute value of hoop compression stress, in ksi (MPa); F_{hc} = critical hoop stress; SF_x = safety factor for axial tension; and SF_h = safety factor for hoop compression.

Axial Compression and Hydrostatic Pressure

The following interaction equation applies if the member is under longitudinal compression tensile stresses and hoop compressive stresses at the same time:

$$\frac{f_a + (0.5f_h)}{F_{xc}}(SF_x) + \frac{f_b}{F_y}(SF_b) \leq 1.0 \tag{3.91}$$

$$SF_h \frac{f_h}{F_{hc}} \leq 1.0 \tag{3.92}$$

Equation (3.91) should reflect the maximum compressive stress combination. The following equation should also be satisfied when $f_x > 0.5F_{ha}$.

$$\frac{f_x - 0.5F_{ha}}{F_{aa} - 0.5F_{ha}} + \left(\frac{f_h}{F_{ha}}\right)^2 \leq 1.0 \tag{3.93}$$

where $F_{aa} = F_{xe}/SF_x$; $F_{ha} = F_{he}/SF_h$; SF_x = safety factor for axial compression; and SF_b = safety factor for bending. $f_x = f_a + f_b + (0.5\ f_h)$ and f_x should reflect the maximum compressive stress combination, where F_{xe}, F_{xc}, F_{he} and F_{hc} are given by Equations (3.34), (3.73), (3.82) and (3.84), respectively.

Note that if $f_b > f_a + 0.5\ f_h$, both Equation (3.88) and Equation (3.91) must be satisfied.

Safety Factors

To compute allowable stresses, the required safety factors presented in Table 3.4 should be used with the local buckling interaction equations.

TABLE 3.4 Safety Factors Based on API RP2A

	Loading			
Design Condition	Axial Tension	Bending	Axial Compression	Hoop Compression
Where the basic allowable stresses would be used (e.g., pressures that will definitely be encountered during the installation or life of the structure)	1.67	F_y/F_b	1.67 to 2.0	2.0
Where the one-third increase in allowable stresses is appropriate (e.g., when considering interaction with storm loads)	1.25	$F_y/1.33F_b$	1.25 to 1.5	1.5

3.7 TUBULAR JOINT DESIGN

In older versions of API RP2A, punching shear governed the design of the tubular joints. The historical development of the API RP2A-WSD provisions clarifies the background of the most recent major updates. According to Marshall and Toprac (1974), the third edition of API RP2A-WSD, issued in 1972, introduced some simple recommendations based on punching shear principles. The fourth edition introduced factors to allow for the presence of load in the chord and the brace-to-chord diameter ratio (β). In the ninth edition, issued in 1977, differentiation was introduced in the allowable stress formulations for the joint and loading configuration, i.e., T/Y, X and K.

Between 1977 and 1983, much work was done, including large-scale load tests to failure, to improve the understanding and prediction of joint behavior. This work culminated in the 14th edition of API RP2A-WSD, in which the punching shear stress formulations were considerably modified and included a more realistic expression to account for the effect of chord loads, as well as providing an interaction equation for the combined effect of brace axial and bending stresses. The 14th edition also introduced the alternative nominal load approach, which gives equivalent results to the punching shear method. The guidance then essentially remained unchanged for all editions up to the 21st, although further recommendations were added on load transfer through the chord in the 20th edition (1993).

Since the 14th edition of API RP2A-WSD was issued, much further knowledge has been gained on the behavior of joints, including both experimental data and results of numerical studies. Over the period 1994 to 1996, MSL Engineering, under the auspices of a joint industry project, undertook an update to the tubular joint database [see the MSL report (1996) and Dier and Lalani (1995)]. This work and more recent studies, notably by API/EWI and the University of Illinois, have formed the basis of the tubular joint strength provisions of ISO 19902. The ISO drafting committee took as its starting point the relevant provisions from the API RP2A-LRFD first edition (similar to the API RP2A-WSD 20th edition) because ISO is in LRFD format. However, the API RP2A-WSD provisions were greatly modified during the drafting process to take account of the increased knowledge.

For the purposes of the supplement to the 21st edition of API RP2A-WSD, the draft ISO provisions, in turn, have been used as a starting basis. Additional studies, not available at the time of the preparation of the draft ISO guideline, have been incorporated into the supplement to the 21st edition of API RP2A-WSD. The major updates between the 21st edition and the supplement to the 21st edition are: a relaxation of the limit on tensile strength, additional guidance on detailing practice, removal of the punching shear approach, new Q_u and Q_f formulations, and a change in the form of the brace load interaction equation.

The old method of calculating the tubular joint capacity, based on the punching shear that was removed from the new API RP2A (2007), is presented here.

3.7.1 Simple Joint Calculation API RP2A (2007)

Joint Classification and Detailing

The first step is to define the joint classification under axial load: the K, X and Y components of load correspond to the three joint types for which capacity equations exist. Such subdivision normally considers all of the members *in one plane* at a joint.

Based on API RP2A (2007), the brace planes within ±15 degrees of each other may be considered as being in a common plane. Each brace in the plane can have a unique classification that could vary with load condition. Note that the classification can be a mix of the three main joint types, which are X, T and K.

Some simple examples of joint classification are presented in Figure 3.29. For a brace to be considered a K joint, the axial load on the brace should be balanced to within 10% by loads in other braces in the same plane and on the same side of the joint. For Y-joint classification, the axial load on the brace transfers as beam shear in the chord. For X-joint classification, the axial load on the brace is transferred through the chord to the opposite side (e.g., to braces, padeyes and launch rails).

Figure 3.29 presents the first three cases (a), (b), (c), which are the main joint classification K, Y, X, respectively.

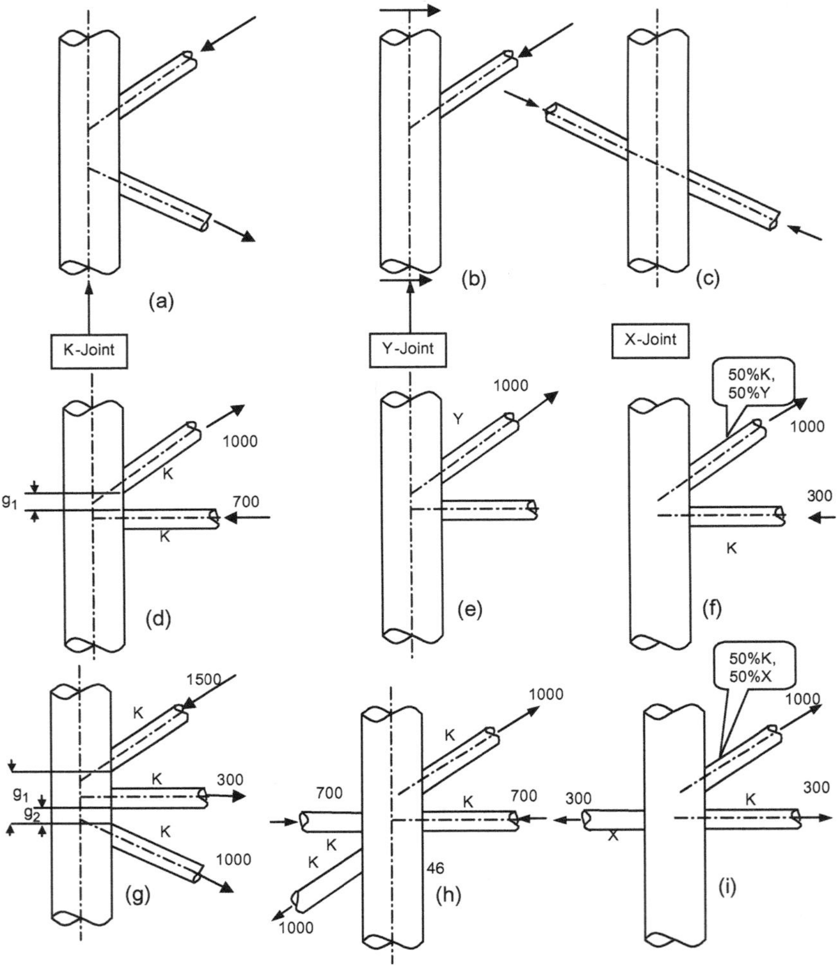

FIGURE 3.29 Joint classification.

Cases (d) and (g) present the gap between the adjacent braces, and case (g) illustrates the existing intermediate brace. The gap (g_2) is the gap between the outer loaded braces.

Case (f) presents a combination of K and Y joints and case (i) presents a combination of 50% K and X joints, while the other 50% of the X joint is the X joint for the brace on the left-hand side. In case (h), because the forces on the member are opposite and equal, the members are considered K joints.

Joint detailing is an essential element of joint design. For unreinforced joints, the recommended detailing nomenclature and dimensions are shown in Figures 3.30 and 3.31. If an increased chord wall thickness (or special steel) is required, it should extend past the outside edge of incoming bracing a minimum of one quarter of the chord diameter or 12 inches (300 mm), whichever is greater. Even greater lengths of increased wall thickness or special steel may be needed to avoid downgrading of joint capacity. If increased wall thickness for a brace or special steel is required, it should extend a minimum of one brace diameter or 24 inches (600 mm), whichever is greater. Neither the cited chord can nor brace stub dimension includes the length over which the 1:4 thicknesses taper occurs. In situations where fatigue has a major effect, tapering on the inside may have the undesirable consequence of fatigue cracking originating on the inside surface, which is difficult to inspect.

The minimum nominal gap between adjacent braces, whether in-plane or out-of-plane, is 2 inches (50 mm). Care should be taken to ensure that overlap of welds at the toes of the joint is avoided.

When overlapping braces occur, the amount of overlap should preferably be at least *d*/4 (where *d* is the diameter of the through brace) or 6 inches (150 mm), whichever is greater. This dimension is measured along the axis of the through member.

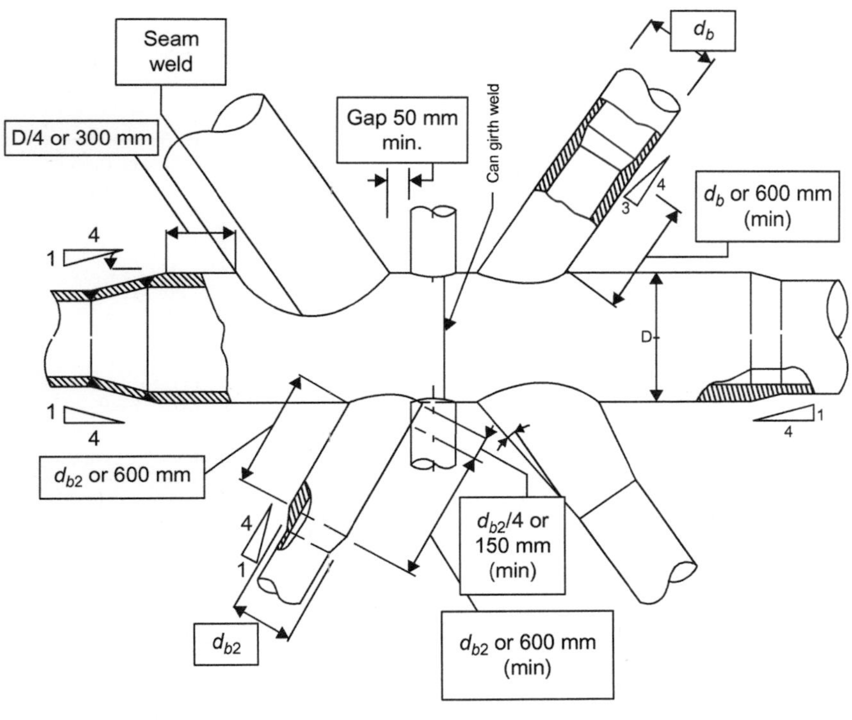

FIGURE 3.30 In-plane joint detailing.

Where overlapping of braces is necessary or preferred, and the braces differ in nominal thickness by more than 10%, the brace with the larger wall thickness should be the through brace and should be fully welded to the chord. Further, where substantial overlap occurs, the larger diameter brace should be specified as the through member. This brace may require an end stub to ensure that the thickness is at least equal to that of the overlapping brace.

Longitudinal seam welds and girth welds should be located to minimize or eliminate their impact on joint performance.

The longitudinal seam weld of the chord should be separated from incoming braces by at least 12 inches (300 mm), as shown in Figure 3.31. The longitudinal seam weld of a brace should be located near the crown heel of the joint.

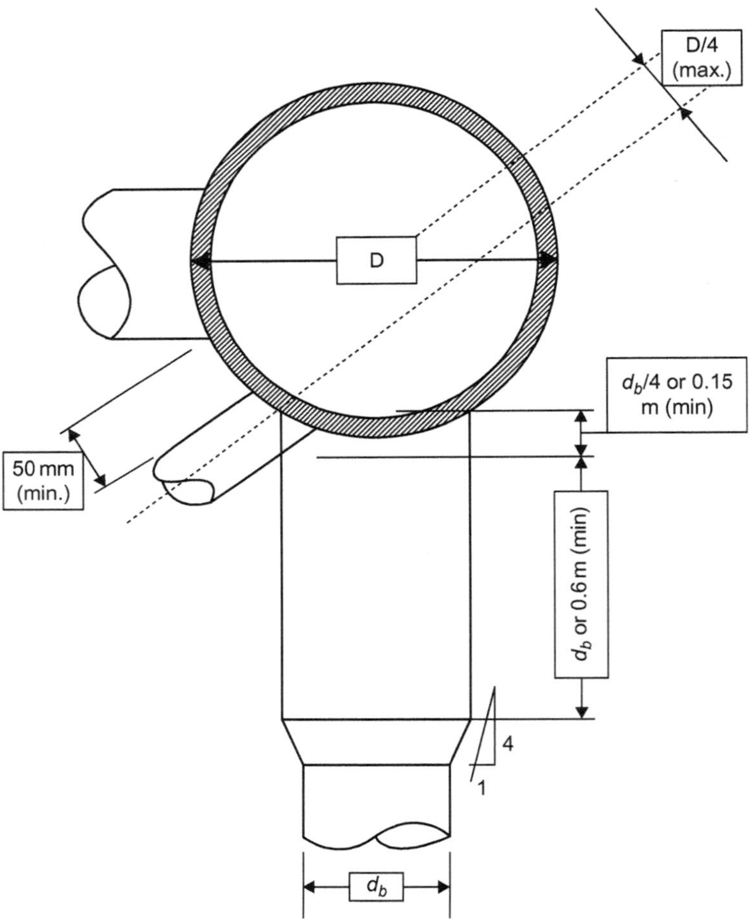

FIGURE 3.31 Out-of-plane joint detailing.

Longer chord cans may require a girth weld. This weld should be positioned at a lightly loaded brace intersection, between saddle and crown locations, as shown in Figure 3.30.

Simple Tubular Joint Calculation

The simple-joint calculation is valid based on the following criteria:

$$0.2 \leq b \leq 1.0$$
$$10 \leq \gamma \leq 50$$
$$30^\circ \leq \beta \leq 90^\circ$$
$$F_y \leq 72\,\text{ksi}\ (500\,\text{MPa})$$
$$g/D > -0.6 (\text{for K joints})$$

Tubular joints without overlap of principal braces and having no gussets, diaphragms, grout or stiffeners should be designed using the following guidelines (plus increase in both cases where applicable).

$$P_a = Q_u Q_f \frac{F_{yc} T^2}{FS \sin \theta} \tag{3.94}$$

$$M_a = Q_u Q_f \frac{F_{yc} T^2 d}{FS \sin \theta} \tag{3.95}$$

where P_a = allowable capacity for brace axial load; M_a = allowable capacity for brace bending moment, F_{yc} = the yield stress of the chord member at the joint (or 0.8 of the tensile strength, if less), in ksi (MPa); FS = safety factor = 1.60.

For joints with thickened cans, P_a should not exceed the capacity limits defined in Equation (3.99).

For axially loaded braces with a classification that is a mixture of K, Y and X joints, take a weighted average of P_a based on the portion of each in the total load.

Strength Factor Q_u

The strength factor Q_u varies with the joint and load type, as shown in Table 3.5. Where the working points of members at a gap connection are separated by more than $D/4$ along the chord centerline, or where a connection has simultaneously loaded branch members in more than one plane, the connection may be classified as a general or multiplanar connection.

Q_β is a geometric factor defined by:

$$Q_\beta = \frac{0.3}{\beta(1 - 0.833\beta)} \quad \text{for } b > 0.6 \tag{3.96}$$

$Q_\beta = 1.0$ for $\beta \leq 0.6$. Q_g is the gap factor defined by $Q_g = 1 + 0.2\,[1 - 2.8\,g/D]^3$ for $g/D \geq 0.05$ and $Q_g = 0.13 + 0.65\,\phi\,\gamma^{0.5}$ for $g/D \leq -0.05$ where $\phi = t\,F_{yb}/(TF_y)$.

TABLE 3.5 Values for Q_u

Joint Classification	Brace Load: Axial Tension	Axial Compression	In-plane Bending	Out-of-plane Bending
K	$(16 + 1.2\gamma)\beta^{1.2}\, Q_g$ but $\leq 40\, \beta^{1.2}\, Q_g$			
T/Y	30β	$2.8 + (20 + 0.8\gamma)\beta^{1.6}$ but $\leq 2.8 + 36\beta^{1.6}$	$(5 + 0.7\gamma)\beta^{1.2}$	$2.5 + (4.5 + 0.2\gamma)\beta^{2.6}$
X	23β for $\beta \leq 0.9$; $20.7 + (\beta - 0.9)(17\gamma - 220)$ for $\beta > 90$	$[2.8 + (12 + 0.1\gamma)\beta]Q_\beta$		

Preferably, the overlap should not be less than $0.25\beta D$. Linear interpolation between the limiting values of the above two Q_g expressions may be used for $-0.05 < g/D < 0.05$ when this is otherwise permissible or unavoidable.

F_{yb} = yield stress of brace or brace stub, if present (or 0.8 times the tensile strength, if less), in ksi (MPa).

The Q_u term for tension loading is based on limiting the capacity to first crack. The Q_u associated with full ultimate capacity of tension-loaded Y and X joints is given in API RP2A.

The X joint, axial tension, Q_u term for $\beta > 0.9$ applies to coaxial braces (i.e., $e/D \leq 0.2$ where e is the eccentricity of the two braces). If the braces are not coaxial ($e/D > 0.2$), then 23β should be used over the full range of β.

Chord Load Factor Q_f

The chord load factor Q_f accounts for the presence of nominal loads in the chord and is calculated from the following equation.

$$Q_f = \left[1 + F_1\left(\frac{FSP_{c\,y}}{P_y}\right) - F_2\left(\frac{FSM_{ipb}}{M_p}\right) - F_3A^2\right] \tag{3.97}$$

The parameter A is defined as:

$$A = \sqrt{\left[\left(\frac{FSP_c}{P_y}\right)^2 + \left(\frac{FSM_c}{M_p}\right)^2\right]} \tag{3.98}$$

where P_c and M_c are the nominal axial load and bending resultant in the chord; P_y is the yield axial capacity of the chord; and M_p is the plastic moment capacity

TABLE 3.6 Values of F_1, F_2 and F_3

Joint Type	F_1	F_2	F_3
T/Y joints under brace axial loading	0.3	0	0.8
K joints under brace axial loading	0.2	0.2	0.3
X joints under brace axial loading[a]			
$\beta \leq 0.9$	0.2	0	0.5
$\beta = 1.0$	−0.2	0	0.2
All joints under brace moment loading	0.2	0	0.4

[a] *Linearly interpolated values between $\beta = 0.9$ and $\beta = 1.0$ for X joints under brace axial loading.*

of the chord. [Where increase is applicable, $FS = 1.20$ in Equations (3.97) and (3.98).]

Table 3.6 presents the coefficients F_1, F_2 and F_3, which depend on joint and load type.

The average of the chord loads and bending moments on either side of the brace intersection should be used in Equations (3.97) and (3.98). Chord axial load is positive in tension, and chord in-plane bending moment is positive when it produces compression on the joint footprint. The chord thickness at the joint should be used in the above calculations.

Joints with Thickened Cans

For simple, axially loaded Y and X joints where a thickened joint can is specified, the joint allowable capacity may be calculated as:

$$P_a = \left[r + (1 - r)(T_n/T_c)^2\right](P_a)_c \tag{3.99}$$

where $(P_a)_c = P_a$ from Equation (3.94) based on chord can geometric and material properties, including Q_f calculated with respect to chord can; $T_n =$ nominal chord member thickness; $T_c =$ chord can thickness; $r = L_c/(2.5D)$ for joints with $\beta \leq 0.9$ and $r = (4\beta - 3)L_c/(1.5D)$ for joints with $\beta > 0.9$; and $L_c =$ effective total length as in Figure 3.30, that r cannot have a value greater than unity.

Alternatively, an approximate closed-ring analysis will be used in the calculation, including plastic analysis with a suitable factor of safety that covers the approximation in the calculation and using an effective chord length up to $1.25D$, where D is the chord diameter, on either side of the line of action of the branch loads at the chord face, but not more than the actual distance to the end of the joint can. The finite element method can be used in case of more complex joints. For multiple branches in the same plane, dominantly loaded in the same sense, the relevant crushing load is $\Sigma_t P_i \sin\theta_t$. A reinforcement can be used, such as diaphragms, rings, gussets or the stiffening effect of out-of-plane members, so it is included in the analysis, although its effectiveness decreases with distance from the branch footprint.

Strength Check

The interaction ratio J_{IR} for the joint with applied axial loads and bending moments in the brace should be calculated using the expression:

$$J_{IR} = \left|\frac{P}{P_a}\right| + \left(\frac{M}{M_a}\right)^2_{ipb} + \left|\frac{M}{M_a}\right|_{opb} \leq 1.0 \quad (3.100)$$

Overlapping Joints

Braces that have in-plane or out-of-plane overlap at the chord member form overlapping joints.

Joints that have in-plane overlap involving two or more braces in a single plane, such as K and KT joints, may be designed using the simple joint equation, using negative gap in Q_g, with the exception that the shear stress parallel to the chord face is a potential failure mode and should be checked. The joint can equation will not apply to overlapping joints with balanced loads. In addition, the axial forces in the overlapping and through braces should be considered to have the same sign, and the combined axial force representing that in the through brace plus a portion of the overlapping brace forces should be used to check the through-brace intersection capacity. The portion of the overlapping brace force can be calculated as the ratio of the cross-sectional area of the brace that bears onto the through brace to the full area.

During calculation, the thicker brace will be considered the through brace if the nominal thickness of the overlapping and through braces differ by more than 10%.

The through-brace capacity should be checked for combined axial and moment loading in the overlapping brace. In this instance, Q_f is calculated based on the through brace.

Joints having out-of-plane overlap may be assessed on the same general basis as in-plane overlapping joints, with the exception that axial load capacity may be calculated as for multiplanar joints.

Grouted Joints

The two traditional types of grouted joints are the fully grouted chord and the double-skin type, where grout is placed in the annulus between a chord member and an internal member. In both cases, the grout is unreinforced and, as far as joint behavior is concerned, no benefit is ascribed to shear keys that may be present.

The value of Q_u for grouted joints is calculated from the following.

In case of axial tension:

$$Q_u = 2.5\gamma\beta K_a \tag{3.101}$$

where $K_a = 0.5(1 + 1/\sin\theta)$.

In case of bending:

$$Q_u = 1.5\gamma\beta \tag{3.102}$$

It should be mentioned that grouted joints cannot fail under compression load and the compression capacity is limited by the braces that carry the compression load. So there is no calculation presented for the axial compression.

Where there are complex joints, for fully grouted and double-skin joints, the Q_u values may be replaced with the values pertinent to grouted joints given in Equations (3.101) and (3.102) in case of axial tension and bending, respectively. In any case, the value of Q_u should not be less than those calculated for simple joints.

For double-skin joints, failure may also occur by chord ovalization. The ovalization capacity can be estimated by substituting the following effective thickness into the simple joint equations.

$$T_e = \sqrt{(T^2 + T_p^2)} \tag{3.103}$$

Figure 3.32 presents an example of calculating the chord length (L_{ch}). For different braces, the chord length will be calculated as:

Brace 1: $L_{ch} = 2x_1 + d_1$
Brace 2: $L_{ch} = 2x_2 + d_2/\sin\theta$
Brace 3: $L_{ch} = 2x_3 + d_3$

where x_1, x_2 and x_3 are obtained as shown in Figure 3.32, T_e = effective thickness (inches or mm), T = wall thickness of chord (inches or mm) and T_p = wall thickness of inner member (inches or mm). T_e should be used in place of T in the simple joint equations, including the γ term.

The Q_f calculation for fully grouted and double-skinned joints should be based on wall thickness of the chord, T; it is presumed that calculation of Q_f has already accounted for load sharing between the chord and inner member, such that further consideration of the effect of grout on that term is unnecessary.

However, for fully grouted joints, Q_f may normally be set equal to 1.0, except in the case of high β ($\geq$0.9), X joints with brace tension and out-of-plane (OPB) or chord axial compression with OPB.

Figure 3.33 presents the configuration of stresses on the tubular joint where t is the brace thickness; g is gap between braces; T is the chord thickness; d is the brace diameter; D is the chord diameter; θ is the brace angle measured from the chord; $\tau = t/T$; $\beta = d/D$; and $\gamma = D/2T$.

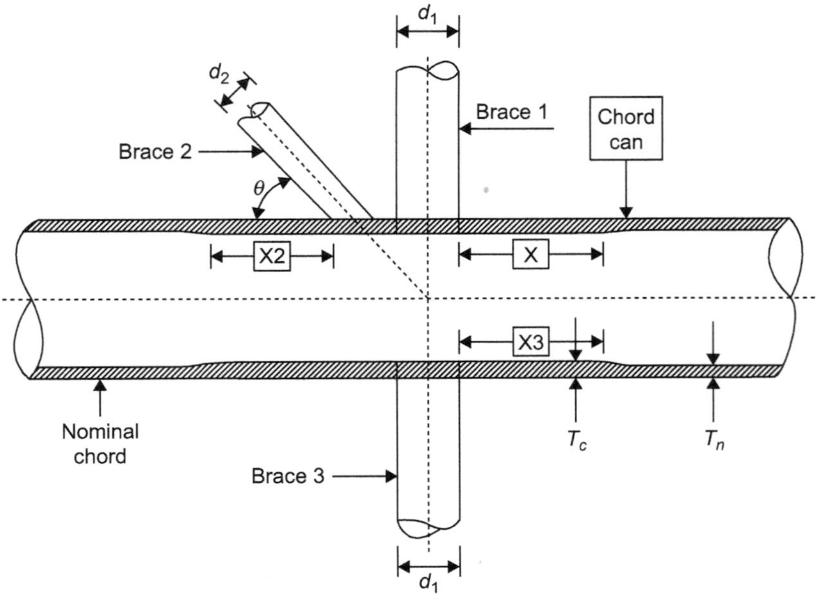

FIGURE 3.32 Chord length, L_{ch}.

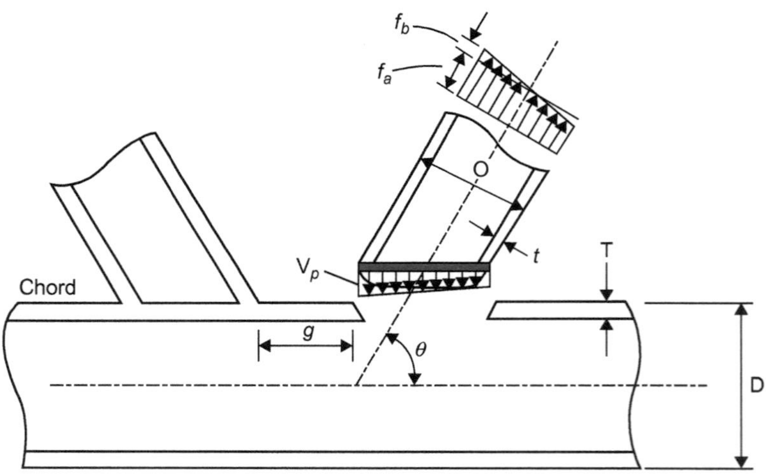

FIGURE 3.33 Geometric parameters for a tubular joint.

3.7.2 Joint Calculation According to API RP2A (2000)

Punching Shear

$$Vp = \tau f \sin\theta \tag{3.104}$$

where f = nominal axial, in-plane bending or out-of-plane bending stressing the brace.

The allowable punching shear stress in the chord wall is the lesser of AISC shear allowable or

$$V_{pa} = Q_q Q_f (F_{yc}/0.6\,\mathrm{g})$$

$$Q_f = 1.0 - \lambda \gamma A^2$$

where $\lambda = 0.030$ for brace axial stress; $\lambda = 0.045$ for brace in-plane bending stress, and $\lambda = 0.021$ for brace out-of-plane bending stress. $A = [(f_{AX})^2 + (f_{IPB})^2 + (f_{OPB})^2]^{0.5}/0.6F_{ye}$. $Q_f = 1.0$ when all extreme fiber stresses in the chord are tensile.

The value of Q_q can be obtained from Table 3.7.

$$Q_b = 0.3/(\beta(1 - 0.833\beta)) \text{ for } \beta > 0.6$$

$$Q_b = 1.0 \text{ for } b \leq 0.6$$

$$Q_g = 1.8 - 0.1(\mathrm{g}/T) \text{ for } \gamma \leq 20$$

$$Q_g = 1.8 - 4(\mathrm{g}/D) \text{ for } \gamma > 20$$

In any case, Q_g should be ≥ 1.0.

Joint classification as K, T, Y or cross-joints should apply to individual braces according to their load pattern for each load case. To be considered a K joint, the punching load in a brace should be essentially balanced by loads on other braces in the same plane on the same side of the joint. In T and Y joints the punching load is reached as beam shear in the chord. In cross-joints, the punching load is carried through the chord to braces on the opposite side.

TABLE 3.7 Values of Q_q

Joint	Axial Compression	Axial Tension	In-plane Bending	Out-of-plane Bending
K (gap)	$(1.10 + 0.2/\beta)Q_g$			
T and Y	$(1.10 + 0.2/\beta)$			
Cross without diaphragm	$(1.10 + 0.20/\beta)$	$(0.75 + 0.20/\beta)Q_\beta$	$(3.27 + 0.67/\beta)$	$(1.37 + 0.67/\beta)Q_\beta$
Cross with diaphragm	$(1.10 + 0.20/\beta)$			

Allowable Joint Capacity

The allowable joint capacity will be calculated as:

$$P_a = Q_u Q_f F_{yc} T^2 / 1.7 \sin\theta \tag{3.105}$$

$$M_a = Q_u Q_f F_{yc} T^2 / 1.7 \sin\theta(0.8d) \tag{3.106}$$

P_a and M_a are the allowable capacity for brace axial load and bending moment, respectively. The values of Q_u are shown in Table 3.8.

Tubular Joint Punching Failure

The experimental test for a frame structure jacket as per the joint industrial study (1999) with K bracing is as shown in Figure 3.34.

TABLE 3.8 Values of Q_u

Joint	Axial Compression	Axial Tension	In-Plane Bending	Out-of-Plane Bending
K (gap)	$(3.4 + 19\beta)Q_g$			
T and Y	$(3.4 + 19\beta)$			
Cross without diaphragm	$(3.4 + 19\beta)$	$(3.4 + 13\beta)Q_\beta$	$(3.4 + 19\beta)$	$(3.4 + 7\beta)Q_\beta$
Cross with diaphragm	$(3.4 + 19\beta)$			

FIGURE 3.34 Failure in the K joint.

Potential for punch through of a compression K-joint member into a chord due to plastic deformations around the intersection under cyclic loads is as shown in the following figures, where punching shear is obvious in Figure 3.35 and the buckling of the tubular joint is illustrated in Figure 3.36.

Figure 3.37 presents tearing in the tubular joint due to direct tensile force.

It is shown in Figure 3.37 as splitting of the paint without cracks in the steel member itself, but with more load the cracks will propagate to the steel member.

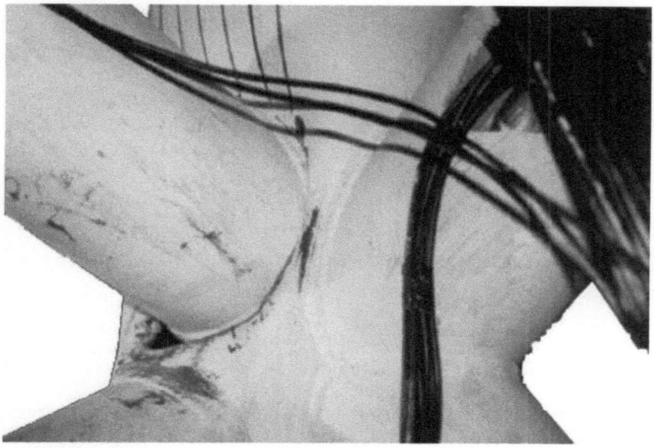

FIGURE 3.35 Punching shear.

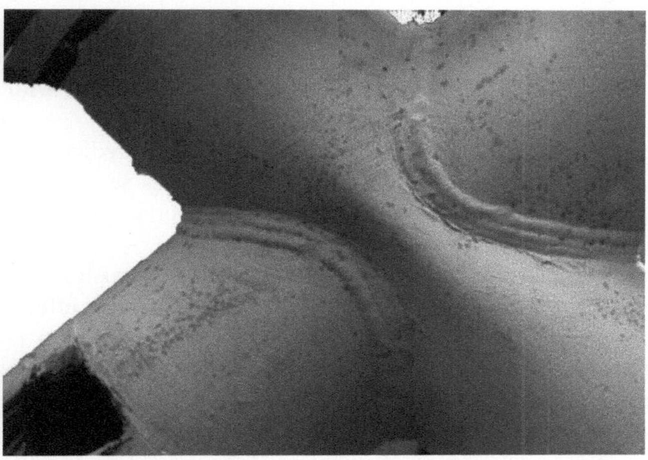

FIGURE 3.36 Buckling of the tubular joint.

FIGURE 3.37 Tearing in the tubular joint.

Therefore, there should be a guideline for tubular joint detailing, so joint can precautions and detailing are as shown in Figure 3.30.

3.7.3 Fatigue Analysis

Experience over the last 60 years and many laboratory tests have proven that a metal may fracture at a relatively low stress if that stress is applied a great number of times. The offshore structure, particularly its tubular joints, must resist progressive damage due to fatigue that results from continuous wave action during the 20- to 30-year design life of the structure. Over the years, platforms are subjected to a wide variety of sea states and within each sea state, the structure experiences many cycles of stress at various levels.

The purpose of fatigue analysis is to account for the fact that the number of cycles of stress that a structural component can withstand varies with the magnitude of the cyclic stress.

Dynamic analysis is used in the fatigue evaluation to predict the number of cycles and magnitude of stresses that occur in various sea states. As in the extreme wave analysis, dynamic effects become increasingly important for deep water structures with a heavy deck load.

Fatigue cracks grow because of tensile stresses; corrosion of a metal is accelerated if the metal is subjected to tensile stress. Thus, the effects of corrosion and fatigue are combined in the case of an offshore platform.

Kuang et al. (1975) described the design of tubular joints for offshore structures as an iterative procedure. The process begins with the sizing of the jacket piles according to the requirements of the specific soil and foundation needs of the platform. Sizing determines the diameter of the jacket legs and allows clearance for the piles to go through them. Once the truss work geometry is selected, the column buckling characteristics determine the diameters of the various jacket braces. The initial wall thicknesses of chords and braces are determined by structural analysis. The next cycle of calculation involves increasing the chord wall thicknesses with heavy joint cans to ensure sufficient static strength to meet code or specification requirements. The next iteration involves calculating the fatigue strength of the joint to determine if it is compatible with the service life requirement of the platform. Depending on the method of fatigue analysis used, allowable stress concentration factors must be either specified for each joint or built into the method of analysis.

For each location around each member intersection of interest in the structure, the stress response for each sea state should be computed, giving adequate consideration to both global and local stress effects.

The stress responses should be combined into the long-term stress distribution, which should then be used to calculate the cumulative fatigue damage ratio, D, where

$$D = \sum(n/N) \tag{3.107}$$

where n = number of cycles applied at a given stress range and N = number of cycles for which the given stress range would be allowed by the appropriate S-N curve.

In most cases, the damage ratio will be calculated for each sea state and combined to obtain the cumulative damage ratio.

In general, the design fatigue life of each joint and member should not be less than the intended service life of the structure multiplied by a safety factor. For the design fatigue life, D should not exceed unity.

For in-situ conditions, the safety factor for fatigue of steel components should depend on the failure consequence and its impact on cost, the environment and the like, and in-service inspectability.

Critical elements are those whose sole failure could be catastrophic. In lieu of a more detailed safety assessment of Category L-1 structures, which are manned and non-evacuated, a safety factor of 2.0 is recommended for inspectable, non-failure-critical connections. For failure-critical and/or noninspectable connections, increased safety factors are recommended, as shown in Table 3.9. A reduced safety factor is recommended for Category L-2 and L-3, which are manned evacuated or unmanned structures, respectively. For conventional steel jacket structures, on the basis of in-service performance data, SF = 1.0 for redundant diver or ROV inspectable framing, with safety factors for other cases being half those in the table.

When fatigue damage can occur due to other cyclic loadings, such as transportation, the following equation should be satisfied.

$$\sum_i SF_i D_i < 1.0 \tag{3.108}$$

where D_i is the fatigue damage ratio for each type of loading and SF_i is the associated safety factor.

TABLE 3.9 Safety Factor for Fatigue Life

Failure Criticality	Inspectable	Not Inspectable
No	2	5
Yes	5	10

For transportation where long-term wave distributions are used to predict short-term damage, a larger safety factor should be considered.

Stress Concentration Factors

The stress concentration factor (SCF) differs from one joint geometry to another. It is known that the applied loads on tubular joints cause stresses at certain points along the intersection weld to be many times the nominal stress acting in the members. The SCF is a multiplier applied to the nominal stress to reach the peak or maximum stress at the hot spot.

The hot spot is the location in the tubular joint where the maximum applied tensile stress occurs. To do a fatigue analysis of certain selected tubular joints in an offshore structure, the stress history of the hot spots in those joints must first be determined. Three basic types of stress contribute to the development of a hot spot. Primary stresses are caused by axial forces and moments resulting from the combined truss and frame action of the jacket. As shown for the in-plane tubular joint in Figure 3.38, hot spots in locations 1, 2, 3, 4 and 6 are most affected by axial forces and in-plane bending moments, while locations 2 and 5 are most affected by the axial forces and circumferential moments in braces.

Secondary stresses are caused by structural details of the connection, such as poor joint geometry, poor fit-up, local variation within the joint due to rigid reinforcement or restraint of the braces by circumferential weld, and these secondary stresses will amplify the primary stresses. Secondary stresses are also caused by metallurgical factors, such as faulty welding practice, insufficient weld penetration, heavy beading, weld porosity or varying cooling rates. These stresses mainly affect location 1, but their effect is also significant at locations 3, 4 and 6. Because metallurgical factors are essential in fatigue stresses on tubular joints, the quality control (QC) procedure for constructing this connection should be given more attention, as is described in Chapter 5.

The most fatigue-sensitive areas in offshore platforms are the welds at tubular joints, because of the high local stress concentrations. Fatigue lives at these locations should be estimated by evaluating the hot-spot stress range (HSSR) and using it as input into the appropriate S-N curve.

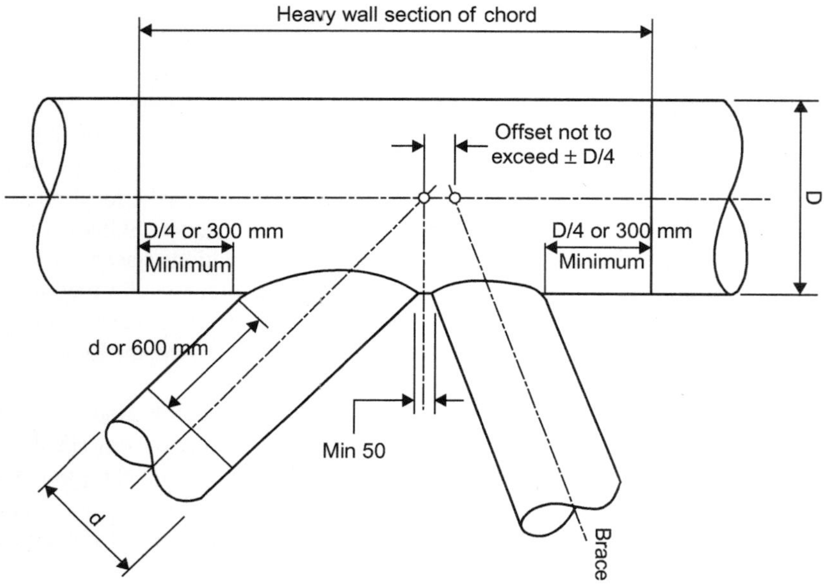

FIGURE 3.38 Hot spots at in-plane tubular joint.

SCFs may be derived from finite element (FE) analyses, model tests or empirical equations based on such methods. When deriving SCFs using FE analysis, it is recommended to use volume (brick and thick shell) elements to represent the weld region and adjoining shell, as opposed to thin shell elements. In such models, the SCFs may be derived by extrapolating stress components to the relevant weld toes and combining them to obtain the maximum principal stress and, hence, the SCF. The extrapolation direction should be normal to the weld toes.

According to Healy and Buitrago (1994) and Niemi et al. (1995), if thin shell elements are used, the results should be interpreted carefully, since no single method is guaranteed to provide consistently accurate stresses.

Extrapolation to the mid-surface intersection generally overpredicts SCFs, but not consistently, whereas truncation at the notional weld toes would generally underpredict SCFs. In place of extrapolation, it is possible to use directly the nodal average stresses at the mid-surface intersection. This will generally overpredict stresses, especially on the brace side.

This last method is expected to be more sensitive to the local mesh size than the extrapolation methods.

The general definition for stress concentration factor for any tubular joint configuration and each type of brace loading is presented by the following formula:

$$\text{SCF} = \text{HSSR/nominal brace stress range}$$

The nominal brace stress range should be based on the section properties of the brace-end under consideration, taking due account of the brace-stub, or a flared member end, if present. Similarly, the SCF evaluation should be based on the same section dimensions. HSSR is the hot spot stress range.

The nominal cyclic stress affecting the chord may also influence the HSSR and should be considered. The SCF should include all stress-raising effects associated with the joint geometry and type of loading, except the local microscopic weld notch effect, which is included in the S-N curve. SCFs may be derived from FE analyses, model tests or empirical equations based on such methods. In general, the SCFs depend on the type of brace cyclic loading if the axial load applied to the brace is in-plane bending or out-of plane bending, the joint type and details of the geometry. The SCF varies around the joint, even for a single type of brace loading.

When combining the contributions from the various loading modes, phase differences between them should be accounted for, with the design HSSR at each location being the range of hot-spot stress resulting from the point-in-time contribution of all loading components.

In general, for all welded tubular joints under all three types of loading, a minimum value of SCF equal to 1.5 should be used.

For unstiffened welded tubular joints, SCFs should be evaluated using the Efthymiou equations, as will be discussed later in thickness effect on the SCF.

The linearly extrapolated hot-spot stress from Efthymiou may be adjusted to account for the actual weld toe position, where this systematically differs from the assumed AWS basic profiles.

The SCF applies also to internally ring-stiffened joints, including the stresses in the stiffeners and the stiffener-to-chord weld. Noting that special consideration should be given to these locations, SCFs for internally ring-stiffened joints can be determined by applying the Lloyd's reduction factors based on the Lloyd's Report (1988) to the SCFs for the equivalent unstiffened joint. For ring-stiffened joints analyzed by such means, the minimum SCF for the brace side under axial or OPB loading should be taken as 2.0.

SCFs in Grouted Joints

Grouted joints are usually used in repairing or strengthening the platform. Grouting joints tends to reduce the SCF of the joint, since the grout reduces the chord deformations. In general, the larger the ungrouted SCF, the greater the reduction in SCF with grouting. Hence, the reductions are typically greater for X and T joints than for Y and K joints. More discussion of the effect of grouting on strengthening is presented in Chapter 7.

S-N Curves for All Members and Connections

For nontubular members and connections in deck structures, appurtenances and equipment and tubular members and attachments to them, including ring stiffeners, may be subject to variations of stress due to wave, wind and other environmental loads or operational loads. Operational loads include loads associated with machine vibration, crane usage and filling and emptying of tanks. Where variations of stress are applied to conventional weld details, identified in ANSI/AWS D1.1- 2002, the associated S-N curves provided in AWS Figure 2.11 should be used, dependent on degree of redundancy. Where such variations of stress are applied to tubular nominal stress situations identified in ANSI/AWS D1.1-2002, the associated S-N curves provided in AWS Figure 2.13 should be used. Stress categories DT, ET, FT, Kl and K2 refer to tubular connections where the SCF is not known.

The hot-spot stress concentration factor can be determined, taking into consideration the exposure of the joint to random variable loads, seawater corrosion, or submerged service with effective cathodic protection system.

The referenced S-N curves in ANSI/AWS D1.1.-2002 Figure 2.11 are class curves. For such curves, the nominal stress range in the vicinity of the detail should be used. Due to load attraction, shell bending, and other, not present in the class-type test specimens, the appropriate stress may be larger than the nominal stress in the gross member. Note that the geometrical stress concentration and notch effects associated with the detail itself are included in the curves.

ET is a simple T-, Y- and K-connection with groove welds or fillet welds as well as complex tubular connections in which the punching shear capacity of the main member cannot carry the entire load. In addition to that, load transfer is accomplished by overlap (negative eccentricity), gusset plates and ring stiffeners. The main stresses are due to tension and compression failure.

FT is a simple T-, Y- or K-connection loaded in tension or bending, having fillet or PJP groove welds and main stress due to shear in the weld.

DT is a connection designed as a simple T-, Y- or K-connection with CJP groove welds, including overlapping connections in which the main member at each intersection meets punching shear requirements. The main stresses are due to tension, compression and bending with reversal action.

K2 is a simple T-, Y- or K-connection in which the gamma ratio R/tc of the main member does not exceed 24 and the main stresses are due to punching.

Reference may alternatively be made to the S-N criteria similar to the other joint (OJ) curves contained within ISO DIS.

The ISO 19902:2004 code proposal uses a weld detail classification system whereby the OJ curves include an allowance for notch stress and a modest geometrical stress concentration.

S-N Curves for Tubular Connections

Design S-N curves are given below for welded tubular and cast joints. The basic design S-N curve is of the form:

$$\mathrm{Log}_{10}(N) = \mathrm{Log}_{10}(k_1) - m\mathrm{Log}_{10}(S) \quad (3.109)$$

where N is the predicted number of cycles to failure under stress range S, k_1 is a constant and m is the inverse slope of the S-N curve.

Table 3.10 presents the basic welded joint (WJ) and cast joint (CJ) curves. These S-N curves are based on steels with yield strength less than 500 MPa (72 ksi).

For welded tubular joints exposed to random variations of stress due to environmental or operational loads, the WJ curve should be used. The brace-to-chord tubular intersection for ring-stiffened joints should be designed using the WJ curve. For cast joints, the CJ curve should be used.

The basic allowable cyclic stress should be corrected empirically for seawater effects, the apparent thickness effect, with the exponent depending on the profile, and the weld improvement factor on S. (An example of S-N curve construction is given in Figure 3.39, below.)

The basic design S-N curves given in Table 3.8 are applicable for joints in air and submerged coated joints. For welded joints in seawater with adequate cathodic protection, if m is equal to the 3 branch of the S-N curve, it should

TABLE 3.10 Basic Design S-N Curves

Curve	$Log_{10}(k_1)$ with S in ksi	$Log_{10}(k_1)$ with S in MPa	m
Welded Joints (WJ)	9.95	12.48	3 for $N < 10^7$
	11.92	16.13	5 for $N > 10^7$
Cast Joints (CJ)	11.80	15.17	4 for $N < 10^7$
	13.00	17.21	5 for $N > 10^7$

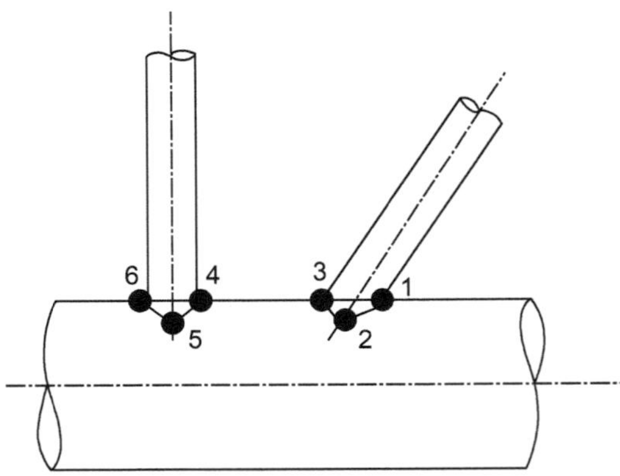

FIGURE 3.39 Allowable fatigue S-N ranges for stress categories for tubular structures in atmospheric service (adopted from AWSD1.1).

be reduced by a factor of 2.0, and if m is equal to the 5 branch, it remains unchanged and the position of the slope change is adjusted accordingly.

Fabrication of welded joints should be in accordance with the quality-control procedure. The curve for cast joints is applicable only to castings having an adequate fabrication inspection plan.

Thickness Effect

The WJ curve is based on a 5/8-inch (16 mm) reference thickness. The following equation will be applied for material thickness above the reference thickness:

$$S = S_o \left(t_{ref}/t\right)^{0.25} \tag{3.110}$$

where t_{ref} = the reference thickness, 5/8-inch (16 mm); S = allowable stress range; S_o = the allowable stress range from the S-N curve; and t = member thickness for which the fatigue life is predicted.

If the weld has profile control, the exponent in the above equation may be taken as 0.20 instead of 0.25. On the other hand, the exponent can be taken as 0.15 if the weld toe has been ground or peened.

The material thickness effect for castings is given by:

$$S = S_o \left(t_{ref}/t\right)^{0.15} \tag{3.111}$$

where the reference thickness t_{ref} is 1.5 inches (38 mm).

No effect should be applied to material thickness less than the reference thickness.

For any type of connection analyzed on a chord hot-spot basis, the thickness of the chord side of the tubular joint should be used in the foregoing equations. For the brace-side hot spot, the brace thickness may be used. The allowable fatigue S-N range for a tubular structure is as in Figure 3.39.

Use of the Efthymiou SCF equations is recommended because this set of equations offers the best option for all joint types and load types and is the only widely vetted set that covers overlapped K and KT joints.

"Mix-and-match" between different sets of equations is not recommended. The Efthymiou equations are also recommended in the "Proposed Revisions for Fatigue Design of Welded Connections" for adoption by the IIW (International Institute of Welding), Eurocode 3 and ISO DIS 14347.

The validity ranges for the Efthymiou parametric SCF equations are: β from 0.2 to 1.0, τ from 0.2 to 1.0, γ from 8 to 32, α (length) from 4 to 40, θ from 20 to 90 degrees and ζ (gap) from $-0.6\beta/\sin\theta$ to 1.0.

For cases where one or more parameters fall outside this range, the following procedure may be adopted:

1. Evaluate SCFs using the actual values of geometric parameters.
2. Evaluate SCFs using the limit values of geometric parameters.
3. Use the maximum of 1 or 2 above in the fatigue analysis.

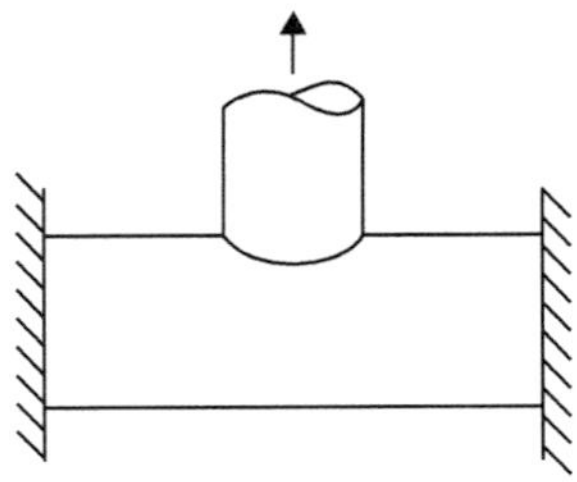

Axial Load—Chord Ends Fixed

Chord saddle

$$SCF = \gamma\tau^{1.1}[1.11 - 3(\beta - 0.52)^2] \sin^{1.6} \tag{3.112}$$

For short chord there is a correction factor F1.
Short chord correction factors (α <12):

$$F1 = 1 - (0.83\beta - 0.56\beta^2 - 0.02)\gamma^{0.23}\exp[-0.21\gamma^{-1.16}\alpha^{2.5}] \tag{3.113}$$

where $\exp(x) = e^x$.
Chord crown

$$SCF = \gamma^{0.2}\tau[2.65 + 5(\beta - 0.65)^2] + \tau\beta(0.25\alpha - 3) \sin\theta \tag{3.114}$$

Brace saddle

$$SCF = 1.3 + \gamma\tau^{0.52}\alpha 0.1[0.187 - 1.25\beta 1.1(\beta - 0.96)] \sin(2.7 - 0.01\alpha)\theta \tag{3.115}$$

For short chord there is a correction factor F1.
Brace crown

$$SCF = 3 + \gamma^{1.2}[0.12\exp(-4\beta) + 0.011\beta^2 - 0.045] + \beta\tau(0.1\alpha - 1.2) \tag{3.116}$$

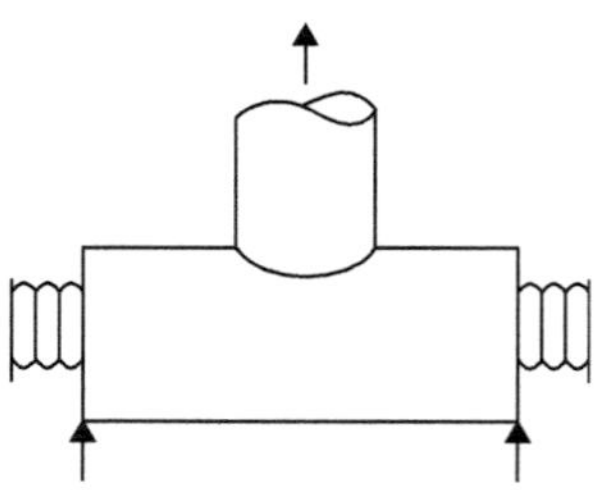

Axial Load—General Fixity Conditions

Chord saddle

$$SCF = [\text{Equation } (3.112)]\, C_1(0.8\alpha - 6)\tau\beta^2(1-\beta^2)^{0.5} + \sin^2 2\theta \quad (3.117)$$

For short chord there is a correction factor F2:

$$F2 = 1 - (1.43\beta - 0.97\beta^2 - 0.03)\gamma^{0.04}\exp[-0.71\gamma^{-1.38}\alpha_{2.5}] \quad (3.118)$$

Chord crown

$$SCF = \gamma^{0.2}\tau[2.65 + 5(\beta - 0.65)^2] + \tau\beta(C_2\alpha - 3)\sin\theta \quad (3.119)$$

Brace saddle: Equation (3.115)
Brace crown

$$SCF = 3 + \gamma^{1.2}[0.12\exp(-4\beta) + 0.011\beta^2 - 0.045] + \beta\tau(C_3\alpha - 1.2) \quad (3.120)$$

Note that, for chord-end fixity parameter C, $0.5 \le C \le 1.0$, and typically $C = 0.7$, $C_1 = 2(C - 0.5)$, $C_2 = C/2$ and $C_3 = C/5$.

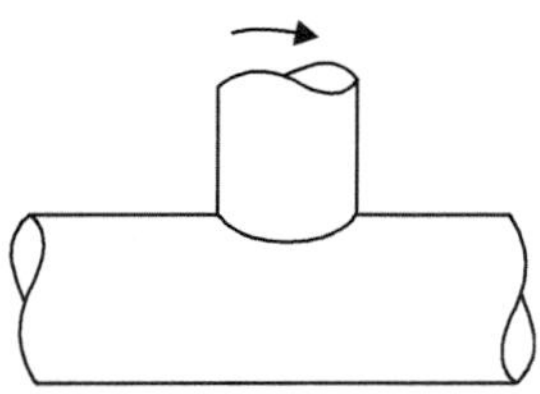

In-Plane Bending

Chord crown

$$SCF = 1.45\beta\tau^{0.85}\gamma(1 - 0.68\beta)\sin^{0.7}\theta \quad (3.121)$$

For short chord there is a correction factor F3, where

$$F3 = 1 - 0.55\beta^{1.8}\gamma 0.16\exp[-0.49\gamma^{-0.89}\alpha^{1.8}] \quad (3.122)$$

Brace crown

$$SCF = 1 + 0.65\beta\tau^{0.4}\gamma(1.09 - 0.77\beta) + \sin^{(0.06\gamma - 1.16)}\theta \quad (3.123)$$

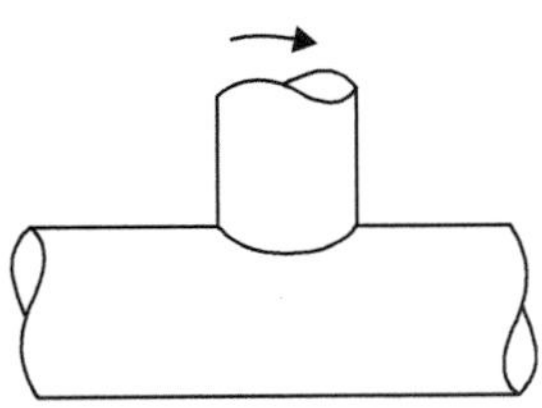

Out-of-Plane Bending

Chord saddle

$$SCF = \gamma\tau\beta(1.7 - 1.05\beta^3)\sin^{1.6}\theta \tag{3.124}$$

Brace saddle

$$SCF = \tau^{-0.54}\gamma^{-0.05}(0.99 - 0.47\beta + 0.08\beta^4) \times [\text{Equation (3.124)}] \tag{3.125}$$

Chord-End Fixity Parameter C

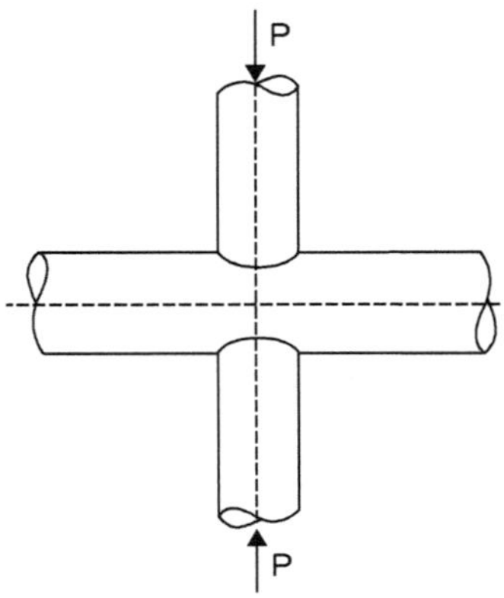

Axial Load (Balanced)

Chord saddle

$$SCF = 3.87\gamma\tau\beta(1.10 - \beta^{1.8})\sin^{1.7}\theta \tag{3.126}$$

Chord crown

$$SCF = \gamma^{0.2}\tau[2.65 + 5(\beta - 0.65)^2] + 3\tau\beta\sin\theta \tag{3.127}$$

Brace saddle

$$SCF = 1 + 1.9\gamma\tau^{0.5}\beta^{0.9}(1.09 - \beta^{1.7})\sin^{2.5}\theta \tag{3.128}$$

Brace crown

$$SCF = 3 + \gamma^{1.2}[0.12\exp(-4\beta) + 0.011\beta^2 - 0.045] \tag{3.129}$$

In joints with short chords (α <12) and closed ends, the saddle SCFs can be reduced by the short chord factors F1 or F2, where:

$$F1 = 1 - (0.83\beta - 0.56\beta^2 - 0.02)\gamma^{0.23}\exp[-0.21\gamma^{-1.16}\alpha^{2.5}] \tag{3.130}$$

$$F2 = 1 - (1.43\beta - 0.97\beta^2 - 0.03)\gamma^{0.04}\exp[-0.71\gamma^{-1.38}\alpha^{2.5}] \tag{3.131}$$

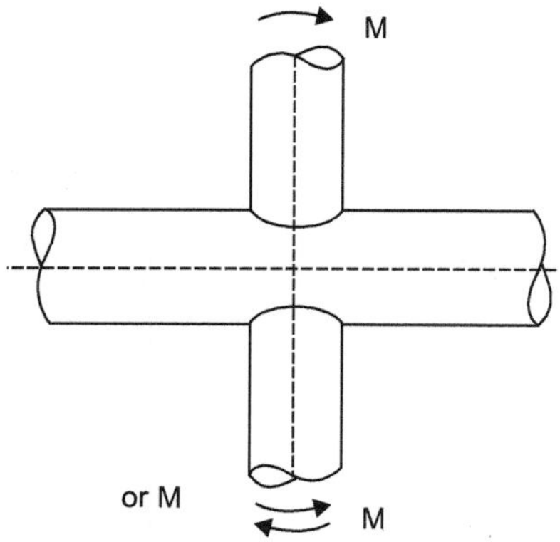

In-Plane Bending

Chord crown: Equation (3.121)
Brace crown: Equation (3.123)

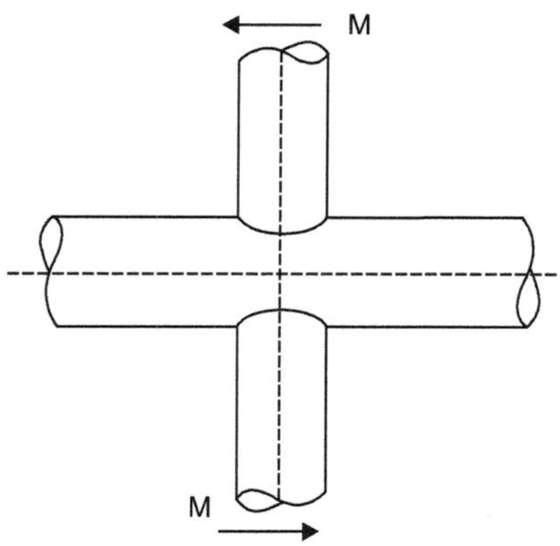

Out-of-Plane Bending (Balanced)

Chord saddle

$$SCF = \gamma\tau\beta(1.56 - 1.34\beta^4)\sin^{1.6}\theta \tag{3.132}$$

Brace saddle

$$SCF = \tau^{-0.54}\gamma^{-0.05}(0.99 - 0.47\beta + 0.08\beta^4) \times [\text{Equation (3.132)}] \tag{3.133}$$

In joints with short chords (α <12) and closed ends, Equations (3.132) and (3.133) can be reduced by the short chord factor F3, where:

$$F3 = 1 - 0.55\beta^{1.8}\gamma^{0.16}\exp[-0.49\gamma^{-0.89}\alpha^{1.8}] \tag{3.134}$$

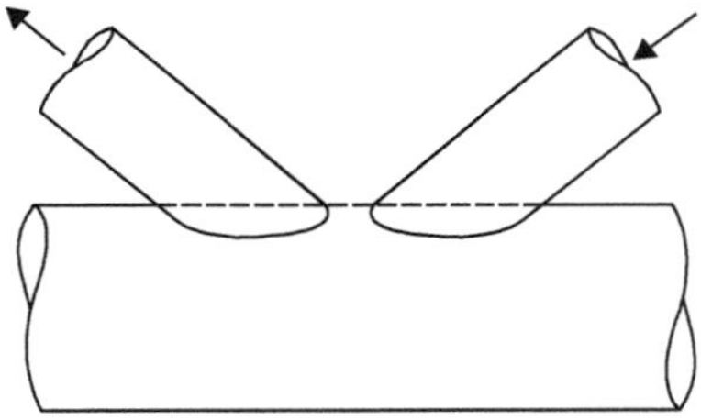

Balanced Axial Load

Chord SCF

$$SCF = \tau^{0.9}\gamma^{0.5}(0.67 - \beta^2 + 1.16\beta)\sin\theta\left[\frac{\sin\theta_{max}}{\sin\theta_{min}}\right]^{0.30}\left[\frac{\beta_{max}}{\beta_{min}}\right]^{0.30} \quad (3.135)$$
$$[1.64 + 0.29\beta^{-0.38}\text{ATAN}(8\zeta)]$$

Brace SCF

$$SCF = 1 + [\text{Equation (3.135)}](1.97 - 1.57\beta^{0.25})\tau^{-0.14}\sin^{0.7}\theta +$$
$$SCF = C\beta^{1.5}\gamma^{0.5}\tau^{-1.22}\sin^{1.8}(\theta_{max} + \theta_{min}) \times [0.131 - 0.084\,\text{ATAN}(14\zeta + 4.2\beta)] \quad (3.136)$$

where $C = 0$ for gap joints, $C = 1$ for the through brace and $C = 0.5$ for the overlapping brace. Note that τ, β, θ and the nominal stress relate to the brace under consideration and ATAN is arctangent evaluated in radians.

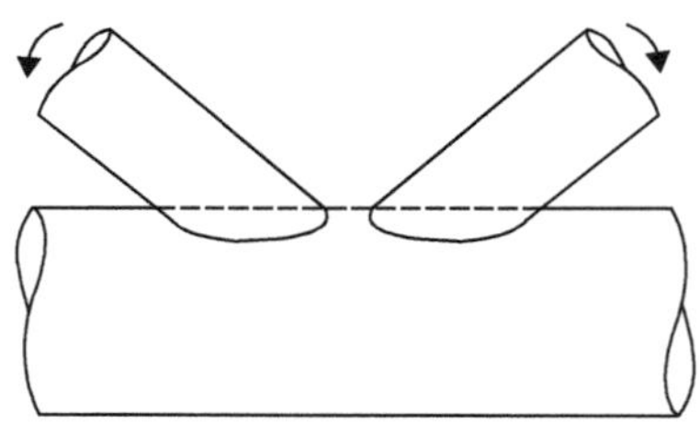

Unbalanced IPB

Chord crown SCF: Equation (3.114). (For overlaps exceeding 30% of contact length, use 1.2 × [Equation (3.121)].)

Gap joint–brace crown SCF: Equation (3.123)

Overlap joint–brace crown SCF: [Equation (3.123)] × (0.9 + 0.4)

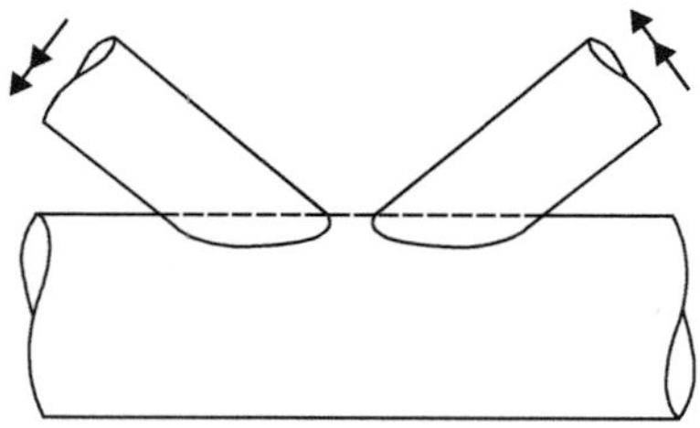

Unbalanced OPB

Chord saddle SCF adjacent to brace A:

$$SCF = [\text{Equation (3.124)}]_A [1 - 0.08(b_{B\gamma})^{0.5} \exp(-0.8x)] + [\text{Equation (3.124)}]_B [1 - 0.08(b_{A\gamma})^{0.5} \exp(-0.8x)][2.05\,(b_{max})^{0.5} \exp(-1.3x)] \quad (3.137)$$

where

$$x = 1 + \frac{\zeta \sin \theta_A}{\beta_A}$$

Brace A saddle SCF:

$$SCF = \tau^{-0.54} \gamma^{-0.05} (0.99 - 0.47\beta + 0.08\beta^4) \times [\text{Equation (3.137)}] \quad (3.138)$$

$$F4 = 1 - 1.07\beta^{1.88} \exp[-0.16\gamma^{-1.06} \alpha^{2.4}] \quad (3.139)$$

Note that $[\text{Equation (3.124)}]_A$ is the chord SCF adjacent to brace A as estimated from Equation (3.124).

The designation of braces A and B is not geometry dependent. It is nominated by the user.

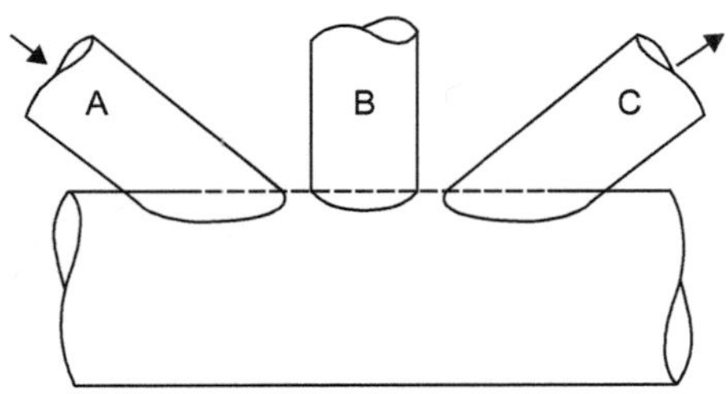

Balanced Axial Load for 3 Bracing

Chord SCF: Equation (3.135)
Brace SCF: Equation (3.136)
For the diagonal braces A and C, use $\zeta = \zeta_{AB} + \zeta_{BC} + \beta_B$
For the central brace B, use $\zeta =$ maximum of ζ_{AB}, ζ_{BC}

In-Plane Bending for 3 Bracing
Chord crown SCF: Equation (3.121)
Brace crown SCF: Equation (3.123)

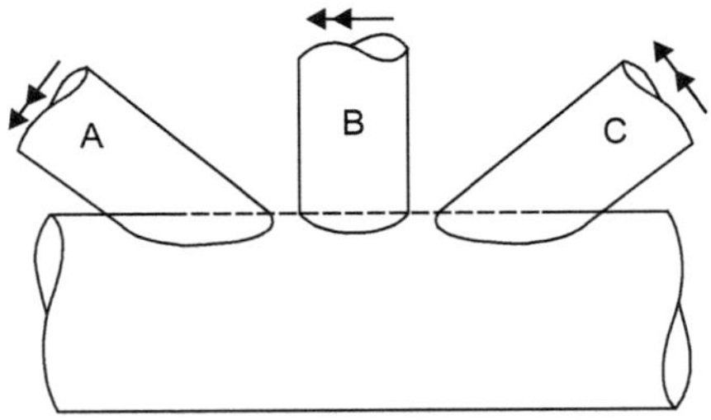

Unbalanced Out-of-Plane Bending for 3 Bracing

Chord saddle SCF adjacent to diagonal brace A:

$$\begin{aligned} SCF = &[\text{Equation (3.124)}]_A[1-0.08(\beta_{B\gamma})^{0.5}\exp(-0.8x_{AB})]\bullet \\ &[1-0.08(\beta_{C\gamma})^{0.5}\exp(-0.8x_{AC})]+[\text{Equation (3.124)}]_B \\ &[1-0.08(\beta_{A\gamma})^{0.5}\exp(-0.8x_{AB})][2.05(\beta_{max})^{0.5}\exp(-1.3x_{AB})] \\ &+[\text{Equation(3.124)}]_C[1-0.08(\beta_{A\gamma})^{0.5}\exp(-0.8x_{AC})] \\ &[2.05(\beta_{max})^{0.5}\exp(-1.3x_{AC})] \end{aligned} \quad (3.140)$$

where

$$x_{AB} = 1 + \frac{\zeta_{AB}\sin\theta_A}{\beta_A} \quad (3.141)$$

$$x_{AC} = 1 + \frac{(\zeta_{AB}+\zeta_{BC}+\beta_B)\sin\theta_A}{\beta_A} \quad (3.142)$$

Chord saddle SCF adjacent to central brace A:

$$\begin{aligned} SCF = &[\text{Equation (3.124)}]_B[1-0.08(\beta_{A\gamma})^{0.5}\exp(-0.8x_{AB})]m\bullet \\ &[1-0.08(\beta C_\gamma)^{0.5}\exp(-0.8x_{BC})]\text{m}+[\text{Equation (3.124)}]_A\bullet \\ &[1-0.08(\beta_{B\gamma})^{0.5}\exp(-0.8x_{AB})]\,[2.05(\beta_{max})^{0.5}\exp(-1.3x_{AB})] \\ &+[\text{Equation (3.124)}]_C\bullet[1-0.08(\beta_{B\gamma})^{0.5}\exp(-0.8x_{BC})] \\ &[2.05(\beta_{max})^{0.5}\exp(-1.3x_{BC})] \end{aligned} \quad (3.143)$$

where

$$m = (\beta_A/\beta_B)^2$$

$$x_{AB} = 1 + \frac{\zeta_{AB}\sin\theta_A}{\beta_B} \text{ and } x_{BC} = 1 + \frac{\zeta_{BC}\sin\theta_B}{\beta_B}$$

Brace saddle SCFs under OPB: obtained from the adjacent chord SCFs using

$$\mathrm{SCF} = \tau^{-0.54}\gamma^{-0.05}(0.99 - 0.47\beta + 0.08\beta^4) \times SCFchord \tag{3.144}$$

where $SCFchord$ = Equation (3.140)KT1 or Equation (3.143).

In joints with short chords ($\alpha < 12$), Equations (3.140), (3.143) and (3.144) can be reduced by the short chord factor F4, where

$F4 = 1 - 1.07\beta^{1.88}\ \exp[-0.16\gamma^{-1.06}\alpha^{2.4}]$.

Effect of weld toe position. Ideally, the SCF should be invariant, given the tubular connection's geometry (γ, τ, β, θ, and ζ). This is how the Efthymiou and all the other SCF equations are formulated. Hot-spot stress is calculated from the linear trend of notch-free stress extrapolated to the toe of the basic standard weld profile, with nominal weld toe position as defined in AWS D1.1 Figure 3.8. When this is done, size and profile effects must be accounted for in the S-N curve, regardless of the underlying cause. This is how the previous API rules were set up.

International thinking tends to suggest that weld profile effects, mainly the variable position of the actual weld toe, should be reflected in the SCF, rather than in the S-N curve. This is consistent with how experimental hot-spot stresses were measured to define the basic international S-N curve for hot-spot fatigue in 16 mm thick tubular joints. One tentative method for correcting analytical SCF for weld toe position was presented in Marshall (1989). Based on Marshall et al. (2005), a more robust formulation is now proposed:

$$SCF_{corr} = 1 - (L_a - L)/L_{mp} \tag{3.145}$$

where SCF_{corr} = the correction factor applied to Efthymiou SCF; L_a = the actual weld toe position for typical yard practice; L = the nominal weld toe position; and L_{mp} = the moment persistence length (distance from nominal toe to reversal of shell bending stress).

Various expressions for L_{mp} are shown in Table 3.11 as a function of joint type, load type and hot-spot orientation.

TABLE 3.11 Expressions for L_{mp}

Circumferential stress at saddle	
All loading modes	$L_{mp} = (0.42 - 0.28\beta)R$
	Angle = $(24 - 16\beta)$ degrees
Longitudinal stress at crown	
Axis symmetric	$L_{mp} = 0.6(RT)^{0.5}$
Gap (g) of K joint	L_{mp} = lesser of $0.6\ (RT)^{0.5}$ or g/2
Outer heel/toe, axial	$L_{mp} = 1.5(RT)^{0.5}$
In-plane bending	$L_{mp} = 0.9(RT)^{0.5}$

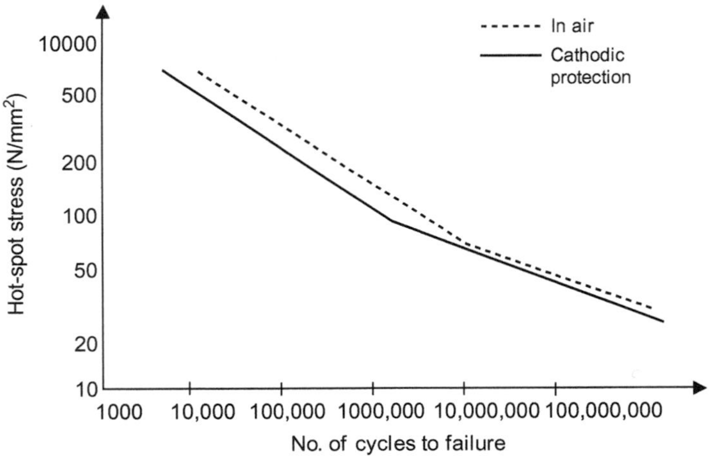

FIGURE 3.40 Tubular joint S-N curve for T = 16 mm (from API RP2A).

R and *T* are radius and thickness, respectively, of the joint-can. Consistency in format with the rules for strain-gauge placement at crown and saddle position may be noted.

Attempts to produce an improved as-welded profile often result in over-welding. As such, high estimates of L_{mp}, which are the low estimates of local stress gradient, will produce conservative corrections. This approach assumes that the weld is not so massive as to change the overall load distribution in the joint-can, and that local hot-spot stresses are dominated by shell bending stress. The relation between the hot-spot stress and the cycles of load until failure is presented in Figure 3.40, where the thickness of the chord is equal to 16 mm.

Failure is expressed as damage or fatigue life damage, so the fatigue life damage is the number of cycles of a particular stress range divided by the allowable number of cycles for that range from the S-N curve.

Table 3.12 is an example of a fatigue analysis, with the stress range and the corresponding number of cycles of stress occurrence and the allowable number of cycles based on the S-N curve. Assuming a point is subject to 5 cyclic stress ranges (due to wave), $D_5 = 10 \times 10^6/50 \times 10^6 = 0.2$, $D_{10} = 0.1$, $D_{20} = 0.2$, $D_{50} = 0.05$ and $D_{90} = 0.025$. Total damage = 0.575.

So, if these waves occurred over 10 years, then the fatigue life = 10/0.575 = 17.4 years.

Jacket Fatigue Design

Dynamic analysis should be carried out to predict the fundamental periods of the platforms in order to confirm the sensitivity of the structure to wave-induced excitation. The fundamental sway periods should be used to derive the dynamic amplification for the in-place analysis loading conditions.

TABLE 3.12 Fatigue Analysis

Stress Range	Number of Cycles of Stress Occurrence	Number of Cycles Allowable
5	10×10^6	50×10^6
10	4×10^5	4×10^6
20	6×10^4	3×10^5
50	5×10^2	1×10^4
90	25	1×10^3

Fatigue analysis should be performed for the jacket structures using methods appropriate to the sensitivity to dynamic loading. A spectral approach will be deemed adequate for platforms with fundamental period less than 3 seconds.

From a practical point of view, the in-service fatigue design life of the joints should be at least two times the service life of the platform (i.e., 50 years).

Fatigue analysis using software will include:

- The environmental loads parameters.
- A sufficient number of wave directions. For each direction, a minimum of four wave heights should be used to compute stress range versus wave height relationship. The directions, wave heights and exceedances selected should be those closest to the directions indicated in the metocean data. A sufficient number of wave crest positions should also be considered.
- If the natural period of the structure exceeds 3 seconds, dynamic amplification effects should be taken into account in calculation of the cycle loading.
- Where significant cyclic stresses may be induced by the action of wind, wave slamming, changes in member buoyancy, etc., such stresses should be combined with those due to wave action to obtain the total effective stress spectrum for a particular member or joint. Fatigue damage accumulated during fabrication and transportation should also be considered.

After that, the analysis procedure takes the following steps.

- For each joint and type of failure under consideration, the stress range spectra should be computed at a minimum of eight positions around the joint periphery.
- Two types of failure should be considered, using the appropriate stress concentration factors (e.g., brace-to-weld failure and chord-to-weld failure).
- For joints other than those between tubular members, individual detailed consideration should be given, with due regard paid to published, reliable experimental data.

SCFs should be determined for tubular to tubular joints, ungrouted and unstiffened. SCF should be calculated using Efthymiou equations as discussed before.

In lieu of a more accurate procedure for analysis, ring-stiffened joints may be checked using the same procedure as for simple joints, but using modified chord thickness.

3.8 TOPSIDE DESIGN

In general, major rolled shapes for offshore structure design should be compact sections, as defined by AISC. The minimum thickness of a structural plate or section should be 6 mm. The minimum diameter to thickness ratio of tubular members should not be less than 20, where the diameter is based on the average of the tubular outside and inside diameters.

Connections design to comply with the codes must fulfill the following minimum requirements.

- The minimum fillet weld should be 6 mm.
- Wherever possible, joints should be designed as simple joints with no overlap.
- Tubular joints should be designed in accordance with API RP2A.

The deflection (discussed in Chapter 2) should be matched with the codes and defined in the owner's specification.

- Deflections should be checked for the actual equipment live loads and casual area live loads. Pattern loading should be considered.
- Deflection of members supporting deflection-sensitive equipment should be not greater than L/500 for beams and L/250 for cantilevers under live loads.
- Deflection of beams in the workrooms and living quarters should be not greater than L/360 for beams and L/180 for cantilevers under live loads.
- Deflection limits for other structures should be L/250 for beams and L/125 for cantilevers under live loads.

In performing structural analysis using software (such as SACS, for example), it is better to define the direction of the model. In most cases, the directions can be:

- $+x$ axis aligned with platform east
- $+y$ axis aligned with platform north
- $+z$ axis aligned vertically upward
- The datum for the axes should be at chart datum.

In general, only the primary structural steel should be modeled. However, secondary members should be modeled where they are necessary for the structural integrity or to facilitate load input. Deck plates should be modeled as shear panel elements. Joint eccentricities should be modeled using discrete elements rather than using the "offset" facility of SACS. When individual elements are used, joint forces can be more easily extracted from the output.

All the differing analyses (in-place, lift, loadout, etc.) should utilize the same base model. That is, the in-place model should form the basis for all the other analyses to be performed.

The in-service analyses should include a basic model of the jacket structure to ensure the correct stiffness interaction between the jacket and the topside structures. Pile foundations to the jackets need not be considered for topside analysis and design—simple supports are sufficient.

3.8.1 Grating Design

Grating is traditionally member used to cover the platform floor (as in onshore facilities). The check list in Table 3.13 should be filled in, to

TABLE 3.13 Grating Design Check List

Item to Be Checked	Yes/No
Grating to be removable	
Vibration performance required	
Corrosive environment	
Grating over stainless steel grating or piping	
Panels to be sized to facilitate fabrication	
Weight and size limitation was defined	
Impact or high local load	
Loads meet the design requirement	
Special operating loads to be considered	
Grating matches the required strength	
Grating within the allowable deflection limit	
Suitable corrosion protection specified	
Anti-slip surface specified	
For fiber-reinforced polymer (FRP) the fire performance requirements were specified	
For FRP the smoke and toxicity requirements were defined	
Grating slope is acceptable	
Adequate lateral restraint was provided	
Tolerance was checked	
Support arrangements provide adequate support at penetrations and cut-outs	
There is an isolation between different materials	

make certain that the design of the grating matches the client's requirements for its function.

Grating should have a 1-inch (25-mm) minimum bearing on supporting steel. Where grating areas are shown as removable on the drawings, the weight of fabricated grating sections for such areas should not exceed 160 kg (350 lb).

Most grating and expanded metal needs to be supported in a specific way. The direction in which the load bars run is the important direction for grating and is usually referred to as the span. For expanded metal, the span is in the direction of the strands. Span is always the least dimension given when referencing a panel size. In most cases, the grating has to be supported in the span direction only and it does not require support on all four sides, unlike the floor plate.

The different grating dimensions are shown in Figure 3.41 and the different types of grating are shown in Figure 3.42.

There is a relation between the span and the deflection. When you select the grating from the manufacturer, you should refer to his type and calculation. To illustrate a grating sheet based on its type and loads, Table 3.14 has been arranged in increasing strength order. The load, L, is a safe superimposed uniformly distributed load in KPa (100 kg/m^2 = 0.98 KPa), where D is the deflection (in mm) for L. Loads are calculated in accordance with an allowable bending stress of 171.6 MPa. Note that the load bars are assumed to be simply supported.

Sometimes pipe support will form a concentrated load on the grating, as shown in Figure 3.43; in this situation, the pipe support is usually a base plate resting directly on the grating, and there is no steel under the support.

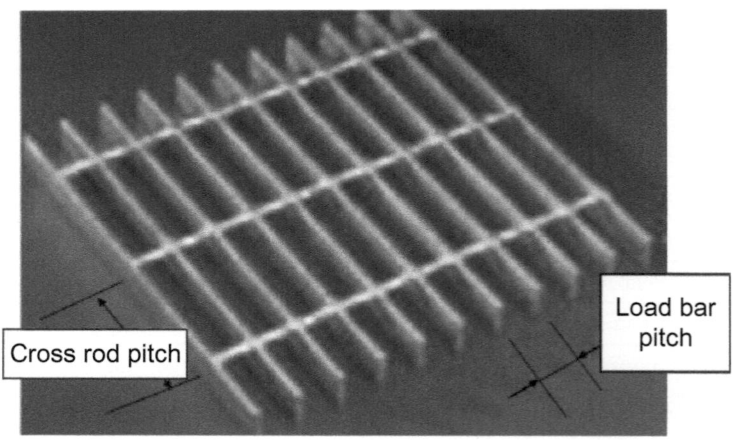

FIGURE 3.41 Grating dimensions.

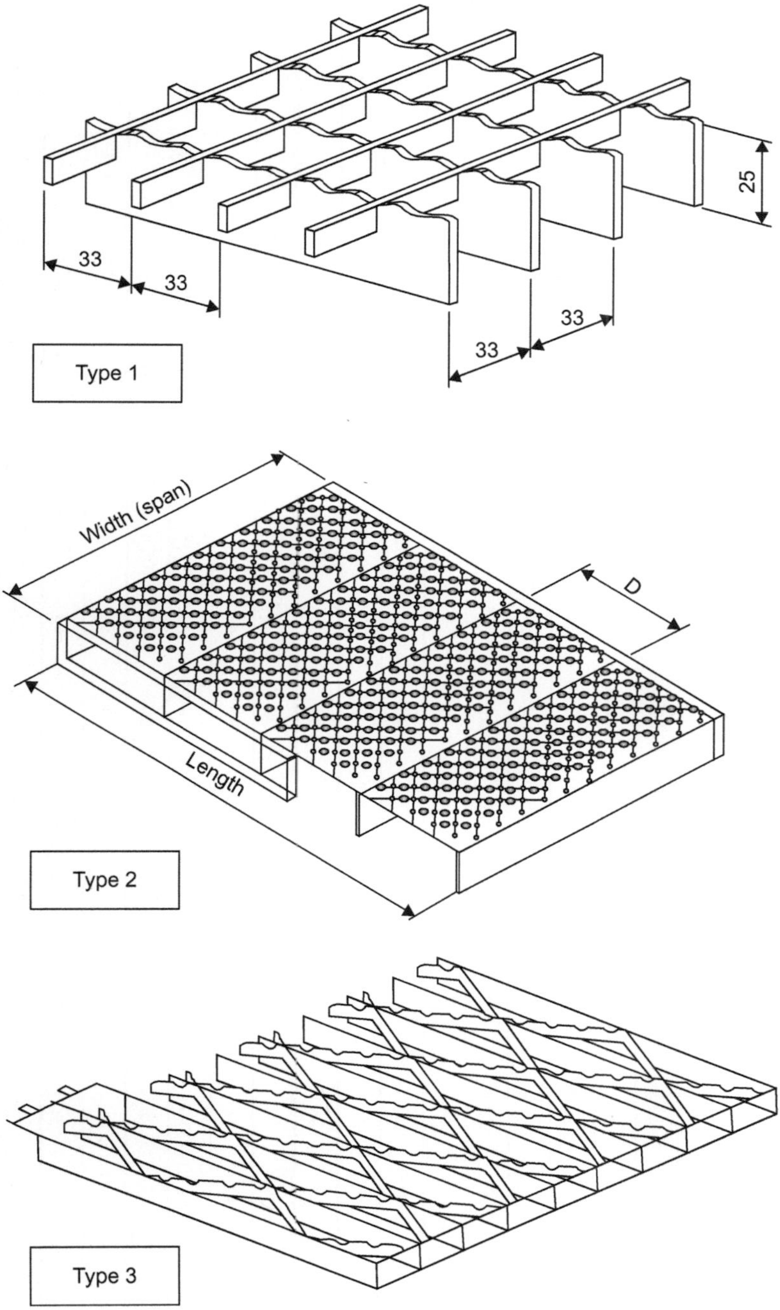

FIGURE 3.42 Types of grating.

TABLE 3.14 Relation between Grating Dimensions, Maximum Span and Maximum Load

Load bar spacing (mm)	Cross bar pitch	Weight (kg/m^2)	Bar size	Load (Kpa)	Spacing between Supports (mm)								
					450	600	750	900	1050	1200	1500	1800	2100
40	100	17.5	25 × 3	L	53	30	19	13	10	7	5	3	2
				D	1.4	2.6	4	5.8	7.8	10.2	16	23.1	31.5
60	50	22.3	25 × 5	L	56	32	20	14	10	8	5	3	2
				D	1.4	2.6	4	5.8	7.8	10.2	16	23.1	31
40	100	26.9	25 × 5	L	70	39	24	17	13	9	6	4	3
				D	1.6	2.9	4.5	6.5	8.8	11.5	18.0	25.9	35.3
30	100	22.8	25 × 3	L	70	39	25	17	13	10	6	4	3
				D	1.4	2.6	4	5.8	7.8	10.3	16.0	23.1	31.5
60	50	26.4	32 × 5	L	76	43	27	19	14	11	7	5	3
				D	1.2	2.2	3.4	4.9	6.7	8.7	13.6	19.7	26.8
30	100	34.7	25 × 5	L	91	51	33	23	17	13	8	6	4
				D	1.6	2.9	4.5	6.5	8.8	11.5	18.0	25.9	35.3
40	100	34	32 × 5	L	120	67	43	30	22	17	11	7	5
				D	1.2	2.2	3.4	4.9	6.7	8.7	13.6	19.7	26.8
30	100	28.4	32 × 3	L	114	64	41	28	21	16	10	7	5
				D	1.1	2	3.1	4.5	6.1	8.0	12.5	18.1	24.6
40	100	42.1	40 × 5	L	226	127	81	56	41	31	20	14	10
				D	0.9	1.6	2.5	3.6	4.9	6.4	10.0	14.4	19.7
30	100	62.9	45 × 5	L	377	212	135	94	67	52	33	23	17
				D	0.8	1.4	2.2	3.2	4.3	5.7	8.9	12.8	17.5

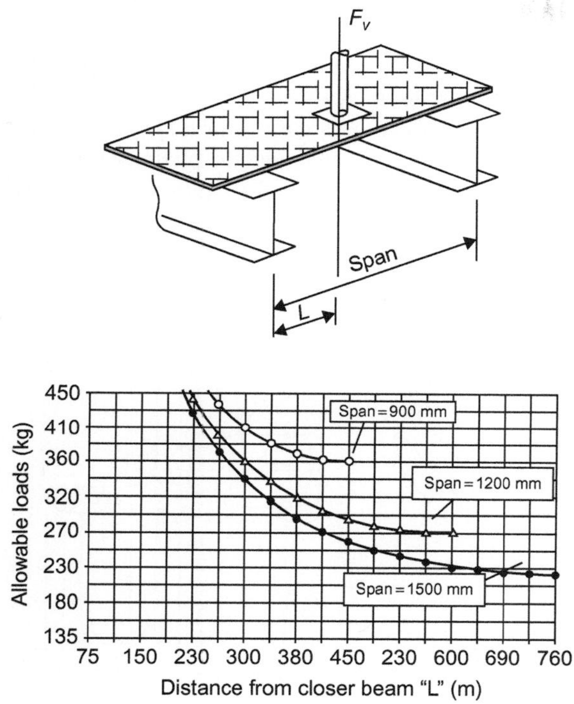

FIGURE 3.43 Relation between allowable loads and location of pipe support for different spans.

3.8.2 Handrails, Walkways, Stairways and Ladders

Handrails, walkways, stairways and ladders should be designed in accordance with OSHA 3124.

Handrails should be provided around the perimeter of all open decks and on both sides of stairways. All handrails should be 1.10 m high and made removable in panels no more than 4.5 m long. Handrail posts should be spaced 1.5 m apart. The gap between panels should not exceed 51 mm. Handrails in the wave zone should also be designed to withstand extreme storm maximal wave loadings.

Walkways, stairways and landings should be designed for load combinations:

1. Dead load plus live loads.
2. Dead loads plus extreme storm 3-second wind gusts and/or extreme storm maximum waves, whichever is applicable.

Stairways should be of structural steel, using a double runner with serrated bar grating treads and handrails.

3.9 BOAT LANDING DESIGN

The boat landing is designed for an impact load from a vessel or boat to the offshore structure. To absorb the impact load, there is usually a fender attached to the boat landing; the fender can be a car tire or a special tire. The connection between the boat landing frame structure and the platform jacket will be through a shock absorber, such as a piston. Figure 3.44 shows the general configuration for the boat landing and its main components and Figure 3.45 shows the boat

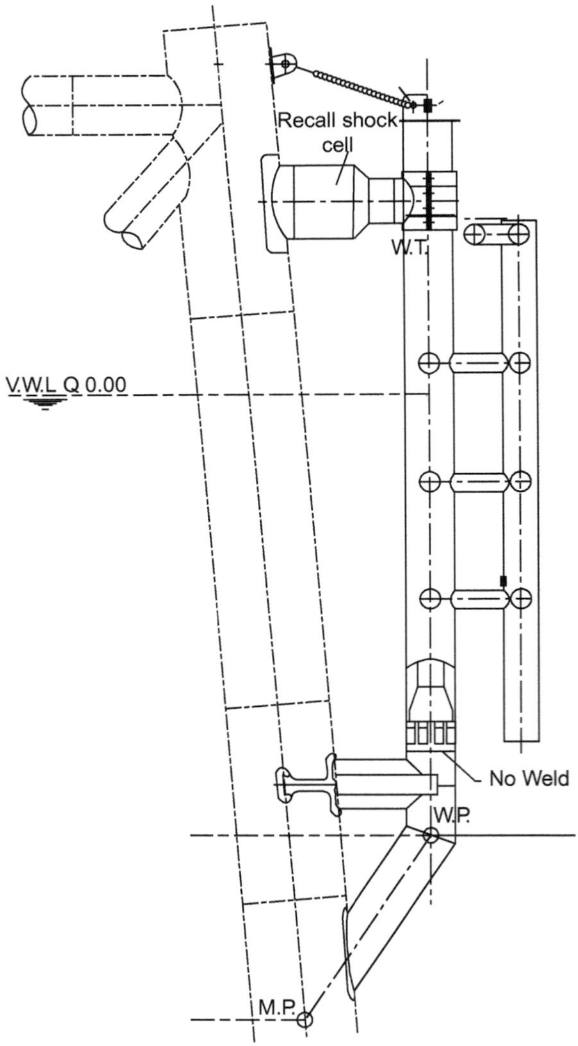

FIGURE 3.44 Boat landing support.

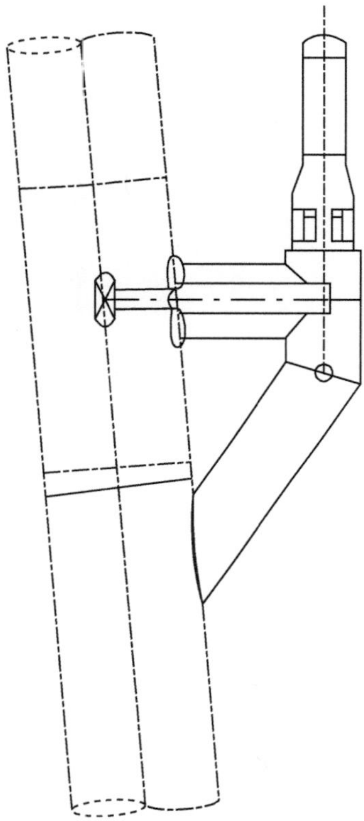

FIGURE 3.45 Connection between the leg and the boat landing.

landing connection to the offshore structure leg, as this mechanisim is traditionally used to absorb the impact load.

Figure 3.46 shows the shock cell.

The barge bumpers and their associated connections to the jacket should be designed for the loading incurred by vessel impact directly in the middle third of the height of the post. Energy to be absorbed in the system should be 560 kJ.

Barge bumpers should be designed to be easily replaced in case of damage. During the detailed design phase, the details considered will include the jacket elevation tolerances. In lieu of specific data, a minimum installation tolerance on jacket elevation of ± 0.3 m will be adopted.

For stab-in guides and bumpers, the following should be considered:

- The aids should be designed so that they fail prior to permanent deformation of any part of the permanent structure. The permanent structural members should be designed so that they can withstand significantly more load than the aids.
- Any deflections must be within the elastic limit of the material.
- A 33% overstress increase in allowable member stresses is permitted.

FIGURE 3.46 Photograph of a shock cell.

3.9.1 Boat Landing Calculation

The calculation of collision force is based on a 1000-ton boat impacting at a velocity of 0.5 m/s using Regal shock cell model SC1830.

Note that the force applied to the boat landing frame has to be defined and the stresses on the frame calculated by the software before beginning the design.

$$m = (1000 \times 9810) \times 2200/2000(9.81 \times 1000) = 1100\,\text{kN}\cdot\text{s}^2/\text{m}$$

$$\nu = \text{approach velocity} = 500\,\text{mm/s}$$

$$E = ½ \times 1100 \times (500)^2 = 13.75 \times 10^7\ \text{N/mm} = 13.75\,\text{kN/m}$$

From the shock cell type, choose the suitable curve for the relation between the energy versus deflection curve, as shown in Figure 3.47.

$$\delta = 360\,\text{mm for } E = 13.75 \times 10^7\ \text{kN/m}$$

The force versus deflection is based on the model curve shown in Figure 3.48.

$$F = 70\,\text{tons} = 686.6\,\text{KN}$$

Some of the energy is absorbed by the vessel (assume 30% absorbed by the vessel and 70% absorbed by the structure).

$$E = 13.75 \times 10^7 \times 0.7 = 9.63 \times 10^7\text{kN/m}$$
$$\delta = 300\,\text{mm}$$
$$F = 510\,\text{kN}$$

This force will not be concentrated at one point because collision occurs across a considerable area, depending on the dimensions of the vessel and its position during impact and also because the fender system distributes the load.

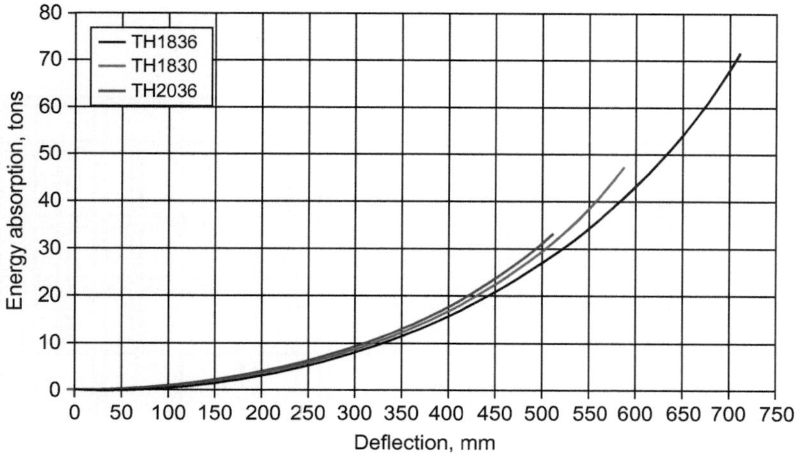

FIGURE 3.47 Energy absorption versus deflection.

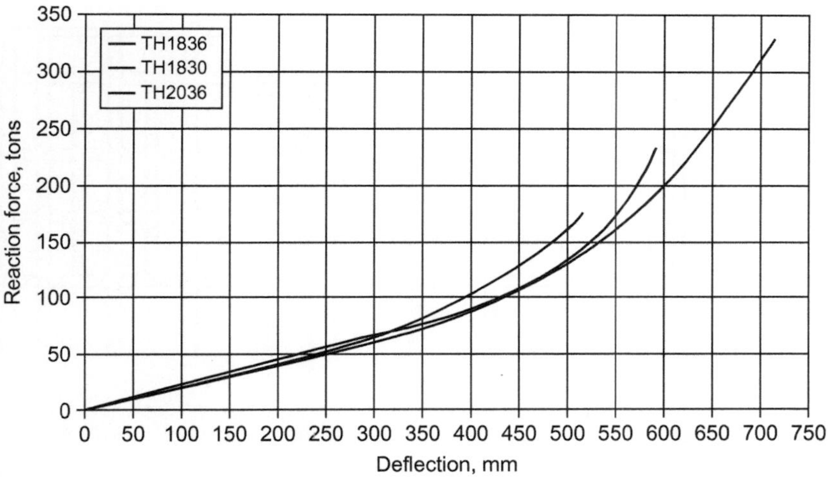

FIGURE 3.48 Reaction force versus deflection.

Cases of Impact Load

Figure 3.49 presents the configuration of the structure of the boat landing. The boat impact length is L, and it is defined at one level in Figure 3.49a and at two levels in Figure 3.49b.

$$F = 9.81 \times 10^5 N$$

$$L = 5560\,mm\,(assumed)$$

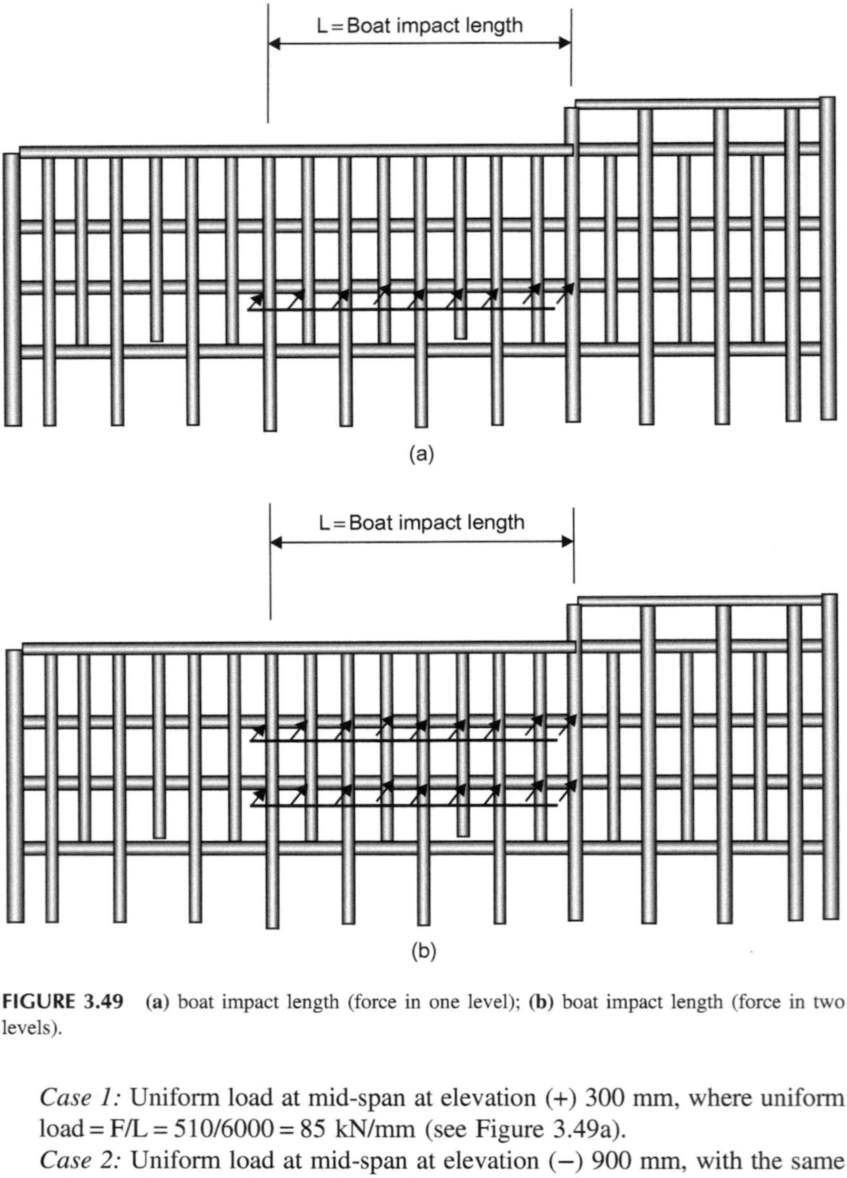

FIGURE 3.49 (**a**) boat impact length (force in one level); (**b**) boat impact length (force in two levels).

Case 1: Uniform load at mid-span at elevation (+) 300 mm, where uniform load = F/L = 510/6000 = 85 kN/mm (see Figure 3.49a).
Case 2: Uniform load at mid-span at elevation (−) 900 mm, with the same values as in case 1 but applied at level −900 mm.
Case 3: Uniform load at mid-span at elevation (+) 300 mm and elevation (−) 900 mm, where uniform load = 0.5 F/L = 42.5 kN/mm (see Figure 3.49b).
Case 4: Assume impact load distributed as a concentrated load at the mid-span at elevation (+) 300 and elevation (−) 900 (6 concentrated loads), where force at each joint = F/6 = 510/6 = 85 kN.

3.9.2 Riser Guard Design

The riser guard design will be the same as that for the boat landing, but a shock cell is not required because boats rarely accidentally collide with the riser guard. Therefore, the riser guard design is based on plasticity, and for the members of the riser guard design to reach the plasticity limit in case of accident they should have a minimum cross-section, in order to reduce the lateral load effect. On the other hand, the design should ensure that maximum deformation of the riser guard will not affect the risers in an accident.

Using a tubular member 12.5 inch × 0.5 inch (323.9 × 12.7 mm), $F_y = 240\ \text{N/mm}^2$ (mild steel).

Figure 3.50 presents the situation of the plasticity as all the pipe cross-sections will be under yield stress.

The plastic modulus = 70.2 inch^3 = 1,150,371.9 mm^3 and $M_p = 240 \times 1{,}150{,}371.9/1000 \times 1000 = 276.09 \times 10^6$ kN/m.

The deformation of the riser guard is presented in Figure 3.51.

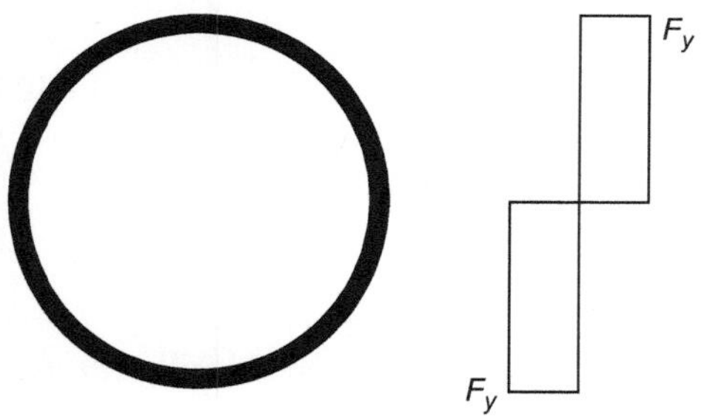

FIGURE 3.50 Member at plastic moment.

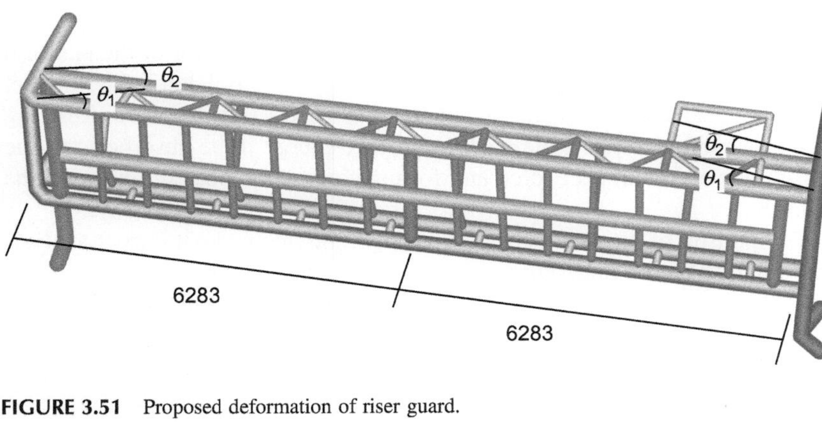

FIGURE 3.51 Proposed deformation of riser guard.

$$P \times \delta = 5 \times M_p \times (\theta_1 + \theta_2)$$

$$\theta_1 = \theta_2 = \delta/6283$$

$$P \times \delta = 5 \times 276.09 \times 10^6 \times 2 \times \delta/6000$$

$$P = 4.6 \times 10^5$$

$$E = 0.9 \times 4.6 \times 10^5 \times d = 0.5mv^2 \times 0.7$$

$$V = 500\ \text{mm/s}$$

$$m = 1100\ \text{N} \cdot \text{s}^2/\text{mm}$$

$$E = 0.7(1/2)1100(500)^2 = 0.9 \times 4.7 \times 10^5 \times \delta$$

The 232 mm deflection is less than the distance between the outside diameter of the risers and the rear face of the riser guard.

Cases of Impact Load

Case 1: Uniform load at mid-span elevation 0.0 (MWL), with L = 7000 mm (assumed), uniform load = 65.7 N/mm.

Case 2: Uniform load at mid-span at elevation (+) 1200 and elevation (−) 1200, so the total load will be distributed to two levels, such that the applied uniform load will be uniform load = 32.9 N/mm (for each elevation).

3.9.3 Boat Landing Design Using the Nonlinear Analysis Method

In some studies, boat landing and riser guard design is done by the nonlinear analysis method, which depends on the strain and denting that will occur on the boat landing member. This analysis is usually performed by special nonlinear analysis software and the member size is reduced as much as possible to reduce the wave load effect on the platform.

A nonlinear analysis is normally used to study the behavior of the platform due to the impact of a 3500 MT vessel at a velocity of 1 knot. The impact energy was calculated and the model was analyzed using software. Resizing the boat landing members allowed the impact energy to be absorbed and the results showed the impact force and formulation of the plastic hinges on the members.

Comparison of the two methods by applying dynamic boat impact analysis shows that dynamic and static (energy) boat impact analyses give a similar impact force for a fixed offshore jacket, and a static boat impact analysis is sufficient for the jacket boat impact analysis. Denting of the member is presented in Figure 3.52.

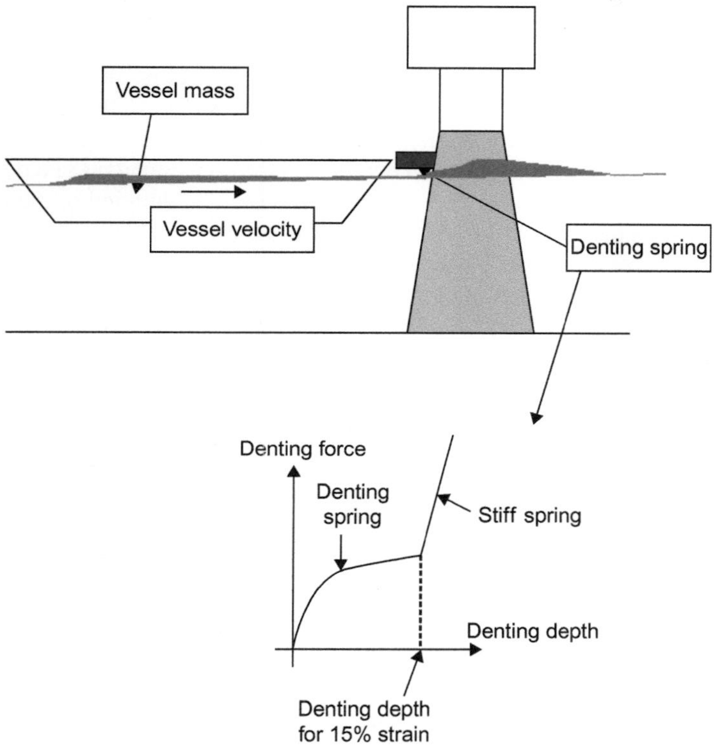

FIGURE 3.52 Denting of the member of a riser guard.

3.9.4 Boat Impact Methods

The static boat impact method utilizes the impact energy to calculate an impact load that is incremented until the impact energy has been dissipated. Fracture control is used in the software to monitor the strain in the members as energy is progressively applied. When the specified strain of 15% is reached in a member, that member is removed and the loading is redirected to other members. If no other load path can be found, the analysis is terminated.

In the dynamic method, the ship is modeled as a mass point connected to the platform through a nonlinear spring. The mass representing the ship is given an initial velocity corresponding to the impact speed and the analysis is carried out as a free vibration problem. The ship force unloads once the spring starts to elongate, i.e. the ship and platform move away from each other. When the contact force has vanished, the ship's mass is disconnected from the model. The dynamic impact analysis is a nonlinear, step-by-step analysis.

For dynamic analysis, the vessel mass and velocity were modeled; at first, the results show a higher (20% to 30%) impact load in dynamic impact analysis than in the static impact condition. Referring to the static impact results reveals that a major portion of the impact energy is absorbed by the denting of the tubular member and the software usually automatically models the denting progress of a pipe during a static impact. The effect of denting is also very important in a dynamic boat impact analysis: in impact between two bodies, the duration of the impact directly affects the impact load. A softer surface in the impact increases the impact time and reduces the impact force. Denting behavior is like a soft spring and makes the impact condition a "soft impact," so it reduces the impact load.

We can use the API formula for modeling a denting spring. According to API RP2A-WSD (21st edition):

$$P_d = 40 F_Y t^2 (X/D)^{0.5} \tag{3.146}$$

where P_d is the denting force and X is the denting depth. F_Y, t and D are the yield stress, wall thickness and diameter of the denting member, respectively.

This formula gives us the relation between denting depth and denting force, and it can be modeled in the software as a nonlinear spring.

After modeling the denting springs, the impact load in the dynamic analysis is reduced and is around 5% less than the static impact load. In some cases, the impact analysis has been done for 12 m and 55 m water depth jackets, and after modeling of the denting, the dynamic impact analysis gave an impact load similar to that in the static impact analysis.

3.9.5 Tubular Member Denting Analysis

Denting of a tubular member has a complicated behavior and there are few formulas to describe the relation between denting force, denting depth and denting energy, some of which are presented in the API RP2A standard. The important point is that, in order to define how much a dent can progress in a tubular member, we need to define the maximum allowable strain in a tubular member; usually this value is 15% in an impact analysis. So the question becomes, how much is the maximum denting depth that causes 15% strain in a tubular member?

The first solution is to model the tubular member in the finite element method (FEM) and to check the strain in the dented tubular; this was done for a 20 inch tubular member. The relation between denting load and denting depth in the FEM analysis is very close to the API formula. This solution requires extensive time for modeling the tube in FEM, so a simplified solution is needed and one is described below.

Referring to FEM deformed shapes, we can conclude that denting deforms a circular shape to approximately an oval shape, as shown in Figure 3.53, and the maximum stress and strain happen at the sides of the deformed tubular, as shown in Figure 3.53, where the radius is minimal.

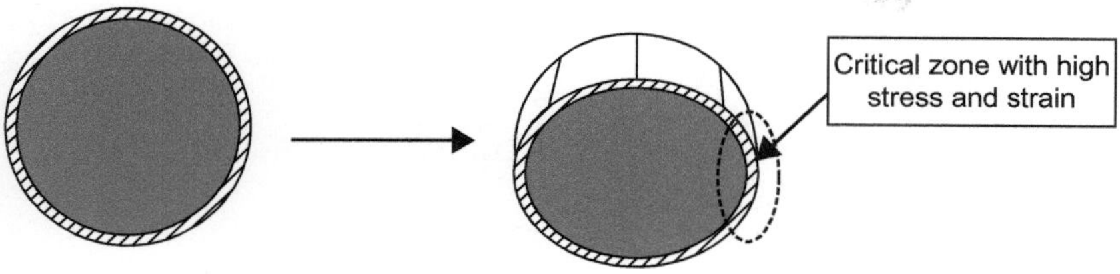

FIGURE 3.53 Critical zone for denting member.

For simplified calculation of the strain at the critical zone of the oval, we need to find the geometry and the minimum radius of the oval. The main assumption is "the circumferences of the circle tubular and deformed elliptic are similar." After finding the oval's minimum radius, the strain can be checked against 15% strain to calculate the denting limit. A simplified method for denting limit calculation is presented in the following section.

This method is valid only for boat landing members with bending behavior. For checking accidental impact on the jacket members (e.g., legs) with axial loads, the simplified method is not valid and more detail analysis needs to be done.

Simplified Method for Denting Limit Calculation

The denting calculation will be calculated based on converting the circle shape of the member to an oval after impact load, as shown in Figure 3.54.

The circle's perimeter is

$$P_c = 2\pi R \tag{3.147}$$

The ellipse's perimeter has a different equation and the following equation has a higher accuracy.

$$P_e \cong 2\pi\left(\frac{a^{1.5}+b^{1.5}}{2}\right)^{2/3} \tag{3.148}$$

Note that in the case of denting of the cylinder section, the perimeter of the circle will be the same after denting and will be like the ellipse.

$$2\pi R = 2\pi\left(\frac{a^{1.5}+b^{1.5}}{2}\right)^{2/3}$$
$$a = (2R^{1.5} - b^{1.5})^{2/3} \tag{3.149}$$

where y is the dent distance

$$b = R - y/2 \tag{3.150}$$

Knowing the dent distance y and the radius of the section R, we can obtain b from Equations (3.149) and (3.150).

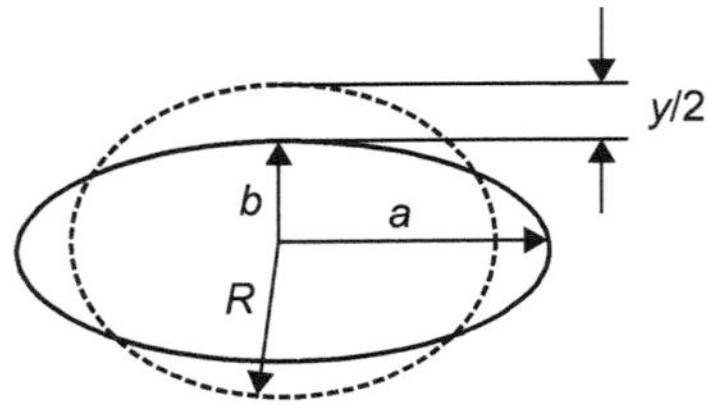

FIGURE 3.54 Calculation parameter.

Assuming that the coordinate system has the origin at the ellipse's center

$$\frac{x^2}{a^2}+\frac{y^2}{b^2}=1$$

we need the radius of curvature at $(x,y)=(a,0)$, which is actually found using calculus. Radius of curvature R is:

$$R=\frac{\left[(x')^2+(y')^2\right]^{1.5}}{x'y''-y'x''} \tag{3.151}$$

where the x and y coordinates can be parameterized as

$$x(\tau)=a\cos(\tau),\ y(\tau)=b\sin(\tau)$$

$$x'(\tau)=-a\sin(\tau),\ y'(\tau)=b\cos(\tau)$$

$$x''(\tau)=-a\cos(\tau),\ y''(\tau)=-b\sin(\tau)$$

Plugging these into the expression for R gives us

$$R=\frac{\left[a^2\sin^2\tau+b^2\cos^2\tau\right]^{1.5}}{ab\left[\sin^2\tau+\cos^2\tau\right]}$$

at $\tau=0$, $R_{min}=b^2/a$.

Figure 3.55 presents the strain ε in the tubular section due to denting.

$$\varepsilon=\frac{\phi\cdot t/2}{R_1\cdot\theta/2}=\frac{t}{R_1}\cdot\frac{\phi}{\theta}$$

$$R_1\theta=2(\theta/2+\phi)R_2$$

$$\phi=\theta/2\left(\frac{R_1}{R_2}-1\right) \tag{3.152}$$

$$\varepsilon=\frac{t}{2R_1}\left(\frac{R_1}{R_2}-1\right)$$

Nonlinear FEM Analysis

Figures 3.56 and 3.57 present the FEM analysis for a pipe 20 inches in diameter with a 0.625 inch thickness. The figures show the effect of a collision of a plate

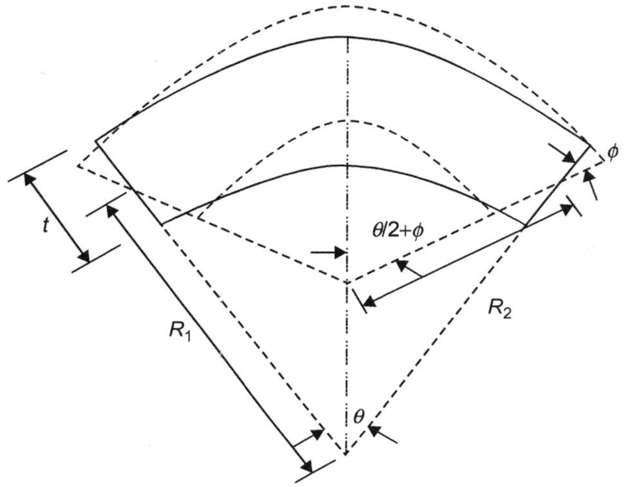

FIGURE 3.55 Strain to the denting member.

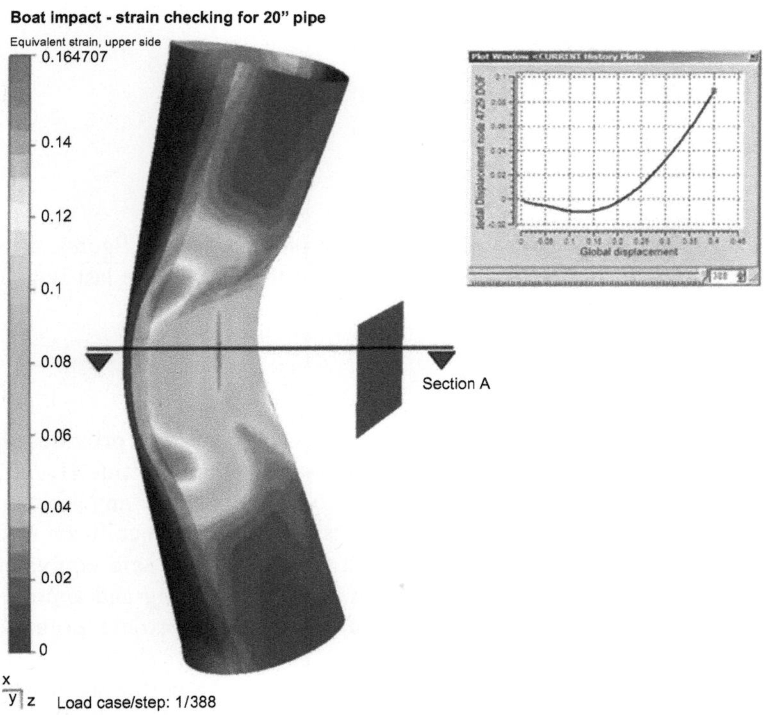

FIGURE 3.56 Strain contour for 400 mm displacement with maximum strain of 16%.

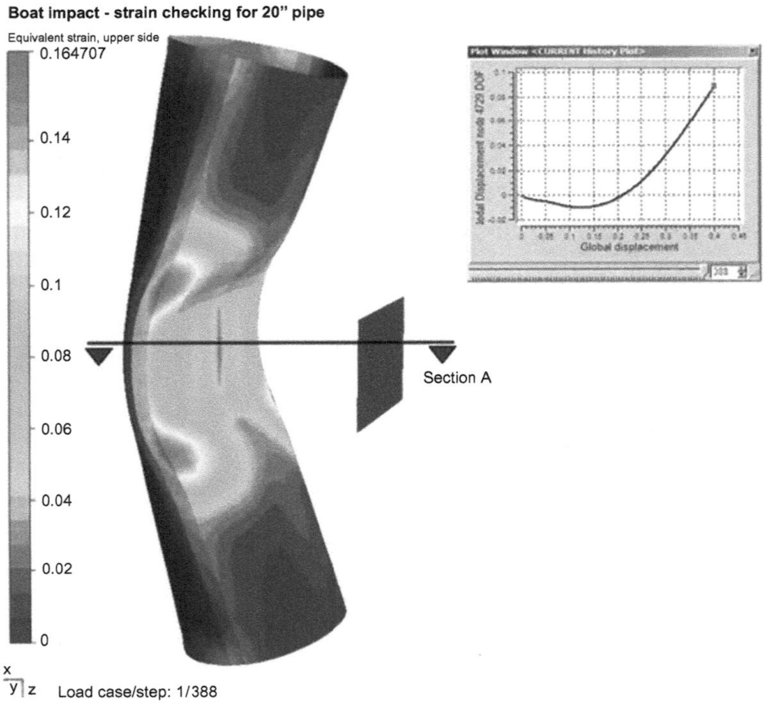

FIGURE 3.57 Strain contour in transverse direction.

and the pipe. The stress and strain contours are presented in the figures, along with the location of higher stresses. As shown in the figures, the last stage of failure occurs at strain value about 0.15.

3.10 RISER GUARD

The riser guard consists of a tubular steel space frame that provides the jacket face between elevations +2.5 m lowest astronomical tide (L.A.T.) and −2.5 m (L.A.T.) to protect the riser from boat collisions or any accidents that may occur. The riser guard should be designed to resist a collision of an equivalent static force acting anywhere on the frame. The said equivalent static force has to be defined by the client or by engineering and approved by the client. This force has to be mentioned in the project premises specification.

A static in-place analysis of the riser guard should be performed using a structural computer program. The model geometry should include all the structural members of the riser guard.

The structural dead loads should be generated by the computer program based on geometry and member properties.

Several load cases should be investigated in the analysis for the equivalent static force. For each case, the program should calculate reactions, deformations and member forces and should check all members for compliance with AISC Code. A 33% stress increase is allowed for all load cases.

3.11 ON-BOTTOM STABILITY

Offshore jacket structures must be temporarily supported by the near-seafloor soils before driving of the foundation piles. The foundation elements that bear on the seafloor soils include the jacket leg extensions, the lowest level of horizontal bracing and mud mats, as shown in Figure 3.5. All these foundation elements must be designed to support the weight of the jacket plus any additional loads imposed by environmental or construction conditions. To achieve a safe yet economical design, the structural and geotechnical engineers must coordinate their activities in sizing the foundation elements.

The first step in the design sequence requires that the geotechnical engineer explore soil conditions at the proposed structure location. By drilling and sampling a boring, geotechnical data are determined for the critical zone, generally located 10 to 20 m below the sea floor. After the structural engineer provides the sizes of the jacket leg extensions and lowest level of horizontal bracing, the geotechnical engineer can then compute the soil resistance developed on these two foundation elements. Depending on the soil resistance, the length of the leg extensions may need to be shortened to ensure full penetration. If the jacket leg extensions and lowest horizontal bracing do not provide adequate temporary support, mud mats must be provided and should be designed based on bearing capacity of the sea-floor soils, as determined by the geotechnical engineer. Although mud mats come into consideration only later, almost all steel jackets are built with mud mats as a means of temporary support. Therefore, mud-mat components are most reliable for on-bottom stability prior to pile installation.

The mud mat is a component designed for most steel jackets. It has become a common practice among engineers to include the mud mat in their jacket design even though the support provided by jacket leg extensions and the lowest level of horizontal bracing is sufficient to stabilize the structure. The fundamental aspects that need to be considered in mud-mat design include the vertical resistance of the soil provided by the bearing capacity of soil and the resistance of the structure against sliding and overturning. Another aspect rarely considered correctly is the settlement of the mud mat.

According to the API code, the bearing capacity analysis should take into account the combined effect of vertical, horizontal and moment loading. More heavily loaded mud mats may experience a lowering of soil stiffness, which can allow load to be transferred to other mud mats.

The safety factors against bearing-capacity failure recommended herein are 2.0 for on-bottom gravity loads alone and 1.5 for the design environmental condition applicable for the installation period.

Allowable steel stresses may be increased by one-third when wave loading is included. In the event of rough seas or if the installation equipment must leave the site for other reasons before the jacket has been adequately secured with piles, the effective weight on bottom may require adjustment to minimize the possibility of jacket movement due to skidding, overturning or soil failure.

Settlement of the mud mat is often predicted by the method of comparison between the applied pressure and the ultimate bearing capacity. This method is simple and easy to apply in the analysis. In the actual case, settlement of the mud mat could be a very important aspect to investigate, especially when the topsoil is weak. Therefore, a proper method should be identified to provide this information as part of the design practice. A check list for on-bottom stability analysis is given in Table 3.15.

TABLE 3.15 Check List for Jacket On-Bottom Stability Analysis

Items	Check Point	Check (yes/no)
I	Computer Model	
1	It is assumed that the model is checked for dimensions, elevations, member group and section properties in the in-service analysis and is upgraded to suit the current analysis. Check for LDOPT, OPTIONS, UCPART in the input file Model modifications to suit the current analysis **(a)** remove pile, appurtenances **(b)** revise Cd, Cm with member and group overrides as for clean members in the entire structure **(c)** Remove marine growth card from the input file **(d)** Check for flooded members and support conditions	
II	Loads	
1	It is assumed that the load calculations are verified in the in-service analysis and only relevant load cases are picked for the current analysis	
2	Wave parameters—Installation wave conditions and directions	
3	Weight of lead and add-on pile sections (before driving in)	
4	Load contingencies	

TABLE 3.15 Check List for Jacket On-Bottom Stability Analysis—cont'd

Items	Check Point	Check (yes/no)
5	Load combinations	
6	**(a)** Load combination without contingencies + environmental forces **(b)** Load combinations with lead piles in (1) Each leg, one at a time + environmental forces (2) All legs at the same time + environmental forces **(c)** Load combination with lead + add-on pile in (1) Each leg, one at a time + environmental forces (2) Two diagonally opposite legs at the same time + environmental forces (3) Two opposite legs, one at a time + environmental forces	
III	Analysis Results	
1	Analyze for Factor of Safety (FoS) determination for sliding and overturning (translate model origin to match load Center of Gravity (CoG); basic load case summary is calculated)	
2	Analyze FoS determination for bearing (translate model origin to match the mud-mat CoG: combined load case summary is considered)	
3	Analyze for bearing pressure check on the jacket structure and mud mat	
4	Enclose sea-state summary (basic/combined load case summary)	
5	Enclose member check report: review overstressed members	
6	Enclose joint check summary: review overstressed joints (check $F_y = \left\{\frac{2}{3}\right\} F_u$ for chords of high-strength members)	
7	Enclose member unity check ratio plot	
8	Enclose FoS calculations	
IV	Factor of Safety and Mud Mat Design	
1	Factor of safety against sliding (>1.50)	
2	Factor of safety against overturning (>1.50)	
3	Factor of safety against bearing (>1.50 without environmental forces)	
4	Factor of safety against bearing (>2.00 with environmental forces)	

3.12 BRIDGES

In cases where two or more platforms are forming a complex or in the case where separate installations are built to support a helideck or a flair, bridges may be required to connect the different structures.

The bridge should be designed to resist the following loads:

- Self-weight of the bridge.
- Uniformly distributed live load equal to 250 kg/m^2 of the walkway area.
- All piping loads carried by the bridge, if any.
- Wind loads acting directly on the bridge.
- Maximum imposed relative displacement between the bridge ends due to the environmental-level loads acting on the two structures connected by the bridge.
- Thermal effect due to temperature changes.

The bearings of the bridge should be designed to allow for the expected displacements and rotations. Normally, the bearings at one end should be hinged, allowing only for rotation. The bearings at the other end should be free for sliding and rotation. Flurogold slide bearings should be adequate to specify the slide bearing.

The design of the bridge should account for a span tolerance of ± 1.0 m liable to result from possible mislocation of the supporting structures. Span length rectification should in this case be accounted for by a possible increase or decrease of the theoretical bridge span or by relocation of the bearings on the structure deck.

The bridge should have an upward camber equal to the deflection supposed to happen under dead loads.

Figure 3.58 presents the hinge support for the bridge, where it can be seen that axial movement is prevented, which is opposite to the situation in Figure 3.59, which shows the roller support that permits axial movement.

FIGURE 3.58 Hinge support for the bridge.

FIGURE 3.59 Roller support for the bridge.

3.13 CRANE LOADS

Normally, the deck crane is installed over a cylindrical pedestal extending to the required level of its fixation. The cylindrical pedestal may contain an inverted, truncated conical transition part in order to reduce the large diameter of the pedestal to the size of the deck leg where the pedestal should be connected.

The cylindrical pedestal, including the conical transition, should be checked for all crane loading conditions. The design engineer should verify that the loads supplied by the crane supplier are rated loads (i.e., including the dynamic amplification) according to the API 2C or Lloyd's Register rules for lifting appliances and that the wind load is considered in the load combinations.

In addition to the static analysis of stresses, a check for fatigue should be considered.

Stress analysis should be carried out according to API RP2A.

The static and dynamic crane loads should be based on data provided by the crane manufacturer

3.14 LIFT INSTALLATION LOADS

Lifting forces are functions of the weight of the structural component being lifted, the number and location of lifting eyes used for the lift, the angle between each sling and the vertical axis and the conditions under which the lift is performed.

All members and connections of a lifted component must be designed for the forces resulting from static equilibrium of the lifted weight and the sling tensions. Moreover, API RP2A recommends that, in order to compensate for any side movements, lifting eyes and the connections to the supporting structural members should be designed for the combined action of the static sling load and a horizontal force equal to 5% of this load, applied perpendicular to the padeye at the center of the pin hole.

It is worth mentioning that, all these design forces are applied as static loads if the lifts are performed in the fabrication yard. On the other hand, if the lifting derrick or the structure to be lifted is on a floating vessel, then dynamic load factors should be applied to the static lifting forces.

In particular, for lifts made offshore, API RP2A recommends two minimum values of dynamic load factors: 2.0 and 1.35. The first factor is for designing the padeyes as well as all members and their end connections framing the joint where the padeye is attached, while the second is for all other members transmitting lifting forces. For load-out at sheltered locations, the corresponding minimum load factors for the two groups of structural components become, according to API RP2A, 1.5 and 1.15, respectively.

Jacket, topside and living quarters lift analyses (onshore and offshore) should be performed based on the requirements of DNV rules. All members and connections should be checked to API RP2A or AISC basic allowable stresses.

It is worth mentioning that the weight contingency factor is a factor that allows for lift weight inaccuracies. For jacket structures, a minimum factor of 1.1 should be used unless the jacket is weighed at the end of the construction by using load cells, in which case this factor may be reduced to 1.03. The weight contingency factor should be applied to the "net weight" and "rigging weight."

Dynamic amplification factors allow for the dynamic effects experienced during the lift. For typical jacket structures, the dynamic amplification factors are as presented in Table 3.16.

Note that the dynamic amplification factors for the offshore lift presented in Table 3.16 should be used for calm sea conditions (Hs <2.5 m). If, for any reason, the lift is carried out in an adverse condition, the factor should be recalculated based on the expected accelerations associated with the sea state.

TABLE 3.16 Dynamic Amplification Factors

	Gross Weight (tons)			
	0–100	100–1000	1000–2500	>2500
Offshore Lift	1.3	1.2	1.15	1.10
Onshore Lift	1.15	1.10	1.05	1.05

3.15 VORTEX-INDUCED VIBRATIONS

In fluid dynamics, vortex-induced vibrations (VIVs) are motions induced on bodies interacting with an external fluid flow and are produced by periodical irregularities in the flow.

A classic example is the VIVs of an underwater cylinder. If you put a pipe into the water and move it through the water in a direction perpendicular to its axis, you can see this vortex. Since real fluids always present some viscosity, the flow around the cylinder will be slowed down while in contact with its surface, forming the so-called boundary layer. At some point, however, the boundary layer can separate from the body because of its excessive curvature. Vortices are then formed, changing the pressure distribution along the surface. When the vortices are not formed symmetrically around the body, different lift forces develop on each side of the body, thus leading to motion transverse to the flow. This motion changes the nature of the vortex formation in a way that leads to a limited motion amplitude.

The tubular members of the flare/vent booms should be checked for VIVs. If the members and booms are found to be dynamically sensitive, they should be checked for fatigue during detailed design.

VIVs are an important source of fatigue damage to offshore platforms, especially for oil exploration and production risers. These slender structures experience both current flow and top-end vessel motions, which give rise to the flow-structure relative motion and cause VIVs. The top-end vessel motion causes the riser to oscillate and the corresponding flow profile appears unsteady.

The possibility of VIVs due to the design current velocity profiles should be considered for all appurtenances, including risers, sump pipes, caissons and any individual members considered potentially susceptible.

One of the classical open-flow problems in fluid mechanics concerns the flow around a circular cylinder, or, more generally, a bluff body. At very low Reynolds numbers, according to the diameter of the circular member, the streamlines of the resulting flow are perfectly symmetrical, as is expected from potential theory.

The Strouhal number, named after Čeněk (Vincent) Strouhal, a Czech scientist, relates the frequency of shedding to the velocity of the flow and a characteristic dimension of the body (diameter, in the case of a cylinder). The Strouhal number is defined as $St = f_{st}D/U$, where f_{st} is the vortex shedding frequency (or the Strouhal frequency) of a body at rest, D is the diameter of the circular cylinder and U is the velocity of the ambient flow. The Strouhal number for a cylinder is 0.2 over a wide range of flow velocities. When the vortex shedding frequency comes close to the natural frequency of vibration of the structure, large and damaging vibrations can occur.

In general, wind, current or any fluid flow past a structural component may cause unsteady flow patterns due to vortex shedding. This may lead to oscillations of slender elements normal to their longitudinal axis. Such vortex-induced oscillations (VIOs) should be investigated.

TABLE 3.17 Relation between Type of Shedding and Reynolds Number

Periodic shedding	$10^2 < Re < 0.6 \times 10^6$
Wide-band random shedding	$0.6 \times 10^6 < Re < 3.0 \times 10^6$
Narrow-band random shedding	$3.0 \times 10^6 < Re < 6.0 \times 10^6$
Quasi-periodic shedding	$Re > 6.0 \times 10^6$

Important parameters governing VIOs are geometry (L/D), damping ratio (ζ), mass ratio [$m^* = m/(\frac{1}{4}\pi\rho D^2)$], Reynolds number ($Re = uD/v$), reduced velocity ($VR = u/f_n D$) and flow characteristics (flow profile, steady/oscillatory flow, turbulence intensity (σ_u/u), etc.), where L = member length (m); D = member diameter (m); m = mass per unit length (kg/m); ζ = ratio between damping and critical damping; ρ = fluid density (kg/m^3); v = fluid kinematic viscosity (m^2/s); u = (mean) flow velocity (m/s); f_n = natural frequency of the member (Hz); and σ_u = standard deviation of the flow velocity (m/s).

The relation between type of shedding and Re is shown in Table 3.17.

3.16 HELIDECK DESIGN

The loads affecting the helideck are discussed in Chapter 2.

- Helidecks are designed in accordance with API RP2A and API RP2L.
- Layout of the helideck should be sufficient for one helicopter.
- One stair for primary access and a ladder for secondary access are required.
- Safety netting should surround the helideck completely, at a minimum width of 1500 mm.
- Paint markings, sizes and colors are in accordance with API RP2L.

The helideck may have a separate platform, as shown in Figure 3.60, and may be connected with the main platform by a bridge.

The helicopter landing area should be clear, without any obstacle, and for obstacles below the landing area, there are some limits from CAP437, as shown in Figure 3.61.

According to CAP437, the following dimensions should be considered.

The safety net's main function is for personnel protection. Therefore, the safety net is usually installed around the landing area, except where adequate structural protection against falls exists. The netting used should be flexible and should be manufactured from non-flammable material, with the inboard edge fastened level with, or just below, the edge of the helicopter landing deck. The net itself should extend 1.5 m in the horizontal plane, so that it

FIGURE 3.60 Helideck as a separate platform.

has an upward and outward slope of at least 10°. Standard dimensions for safety nets are given in Table 3.18.

The net should be strong enough to withstand and contain, without damage, a 100 kg weight being dropped from a height of 1 m.

According to API RP2L, a safety net designed to meet these criteria should not act as a trampoline, giving a "bounce" effect. Where lateral or longitudinal center bars are provided to strengthen the net structure, they should be arranged and constructed to avoid causing serious injury to persons falling on them. The ideal design has a "hammock" effect, which should securely contain a body falling, rolling or jumping into it, without serious injury. When considering the securing of the net to the structure and the materials used, care should be taken that each segment will meet adequacy of purpose considerations. Polypropylene deteriorates over time; various wire meshes have been shown to be suitable if properly installed.

For the helideck markings, the color of the helideck should be dark green or gray, as shown in Figure 3.62. In addition, the perimeter should be clearly marked with a white painted line 300 mm wide.

Figure 3.62 also presents the dimensions of the yellow circle and the H marking on the helideck.

TABLE 3.18 Helicopter weights, dimensions and D value for different types as per CAP437

Type	D value (m)	Perimeter "D" marking	Max weight (ton)	"t" value	Landing net size	Landing minimum size, (m)
Bolkow bo 105D	11.81	12	2.300	2.4t	Not required	-
Bolkow 117	13.00	13	2.300	3.2t	Not required	-
Augusta A109	13.05	13	2.600	2.6t	Small	9 m × 99 m
Dauphin SA365N2	13.68	14	4.250	4.3t	Small	99 m × 9 m
Sikorsky S76B&C	16	16	5.307	5.3t	Medium	-
Bell 212	17.46	17	5.080	5.1t	Not required	-
Super Puma AS 332L2	19.50	20	9.300	9.3t	Medium	12 m × 12 m
Super Puma AS 332L	18.70	19	8.599	8.6t	Medium	12 m × 12 m
Bell 214ST	18.95	19	7.936	8.0t	Medium	12 m × 12 m
Sikorsky S61N	22.20	22	9.298	9.3t	Large	15 m × 15 m
EH101	22.80	23	14.600	14.6t	Large	15 m × 15 m
Boeing BV234LR Chinook**	30.18	30	21.315	21.31t	Large	15 m × 15 m

Note: With a skid-fitted helicopter, the maximum height may be increased with ground-handling wheels fitted.
***Note: The BV234 is a tandem rotor helicopter and in accordance with ICAO the helicopter size is 0.9 of the helicopter D value, i.e., 27.16 m.*

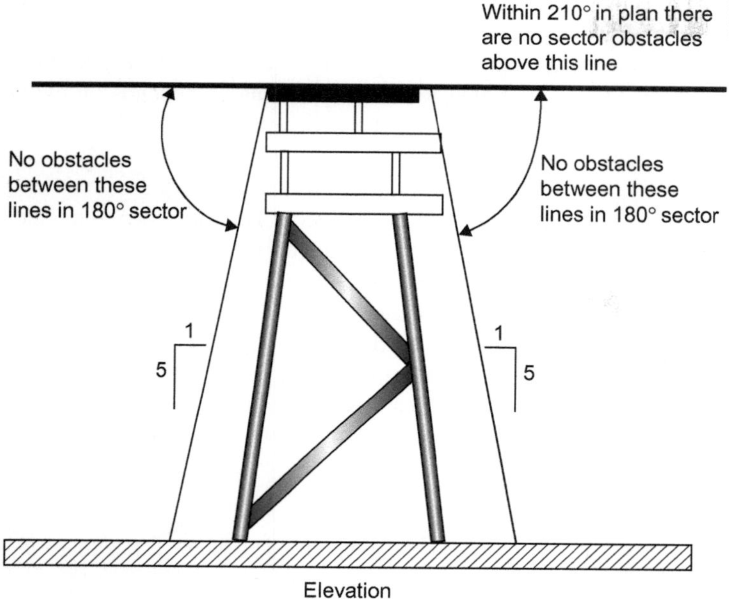

FIGURE 3.61 Limits of free obstacles below landing area level.

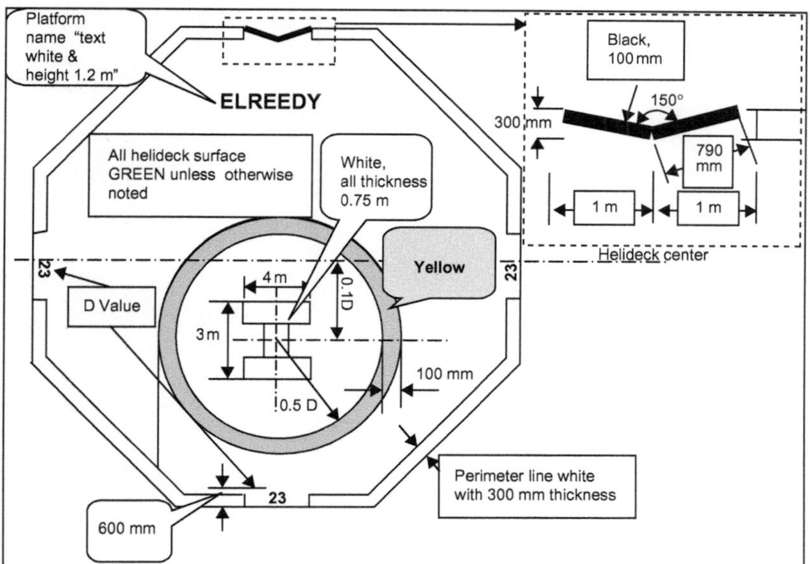

FIGURE 3.62 Helideck marking specifications and dimensions.

The letter H on the helideck has standard dimensions, as shown in Figure 3.62, and it is painted white.

Figure 3.63 presents the obstacle limitation in the helideck plan view.

Based on CAP 437, the helideck netting dimension depends on the landing net size, which, in turn, depends on the helicopter type, the D value and the maximum weight (see Table 3.18).

Figure 3.64 presents the configuration of the safety net.

The helicopter should be tied to the helideck. This tie-down point should be located and be of such strength and construction to secure the helicopter when subjected to weather conditions pertinent to the installation design considerations. The tie-down rings should match the tie-down strap attachments. Note that the maximum bar diameter of the tie-down ring should be 22 mm in order to match the strap hook dimension of the tie-down strap in most helicopters. Advice on the recommended safe working load should be obtained from the helicopter operator. Figure 3.65 presents a suitable tie-down configuration;

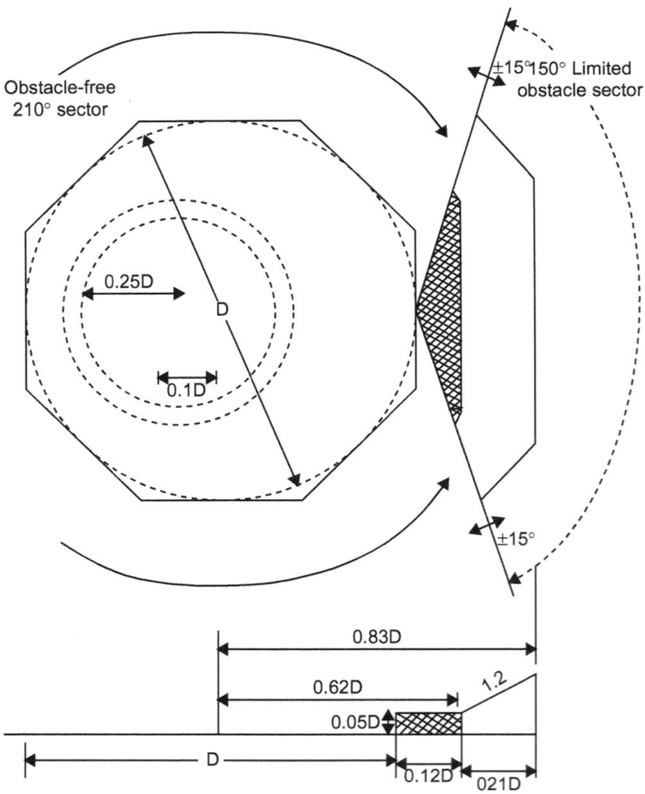

FIGURE 3.63 Obstacle limitation for single main rotor as per CAP437.

FIGURE 3.64 The main configuration of the traditional safety net.

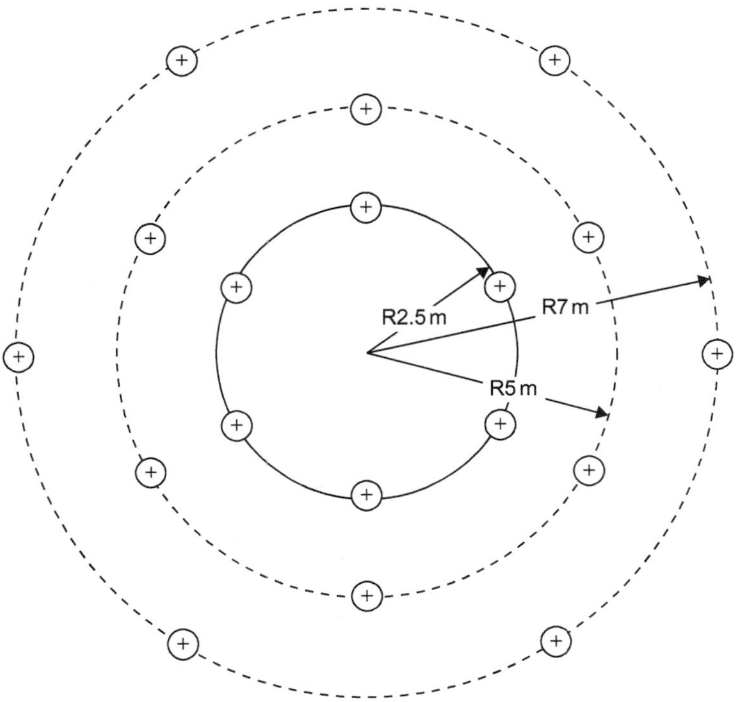

FIGURE 3.65 Suitable tie-down configuration.

note that the outer circle is not required if the D value is less than 22.2 m. The centers of all the circles coincide with the center of the marking circle shown in Figure 3.62.

3.17 STRUCTURE ANALYSIS AND DESIGN QUALITY CONTROL

An offshore structure is complicated and not an easy structure to design, so it requires many quality checks. One method is to use check lists to ensure that all the factors are taken into account for design.

Table 3.19 is a check list for jacket in-place analysis and Table 3.20 is a check list for topside in-place analysis.

TABLE 3.19 Check List for Jacket In-Place Analysis

Items	Check Point	Check (yes/no)
	Computer Model	
1	Framing dimensions according to the drawings or sketches	
2	Framing elevations according to the drawings or sketches	
3	Water depth and mud line elevation match the BOD	
4	Member group properties: E, G, density match the BOD	
	(a) Section details, segment lengths, member offsets	
	(b) Corrosion allowance in splash zone	
	(c) Zero density for wishbone elements in ungrouted jackets	
	(d) Grout density corrections for grouted leg and pile	
5	Member properties: Kx, Ky, Lx, Ly	
6	Member end releases where applicable	
7	Flooded members (legs, risers, J tubes, caissons, etc.)	
8	Dummy members (relevant joints kept, rest deleted)	
9	Plate/membranes modeled correctly	
10	Boundary conditions	
11	Drag and inertia coefficients (smooth, rough members)	
12	Marine growth data as per BOD	
13	Member and group overrides	

TABLE 3.19 Check List for Jacket In-Place Analysis—cont'd

Items	Check Point	Check (yes/no)
	(a) No wave load and marine growth on piles and wish bones in jacket legs	
	(b) Enhancement of Cd, Cm for anode supported members	
	(c) Enhancement of Cd, Cm for jacket walkway members	
	(d) Enhancement of Cd, Cm	
15	Hydrostatic collapse check selected with redesign option	
16	Allowable stress modifiers for extreme storm load cases	
17	Unity check ranges in the analysis input file	
	Loads	
18	Load description, calculations and distribution	
19	Wave theory, wave, current and wind directions	
	(Non-linear current stretch with apparent wave period calculation)	
20	Equivalent Cd, Cm calculations for items mentioned above	
21	Load contingencies match the BOD	
22	Load combinations for operating and extreme cases	
23	Load summations	
24	Load summation verification against weight control data	
	PSI Data and Input File	
25	Units for T-Z, Q-Z and P-Y data from the geotechnical data	
26	Pile segmentation data, end bearing area	
27	With reference to the input format, check T, Q and P factor	
	Analysis Results	
28	Enclose sea-state summary to be checked against load summation	
29	Enclose member check summary: review for overstressed members	
30	Enclose joint check summary: review overstressed joints (check $F_y = 2/3F_u$ for chords of high-strength members)	
31	Check maximum pile compression and tension	

(Continued)

TABLE 3.19 Check List for Jacket In-Place Analysis—cont'd

Items	Check Point	Check (yes/no)
32	Enclose model plots: Joints/group/section names, Kx, Ky, Lx, Ly, and loading	
33	Enclose deflection plots, member unity check ratio plots	
34	Enclose hydrostatic collapse check reports and check for need of rings	
35	Review pile factor of safety calculations	
36	Review permissible deflection calculations	
37	Review plot plan and latest structural drawings	
38	Review relevant sections of weight control report	
	Dynamic In-Place Analysis (if required)	
39	Determine DAF based on single Degree of Freedom DoF concept and apply on wave load cards.	
40	Determine DAF based on inertia load distribution and apply on total structure	
	General	
41	Joint name range identified for each framing level and sequential	
42	Member group name specific to each framing level and sequential	
43	Check for future loads and loads due to specific requirement, such as rigless interventions	

Abbreviations: BOD = basis of design; DAF = dynamic amplification factor

The calculation report delivered to the client should include, as a minimum, the following data:

- Calculation cover sheet
- Contents page
- Description of the analysis methodology and techniques
- Explanation of the model geometry and axis system
- Explanation of boundary conditions
- Explanation of the input loads and any supporting calculations showing the development of the loads
- Load combination matrix
- Explanation of the stress analysis assumptions, including member effective lengths philosophy

TABLE 3.20 Check List for Topside In-Place Analysis

Items	Check Points	Check (yes/no)
	Computer Model	
1	Framing dimensions match with the drawings	
2	Framing elevations match with the drawings	
3	Member properties: Kx, Ky, Lx, Ly	
4	Member end releases where applicable	
5	Plate and membranes modeled correctly	
6	Boundary conditions	
7	Review the loads in the analysis input file	
8	Allowable stress modifiers for extreme storm load cases	
9	Unity check ranges in the analysis input file	
	Loads	
10	Load description and calculations	
11	Secondary structural item dead load calculations	
12	Equipment, piping operating and dry load, E&I bulk load calculations	
13	Wind load calculations, wind area considered	
14	Earthquake load calculation	
15	Load contingencies	
16	Load combinations for operating and extreme cases	
17	Load combinations for local checks	
18	Load summations	
	Installation and pre-service loads may be applied as a separate load case	
	Crane load cases may be magnified and applied for local checks	
19	Check for future loads and loads due to specific requirement, such as rigless interventions	
	Analysis Results	
20	Enclose sea-state summary and check against load summation	
21	Enclose member check summary: review for overstressed members	

(Continued)

TABLE 3.20 Check List for Topside In-Place Analysis—cont'd

Items	Check Points	Check (yes/no)
22	Enclose joint check summary: review overstressed joints (check $F_y = 2/3F_u$ for chords of high-strength members)	
23	Enclose model plots: Joint/group/section names, K_x, K_y, L_x, L_y, F_y	
24	Enclose deflection plots, member unity check ratio plots	

TABLE 3.21 Report Check List

	Task	LE	AE	CE
Model Set-up	Latest data/information used			
	Unique model/run name used			
	Geometry			
	Support conditions			
	Member effective lengths			
	Loads			
	Load combinations[a]			
	Line by line check of input file			
Self Checking	Review model geometry plots			
	Check load sums against load combination matrix			
	Program-generated loads reasonable (e.g., dead, wave, wind, etc.)			
Post-run Verification	Review errors and warning messages			
	Check reaction totals match load combination matrix			
	Check deflections are consistent with expected loadings, storm directions, etc.			
	Check pile convergence, stresses and safety factors			
	[b]Review member stresses			
	Review tubular joint stresses			
	Model and results saved on the server			
	All checks complete and satisfactory			

Abbreviations: LE = lead engineer; AE = analyst engineer; CE = checking engineer.
[a]*The combination of different loads is one case of loading.*
[b]*For local design, hydrostatic test weights should be considered where applicable.*

- Discussion of the results, including deflections, member stresses, joint stresses, reactions, pile capacities, etc.
- Additional calculations supporting the results, including calculations showing the justification of overstressed members, etc.
- Attachments should contain the following:
  - Any reference information used for the model
  - Drawings
  - Input files (including the model, PSI input file, joint-can input, etc.)
  - Model plots showing joint numbers and member groups. Member effective lengths, etc., may also be presented if required.
  - Selected results, including reactions, summary of maximum member U_c, summary of maximum joint U_c, summary of maximum joint deflections, pile and safety factors and other relevant results
  - Calculation check list and any other check list that ought to be included

A summary report check list to be used by the lead engineer, analyst engineer and checking engineer is given in Table 3.21.

BIBLIOGRAPHY

Dier, A.F., Lalani, M., 1995. Strength and stiffness of tubular joints for assessment/design purposes, Offshore Technology Conference, Houston, TX, Paper OTC 7799.

EDI, SACS IV. User's Manual, release 5.1, Published by SACS-Bentely, USA 2003.

Healy, R.E., Buitrago, J., 1994. Extrapolation procedures for determining SCFs in mid-surface tubular joint models. Sixth International Symposium on Tubular Structures. Monash University, Melbourne, Australia.

International Organization for Standardization, Petroleum and Natural Gas Industries—Offshore Structures—Part 2: Fixed Steel Structures, ISO/DIS 19902:2004.

Kuang, J.G., Potvin, A.B., Leick, R.D., 1975. Stress concentration in tubular joints. Paper presented at the Offshore Technology Conference, OTC paper No. 2205, Houston, TX.

Lloyd's Register of Shipping, 1988. Stress Concentration Factors for Tubular Complex Joints, Complex Joints JIP. Final Report No. 3 of 5 of Simple Unstiffened Joint SCFs, The Health and Safety Executive, London.

Marshall, P.W., 1989. Recent developments in fatigue design rules in the U.S.A. In: Fatigue Aspects in Structural Design. Delft University Press, Delft, Netherlands.

Marshall, P.W., Bucknell, J., Mohr, W.C., 2005. Background to new RP2A-WSD fatigue provision. Paper OTC 17295. In: Proceedings of the Offshore Technology Conference, Houston, TX, May 2005.

Marshall, P.W., Toprac, A.A., 1974. Basis for tubular joint design. Weld J 53 (5), 192–201.

MSL Engineering Limited, 1996. Assessment Criteria, Reliability and Reserve Strength of Tubular Joints, Doc. Ref. C14200R018, Ascot, England.

Niemi, E., Fricke, W., Wolfgang Fricke, Maddox, S.J. et al., 1995. Stress Determination for Fatigue Analysis of Welded Components. IIW-1221-93, Abington Publishing, Cambridge, UK.

Report for Joint Industry Typical Frame Projects, 1999. Bomel Engineering Consultants. Maidenhead, Berkshire, England.

Yettram, A., Husain, H., 1966. Plane framework methods for plates in extension. J. Eng. Mech. Div. ASCE 91 (EM3), 53–64.

Chapter 4

Geotechnical Data and Pile Design

4.1 INTRODUCTION

To begin the design of any structure, the designer must know the type of soil on which the structure will be built. So the first step in design of an offshore structure platform is to perform a soil investigation, but the offshore soil investigation technique is not like onshore soil investigation.

A fixed offshore platform is defined as a jacket-type structure extending above water level and supported by relatively long piles, because this type of structure has been an industry standard for many years. Of course, nowadays other types of structures are being built and many different designs are being proposed, as presented in Chapter 1, and the soil investigation requirements and its depth depend on the type of structure, since they could vary considerably for different types of structures.

In a traditional fixed offshore platform, the soil investigation usually consists of a single boring sampled to a penetration below the expected pile-tip penetration depth.

4.2 INVESTIGATION PROCEDURE

The final design of a fixed offshore platform is based on the best soil field test data that can be obtained at the exact structure location. Therefore, an accurate survey should be implemented with the soil investigation. In general, the soil data are usually needed well in advance of any construction at a site.

The marine vessel and drilling equipment that must be provided for conducting and supporting the soil investigation should have a mooring or positioning system and should be capable of drilling and sampling well below the maximum expected penetration of the piles. Furthermore, offshore soil investigation is typically more expensive than similar work onshore. This is due to the cost of the marine vessel itself, as its fee is calculated per day.

Provision should be made for on-site determinations of pile capacity versus depth for the pile size proposed to be used in the structure. Thus, an adequate but not excessive depth of investigation can be ensured for the pile load anticipated.

Therefore, the objectives of an investigation should always be kept in mind and the planning effort and selection of equipment and techniques should be aimed at achieving the objectives at the lowest cost. If there is leeway in scheduling of the work, the work can be undertaken in suitable weather; if there is no leeway, larger equipment may be needed to cope with more severe conditions. The drilling and sampling techniques must be adapted to the marine environment in order to produce the desired results within a reasonable time.

Once the desired field investigation has been completed, further investigation is needed in the laboratory to evaluate and to obtain soil properties and to apply the field and laboratory results to the design problem.

It is appropriate to mention at this point that a fairly wide range of results can be obtained, particularly in the field, depending upon several factors. It is essential that the factors influencing the results be evaluated in interpreting the data and in applying the results to the design.

Offshore soil investigation provides useful information in connection with platform installation as well as design. An accurate water-depth determination, made in connection with obtaining a sample exactly at the seabed, gives the engineering office the required information, and high-quality samples and data near the seabed provide very useful information relative to jacket support before piles are driven, jacket leg penetration below the seabed and soil-pile interaction under lateral load.

The observations made during drilling and sampling should be recorded, because they can also provide some indication of potential problems in pile installation. The scope of work and execution of an investigation should encompass all possible elements important to design and construction.

4.2.1 Performing an Offshore Investigation

An investigation at sea must begin by identifying the location, which is ordinarily done by a survey boat. After the correct location has been established, a buoy may be dropped to mark the location for drilling. The survey vessel also assists in running the anchors for the drilling vessel. In most cases, while drilling and sampling are in progress, the survey vessel remains in the area to serve as an emergency stand.

Most offshore investigations done at the present time utilize self-propelled vessels, usually of the oilfield-supply class. Such vessels have adequate deck space and can easily be outfitted for drilling. A typical vessel length is in the range of 40 to 70 m. The vessel is equipped with winches, cables and anchors for four-point mooring. Anchor lines usually approach about eight times water depth.

Anchors may weigh as much as, or more than, 2.5 tons, to give reasonable assurance of maintaining station in adverse conditions of wind, wave, current and mud-line soils. As much as possible, the bow of the vessel is oriented into the prevailing surface sea and wind to minimize vessel roll, pitch and heave. Satisfactory drilling and sampling can be done in seas of 1.8–2.4 m (6–8 ft) height.

The supply vessel should have sufficient quarters for both the vessel and drilling crews, and two drilling crews are provided so that work can be conducted on a 24-hour basis once a boring is started.

Many vessels have been modified to include a centrally located drilling well through the hull. The central location minimizes the amount of vessel motion and provides ample work space around the drilling rig. If such a well is not available, drilling can be done from a temporary deck cantilevered over the side. The best location for the platform, from the standpoint of vessel motion, is at the mid-point of the side.

4.2.2 Drilling Equipment and Method

Nowadays, the offshore industry has moved to using deep-water platforms designed and built for depths approaching 150 m (500 ft); one has even been designed for a water depth of about 300 m. Pile foundations frequently penetrate 90–120 m into the sea bottom. The combination of water depth and bottom penetration means that drilling equipment for soil borings should have a depth capability of 300 m or more.

A convenient rig has utilities that include hoisting drums and a rotary pump.

A drilling operation may be improvised from a power swivel or tongs to provide rotation, a crane to provide hoisting capacity and a pump for circulation.

Weight on the bit must be provided through drill collars. Telescoping bumper subs are sometimes used to keep the bit on the bottom as the vessel heaves, or a motion compensator can be used for the same purpose. A weight indicator on the drilling line permits the driller to observe and to control the bit weight and to prevent damage to the drill pipe.

In the early days of offshore soil exploration, it was customary to set a casing or conductor from the drilling deck into the sea floor. This provided a means of recirculating drilling fluid and allowed repeated entry into the hole to drill and sample by conventional land methods. However, as work moved into deeper water, this procedure became quite slow and costly. The longer time on site invited more weather interference. Therefore, it became desirable to develop new and faster drilling and sampling techniques; the so-called "wire-line" methods were the result.

4.2.3 Wire-Line Sampling Technique

In the wire-line method, the drill pipe itself serves as the conductor; no other pipe is used. An open-center drill bit is used on the lower end. A boring is advanced by ordinary rotary methods; drilling fluid pumped through the drill pipe prevents soil from entering the open-center bit. Since there is no provision for recirculation of drilling fluid, all fluid pumped is expended as it emerges from the boring at the sea floor; thus a continuous supply of new fluid is required. As discussed below, a fluid of controlled weight and viscosity is used to counteract formation pressures and to stabilize the drilled hole.

The wire-line sampler consists of a thin-walled tube attached to the lower end of a device incorporating a down-hole hammer or set of mechanical parts. The device is run through the bore of the drill pipe, so its size is limited by the inside diameter.

The driving operation results in some disturbance to clay samples. Shear strengths measured on driven samples will be generally lower than on pushed samples. Comparative borings and tests have been made to evaluate this effect.

The wire-line procedure for sampling sand is obviously the same as the standard penetration test (SPT), in that drill rods attached to the sampler are eliminated and driving is by means of the down-hole hammer, which imparts a fairly uniform driving energy at any depth. Correlations have been made between driving resistances and soil strength; it is doubtful that the SPT method has any validity at the great sampling depths offshore. The wire-line procedure provides a qualitative indication of density and permits close observations of variations with depth.

The advantages of using the wire-line procedure are: 1) the sampler is completely independent of the drill pipe, and thus from vessel-imparted motion, and 2) sampling can be done rapidly, in that a boring that might take 4–5 days to complete using conventional methods can be finished in about one day with wire-line techniques. This results in reduced weather exposure and, obviously, reduced cost.

4.2.4 Offshore Soil Investigation Problems

The effects of weather upon marine operations are well known. The smaller equipment used in soil-boring operations is considerably more vulnerable than the large equipment used for offshore well drilling and construction.

Boring sites are frequently 100 miles or more from land or safe anchorage. It is therefore necessary that marine equipment used for soil-boring operations be able to endure sea conditions much worse than those in which drilling can be conducted. If adverse conditions are encountered after a boring is started, it may be necessary to suspend drilling temporarily or even to withdraw the drill pipe, reposition anchors and redrill the boring to continue sampling at greater depth.

As stated above, it is important to define the water depth, as the design of the jacket is based on this value. The matter of determining water depth may appear simple, but it can be very difficult because of currents, tide and very soft seabed soils, yet an exact water depth is essential, so that sampling can be begun at the seabed.

Probably the best way to determine water depth is to use a sounding weight on the small wire line used for sampling and to use a wire-line counter to measure the length of line. The weight must be adequate to minimize current effects but must be designed so that it defines the seabed and does not sink into a very soft bottom. The water depth can be confirmed on the first sampling attempt.

World-wide predictions of seasonal weather conditions and sea states are available from several sources. Short-term forecasts are also available in most offshore areas. Predicted tide and current tables can be obtained for almost any area. All of this information is useful in design planning and jacket installation.

If the platform is located in an area where significant tide changes occur, it is necessary to make frequent observations to define the tide cycle and thereby control the sampling depth.

A plot of time versus water depth can be used to reduce water depth to any desired datum. The time, date and measured water depth should be recorded at the start of a boring. Tide variations can sometimes be recorded by a suitable fathometer.

In areas of very soft, underconsolidated soils, it is necessary to exercise very careful control of drilling fluid weight to counteract the tendency of the material to squeeze into the drilled hole and up into the drill pipe. Failure to do so can result in very severe sample disturbance. Furthermore, in such areas, the problem of handling disturbance on recovered samples is also quite severe.

Once a boring has penetrated a granular soil formation, it is essential that drilling mud having suitable viscosity and gel properties be used to stabilize the drilled hole and to prevent caving; commercial saltwater gel is excellent for this purpose. Particularly in glacial deposits, coarse granular material such as cobbles or boulders may be encountered and may make drilling extremely difficult. The presence of rock formations within the depth of investigation requires that special tools and procedures be used.

In most cases, gas may be present within formations penetrated by soil borings and flows of water may also be encountered. The normal procedure of using a blowout preventer on cased holes is not easily applied to the wire-line method.

Sometimes a large mobile rig may be used to drill at a location where a platform may be installed later, and it may be desirable to make a soil boring from the mobile rig. If space is available, the boring can be made without interference with the normal rig activity by placing a soil-boring rig on board. The drilling and sampling procedures are identical to those used from a floating vessel. The large rigs, either jack-ups or floaters, provide a relatively stable base for soil-boring operations.

Sometime the soil boring is performed using a large oilfield rig. If the soil-boring rig cannot be accommodated, it is possible to use the large drilling rig and its crews, working under technical supervision, to make the soil boring. The large mobile rig and its support equipment may cost \$25,000 to \$50,000 per day; therefore, the soil boring may be quite expensive. This approach should generally be avoided if other means are available for making a boring. If the large rig must be used, wire-line procedures will minimize the rig time devoted to soil-boring operations.

Existing diver-operated equipment consists of fairly conventional drilling equipment adapted to work on the sea floor with support from a surface vessel. The diver-operated approach to drilling and sampling may become more competitive with other traditional methods, but, diver training must also include experience in soil sampling.

Several pieces of equipment have been developed to operate on the sea floor by remote control from a surface vessel. To date, this equipment, known as a remotely operated vehicle (ROV), does not have the capability to sample at the depths required to investigate deep-pile foundations. As development continues, the use of such equipment may become more feasible.

Much attention has been given to use of small manned submersibles to perform various underwater tasks without exposing personnel to the pressures associated with diving. These devices have been equipped with manipulators of various kinds, and they have been used to perform in-situ tests at shallow penetrations while resting on the ocean bottom. A logical extension of this technology would be to equip a submersible to drill and sample at significant depth; no such equipment has yet been built, although a design has existed since 1996.

4.3 SOIL TESTS

The wet rotary process, which is commonly used with onshore soil boring, is also used for offshore soil testing. In this respect, there is little difference between onshore and offshore practice except for some details in the way the objective is accomplished. However, the different offshore environmental conditions have necessitated changes in soil-sampling procedures. An understanding of onshore sampling techniques, tools and results will aid in understanding the required alterations, the concessions made and the advantages of the offshore procedures.

Most onshore sampling is done by what is termed conventional means, which are also used in offshore sampling. At the desired sampling depth, the drill pipe is pulled from the hole and the drill bit is replaced by a soil sampler. The sampler is run to the bottom on the drill pipe. After the sample has been taken, the drill pipe is again pulled to retrieve the sample, then the soil sampler is replaced by the bit and the drill pipe is run back into the hole to advance the boring to the next sampling interval.

The cost of soil investigation offshore is very high, as is the case with preliminary engineering. There are many sources of useful information and data about soil characteristics as they relate to the geologic information for the platform location, drilling records, acoustic data or weather and sea state.

The most common method of sampling cohesive soils (clay) is to push a thin-walled tube into undisturbed material below the bottom of the drilled hole; this procedure is standardized in ASTM D-1587. Penetration of the tube is normally achieved by a rapid, continuous push from a pull-down system on the drilling rig. The sampling tube is vented to permit escape of fluid as soil enters.

The tube wall must be clean and smooth to minimize friction. The tube diameter may range from 50 to 100 mm and the wall thickness may be 1.5–3 mm. The lower end of the tube is sharpened and is swaged to give a slight inside clearance. The tube length to diameter ratio is usually between about 10 and 15. The length of sample as compared to the length of push is ideally near 100%. All of these factors have a bearing on obtaining an undisturbed sample and, therefore, upon certain soil parameters that may be determined from tests on samples.

As mentioned above, for granular soils (sands), the standard penetration test (SPT) has been widely used in onshore projects; the procedure is described in ASTM D-1586. The SPT is a well-established and unsophisticated method, developed in the United States around 1925, that has undergone refinements with respect to equipment and testing procedure. The testing procedure varies in different parts of the world. Therefore, standardization of SPT was essential in order to facilitate the comparison of results from different investigations. The equipment is simple, relatively inexpensive and rugged. Another advantage is that representative but disturbed soil samples are obtained. The reliability of the method and the accuracy of the result depend largely on the experience and care of the engineer on site.

A split-barrel sampler is driven from the bottom of a prebored hole into the soil by means of a 63.5 kg hammer, dropped freely from a height of 0.76 m. The diameter of the prebored hole varies normally between 60 and 200 mm. If the hole does not stay open by itself, casing or drilling mud should be used. The sampler is first driven to a depth of 150 mm below the bottom of the prebored hole, then the number of blows required to drive the sampler another 300 mm into the soil, the so-called N30 count, is recorded. The rods used for driving the sampler should have sufficient stiffness. Normally, when sampling is carried out to depths greater than around 15 m, 54 mm rods are used.

The quality of test results depends on several factors, such as energy actually delivered to the head of the drill rod, the dynamic properties of the drill rod, the method of drilling and borehole stabilization. The energy actually delivered can vary between 50% and 80% of the theoretical free-fall energy. Therefore, correction factors for rod energy (60%) are commonly used (Seed and De Alba, 1986). The SPT can be difficult to perform in loose sands and silts below ground-water level (typical for land reclamation projects), as the borehole can collapse and disturb the soil to be tested. The following factors can affect the test results: nature of the drilling fluid in the borehole, diameter of the borehole, the configuration of the sampling spoon and the frequency of delivery of the hammer blows.

Therefore, it should be noted that drilling and stabilization of the borehole must be carried out with care. The measured N-value (blows/0.3 m) is the so-called standard penetration resistance of the soil. The penetration resistance is influenced by the stress conditions at the depth of the test. Based on settlement observations of footings, Peck et al. (1974) proposed the following relationship for correction of confinement pressure. The measured N-value is to be multiplied by a correction factor C_N to obtain a reference value, N_1, corresponding to an effective overburden stress of 1 t/ft^2 (approximately 107 kPa)

$$N_1 = N \cdot C_N \tag{4.1}$$

where C_N is a stress correction factor or standard penetration test and p' is the effective vertical overburden pressure (see Equation 4.2).

$$C_N = 0.77 \cdot \log 10(20/p') \tag{4.2}$$

Preliminary knowledge of the geologic conditions in an offshore area will aid in selection of exploration tools best suited to the job. However, at the present time, sub-bottom profiles are the main source of information about soils and rock that may be encountered.

In many areas of offshore oilfield activity, previous investigations have been made. Although information may be for a location miles away from the desired location, it can provide insight into foundation conditions. Furthermore, general information about the character of materials within the probable depth of a soil boring may sometimes be obtained from well records.

Geophysical exploration precedes drilling activity in offshore areas. The depth of a geologic structure important from a production standpoint is generally well below the zone of interest for foundations; records from seismic exploration are probably of very limited value for foundation purposes. However, shallow-penetration surveys are frequently available and do provide useful data on soil stratigraphy.

Many in-situ testing devices have been developed and are in use to determine soil properties and conditions in the ground. Among them are the cone penetrometer, commonly known as the "Dutch" cone, the pressure meter and the vane shear device. None of these provides a sample for other tests; if samples are required, sampling must be done between the in-situ tests or in a companion boring.

The Dutch cone was developed principally to define granular soil strata that would serve to support point-bearing piles. The cone is modified to measure both side friction and point-bearing resistance. Experienced operators claim to be able to identify soil types by the ratios of these resistances. Modern cone equipment has been used successfully only to shallow penetrations at sea. Remotely operated sea floor equipment presently being developed in Europe may have a substantially greater depth capability.

The pressure meter is designed to determine soil behavior by using an expanding pressure cell to measure load-deformation characteristics. The pressure cell is either driven or pushed into undisturbed soil to begin testing or is installed in the bottom of a drilled hole of carefully controlled dimensions. Readings, interpreted in terms of modulus of deformation and limit pressure, are used to determine bearing capacity, settlement and other data. This tool has seen little success in connection with deep exploration for pile foundations at sea.

The remote vane, which is the vane shear device, has been used for years to measure in-place shear strengths of soils; however, soil shear failure was produced through torque applied to rods extending to the surface.

Remotely operated vane equipment has been developed in recent years and has been used successfully in a number of offshore investigations in conjunction with wire-line sampling operations.

Tests can be conducted fairly quickly and economically in material having a shear strength up to about 192 kN/m^2. The present vane probe is powered electrically through a conductor cable extending to the surface but has no connection to the drill pipe. Continuous torque readings throughout a test are monitored on a readout at the surface. Development is under way on another device that has no electrical connection to the surface; readings will be stored in the down-hole device and will be read when it is brought to the surface after a test.

Results of the remote vane tests show strengths consistently higher than can be measured on samples recovered. This is to be expected because of disturbance created by sampling and by sample handling and because of the pressure relief experienced by samples brought to the surface. The vane shear device probably better measures the true in-situ shear strength of a cohesive soil than any other device.

Until the principal emphasis shifts from pile-supported platforms, which is not expected any time soon, and until offshore activity extends into much deeper water, the present methods and techniques of offshore soil investigation will continue to be employed. Boat drilling and wire-line sampling methods offer the most versatile and economical means of investigation. Use of remote devices, such as the vane probe and cone penetrometer, in drilled holes will increase and will add much to our knowledge of in-situ soil properties. Dynamic positioning of surface drilling vessels or support vessels for other operations will become more important as operations move into deeper waters beyond the continental shelf. There will be much more activity in the development of remotely operated sea floor drilling and sampling devices. Manned submersibles should also play an increasingly important role with further development.

4.4 IN-SITU TESTING

As mentioned above, offshore soil investigation is much more expensive than onshore investigation. Therefore, the NORSOK (2004) report recommended that all in-situ test equipment systems prescribed should be checked for functionality during mobilization of equipment on board the survey vessel. The functionality checks should include, but not be limited to, the signal response of sensors, the data-acquisition system and a wet test of essential subsea equipment.

In-situ equipment with electronic transmission should be designed to sustain the water pressures expected in the field.

During testing, zero readings of all sensors should be recorded before and after each test. The specifications for in-situ test equipment are made for the most commonly used tests. For other in-situ testing, equipment specifications and procedures should be established prior to mobilization.

Records of experience with the use of the equipment, routines and procedures for interpretation of measurements for assessment of soil parameters should be documented and should be made available on request.

In most cases, the in-situ test tool can be inserted into the soil from the seabed to either a pre-set depth or to refusal due to limitation in pushing force, capacity of load sensor(s) or other factors, which is called *seabed mode,* or from the bottom of the borehole to either a pre-set depth or to refusal due to limitation in pushing force, capacity of load sensor(s) or other factors, which is called *drilling mode*.

It is very important that the rig does not interact with the seabed soil in such a way that the result of the in-situ test is influenced by its presence. Ideally, the footprint and weight of the seabed rig should be such that the topmost soil where the in-situ test is carried out is not disturbed or influenced by stresses from the seabed rig.

The effect of the seabed rig on the in-situ test results may be reduced by careful consideration of whether

- the contact area (footprint) is ring- or rectangle/square-shaped, with an open space where the in-situ tool is pushed into the seabed
- skirts are used on the periphery of the rig to transfer forces to stiffer soil
- the weight of the rig is balanced so that it is no larger than what is required to provide sufficient reaction force.

In order to be able to evaluate any effects of the rig on the in-situ measurements, any penetration into the sea bottom should be monitored, using a TV camera mounted on the rig.

Use of a remotely operated vehicle (ROV) is a possible alternative for performing shallow testing, depending on soil type and required penetration depth.

The drilling of the borehole should be carried out in such a way that the disturbance to the soil below the drill bit is minimized. In order to avoid any disturbed zone below the drill bit, the in-situ test tool should penetrate at least 1 m if soil strength and density allow. The disturbed zone can be assessed by continuous CPT/CPTU (see next section) penetrated to approximately 3 m below the drill bit.

4.4.1 Cone Penetration Test (CPT)

The CPT (Dutch cone) is modified to piezocone penetration depth. The piezocone test (CPT testing that also gathers piezometer data, called CPTU testing) is a CPT with additional measurement of the pore water pressure at one or more locations (U_1, U_2 and U_3) on the penetrometer surface, as shown in Figure 4.1.

The CPT is an in-situ testing method for determining the geotechnical engineering properties of soils and for delineating soil stratigraphy. It was initially developed in the 1950s at the Dutch Laboratory for Soil Mechanics in Delft to investigate soft soils (which is why it has also been called the Dutch cone test). Today, the CPT is one of the most used and accepted in-situ test methods for soil investigation world-wide.

The test method consists of pushing an instrumented cone, with the tip facing down, into the ground at a controlled rate, usually 20 mm/s. The resolution of the CPT in delineating stratigraphic layers is related to the size of the cone tip, with typical cone tips having a cross-sectional area of either 1000 or 1500 mm², corresponding to diameters of 36 and 44 mm.

American Society of Testing Materials (ASTM) presents the apparatus and test procedure for the CPT and the measurement of q_c. In particular, ISO (2005) prescribes cones with a base area in the range of 500 mm^2 to 2000 mm^2 and a penetration rate of 20 ±5 mm/s.

It is noted that the CPT-based design methods were established for cone resistance values up to 100 MPa. Caution should be used when applying the methods to sands with higher resistances.

The early applications of CPT mainly determined soil bearing capacity. The original cone penetrometers involved simple mechanical measurements of the total penetration resistance to pushing a tool with a conical tip into the soil. Different methods were employed to separate the total measured resistance into components generated by the conical tip due to the tip friction and friction generated by the rod string.

In the 1960s, a friction sleeve was added to quantify the friction component and to aid in determining soil cohesive strength (Begemann, 1965). Electronic

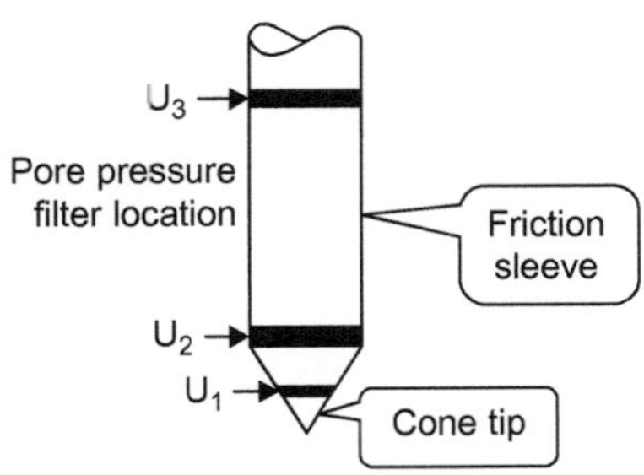

FIGURE 4.1 Sketch for cone penetration test (CPT).

measurements began in 1948 and improved further in the early 1970s (de Reister, 1971).

Most modern electronic CPT cones now also employ a pressure transducer with a filter to gather pore water pressure data. The filter is usually located either on the cone tip (the so-called U_1 position), immediately behind the cone tip in the most common U_2 position or behind the friction sleeve, which is the U_3 position. Pore water pressure data aids in stratigraphy and is primarily used to correct tip friction values. CPT and CPTU testing equipment generally advance the cone using hydraulic rams mounted on a heavily ballasted vehicle or using screwed-in anchors as a counterforce. One advantage of CPT over the SPT is a more continuous profile of soil parameters, with CPTU data recorded typically at 20 mm intervals.

CPT for geotechnical applications was standardized in 1986 by ASTM Standard D-3441 (ASTM, 2004). ISSMGE provides international standards for CPT and CPTU. Later, ASTM standards addressed the use of CPT for environmental site characterization and groundwater monitoring activities. Particularly for geotechnical soil investigations, CPT is gaining popularity over SPT because of its increased accuracy, speed of deployment, more continuous soil profile and reduced cost in relation to other soil-testing methods. The ability to advance additional in-situ testing tools using the CPT direct-push drilling rig, including the seismic tools, is accelerating this process.

The arrangement for the CPT on the drilling vessel is shown in Figure 4.2.

After performing the geotechnical investigations, there will be several vertical CPT profiles; for example, one per platform leg. Therefore, it is recommended that at least two approaches be calculated. The capacity should be first, based on the combined averaged resistance capacity q_c profile and then based on individual q_c profiles. After that, experience is important in providing judgment for selecting the most appropriate resistance capacity profile and associated final axial capacity.

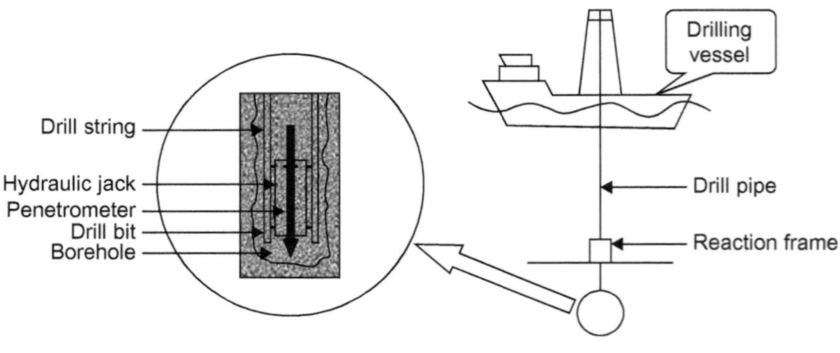

FIGURE 4.2 Arrangement for cone penetrometer tests.

Equipment Requirements

The geometry of the cone penetrometer, with tip, sleeve and pore pressure filters, should follow IRTP (1999), from which the following is extracted: 1) the penetrometer tip and adjoining rods should have the same diameter for at least 400 mm behind the tip, and 2) the cone should have a nominal cross-section area, Ac, of 1000 mm^2, with 35.3 mm $\leq d_c \leq$ 36.0 mm and 24.0 mm $\leq h_c \leq$ 31.2 mm, where d_c is the cone diameter and h_c is the height of the conical part.

According to the IRTP (1999), cone penetrometers with a diameter between 25 mm ($Ac = 500$ mm^2) and 50 mm ($Ac = 2000$ mm^2) are permitted for special purposes, without the application of correction factors. The recommended geometry and tolerances given for the 1000 mm^2 cone penetrometer should be adjusted proportionally to the diameter.

CPT Testing Procedure

The testing procedure should comply with the IRTP (1999) as follows:

- The nominal rate of penetration should be 20 mm/s with an accuracy of ±5 mm/s.
- The length of each stroke should be as long as possible, with due consideration of the mechanical and strength limitations of the equipment, noting that continuous penetration is preferred.
- Readings of all channels should be taken at least once per second (for every 20 mm of penetration).
- For quality control, zero readings should be recorded before and after each test.

For CPTU testing, the filter stones should be fully saturated and the pore pressure measurement system should give an instantaneous response to changes in pressure. Documented procedure for saturation of filter stones should be available.

Pore pressure dissipation tests should be carried out to at least 50% consolidation (i.e., halfway between pore pressure after stopping penetration of the cone and the assessed in-situ pore pressure). In cases where this may imply long measurement periods, the client and the contractor should agree on the maximum test duration, depending on the importance of the results.

The sampling rate during a dissipation test should be at least:

- During the first minute, 2 times each second
- Between the first minute and tenth minute, once each second
- Between 10th and 100th minutes, once every 2 seconds
- After 100 minutes, once every 5 seconds

The data-acquisition system should be such that overall accuracy is maintained.

The resolution of the measured results should be within 2% of the measured value.

During actual testing with wire-line tools, the data-acquisition system should allow for real-time inspection of the measured results, in both digital and graphical form.

The measured results should be stored digitally for subsequent processing.

Calibration Requirements

For each cone penetrometer, an accurate calibration should be made of the area ratios of the cone and the friction sleeve, as given in IRTP (1999). These values are characteristics for each cone penetrometer and should be documented in each field report, because they are very important for data reduction. The calibrations should be checked at least once a year.

Calibrations of each sensor should be made prior to each project and at least every third month or after about 100 soundings. If the load cells have been loaded close to maximum capacity, new calibrations should be carried out.

During field work, regular function checks of the cone penetrometer and measuring system should be carried out. As discussed above, offshore soil investigation is very expensive, so it is important to follow the calibration procedure for any tools before use. The owner's representative should check this with the contractor.

A calibration certificate for each cone penetrometer should be presented before mobilization.

For each cone or CPTU representative of each type (i.e., of the same load cell capacity), the following temperature calibration should be documented to have been done at least once:

- Variation of response to zero load for temperatures varying from 0°C to 40°C
- Calibration factor of each sensor at a temperature of +5°C; the method used to obtain the calibration factors should be explained.

The use of accuracy classes, as required in IRTP (1999), should be adopted. Equipment and procedures to be used should be selected according to the required accuracy class given in Table 4.1. These precautions are very critical and it is important that they be considered and monitored by the project quality team.

If all possible sources of errors are added, the accuracy of the recordings should be better than the largest of the values given in Table 4.1. The relative or percent accuracy applies to the measured value and not the measuring range or sensor capacity.

Class 1 is meant for situations where the results will be used for precise evaluation of stratification and soil type as well as parameter interpretation in profiles including soft or loose soils.

For Class 3, the results should only be used for stratification and for parameter evaluation in stiff or dense soils. Class 2 may be considered more appropriate for stiff clays and sands.

TABLE 4.1 Accuracy Classes

Class	Measured Parameter	Minimum Allowable Accuracy	Maximum Length between Measurements (mm)
1	Cone resistance	50 kPa or 3%	20
	Sleeve friction	10 kPa or 10%	
	Pore pressure	5 kPa or 2%	
	Inclination	2 degrees	
	Penetration	0.1 m or 1%	
2	Cone resistance	200 kPa or 3%	20
	Sleeve friction	25 kPa or 15%	
	Pore pressure	25 kPa or 3%	
	Inclination	2 degrees	
	Penetration	0.2 m or 2%	
3	Cone resistance	400 kPa or 5%	50
	Sleeve friction	50 kPa or 15%	
	Pore pressure	50 kPa or 5%	
	Inclination	5 degrees	
	Penetration	0.2 m or 2%	

During the sounding, zero readings should be taken with the probe temperature as close as possible to the ground temperature, and all sensors and other electronic components in the data-acquisition system should be temperature stabilized.

CPT Results

The reporting of results from CPTs should comply with IRTP (1999).

For each cone test, the following information should be reported offshore after each test and in the field report (either in tabular form or in the CPT profiles):

- location
- test number
- coordinates of test location
- date of test
- cone serial number
- cone geometry and dimensions, with position and dimensions of filter stone

- capacity of sensors (tip, pore pressure and sleeve friction)
- calibration factors used
- zero readings of all sensors before and after each test, either at sea floor or bottom of borehole
- observed wear or damage on tip or sleeve
- penetration rate
- any irregularities during testing
- theoretical effective (net) area ratio of tip and friction sleeve
- water depth to sea floor during test
- corrections due to tidal variations, if any
- observed sinking in of the frame
- the inclination of the cone penetrometer to vertical axis, for a maximum penetration depth spacing of 1 m (for seabed testing only).

The measured results (in engineering units) should be presented in digital form, as:

- depth of penetration (in m)
- cone tip resistance (in Mpa)
- pore pressure(s) (in Mpa)
- sleeve friction (in kPa)
- total thrust during test (in kN)

The graphical representation of the results from CPTs in the field (offshore) should be presented. If not otherwise agreed upon, the depth scale should be 1 m (field) = 10 mm (plot).

The zero reference for seabed CPTs should be the sea bottom, and for down-hole CPT, the bottom of the borehole. The selection of the scale for presenting the measured cone resistance, pore pressure and friction should be reasonable to suit the soil conditions.

In addition to the measured CPT/CPTU values, the following corrected or derived parameters should be presented:

- corrected cone penetration resistance, $q_t = q_c + (1 - a)u$ (with u measured behind the cone)
- corrected sleeve friction, f_t, only if pore pressures have been measured at both ends of the friction sleeve; friction ratio, $R_f = (f_s/q_t) \times 100\%$ or, if relevant, $(f_t/q_t) \times 100\%$, where f_t is the sleeve friction corrected for pore pressure effects, noting that this requires measurement of the pore pressure at both ends of the friction sleeve
- pore pressure ratio, with pore pressure recorded behind the cone

For down-hole tests, it is important that the derived parameters are corrected to be referenced to sea.

The output data are presented in a graph, as shown in Figure 4.3.

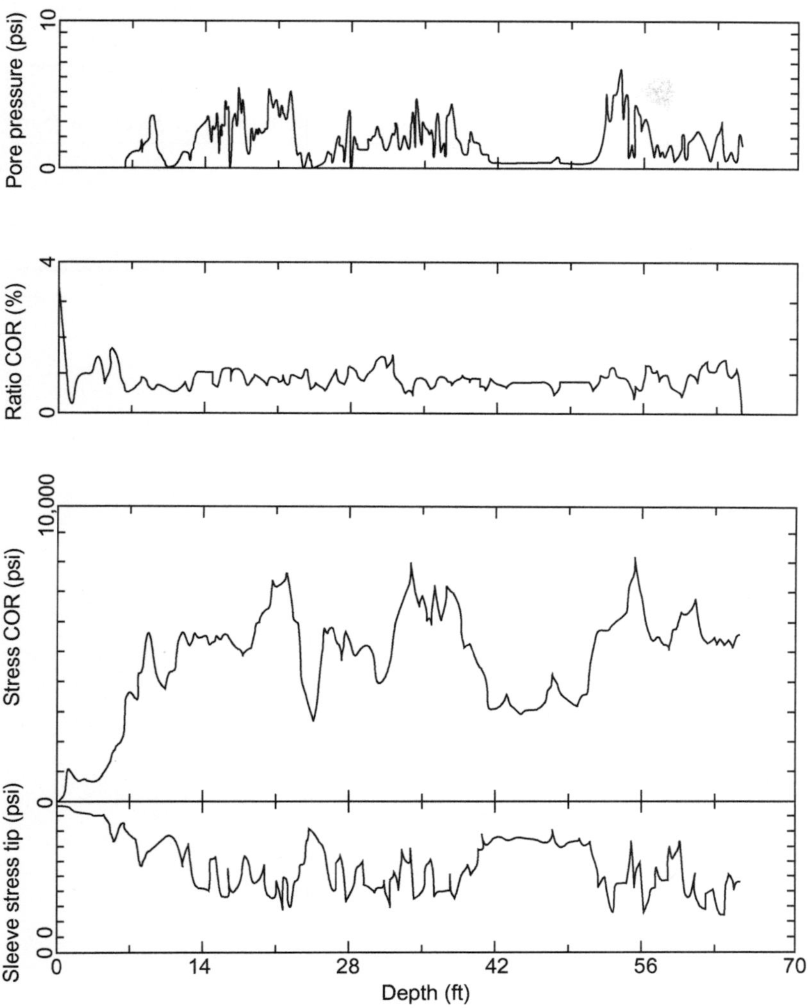

FIGURE 4.3 CPTU output results.

4.4.2 Field Vane Test

Vane blades should be rectangular, as defined in ASTM D-2573-01 or BS 5930:1999.

Shear strengths are given in Table 4.2, based on the geometrical dimensions of the vane blades. The vane test in the borehole offshore is presented in Figure 4.4.

TABLE 4.2 Measuring Range with Different Vane Blade Dimensions

Measuring Range of s_u	Vane Height (mm)	Vane Width (mm)	Vane Blade Thickness (mm)
0–50	130	65	2
30–100	110	55	2
80–250	80	40	2

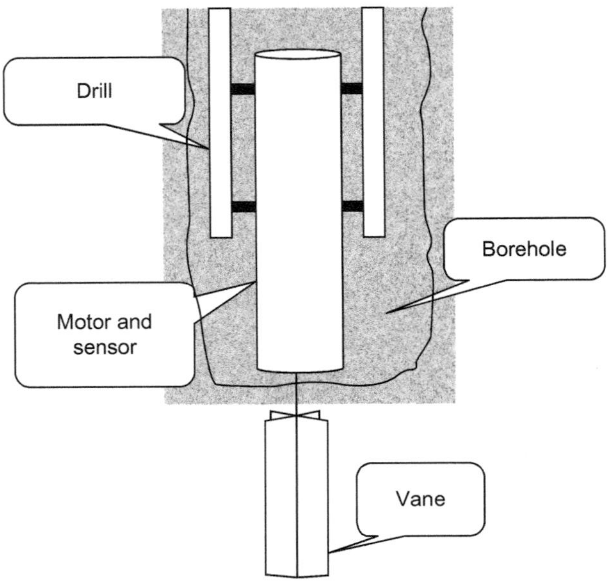

FIGURE 4.4 Sketch for vane test.

Testing Procedure

As a normal procedure before a vane test is started, the vane blade should be pushed at least 0.5 m below the bottom of the borehole. The pushing rate should be less than 25 mm/s. The time from the instant when the desired test depth has been reached to the beginning of the test (waiting time) should be 2 min to 5 min.

The rotation of the vane should be smooth and for the initial test (undisturbed) it should be 6 degrees to 12 degrees per minute.

To measure remolded shear strength, the vane should be rotated at least 10 times at a rate ≥4 RPM (rev/min) and until a constant torque over 45 degrees of continuous rotation has been reached. At the end of the rapid rotations, the remolded shear strength should be measured without delay, with a rotation rate equal to that used for intact shear strength.

It is also possible to do vane tests in the seabed mode. The depth intervals between tests should be at least 0.5 m.

The insertion method and test procedure to be used should be described, giving particular information about the method for insertion and penetration of vanes below the bottom of the borehole, possible rotation rates available and the method for providing torque and reaction.

The data-acquisition system should be such that the overall accuracy is maintained. Taking into account all sources of error, including the data-acquisition system, the uncertainty in the measured torque should not exceed the smaller of 5% of measured value or 2% of the maximum value of the measured torque of the layer under consideration.

The resolution of the measured result should be within 2% of the measured value.

During testing, the data-acquisition system should allow for real-time inspection of measured results in both digital and graphical form.

It is very important that, at least every year and before each project, the sensor for measuring the torque during vane testing should be calibrated. If the sensor is loaded close to its maximum or any damage is suspected, it should be checked and recalibrated. In addition, function checks should be carried out in the field.

For each vane test, the following information should be given:

- site area
- date of test
- operator
- boring number/test identification
- water depth at test location
- dimensions of vane
- depth below sea bottom to vane tip
- depth below bottom of borehole to vane tip
- rate of rotation
- the complete curve of torque versus rotation (degrees)
- time to failure
- any irregularities during testing
- formula used to calculate the vane undrained shear strength, s_{uv}, including the assumption made for shear stress distribution on ends of the vane.

4.5 SOIL PROPERTIES

In soil mechanics, the equilibrium and movement of soil bodies is studied, where soil is understood to be the weathered natural material in the upper layers (the upper 20 to 100 m) of the Earth's crust. This material may be gravel, sand, clay, peat, or some other soft and loose granular medium. The characteristics of these materials are quite different from those of artificial (man-made) materials, such as steel, concrete, etc. Man-made materials usually are much more consistent than soils and exhibit relatively simple, linear, mechanical behavior, at least if the deformations are not too large.

The mechanical properties of soils are usually strongly nonlinear, with the material exhibiting irreversible plastic deformations when loaded and unloaded, even at low stress levels, and often showing anisotropic behavior, creep and such typical effects as a volume change during shear. This mechanical behavior of soil is also difficult to predict, because the structure of the soil may be highly inhomogeneous, as a result of its geological history, and it is often not possible to determine the detailed behavior of the soil by tests in the laboratory or in situ.

The behavior of soils may be further complicated by the presence of water in the pores. This relatively stiff fluid in the pores may prevent or retard volume deformations.

For all these reasons, the characterization of the mechanical behavior of soils is often done in a schematic way only, and its form is adapted to the particular type of problem under consideration.

The soil layers have the dominating effect, whereas the settlement of an embankment is mainly governed by the deformation properties of the soil, including creep.

Thus, in soil mechanics, the range of applicability of a certain parameter is often restricted to a limited class of problems. Many properties cannot be used outside their intended field of application. Nevertheless, various properties may all derive from such common basic phenomena as interparticle friction, or the structure of a granular medium, so that there may well exist good correlations between certain properties. In this chapter, some of these properties are reviewed, and some correlations are discussed. It should be noted that, in engineering practice, nothing can beat the results of experimental determination of the soil parameters in situ or in the laboratory. A correlation may at best give a first estimate of the order of magnitude of a parameter.

Globally, it is increasingly evident that there is no unique combination of laboratory and in-situ testing programs that is likely to provide all the appropriate parameters for design of foundations in carbonate soils, but some laboratory and in-situ tests have been found useful. At a minimum, a laboratory testing program for carbonate soils should determine:

- Material composition, particularly carbonate content
- Material origin, to differentiate between skeletal and nonskeletal sediments
- Grain characteristics, such as particle angularity, porosity and initial void ratio
- Compressibility of the material
- Soil strength parameters, particularly friction angle
- Formation cementation, at least in a qualitative sense.

For site characterization, maximum use of past local experience is important, particularly in the selection of an appropriate in-situ program. In new, unexplored territories where the presence of carbonate soils is suspected, selection of an in-situ test program should draw upon any experience with carbonate soils where geographical and environmental conditions are similar.

4.5.1 Strength

Soils usually cannot transfer stresses beyond a certain limit. This is called the strength of the soil. The shear strength of soils is usually expressed by Coulomb's relation between the maximum shear stress and the effective normal stress:

$$\tau_{\max} = c + \sigma' \tan \phi \tag{4.3}$$

where c is the cohesion and ϕ is the friction angle. For sands, c is usually negligible, so that ϕ is the only strength parameter.

For clays it is often most relevant to consider the strength in undrained conditions, during which the effective stress remains constant. The undrained shear strength is usually denoted by s_u, and it is often considered irrelevant to what degree this is to be attributed to cohesion or friction. The shear strength parameters can be determined in the laboratory, for instance by tri-axial testing.

A simple and useful in-situ test is the cone penetration test (CPT), in which a cone is pushed into the ground using hydraulic pressure equipment, while recording the stress at the tip of the cone, and often also the friction along the lower part of the shaft. The test is used in The Netherlands as a model test for a pile foundation, and the results are used directly to determine the bearing capacity of end-bearing piles, using simple scale rules.

The CPT can also be used to estimate the strength of a soil, however, by using certain correlations. For a penetration test in sand, for instance, the bearing capacity of the cone, q_c, is, according to Brinch Hansen's formula,

$$q_c = s_q N_q \sigma'_v = s_q N_q \gamma' z \tag{4.4}$$

where s_q is a shape factor to express effective weight of the overburden, for which one may use the formula:

$$s_q = 1 + \sin \phi \tag{4.5}$$

z is the depth, and N_q is a dimensionless constant for which theoretical analysis has given the value as in the following equation:

$$N_q = \frac{1 + \sin \phi}{1 - \sin \phi} \exp(\pi \tan \phi) \tag{4.6}$$

The predicted cone resistances (q_c) for various types of sand at a depth of 10 m and 20 m are shown in Table 4.3, assuming that $\gamma = 10$ kN/m^3. The values

TABLE 4.3 Guidance for Cone Resistance in Sand

Soil type	ϕ	N_q	q_c (z =10 m) (MPa)	q_c (z = 20 m)
Loose sand	30°	18.4	2.8	5.5
Medium dense sand	35°	33.3	5.2	10.5
Very dense sand	40°	64.2	10.5	21.1

in the table are indeed often observed for sand layers at these depths. They may also be used inversely: if a certain cone resistance is observed, it is indicative for a certain type of material.

For a CPT in clay soil, the Brinch Hansen formula can be used to correlate the CPT result to the undrained shear strength.

The general Brinch Hansen formula is

$$q_c = s_c N_c c + s_q N_q \sigma'_v \tag{4.7}$$

Because the test is performed very quickly, the soil behavior can be considered to be undrained. The values for the coefficients can then be taken as $N_c = 5.14$, $N_q = 1$, $s_c = 1.3$, $s_q = 1$. Equation (4.7) now reduces to

$$q_c = 6.7 s_u + \sigma'_v \tag{4.8}$$

where the cohesion c has been interpreted as the undrained shear strength s_u.

The undrained shear strength of normally consolidated clays depends upon the vertical stress σ'_v. A relationship that is often used is the correlation proposed by Ladd,

$$s_u = 0.22 \sigma'_v \tag{4.9}$$

Substitution of this result into Equation (4.8) gives

$$q_c \approx 11 s_u \tag{4.10}$$

For a soft clay, with $s_u = 20$ kPa, the order of magnitude of the cone resistance would be 220 kPa $\approx$ 0.2 MPa. Such values are indeed often observed. Again, they may also be used to estimate the undrained shear strength from CPT data.

Table 4.4 provides guidance to cohesive soil characteristics based on the rule of thumb.

The relative density for cohesionless soil can be predicted using Table 4.5 as a guideline.

Tables 4.6 and 4.7 present terminology for soil structure and characteristics and the common names of various soil types.

TABLE 4.4 Consistency of Cohesive Soil

Consistency	Unconfined Compressive Strength (tons/ft^2)	Rule-of-Thumb Test
Very soft	0–0.25	Core (height twice diameter) sags under own weight
Soft	0.25–0.5	Can be pinched in two by pressing between thumb and finger
Firm	0.5–1.0	Can be imprinted easily with finger
Stiff	1.0–2.0	Can be imprinted with considerable pressure from fingers
Very stiff	2.0–4.0	Barely can be imprinted by pressure from fingers
Hard	>4.0	Cannot be imprinted by fingers

TABLE 4.5 Degree of Compactness for Cohesionless Soil

Degree of Compactness	Relative Density (%)
Very loose	0–15
Loose	15–35
Medium dense	35–65
Dense	65–85
Very dense	85–100

TABLE 4.6 Terminology for Soil Structure Characteristics

Slickensided	Cut by old fracture planes that are slick and glossy in appearance and constitute planes of weakness; generally of random orientation.
Fissured	Containing old shrinkage cracks that are frequently filled with fine sand or silt; generally predominantly vertical.
Laminated	Composed of thin layers of varying color and texture.
Interbedded	Composed of alternate layers of different soil types.
Calcareous	Containing appreciable quantities of calcium carbonate.
Well graded	Having a wide range of particle sizes and substantial amounts of all intermediate particle sizes.
Poorly graded	Predominantly of one particle size, or having a range of sizes with some intermediate sizes missing.

TABLE 4.7 Common Soil Types

Topsoil	Surface formation, generally black or gray due to organic content or degree of weathering; the top portion of the soil profile that supports vegetation.
Hard pan	Hard, cemented conglomerate that will not soften when wet.
Fill	Any man-made soil deposit.
Caliche	Layers of soil cemented together by calcium carbonate deposited by evaporation of ascending or descending ground waters.
Adobe	Heavy-textured, light-colored, alluvial clay soils occurring in the south-western part of the United States.
Gumbo	Fat clays with little sand that, when saturated with water, are impervious and have a waxy or soapy appearance and feel.
Muck	Highly organic soil of very soft consistency.

TABLE 4.8 Approximate Property Values for Different Soil Types

Soil Type	Undrained Shear Strength (S_u, in kPa)	Effective Cohesion (c', in kPa)	Friction Angle (f', in degrees)	Saturated Density (D_s, in t/m^3)	Voids Ratio
Soft to firm clay	10–50	5–10	19–24	1.4–1.8	
Stiff clay	50–100	10–20	22–29	1.8–1.9	
Very stiff to hard clay	100–400	20–50	27–31	1.9–2.2	
Silt	10–50	—	27–35	1.7–2.3	1.1–0.3
Loose sand	—	—	29–30	1.7–1.8	1.1–0.8
Medium dense sand	—	—	30–40	1.8–2.0	0.8–0.5
Dense sand	—	—	35–45	2.0–2.3	0.5–0.2
Gravel	—	—	35–55	1.7–2.4	1.1–2.2

The approximate soil parameters for different types of soil are illustrated in Table 4.8.

4.5.2 Soil Characterization

A key part of developing realistic analytical models to evaluate cyclic loading effects on piles is the characterization of soil-pile interaction behavior.

High-quality in-situ, laboratory, and model-prototype pile-load tests are essential in such characterizations. In developing soil-pile interaction characterizations, it is important that pile installation and pile-loading conditions be integrated into the testing programs.

According to McClelland et al. (1971), in-situ tests, such as vane shear, cone penetrometer, pressure meter and other tests, can provide important insights into in-place soil behavior and stress-strain properties.

Both low- and high-amplitude stress-strain properties can be developed. Long-term static and creep loadings or short-term dynamic, impulsive and cyclic loadings sometimes can be simulated with in-situ testing equipment.

Laboratory tests on representative soil samples permit a wide variety of stress-strain conditions to be simulated and evaluated. Soil samples can be modified to simulate pile-installation effects, such as remolding and reconsolidating, to estimate in-situ stresses. The samples can be subjected to different boundary conditions, such as tri-axial, simple shear and interface shear, and to different levels of sustained and cyclic shear time histories to simulate in-place loading conditions.

Another important source of data to develop soil characterizations for cyclic loading analyses are tests on model and prototype piles. Based on Bogard et al. (1985) and Karlsrud and Haugen (1985), model piles can be highly instrumented, and repeated tests can be performed in soils and for a variety of loadings.

Geometric scale, time scale and other modeling effects should be carefully considered in applying results from model tests to prototype behavior analyses. As discussed by Pelletier and Doyle (1982) and Arup et al. (1986), the data from load tests on prototype piles are useful for calibrating analytical models.

Such tests, even if not highly instrumented, can provide data to guide development of analytical models. These tests can also provide data to verify results of soil characterizations and analytical models.

Prototype pile-load testing, coupled with in-situ and laboratory soil testing and realistic analytical models, can provide an essential framework for making realistic evaluations of the responses of piles to cyclic axial loadings.

The foundation should be designed to carry static, cyclic and transient loads without excessive deformations or vibrations in the platform. Special attention should be given to the effects of cyclic and transient loading on the strength of the supporting soils as well as on the structural response of piles.

It is very important to consider the possibility of movement of the sea floor against the foundation members, and the forces caused by such movements, if anticipated, should be considered in the design.

4.6 PILE FOUNDATIONS

Offshore structure platforms commonly use open driven piles. These piles are usually driven into the seabed with impact hammers, which use steam, diesel fuel or hydraulic power as the source of energy. Therefore, the pile wall thickness should be adequate to resist axial and lateral loads as well as the stresses during pile driving.

According to Smith (1962), it is possible to predict approximately the stresses during pile driving using the principles of one-dimensional elastic stress wave transmission by carefully selecting the parameters that govern the behavior of soil, pile, cushions, capblock and hammer.

The above approach may also be used to optimize the pile hammer cushion and capblock using the computer analysis commonly known as wave equation analyses. The design penetration of driven piles should be determined, rather than correlation of pile capacity with the number of blows required to drive the pile a certain distance into the seabed.

If a pile stops before it reaches design penetration, one or more of the following actions can be taken:

1. Review of all aspects of hammer performance, possibly with the aid of hammer and pile-head instrumentation, which may identify problems that can be solved by improving hammer operation and maintenance or by the use of a more powerful hammer.
2. Reevaluation of design penetration by reconsideration of loads, deformations and required capacities of both individual piles and the foundation as a whole, which may identify reserve capacity available. An interpretation of driving records in conjunction with the instrumentation mentioned above may allow design soil parameters or stratification to be revised and pile capacity to be increased.
3. Modifications to piling procedures (usually the last course of action), which may include one of the following: The soil plug inside the pile is removed by jetting and air lifting or by drilling to reduce pile-driving resistance. If plug removal results in inadequate pile capacities, the removed soil plug should be replaced by a gravel grout or concrete plug having sufficient load-carrying capacity to replace that of the removed soil plug. Attention should be paid to plug/pile load-transfer characteristics. Note that plug removal may not be effective in some circumstances, particularly in cohesive soils.

Soil below the pile tip is removed by drilling an undersized hole or by lowering jetting equipment through the pile, which acts as the casing pipe for the operation. The effect on pile capacity of drilling an undersized hole is unpredictable unless there has been previous experience under similar conditions.

According to API RP2A (2007), jetting below the pile tip should be avoided because of the unpredictability of the results.

A first stage or outer pile is driven to a predetermined depth, the soil plug is removed, and a second stage or inner pile is driven inside the first-stage pile. In this case, grouting will be inserted in the annulus between the two piles to provide load transfer between the two piles and to develop composite action.

The piles of offshore structures are exposed to static and cyclic loading, and the loads will be axial and lateral. So all of these load effects should be considered in the pile design.

4.6.1 Pile Capacity for Axial Loads

On the basis of API RP2A (2007), the ultimate static axial capacity (*QI*) of an open-ended pipe pile in compression is given by the equation:

$$Q_t = Q_f + Q_s + \{\text{small values from } Q_{fi} \text{ or } Q_{sp}\} \tag{4.11}$$

$$Q_t = f \bullet A_s + A_a \bullet Q + \mathrm{A}_p . q \tag{4.12}$$

Where A_p is the area of pile or

$$Q_t = f \bullet A_s + A_a \bullet Q + fA_{si} \tag{4.13}$$

where Q_t = the total pile resistance, Q_f = total outside shaft resistance, Q_s = end-bearing capacity of the annulus, Q_{fi} = total inside shaft resistance, Q_{sp} = bearing capacity of the soil beneath the plug, f = unit skin friction capacity (in kPa), A_s = outside surface area of pile (in m^2), q = unit end-bearing capacity (in kPa), A_a = the area of the pile annulus (in m^2) and A_{si} = inside surface area of pile (in m^2).

In computing pile loading and capacity, the weight of the pile-soil plug system and hydrostatic uplift should be considered.

In determining the load capacity of a pile, consideration should be given to the relative deformations between the soil and the pile, as well as the compressibility of the soil-pile system.

Coyle and Reese (1966), Murff (1980) and Randolph (1983) discussed skin friction and assumed that the maximum skin friction along the pile and the maximum end bearing are mobilized simultaneously. However, the ultimate skin friction increments along the pile are not necessarily directly additive, nor is the ultimate end bearing necessarily additive with the ultimate skin friction. In some circumstances, this effect may result in the capacity being less than that given by Equation (4.11).

In such cases, a more explicit consideration of axial pile performance effects on pile capacity may be warranted. Pile sizing should be based on what experience has shown can be installed consistently, practically and economically under similar conditions with the installation equipment being used. Alternatives for possible remedial action in the event design objectives cannot be obtained during installation should also be investigated and defined prior to construction.

For the pile system, the pile-capacity factor of safety is defined in Table 4.9 according to API RP2A (2007). The allowable skin friction values on the pile section on the upper surface of the pile should be discounted in computing skin friction resistance, Q_f. The end-bearing area of a pilot hole, if drilled, should be discounted in computing total bearing area.

TABLE 4.9 Design Parameter Guide for Cohesionless Siliceous Soil (Based on API RP2A)

Soil Description	Soil Condition	Shaft Friction Factor	Limited Shaft Friction Values (in kPa)	End-Bearing Factor N_q	Limited Unit End-Bearing Values (in Mpa)
Sand	Very loose				
Sand	Loose				
Sand-silt	Loose				
Silt	Medium dense				
Silt	Dense				
Sand-silt	Medium dense	0.29	67	12	3
Sand	Medium dense	0.37	81	20	5
Sand-silt	Dense	0.37	81	20	5
Sand	Dense	0.46	96	40	10
Sand-silt	Very dense	0.46	96	40	10
Sand	Very dense	0.56	115	50	12

Skin Friction and End Bearing in Cohesive Soils

Traditionally, piles for an offshore structure platform are pipe pile. If the pipe pile penetrates cohesive soils, the shaft friction, f (in kPa), at any point along the pile may be calculated by:

$$f = \alpha c \tag{4.14}$$

where α = a dimensionless factor and c = undrained shear strength of the soil at the point in question.

The factor α can be computed by:

$$\begin{aligned} \alpha &= 0.5\psi - 0.5\psi \leq 1.0 \\ \alpha &= 0.5\psi - 0.25\psi > 1.0 \end{aligned} \tag{4.15}$$

with the constraint that $\alpha \leq 1.0$, where $\psi = c/p$, for the point in question and p' is the effective overburden pressure at the point in question (in kPa). For under-consolidated clays, clays with excess pore pressures undergoing active consolidation, α can usually be taken as 1.0.

The appropriate methods for determining the undrained shear strength, c, and effective overburden pressure, p', including the effects of various sampling and testing procedures, are important. As the number of pile-load tests is not enough in soils having c/p' ratios greater than three, Equation (4.15) should be applied with some engineering judgment for high c/p' values. The same engineering judgment should be applied for deep-penetrating piles in soils with high undrained shear strength, c, where the computed shaft frictions, f, using Equation 4.14 above, are generally higher than previously specified in API RP2A. In the case of very long piles, some reduction in pile capacity occurs, because the shaft friction may reduce to some lesser residual value on continued displacement.

For piles end bearing in cohesive soils, the unit end bearing, q (in kPa), may be computed by:

$$q = 9c \tag{4.16}$$

It is obvious that in open-driven piles the shaft friction, f, acts on both the inside and outside of the pile. The total resistance is the sum of the external shaft friction, the end bearing on the pile wall annulus and the total internal shaft friction or the end bearing of the plug, whichever is less.

If the pipe pile is considered to be plugged, the bearing pressure may be assumed to act over the whole cross-section of the pile. For unplugged piles, the bearing pressure will be calculated on the pile wall annulus only. Whether a pile is considered plugged or unplugged may be based on static calculations. For example, a pile could be driven in an unplugged condition but act plugged under static loading.

In some cases, piles are driven in undersized drilled holes, piles are jetted in place or (in some minor projects) the piles are drilled and grouted in place. In these situations, the soil disturbance resulting from installation will affect the shaft friction values. In general, f should not exceed values for driven piles; however, in some cases for drilled and grouted piles in overconsolidated clay, f may exceed these values.

In determining f for drilled and grouted piles, the strength of the soil-grout interface, including potential effects of drilling mud, should be considered. As discussed by Kraft and Lyons (1974), a further investigation and check should be made of the allowable bond stress between the pile steel and the grout.

The shaft friction values, f, in the cohesive layers should be as given in Equation (4.14). End-bearing values for piles tipped in cohesive layers with adjacent weaker layers may be as given in Equation (4.16), assuming that the pile achieves penetration of two to three pile diameters or more into the layer in question and the tip is approximately three pile diameters above the bottom of the layer, to avoid punch through.

Some modification in the end-bearing resistance may be necessary if these distances are not achieved.

Shaft Friction and End Bearing in Cohesionless Soils

A simple method for assessing pile capacity in cohesionless soils will be discussed. There are reliable methods for predicting pile capacity that are based on direct correlations of pile unit friction and end-bearing data with cone penetration test (CPT) results. The CPT-based methods have been discussed in depth recently and have been found to provide good results from a statistical point of view as the results coincide with pile-load test results. Although they are not required, they are, in principle, the preferred methods. CPT-based methods also cover a wider range of cohesionless soils. Because experience with CPT in offshore structures is limited, it should be applied only by competent engineers who are experienced in the interpretation of CPT data and who understand the limitations and reliability of the methods. A pile-driving instrumentation data system may be necessary to provide more accurate calculations.

For pipe piles in cohesionless soils, the unit shaft friction at a given depth, f, may be calculated by:

$$f = \beta p' \tag{4.17}$$

where β = a dimensionless shaft friction factor and p' = effective overburden pressure at the depth in question.

Table 4.10 may be used for selection of β values for open-ended pipe piles driven unplugged if other data are not available. Values of β for full-displacement piles (i.e., driven fully plugged or closed ended) may be assumed to be 25% higher than those given in Table 4.10. For long piles, f may not increase linearly with the overburden pressure, as implied by Equation (4.17). In such cases, it may be appropriate to limit f to the values given in Table 4.10.

For piles end bearing in cohesionless soils, the unit end bearing q may be calculated by:

$$q = Nqp' \tag{4.18}$$

where Nq = a dimensionless bearing capacity factor and p' = effective overburden pressure at the depth in question.

Recommended Nq values are also presented in Table 4.9.

TABLE 4.10 Pile-Capacity Factor of Safety in API RP2A (2007)

Load Condition	Factor of Safety
Design environmental conditions with appropriate drilling loads	1.5
Operating environmental conditions during drilling operations	2.0
Design environmental conditions with appropriate producing loads	1.5
Operating environmental conditions during producing operations	2.0
Design environmental conditions with minimum loads (for pullout)	1.5

In the case of long piles, q may not increase linearly with the overburden pressure, which is different from what is stated in Equation (4.18). In such cases, it may be appropriate to limit q to the values given in Table 4.9. For plugged piles, the unit end bearing q acts over the whole cross-section of the pile. For unplugged piles, q will be calculated by considering the area of the pile annulus only. In this case, additional resistance is offered by friction between soil plug and inner pile wall.

Whether a pile is considered to be plugged or unplugged may be based on static calculations using a unit skin friction on the soil plug equal to the outer skin friction. The design parameters in Table 4.9 are just a guide from API RP2A, and detailed information must be obtained from the CPT results, strength tests and other soil and pile response tests.

The soil condition is identified based on the relative density, as shown in Table 4.5.

Olson (1984) compared the load test data for piles in sand (obtained by measuring the axial load capacities for open steel piles) and the calculated capacity from API RP2A. Studies done in 2005 by Lehane indicate that variability in capacity predictions using the API calculation method may exceed those for piles in clay. The research also indicated that the calculation method is conservative for short offshore piles [short = piles less than 45 m (150 ft) long] in dense to very dense sands loaded in compression and may be unconservative in all other conditions. In unfamiliar situations, the designer may want to account for this uncertainty through a selection of conservative design parameters or by going toward higher factors of safety.

In cases of soil types that do not have characteristic values that fall within the ranges of soil density and the description given in Table 4.9, or for materials with unusually weak grains or compressible structure, Table 4.9 will be not suitable for use in selecting design parameters. A special laboratory or field test is required to obtain the design parameters, as in the case of very loose silts, or soils containing large amounts of mica or volcanic grains, or sands containing calcium carbonate, which are found extensively in many areas of the oceans. From a practical point of view, it suggests that driven piles in these types of soils may provide lower design strength parameters than are described in Table 4.9.

Drilled and grouted piles in carbonate sand may have significantly higher capacities than driven piles and have been used successfully in many areas with carbonate soils. The characteristics of carbonate sands are highly variable, and experience with the behavior of this type of soil in the location of the platform will be the main thing that governs the design parameters that will be used.

It is worth mentioning that pile capacity is improved in carbonate soils of high densities and higher quartz contents.

Note that cementation may increase end-bearing capacity but result in a loss of lateral pressure and a corresponding decrease in frictional capacity.

As discussed before, in cohesion soil, for piles driven in undersized drilled or jetted holes, the values of f and q should be determined by some method that considers the effect of soil disturbance due to installation, but they should not exceed values for driven piles. Except in unusual soil types, such as described above, the f and q values given in Table 4.9 may be used for drilled and grouted piles, with consideration given to the strength of the soil-grout interface.

The unit shaft friction values in cohesionless layers and the end-bearing values for piles tipped in cohesionless layers with adjacent layers of lower strength may also be taken from Table 4.9, provided that the pile achieves penetration of two to three diameters or more into the cohesionless layer, and the tip is at least three diameters above the bottom of the layer, to preclude punch through.

4.6.2 Foundation Size

In most cases, in the FEED engineering phase, the pile configuration will be defined based on past experience. During selection of the size of the pile foundation, the following should be considered: diameter, penetration, wall thickness, type of tip, spacing, number of piles, geometry, location, seabed restraint, material strength, installation method and other parameters as may be appropriate.

A number of different analysis procedures may be utilized to determine the requirements for a foundation. At a minimum, the procedure used should properly simulate the nonlinear response behavior of the soil and ensure load-deflection compatibility between the structure and the pile-soil system.

For deflections and rotations of individual piles, the total foundation system should be checked at all critical locations, which may include pile tops, points of contraflecture, mud line, etc. Deflections and rotations should not exceed serviceability limits, which would render the structure inadequate for its intended function.

Pile Penetration

The design pile penetration should be sufficient to develop adequate capacity to resist the maximum computed axial bearing and pullout loads with an appropriate factor of safety.

The ultimate pile capacities can be computed in accordance with previous sections, or by other methods that are supported by reliable comprehensive data. API RP2A (2007) defined the minimum factor of safety by dividing the ultimate pile capacity into the actual load, as shown in Table 4.10.

There are two safety factors in API RP2A that depend on considering the design environmental conditions with the 100-year storm wave effect and the operating environmental conditions, including the maximum wave height per year.

The provisions of API RP2A (2007) for sizing the foundation piles are based on an allowable stress (working stress) method, except for pile penetration. In this method, the foundation piles should conform to the requirements of specification and design. Any alternative method supported by sound engineering principles and empirical evidence may also be utilized. Such alternative methods include the limit state design approach or ultimate strength design of the total foundation system.

4.6.3 Axial Pile Performance

Static Load-Deflection Behavior

The static pile axial deflection should be compatible with the structural forces and deflection, so it should be within the service limit. An analytical method for determining axial pile performance is provided in Meyer et al. (1975).

This method makes use of axial pile shear transition versus local pile deflection (t-z) curves to model the axial support provided by the soil along the sides of the pile. An additional (Q-z) curve is used to model the tip and bearing versus the deflection response. (Methods for constructing t-z and Q-z curves are given below.) Pile response is affected by load directions, load types, load rates, loading sequence, installation technique, soil type, axial pile stiffness and other parameters.

Some of these effects for cohesive soils have been observed in both laboratory and field tests.

In cases where soils exhibit strain-softening behavior or where the piles are axially flexible, the actual capacity of the pile may be less than that given by Equation (4.11). Note that other factors, such as increased axial capacity under loading rates associated with storm waves, may counteract the above effects, as discussed by Dunnavant et al. (1990).

Cyclic Response

The definition of cyclic loading, as the load that includes inertial loadings developed by environmental conditions such as storm waves and earthquakes, can have two opposite effects on the static axial capacity, as the pile can be under compression and sometimes under less compression or under tension. On the other hand, repetitive loadings can cause a temporary or permanent decrease in load-carrying resistance and may cause an accumulation of deformation. In cyclic loading, rapidly applied loading may occur, which causes an increase in the load-carrying resistance and stiffness of the pile, while very slowly applied loading can cause a decrease in load-carrying resistance and stiffness of the pile.

The design pile penetration selected should be enough to develop an effective pile capacity to resist the design static and cyclic loadings.

The design pile penetration can be confirmed by performing pile response analyses of the pile-soil system subjected to static and cyclic loadings. The pile-soil resistance-displacement *t-z* and *Q-z* characterizations are discussed below.

When any of the above effects are explicitly considered in pile-response analysis, the design static and cyclic loadings should be imposed on the pile top and the resistance-displacements of the pile determined. At the completion of the design loadings, the maximum pile resistance and displacement should be determined. Pile deformations should meet structure serviceability requirements.

Axial Load-Deflection (t-z and Q-z) Data

The fixed offshore platform pile foundation should be designed to resist static and cyclic axial loads. The axial resistance of the soil is provided by a combination of axial soil-pile unit skin friction or load transfer along the sides of the pile and end-bearing resistance at the pile tip. The plotted relationship between mobilized soil-pile shear transfer and local pile deflection at any depth is described using a *t-z* curve. Similarly, the relationship between mobilized end-bearing resistance and axial tip deflection is described using a *Q-z* curve.

Axial deformation of piles may be modeled in a similar way to the lateral case, to permit stress transfer to be computed and axial pile stiffness to be assessed. For axial loading, *t-z* curves are used to represent the resistance along the pile shaft, and *Q-z* curves are introduced to model end bearing. Characteristic shapes of the curves are shown in Figure 4.5(a) and (b) for clay and sand soil, respectively.

Axial support to a pile is provided by surrounding soil, and axial pile deformation at the end of pile depth may be considered to consist of four components: elastic pile deformation, elastic soil deformation, plastic soil deformation and plastic soil-pile slip deformation. The purpose of the *t-z* curves is to model the latter three components. *Q-z* curves model elastic and plastic soil deformation around the pile tip. Elastic pile deformation is not directly related to soil characteristics and is modeled in the beam-column representation of the pile.

In the past, *t-z* curves were based directly on experimental evidence from Coyle and Reese (1966). This led to an adopted standard that peak shaft resistance was mobilized at a vertical relative pile-soil movement (*Zc*) of 2.54 mm (0.1 inch) in sands. For clays, the value is 1% of pile diameter. It has been shown theoretically and experimentally that the form of the *t-z* curve will be a function of the pile length and diameter, soil stiffness and shaft resistance, as discussed by Kraft et al. (1981). However, to account for these characteristics, the average shear modulus of the soil must be known. In sand, the appropriate strain-level shear modulus is only known within an order of magnitude,

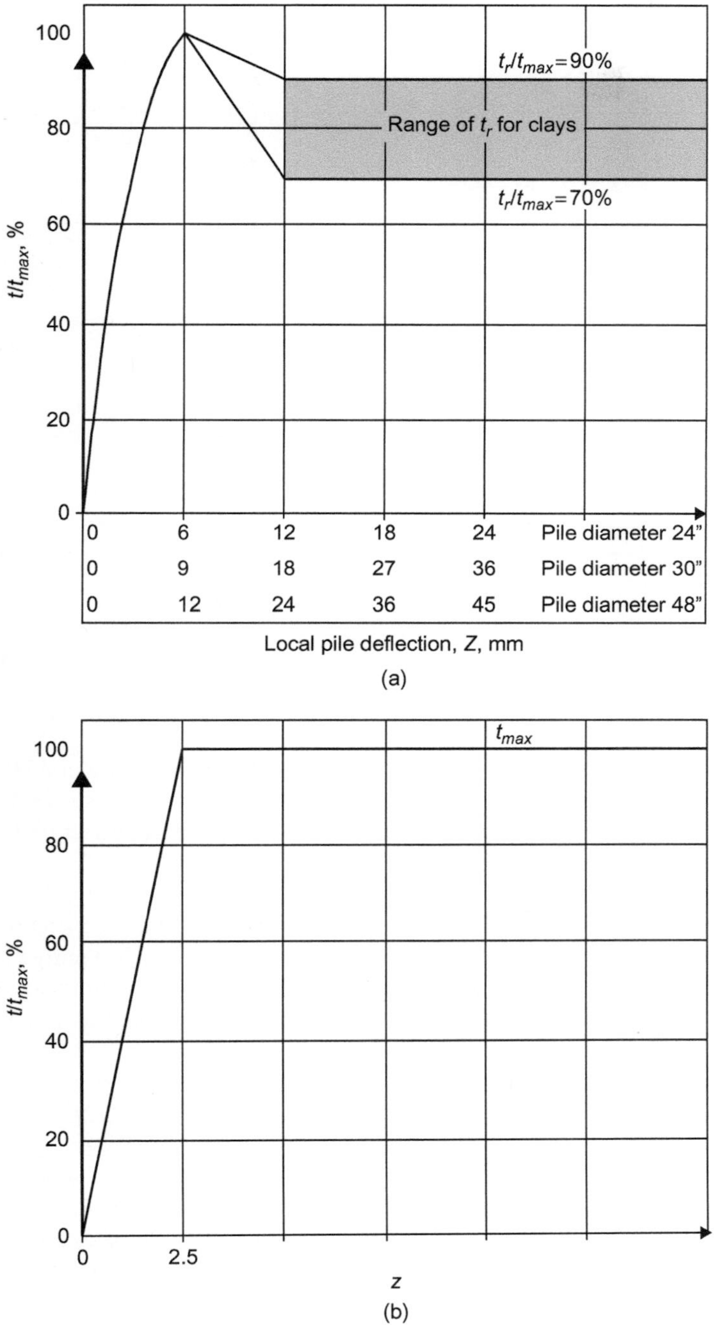

FIGURE 4.5 (a) Sketch of the shape of the *t-z* curve for clay soil with different pile diameters, (b) Sketch of the shape of the *t-z* curve for sand with different pile diameters.

(Continued)

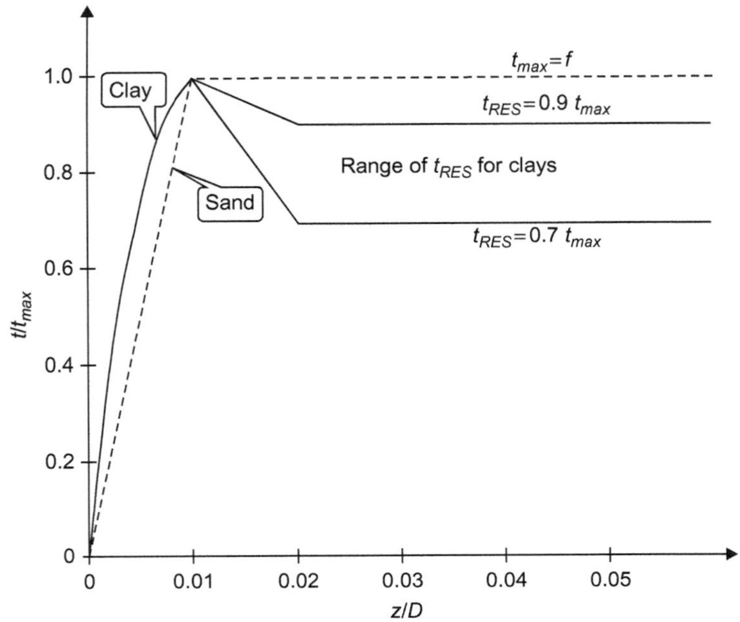

FIGURE 4.5 (c) *t-z* curves recommended for noncarbonate soils, according to API RP2A.

and thus *t-z* curves generated by this method will contain considerable uncertainty. However, a parametric study by Meyer et al. (1975) showed that a six-fold variation in soil yield displacement had only a small effect on the predicted pile head displacement.

The data include a peak-residual behavior for *t-z* curves in clay, the governing parameter of which is the ratio of peak to residual unit skin friction. The recommended range for this parameter is 70% to 90%. Vijayvergiya (1977) indicates that this parameter decreases with increasing overconsolidation ratio. The peak residual behavior in sand has been adopted as per Wiltsie et al. (1982).

For some projects, in the absence of more definitive criteria, the recommended *t-z* curves for noncarbonate soils, according to API RP2A, are shown in Figure 4.5(c).

Table 4.11 presents the relation between the vertical displacement of the pile to the pile diameter and the skin friction between the pile and the soil as a percentage of the total friction capacity diameter. The shape of the *t-z* curve at displacements greater than z_{max}, as shown in Figure 4.5, should be carefully considered.

TABLE 4.11 Relation between the Ratio of Pile Deflection to the Diameter and the Skin Friction Capacity

Soil Type	z/D	*Pile Axial Deflection for Different Pile Diameters, mm*			t/t_{max}
		24″	*36″*	*48″*	
Clays	0.0016	0.98	1.46	2.0	0.30
	0.0031	1.89	2.83	3.8	0.50
	0.0057	3.5	5.21	7.0	0.75
	0.0080	4.9	7.32	9.8	0.90
	0.0100	6.1	9.14	12.2	1.00
	0.0200	12.2	18.3	24.4	0.70–0.90
	∞				0.70–0.90
Sands	z (mm)				t/t_{max}
	0.000	0	0	0	0.00
	2.5	2.5	2.5	2.5	1.00
	∞				1.00

z = local pile deflection (in mm), D = pile diameter (in mm), t = mobilized soil-pile unit skin friction (in lb/ft^2 or kPa) and t_{max} = maximum soil-pile unit skin friction capacity computed (in lb/ft^2 or kPa).

The ratio between the residual unit skin friction to the maximum skin friction between pile and soil t_r/t_{max} at the axial pile displacement at which it occurs (z_r) is affected by the soil stress-strain behavior, stress history, pipe installation method, pile load sequence and other factors.

The value of t_r/t_{max} can range from 70% to 90%. Laboratory, in-situ or model pile tests can provide valuable information for determining values of t_r/t_{max} and z_r for various soils.

The end-bearing or tip-load capacity should be determined. However, relatively large pile-tip movements are required to mobilize the full end-bearing resistance. A pile-tip displacement up to 10% of the pile diameter may be required for full mobilization in both sand and clay soils. In the absence of more definitive criteria, the curve in Figure 4.6 is recommended for both sands and clays.

Table 4.12 presents the relation between the axial displacement of the pile relative to the pile diameter and the end-bearing capacity as a percentage of the total end-bearing capacity.

The recommended curve is shown in Figure 4.6

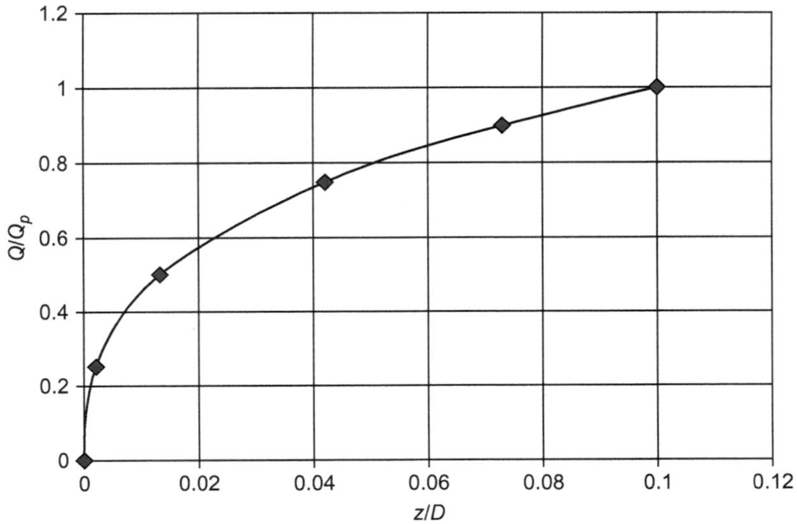

FIGURE 4.6 Pile-tip load displacement (Q-z) curve.

TABLE 4.12 Relation between the Axial Deflection to Pile Diameter Ratio and Percentage of End-Bearing Capacity

z/D	*Pile Axial Deflection for Different Pile Diameters, mm*			*Q/Q$_p$*
	24″	*36″*	*48″*	
0.002	1.2	1.8	2.4	0.25
0.013	7.9	11.9	15.8	0.50
0.042	25.6	38.4	51.2	0.75
0.073	44.5	66.8	89	0.90
0.100	61	91.4	121.9	1.00

z = axial tip deflection (in mm), D = pile diameter (in mm), Q = mobilized end-bearing capacity (in lb or KN) and Q_p = total end bearing (in lb or KN).

Axial Pile Capacity

A number of studies were performed that were aimed at collecting and comparing axial capacities from relevant pile load tests to those predicted by traditional offshore pile design procedures. Studies like these can be very useful in tempering one's judgment in the design process. It is clear, for example, that there is considerable scatter in the various plots of measured versus predicted capacities.

The designer should be aware of the many limitations of such comparisons when making use of the results. Limitations of particular importance include:

1. There is considerable uncertainty in the determination of both predicted capacities and measured capacities. For example, determination of the predicted capacities is very sensitive to the selection of the undrained shear strength profile, which itself is subject to considerable uncertainty. The measured capacities are also subject to interpretation as well as possible measurement errors.
2. Conditions under which pile load tests are conducted generally vary significantly from design loads and field conditions. One clear limitation is the limited number of tests on deeply embedded, large-diameter, high-capacity piles. Generally, pile load tests have capacities that are 10% or less of the prototype capacities. Briaud et al. (1984) mentioned that another factor is that the rate of loading and the cyclic load history are usually not well represented in load tests. According to Clarke (1993), pile load tests are often conducted before full set-up occurs, for practical reasons. Furthermore, pile-tip conditions (closed versus open-ended) may differ from offshore piles.
3. In most of the studies, an attempt has been made to eliminate the factors that are thought to be significantly affected by extraneous conditions in load testing, such as protrusions on the exterior of the pile shaft (weld beads, cover plates, etc.), installation effects (jetting, drilled out plugs, etc.), and artesian conditions, but it is not possible to be absolutely certain in all cases.

The tests most relevant to offshore applications have all been conducted in the United States or in Europe. As regional geology and particularly operating experience are considered very important in foundation design, care should be exercised in applying these results to other regions of the world. In addition, the designer should note that certain important tests in silty clays of low plasticity indicate overprediction of frictional resistance by Equations (4.14) and (4.15). The reason for the overprediction is not well understood and has been an area of active research. The designer is thus cautioned that pile design for soils of this type should be given special consideration.

Additional considerations that apply to drilled and grouted piles are discussed by Kraft and Lyons (1974) and O'Neill and Hassan (1994).

Pile load tests are commonly used as a basis for determining pile load-movement characteristics. In clay, the ultimate capacity of the pile, as shown in Figure 4.7, reaches a maximum value at some movement, beyond which there is a gradual drop to a residual value.

The frictional resistance increases rapidly and reaches a maximum value at a very small displacement, referred to as the *critical movement*. However, the point resistance continues to increase beyond this critical movement and tends to reach a maximum value at a relatively larger movement. This maximum value is referred to as the *end-bearing capacity*.

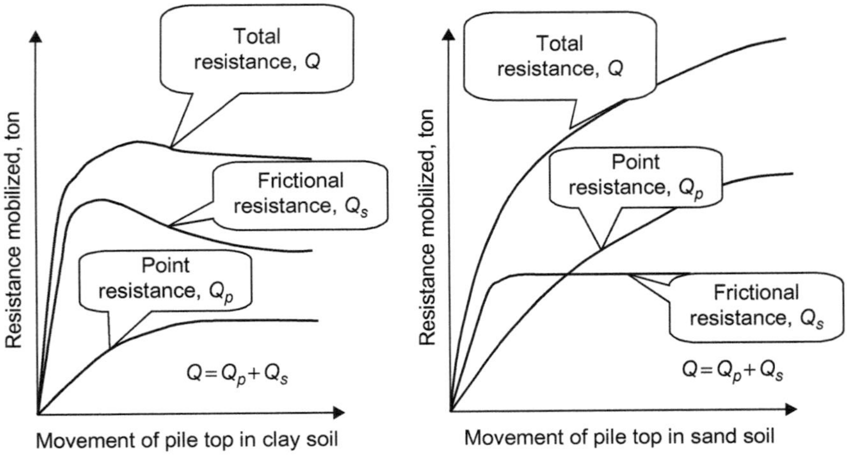

FIGURE 4.7 Typical load-movement characteristics of an axial loaded pile.

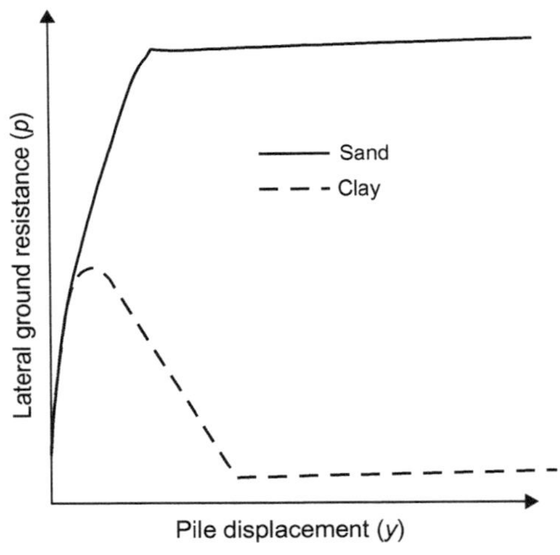

FIGURE 4.8 Example of a typical *p-y* curve for a 36-inch pile.

In sand, as presented in Figure 4.7, the ultimate capacity seems to increase and reach a constant value. The point resistance in sand continues to increase gradually. This is probably why a pile in sand does not usually reach a plunging failure during a load test. The difference between pile behavior in sand and that in clay is attributed to the different point and frictional resistances as a function of pile movement.

The relation between the lateral resistance and displacement for a 36-inch-diameter pile in clay and sand is shown in Figure 4.8.

Laterally Loaded Piles Reaction

The pile foundation in a fixed offshore structure should be designed to carry lateral loads, whether static or cyclic. Additionally, the designer should consider overload cases in which the design lateral loads on the platform foundation are increased by an appropriate safety factor.

The designer should be sure that the overall structural foundation system will not fail under overloads. Note that the lateral resistance of the soil near the surface is significant to pile design and the effects of scour and soil disturbance on this resistance during pile installation should be considered. Under lateral loading, clay soils behave as a plastic material, which makes it necessary to relate pile-soil deformation to soil resistance. Therefore, lateral soil resistance deflection (*p-y*) curves should be obtained using stress-strain data from laboratory soil samples and should be one of the deliverables in the geotechnical report. The ordinate for these curves is soil resistance, p, and the abscissa is soil deflection, y. By iterative procedures, a compatible set of load-deflection values for the pile-soil system can be developed.

Matlock (1970) performed a comprehensive study of the design of laterally loaded piles in soft clay, and Reese and Cox (1975) performed a study of laterally loaded piles in stiff clay.

In the absence of more definitive criteria, Figure 4.11 and Table 4.14 may be used for constructing ultimate lateral-bearing-capacity curves and *p-y* curves.

It is noted that these *p-y* curves are recommended for estimating pile bending moment, displacement and rotation profiles for various (static or cyclic) loads. Different criteria may be applicable for fatigue analysis of a pile that has previously been subjected to loads larger than those used in the fatigue analysis and that resulted in "gapping" around the top of the pile.

Lateral Bearing Capacity for Soft Clay

According to API RP2A (2007), for static lateral loads the ultimate unit lateral bearing capacity of soft clay p_u has been found to vary between $8c$ and $12c$, except at shallow depths, where failure occurs in a different mode due to minimum overburden pressure. Cyclic loads cause deterioration of lateral bearing capacity below that for static loads.

In the absence of more definitive criteria, the following is recommended: p_u increases from $3c$ to $9c$ as X increases from 0 to X_R according to:

$$p_u = 3c + \gamma X + J(cX/D) \tag{4.19}$$

and

$$p_u = 9c \quad \text{for} \quad X \geq X_R \tag{4.20}$$

For a condition of constant strength with depth, Equations (4.19) and (4.20) are solved simultaneously to have the following equation:

$$X_R = 6D/[(\gamma D/c) + J] \tag{4.21}$$

where p_u = ultimate resistance (in kPa), c = undrained shear strength for undisturbed clay soil samples (in kPa), D = pile diameter (in mm), γ = effective unit weight of soil (in MN/m^3), J = a dimensionless empirical constant with values ranging from 0.25 to 0.5 (having been determined by field testing; a value of 0.5 is appropriate for Gulf of Mexico clays, so it should be checked in another areas), X = depth below soil surface (in mm) and X_R = depth below soil surface to bottom of reduced resistance zone (in mm).

Where the strength varies with depth, Equations (4.19) and (4.20) may be solved by plotting the two equations, i.e., p_u versus depth. The point of first intersection of the two equations is taken to be X_R.

These empirical relationships may not apply where strength variations are erratic. In general, minimum values of X_R should be about 2.5 pile diameters.

On the other hand, in soft clay, the load-deflection (*p-y*) curves for lateral soil resistance-deflection relationships for piles are generally nonlinear. The *p-y* curves for the short-term static load case may be generated from Table 4.13, where p = actual lateral resistance (in kPa), y = actual lateral deflection

$$y_c = 2.5\varepsilon_c D \tag{4.22}$$

and ε_c is the strain that occurs at one-half the maximum stress on laboratory unconsolidated, undrained, compression tests of undisturbed soil samples.

For the case where equilibrium has been reached under cyclic loading, the *p-y* curves may be generated from Table 4.14.

Lateral Bearing Capacity for Stiff Clay

For static lateral loads, the ultimate bearing capacity p_u for stiff clay ($c >$ r 96 kPa), as for soft clay, would vary between $8c$ and $12c$. Due to rapid

TABLE 4.13 Relation between Pile Lateral Load and Lateral Deflection

p/p_u	y/y_c
0.00	0.00
0.23	0.1
0.33	0.3
0.5	1.0
0.72	3.0
1.00	8.00
1.00	∞

TABLE 4.14 Relation between Pile Load and Lateral Displacement

$X > X_R$		$X < X_R$	
p/p_u	y/y_c	p/p_u	y/y_c
0.00	0.00	0.00	0.0
0.23	0.1	0.23	0.1
0.33	0.3	0.33	0.3
0.50	1.0	0.50	1.0
0.72	3.0	0.72	3.1
0.72	∞	0.72 X/X_R	15.0
		0.72 X/X_R	∞

deterioration under cyclic loading, the ultimate resistance will be reduced to something considerably less and should be so considered in cyclic design. Furthermore, because stiff clays also have nonlinear stress-strain relationships, they are generally more brittle than soft clays. In developing stress-strain curves and subsequent *p-y* curves for cyclic loads, good judgment should reflect the rapid deterioration of load capacity at large deflections for stiff clays.

For a more detailed study of the construction of *p-y* curves, see Matlock (1970) for soft clay, Reese and Cox (1975) for stiff clay, O'Neill and Murchison (1983) for sand and Georgiadis (1983) for layered soils.

Lateral Bearing Capacity for Sand

A series of studies verified the theoretical studies with the field-test results during lateral loading of a 24-inch diameter test pile installed at sites with clean, fine sand and silty sand. (The studies were funded by Amoco's production company, Chevron oil field research, Esso's production research company, Mobil Oil Corporation and Shell's development company.) The results suggest a shape for the *p-y* curve as shown in Figure 4.9, where the initial part is a straight line representing the elastic behavior and the horizontal straight line represents the plastic behavior, with the straight lines connected by a parabola.

The values of *Pm* and *pc* are a function of the ultimate soil resistance. A difference between the ultimate resistance from theory and that from experiments was observed that was covered by empirical factors. Another study was performed by O'Neill and Murchison (1983), who evaluated *p-y* relationships in

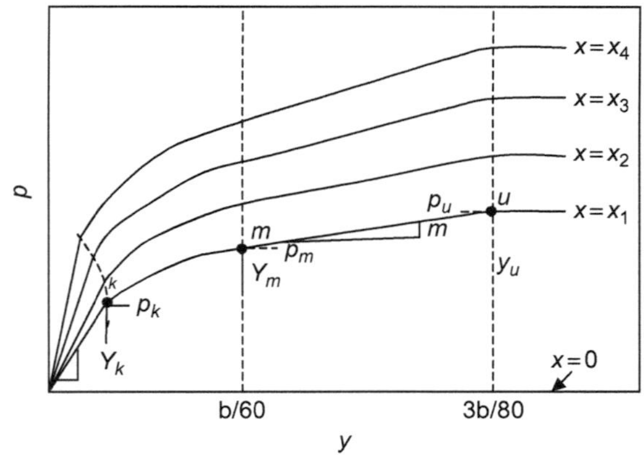

FIGURE 4.9 Typical family of p-y curves for the proposed criteria.

sands. API RP2A (2007) recommends that the p-y curve be calculated using the information from that study.

The ultimate lateral bearing capacity for sand has been found to vary from a value at shallow depths determined by Equation (4.21) to a value at deep depths determined by Equation (4.23). At a given depth, the equation giving the smallest value of p_u should be used as the ultimate bearing capacity.

$$p_{us} = (C_1 H + C_2 D)\gamma H \tag{4.23}$$

$$p_{ud} = C_3 D \gamma H \tag{4.24}$$

where p_u = ultimate resistance (force/unit length) (in kN/m) (s = shallow, d = deep); γ = effective soil weight (in KN/m^3); H = depth (in m); ϕ' = angle of internal friction of sand (in degrees); C_1, C_2, C_3 = coefficients determined from Table 4.15 as a function of ϕ'; and D = average pile diameter from surface to depth (in m).

The relationship between lateral soil resistance and deflection (p-y curve) for sand is nonlinear. If there is no definitive information available, the curve may be approximated at any specific depth H, according to API RP2A, by the following equations:

$$P = Ap_u \tanh\left[\frac{kH}{Ap_u} y\right] \tag{4.25}$$

where A is a factor to account for the cyclic or static loading condition, evaluated by $A = 0.9$ for cyclic loading and by $A \geq 0.9$ for static loading, so that:

$$A = [3.0 - 0.8(H/D)] \tag{4.26}$$

TABLE 4.15 Coefficient C_1, C_2, C_3

Angle of internal friction, ϕ'	C_1	C_2	C_3
20	0.6	1.5	8.5
21	0.7	1.6	9.6
22	0.8	1.7	10.8
23	0.9	1.8	12.2
24	1.0	1.9	13.8
25	1.1	2.0	15.6
26	1.2	2.1	17.6
27	1.3	2.2	19.9
28	1.4	2.3	22.5
29	1.6	2.5	25.4
30	1.7	2.6	28.7
31	1.9	2.7	32.4
32	2.1	2.9	36.6
33	2.3	3.0	41.4
34	2.5	3.2	46.7
35	2.8	3.4	52.8
36	3.1	3.6	59.6
37	3.4	3.8	67.4
38	3.8	4.0	76.1
39	4.2	4.2	86.0
40	4.6	4.4	101.5

where p_u = ultimate bearing capacity at depth H (in kN/m); k = initial modulus of subgrade reaction (in kN/m^3), as determined from Table 4.16 as a function of the angle of internal friction, ϕ', and the relative density for sand under the water table; y = lateral deflection (in inches or m); and H = depth (in m).

Alternative Methods of Determining Pile Capacity

In clay soil, there are alternative methods described by API RP2A (2007) for determining pile capacity that comply with industry experience. One such method is summarized below.

TABLE 4.16 Relation between Subgrade Reaction, Angle of Internal Friction and Relative Density for Sand below Water Table

Soil Type	Angle of Internal Friction, ϕ'	Relative Density, %	Subgrade Reaction, kN/m^3
Very loose	<29	20	265.7
Loose sand		25	426.3
		30	553.6
		35	744.6
	29–30	40	996.4
Medium dense		45	1356.3
		50	1716.1
		55	2026.1
	30–36	60	2491.1
Dense		65	2850.9
		70	3293.8
		75	3792.0
	36–40	80	4262.6

For piles driven through clay, the skin friction, f, is equal to or less than, but should not exceed, the undrained shear strength of the clay c_u, as determined by unconsolidated-undrained (UU) tri-axial tests and miniature vane shear tests.

Unless test data indicate otherwise, f should not exceed c_u or the following limits:

1. For highly plastic clays, f may be equal to c_u for underconsolidated and normally consolidated clays. For overconsolidated clays, f should not exceed 48 kPa for shallow penetrations or the equivalent value of c_u for a normally consolidated clay for deeper penetrations, whichever is greater.
2. For other types of clay:

$$f = c_u \quad \text{for} \quad c_u < 24\,\text{kPa} \tag{4.27}$$

$$f = c_{u/2} \quad \text{for} \quad c_u < 72\,\text{kPa} \tag{4.28}$$

f varies linearly for values of c_u between the above limits.

It has been shown that the above method provides, on the average, a reasonable accuracy with the pile load test database results as discussed by Olson (1984). The method you use should take into consideration previous experience of the soil for the specific site and industrial experience in similar soil.

Establishing Design Strength and Effective Overburden Stress Profiles

The undrained shear strength and effective overburden stress profiles are the two factors most affecting the pile axial capacity calculation. The wide variety of sampling techniques and the potentially large scatter in the strength data from the various types of laboratory tests complicate appropriate selection. Therefore, the undrained unconsolidated tri-axial compression tests should be done carefully with the full quality-control system from the field sample and the laboratory tests as described previously.

The following factors can cause capacity degradation of long piles driven in clay soils:

1. Continued shearing of a particular soil horizon during pile installation
2. Lateral movement of soil away from the pile due to “pile whip” during driving
3. Progressive failure in the soil due to strength reduction with continued displacement (softening)

The occurrence of degradation due to these effects depends on many factors related to both installation conditions and soil behavior.

Time Affects Changes in Axial Capacity in Clay Soil

The pile capacity calculated from the previous equation doesn’t consider the effect of time on the pile capacity, such as in old platforms constructed 40 or more years ago. A recent study was performed to define the behavior of axial capacity in clay soil with time.

Clarke (1993) and Bogard et al. (1990) conducted field measurement studies in which it was shown that the time required for driven piles to reach ultimate capacity in a cohesive soil can be relatively long, as much as 2–3 years.

It is worth mentioning that the rate of strength gain is highest immediately after driving, and the rate decreases during the dissipation process. Thus a significant strength increase can occur in a relatively short time.

During pile driving in normally to lightly overconsolidated clays, the soil surrounding a pile is significantly disturbed, the stress state is altered, and large excess pore pressures can be generated. After installation, the excess pore pressures begin to dissipate, which means that the surrounding soil mass around the piles begins to consolidate, so the pile capacity increases with time. This process is usually referred to as *set-up*. The rate of excess pore pressure dissipation is a function of the coefficient of radial consolidation, pile radius, plug characteristics and soil layering.

In the most popular case, where the driven pipe piles supporting a structure have design loads applied to the piles shortly after installation, the time-consolidation characteristics should be considered in pile design. Note that in traditional fixed offshore structure installation, the time between pile installation and the platform's becoming totally loaded is 1–3 months, but if the commissioning and start-up come early, this information should be transferred to the engineering office, as the expected increase in capacity with time is an important design variable that can affect the safety of the foundation system during the early stages of the consolidation process.

The behavior of piles subjected to significant axial loads in highly plastic, normally consolidated clays was studied using a large number of model pile tests and some full-scale pile load tests.

From the study of pore-pressure dissipation and load test data at different times after pile driving, empirical correlations were obtained between the degree of consolidation, degree of plugging and pile shaft shear-transfer capacity. This study revealed that there is no significant change in capacity with time for closed-ended steel piles in heavily overconsolidated clay. This is contrary to tests on 0.273 m (10.75 inch) diameter closed-ended steel piles in overconsolidated Beaumont clay, where considerable and rapid set-up in 4 days was found, so the pile capacity at the end of installation was never fully recovered.

So it is very important to highlight that the axial capacity of the pile with time is under research and development and there is no solid formula or equation to follow. The focus should be on research done on the specific site location and on the previous history of the location.

4.6.4 Pile Capacity Calculation Methods

API RP2A (2007) presents new methods for calculating pile capacity based on the cone penetration test (CPT).

As outlined previously, a simple method for assessing pile capacity in cohesionless soils is the recommended method in previous editions of API RP2A-WSD. There are reliable CPT-based methods for predicting pile capacity. These methods are all based on direct correlations of pile unit friction and end-bearing data with cone tip resistance (q_c) values from CPT. These CPT-based methods cover a wider range of cohesionless soils, are considered fundamentally better, and have shown statistically closer predictions of pile load test results.

The new CPT-based methods for assessing pile capacity in sand are preferred to the previous method. However, more experience is required with all these new methods before any single one can be recommended for routine design instead of the previously presented method. API stated clearly that the new CPT-based methods should be used only by qualified engineers who are experienced in interpreting CPT data and who understand the limitations and reliability of the CPT-based methods.

The assumption is made that friction and end-bearing components are uncoupled. Hence, for all methods, the ultimate bearing capacity in compression (Q_d) and tensile capacity (Q_t) of plugged open-ended piles are determined by:

$$Q_t = Q_f + Q_p = P_o \int f_{c,z} dz + A_q q_p \tag{4.29}$$

$$Q_f = P_o \int f_{t,z} dz \tag{4.30}$$

As the friction component, Q_f, is calculated by numerical integration, so the results are sensitive to the depth increment used, especially in the case of CPT-based methods. The depth increments for CPT-based methods should be in the order of 1% of the pile length (or smaller) and, in any case, the depth increment should not exceed 250 mm.

There are four recommended CPT-based methods mentioned in API RP2A:

1. Simplified ICP-05
2. Offshore UWA-05 (Lehane et al., 2005a,b)
3. Fugro-05 (Lehane et al., 2005a; Kolk et al., 2005)
4. NGI-05 (Lehane et al., 2005a; Clausen et al., 2005)

The first method is a simplified version of the design method recommended by Jardine et al. (2005), whereas the second is a simplified version of the UWA-05 method applicable to offshore pipe piles. The other three methods are summarized by Lehane et al. (2005a). It is important to avoid calculating friction and end-bearing components from different methods.

The unit skin friction formulae for open-ended steel pipe piles for the first three recommended CPT-based methods (Simplified ICP-05, Offshore UWA-05 and Fugro-05) can all be considered as special cases of the general formula:

$$f_z = u.q_{cz}\left(\frac{\sigma'_{vo}}{P_a}\right) A_r^b \left[\max\left(\frac{L-z}{D}, v\right)\right]^{-c} [\tan \delta_{cv}]^d \left[\min\left(\frac{L-z}{D}\frac{1}{v}, 1\right)\right] \tag{4.31}$$

where f_z is the unit skin friction, δ_{cv} is pile–soil constant-volume interface friction angle; L is the pile length underneath the seabed, $A_r = 1 - (D_i/D)^2$; D_i is the pile inner diameter ($D_i = D - 2t$); z is the depth under the seabed; q_{cz} is the CPT tip resistance at depth z; D is the outer diameter; t is the wall thickness and P_a is the atmospheric pressure equal to 100 kPa.

TABLE 4.17 **Unit Skin Friction Parameter Values for Driven Open-Ended Steel Pipes (Simplified ICP-05 and Fugro-05 Methods)**

Method	Load Direction	a	b	c	d	e	u	v
Simplified ICP-05	Compression	0.1	0.2	0.4	1	0	0.023	$4\sqrt{A_r}$
	Tension	0.1	0.2	0.4	1	0	0.016	$4\sqrt{A_r}$
Fugro-05	Compression	0.05	0.45	0.90	0	1	0.043	$2\sqrt{A_r}$
	Tension	0.15	0.42	0.85	0	0	0.025	$2\sqrt{A_r}$

Table 4.17 provides the recommended values for parameters a, b, c, d, e, u and v for compression and tension, which are the unit skin friction parameter values for driven open-ended steel piles for the Simplified ICP-05, Offshore UWA-05 and Fugro-05 methods.

Additional recommendations for computing unit friction and end bearing for all four CPT-based methods are presented in the following subsections.

Simplified ICP-05

A comprehensive overview of research work performed at Imperial College on axial pile design criteria of open- and closed-ended piles in clay and sand is presented by Jardine et al. (2005). The design equations for unit friction in sand in this publication include that unit friction is favorably influenced by soil dilatancy. This influence reduces with increasing pile diameter.

The Simplified ICP-05 method as presented by API RP2A for unit skin friction of open-ended pipe piles and the parameter values in Table 4.17 are a conservative approximation of the full ICP-05 method, since dilatancy is ignored and some parameter values were conservatively rounded up and down.

Use of the original design equations in Jardine et al. (2005) may be considered, particularly for small-diameter piles ($D < 0.76$ m), provided that larger factors of safety are considered in the WSD design.

The ultimate unit end bearing for open-ended pipe piles, Q_p, follows the recommendations of Jardine et al. (2005), which specify an ultimate unit end bearing for plugged piles given by:

$$Q_p = q_{ca}\{[0.5 - 0.25\log 10(D/D_{CPT})] \geq 0.15\, q_{c,av,1.5D}\} \tag{4.32}$$

where q_{ca} is the average q_{cz} value between 1.5 D_o above oil tip to 1.5 D_o below pile tip level and D_{CPT} is the CPT tools diameter (about 36 mm for a standard 1000 mm^2 base net cone).

Jardine et al. (2005) specify that plugged pile end-bearing capacity applies; that is, the unit end-bearing Q_p acts across the entire tip cross-section, only if both the following conditions are met:

$$D_i < 2(D_r - 0.3)$$

[Note: D_i units are inches or m and D_r units are not percentages, but fractions.]

and

$$D_i/D_{CPT} < 0.083\, q_{c,z} < p_a \tag{4.33}$$

Should either of the above conditions not be met, then the pile will behave unplugged and the following equation should be used for computing the end-bearing capacity:

$$Q_p = \pi\, t(D - t) q_{c,z} \tag{4.34}$$

The full pile end bearing calculated by using Equation (4.32) for a plugged pile should not be less than the end-bearing capacity of an unplugged pile calculated from Equation (4.34).

Offshore UWA-05

For friction, Lehane et al. (2005a) summarize the results of recent research work at the University of Western Australia on development of axial pile design criteria of open- and closed-ended piles driven into silica sands. The full design method presented by Lehane et al. (2005a,b) for unit friction on pipe piles includes a term allowing for favorable effects of soil dilatancy (similar to ICP-05) and an empirical term allowing for partial soil plugging during pile driving. Lehane et al. recommend for offshore pile design to ignore the two favorable effects of dilatancy and partial plugging. The original full design equations in Lehane et al. (2005a) will be used for small-diameter piles, $D < 750$ mm (30 inches), provided that larger factors of safety are considered in the WSD design.

For end bearing, Lehane et al. (2005a,b) present design criteria for ultimate unit end bearing of plugged open-ended pipe piles. Their full design method for pipe piles includes an empirical term allowing for the favorable effect of partial plugging during pile driving. For offshore pile design, Lehane et al. (2005a,b) recommend this effect be ignored, resulting in the recommended design equation for plugged piles in the Offshore UWA-05 method:

$$Q_p = q_{ca}(0.15 + 0.45 A_r) \tag{4.35}$$

Since the UWA-05 method considers nonplugging under static loading to be exceptional for typical offshore piles, the method does not provide criteria for unplugged piles.

The unit end-bearing q_p calculated in Equation (4.35) is therefore acting across the entire tip cross-section. The use of $q_{c,av15D}$ in Equation (4.35) is not recommended in sand profiles, where the CPT q_c values show significant variations in the vicinity of the pile tip or when penetration into a founding stratum is less than five pile diameters. For these situations, Lehane et al. (2005a) provide guidance on the selection of an appropriate average q_c value for use in place of q_c.

The unit skin friction in compression and tension will be obtained as follows:
Compression:

$$f_{zc} = 0.03.q_{cz}\left(\frac{\sigma'_{vo}}{P_a}\right)A_r^{0.3}\left[\max\left(\frac{L-z}{D},2\right)\right]^{-0.5}[\tan\,\delta_{cv}]\left[\min\left(\frac{L-z}{2D},1\right)\right] \quad (4.36)$$

Tension:

$$f_{zt} = 0.022.q_{cz}\left(\frac{\sigma'_{vo}}{P_a}\right)A_r^{0.3}\left[\max\left(\frac{L-z}{D},2\right)\right]^{-0.5}[\tan\,\delta_{cv}]\left[\min\left(\frac{L-z}{2D},1\right)\right] \quad (4.37)$$

Fugro-05

For friction, the Fugro-05 method is a modification of the ICP-05 method. This method was studied by Fugro (2004) and Kolk et al. (2005), as well as by Lehane et al. (2005a), and in Equation (4.31) and the parameter values in Table 4.17. When using ICP-05 and the UWA-05 or Fugro-05 methods, it is recommended to consider larger factors of safety, as discussed by CUR (2001) in the reliability of design using these methods.

For end bearing, the basis for the ultimate unit end bearing on pipe piles according to Fugro-05 is presented in the research report to API (Fugro, 2004) and is summarized by Kolk et al. (2005). The recommended design criterion for plugged piles is given by:

$$Q_p = 8.5\sqrt{p_a q_{ca}}A^{0.25}r \quad (4.38)$$

Note that the UWA-05 and Fugro-05 methods do not specify unplugged end-bearing capacity because traditional fixed offshore platform piles behave in a plugged end mode during static loading, as discussed by CUR (2001). It can be shown that plugged behavior applies in the following cases:

1. If the cumulative thickness of sand layers within a soil plug is in excess of $8D$, or
2. The total end-bearing Q_p is limited, as follows:

$$Q_p \leq Q_{f,I,clay}e^{Ls/D} \quad (4.39)$$

The value of the friction capacity ($Q_{f,i,clay}$) inside the pile in plugged soil can be calculated using the same methods used in calculating the pile friction in clay.

The above criteria apply for fully drained behavior of sand within the pile plug. Criteria for undrained and partially drained sand plug behavior are presented by Randolph et al. (1991).

For the exceptional case of unplugged end-bearing behavior in fully drained conditions, reference is made to CUR (2001) and Lehane and Randolph (2002) for estimating end-bearing capacity.

NGI-05

For friction calculation in NGI-05, Clausen et al. (2005) provided the ultimate unit skin friction values for tension (f_{tz}) and compression (f_{cz}) for driven open-ended steel pipe piles:

$$f_{tz} = (z/L)p_a(\sigma_{vo}/p_a)^{0.25}F_{Dr} > 0.1\sigma'_{vo} \quad (4.40)$$

$$f_{cz} = 1.3(z/L)p_a(\sigma'_{vo}/p_a)^{0.25}F_{Dr} > 0.1\sigma'_{vo} \quad (4.41)$$

where:

$$F_{Dr} = 2.1(D_r - 0.1)^{1.7} \quad (4.42)$$

$$D_r = 0.4 \ln\left(\frac{q_{c,z}}{22(\sigma'_{vo}p_a)^{0.5}}\right) > 0.1 \quad (4.43)$$

Note: $D_r > 1$ should be accepted and used.

As for the "full" ICP-05, "full" UWA-05 and the Fugro-05 methods, higher factors of safety are recommended when using the NGI-05 method.

For end bearing, the recommended design equation for ultimate unit end bearing of plugged open-ended steel pipe piles in the NGI-05 method, according to Clausen et al. (2005), is:

$$Q_p = \frac{0.7q_{ca}}{1 + 3D_r^2} \quad (4.44)$$

where

$$D_r = 0.4 \ln\left([q_{ca}/(22(\sigma'_{vo}p_a)^{0.5})] > 0.1\right) \quad (4.45)$$

Note that $D_r > 1$ should be accepted and used.

The resistance of nonplugging piles should be computed using an ultimate unit wall end-bearing value (q_{wz}) given by:

$$q_{wz} = q_{cz} \quad (4.46)$$

and an ultimate unit friction (f_{pz}) between the soil plug and inner pile wall given by:

$$f_{pz} = 3f_{cz} \quad (4.47)$$

The lower value of the plugged resistance by Equation (4.46) and unplugged resistance by Equations (4.46) and (4.47) should be used in design.

Application of CPT

When using CPT measurement as the basis for the previous methods to calculate the unit skin friction and end bearing for the pile, some precautions should be taken, the peak unit skin friction in compression and tension at a given depth, f_{cz} and f_{tz}, are not unique and are both dependent on pile geometry. In general, the axial load and deformation response are affected by the pile penetration depth, the pile diameter and its wall thickness. Note that increased pile penetration will decrease these ultimate values at a given depth.

In the case of doing the test to obtain the q-z data for axial load-deformation response, the end-bearing Q_p is assumed to be fully mobilized at a pile tip-displacement value of $0.1D$.

Soil types such as carbonate sands, micaceous sands, glauconitic sands, volcanic sands, silts and clayey sands have unusually weak structures with compressible grains. These types require special consideration in situ and in the laboratory tests for selection of an appropriate design method and design parameters according to Thompson and Jardine (1998) and Kolk (2000) for pile design in carbonate sand, and according to Jardine et al. (2005).

When using CPT in cohesionless soil, such as gravels, when particle sizes are in excess of 10% of the CPT cone diameter, one possible approach could be to use the lower-bound q_c profile. In this case, one can estimate the end-bearing capacity profile from the adjacent sand layers.

In the case of using CPT in weaker clay layers near the pile tip, it is recommended to obtain q_c data averaged between $1.5D$ above the pile tip and $1.5D$ below the pile tip level, provided q_c does not vary significantly. The UWA method should be used, if significant q_c variations occur, to compute $q_{c,av}$.

A thin clay layer, less than around $0.1D$ thick, is a problem especially when CPT data are discontinuous vertically or not all pile locations have been investigated. From a practical point of view, the offshore piles usually develop only a small percentage of q_p under extreme loading conditions. So, the finite element method can be used in calculating the pile capacity and settlement of a pile tip on sand containing weaker layers. It also may be considered in assessing the axial pile response under such conditions.

It is recommended that the end-bearing component be reduced in the case of the pile tip being within a zone up to $\pm 3D$ from such layers. When q_c data averaging is also applied to this $\pm 3D$ zone, the combined effects may be unduly conservative and such results should be critically reviewed. This rule also applies to large pile diameters ($D > 2$ m).

4.6.5 Pile Capacity under Cyclic Loadings

Environmental loadings are developed by winds, waves, currents, earthquakes and ice floes. These loadings are consider the source of cyclic loading, as these loads can have both low- and high-frequency cyclic components in which the rates of load application and duration are measured in seconds. Storm and ice loadings can have several thousand cycles of applied forces, while earthquakes can induce several tens of cycles of forces.

For most fixed offshore platforms supported on piles, experience has proven the adequacy of determining pile penetration based on static capacity evaluations, static ultimate design loads and commonly accepted factors of safety that, in part, account for the cyclic loading effects.

Cyclic Loading Effects

A study by Briaud et al. (1984) on the axial capacity and performance of piles showed that, as compared with long-term, static loadings, cyclic loading may have the following important influence on pile axial capacity and stiffness: it may decrease capacity and stiffness due to the repeated loading. On the other hand, it may increase capacity and stiffness due to the high rates of loading.

The cyclic loading effect on pile capacity is caused by resistance from the pile and soil and the load itself.

Cyclic loading can provide a positive effect by stiffening and strengthening or, on the contrary, may may be a reason for softening and weakening of the soil around the pile; also, it may cause accumulation of pile displacements.

For earthquakes, the free-field ground motions, which are natural phenomena, are not affected by the piles and structure but, due to their vibration, cyclic straining develops in the soil; these effects may influence pile capacity and stiffness. The soil will damp the earthquake vibration and absorb the loading energy.

Analytical Models

A primary objective of these analyses is to ensure that the pile and its penetration under static and cyclic loading are adequate to meet the structure's serviceability and capacity based on ultimate limit state requirements.

Different analytical models have been developed and applied to determine the cyclic axial response of piles, as presented in API RP2A. In general, these models can be categorized into two types, discrete element models and continuum models, noting that the two models use the finite difference method or finite element method. The two types of analytical models can be described as follows.

Discrete Element Models

The soil around the pile is modeled by a series of uncoupled springs or elements attached between the piles, and the far-field soil in most cases is assumed to be rigid. The material behavior of these elements may vary from linearly elastic to nonlinear, hysteretic, and rate dependent. According to Bea and Audibert (1979) and Karlsrud et al. (1986), the soil elements are commonly referred to as shaft resistance-displacement (*t-z*) and tip resistance-displacement (*Q-z*) elements.

The pile can be modeled as a series of discrete elements; for example, rigid masses interconnected by springs, or it can be modeled as a continuous element (rod), either linear or nonlinear. In these models, material properties for soil and pile can vary along the pile.

This type of model is the most famous in fixed offshore structure analysis.

The primary steps in performing an analysis of cyclic axial loading effects on a pile using discrete element models are summarized in the following:

1. The pile-head loadings should be characterized in terms of the magnitudes, durations and numbers of cycles. This includes both long-term loadings and short-term cyclic loadings. Typically, the design static and cyclic loadings expected during a design event are chosen.
2. The properties of the pile, including its diameter, wall thickness, stiffness properties, weight and length, must be defined. Therefore, this will need an initial estimate of the pile penetration that might be appropriate for the design loadings. Note that an empirical, pseudostatic method based on pile load tests or soil tests might be used to make such estimates.
3. Soil properties should be defined, as any analytical approaches will require different soil parameters.

For practical reasons, discrete element models solved numerically have seen the most use in evaluation of piles subjected to high-intensity cyclic loadings. Based on Poulos (1983) and Karlsrud et al. (1986), results from these models are used to develop information on pile accumulated displacements and on pile capacity following high-intensity cyclic loadings.

Continuum Models

The soil around the pile is idealized as a continuum attached continuously to the pile.

The pile is typically modeled as a continuous rod, either linear or nonlinear. According to Novak and Sharnouby (1983) and Desai and Holloway (1972), the model material properties can vary in any direction. A wide range of assumptions can be used regarding boundary conditions, solution characteristics and others that lead to an unlimited number of variations for either of the two approaches.

The main key to establishing the model is to define the elastic properties of the soil (E, G, ν, D). However, in the discrete element model, soil resistance-displacement relationships along the pile shaft (*t-z*) and at its tip (*Q-z*) should be determined. In-situ and laboratory soil tests, and model and prototype pile load tests can provide a basis for such determinations. These tests should at least implicitly include the effects of pile installation, loading and time. Another important key to the model is to establish the boundary condition of the model.

The finite element models have been used for specialized analyses of piles subject to monotonic axial loadings, as discussed by Novak and Sharnouby (1983).

Because elastic continuum models solved analytically are similar to machine vibration analyses, the third model is useful in evaluating piles subjected to low-intensity, high-frequency cyclic loadings at or below design working loadings. In the case of higher-intensity loading, Lysmer (1978) suggested that, where material behavior is likely to be nonlinear, the continuum model solved analytically can still be used by employing equivalent linear properties that approximate the nonlinear, hysteretic effects.

4.7 SCOUR

Seabed scour affects both lateral and axial pile performance and capacity. Scour prediction remains an uncertain art. Sediment transport studies may assist in defining scour design criteria, but local experience, from ROV inspection of the existing platform, is the best guide. The uncertainty in design criteria should be handled by robust design, or by an operating and maintenance strategy of monitoring the scour and performing a proper remedy maintenance action. Typical remediation experience, as documented in scour design criteria, will usually be a combination of local and global scour.

Scour (seabed erosion due to wave and current action) can occur around offshore piles. Common types of scour are:

- General or global scour, which affects the area of the piles and is usually twice the area that is covered by the platform
- Local scour, which occurs around specific areas of the structure, such as the piles.

Scour occurs if the water velocity is high enough to lift and carry the seabed sediments in suspension from the area. Turbulence assists this process by breaking up consolidated sediments. Scour is a particular problem in the southern end of the North Sea, which has high tidal currents and a loose sand seabed. Significant scour may occur during a single storm.

Bijaker (1980) defines three mechanisms that cause scour to occur:

- An increase in water velocity around the object
- A vortex trail shed on the downstream side of the object
- A vertical component of the water velocity caused by the presence of the object.

Niedoroda et al. (1981) discussed the process of scour formation in more detail, and Chow and Herbitch (1978) have studied scour patterns around pile groups.

In general, in design of the offshore jacket structure it is assumed that local scour is 1.5 times the leg diameter, and depth of global scour is assumed to be 1 m.

There is no generally accepted method to account for scour in axial capacity for offshore piles. Whitehouse (1998) gave techniques for scour depth assessment. In addition, general scour data may be obtained from national authorities.

Scour decreases axial pile capacity in sand. Both friction and end-bearing components may be affected. This is because scour reduces both q_c and σ'_v (vertical effective stress), whereas q_c is simply proportional to σ'_v

$$q_{cf} = \phi q_{co} \tag{4.48}$$

where $q_{c,f}$ is the final q_c value after general scour, q_{co} is the original q_c value before general scour, ϕ is a dimensionless scour reduction factor $= \sigma'_{vf}/\sigma'_{vo}$, where σ'_{vf} = final vertical effective stress value and σ'_{vo} = original vertical effective stress value.

The scour reduction factor is presented by the following equation based on a conservative approach (Fugro, 1995) for normally consolidated sands.

$$\phi = \left(\frac{1}{1+2k_o} \sqrt{\frac{H' + 2k_o\sqrt{H_{GS}H' + H^2 H_{GS}}}{H_{GS} + H'}} \right) \tag{4.49}$$

where H' = depth below final seabed level $= H - H_{GS}$ and H_{GS} = general scour depth.

Scour reduces lateral soil support, which leads to an increase in pile maximum bending stress. Scour is generally not a problem for cohesive soils, but should be considered for cohesionless soils.

As shown in Figure 4.10, H_{GS} is the general scour depth, H_O is the overburden reduction depth ($6.0 \times D$ is typical), H_{LS} is the local scour depth ($1.5 \times D$ is typical), p' is the vertical effective stress, H is the depth below original sea floor and H' is the depth below final general sea floor.

E_S is the initial modulus of subgrade reaction and is calculated from the following equation:

$$E_S = kH$$

Therefore, scour will decrease the modulus of the subgrade, which is the lateral soil support, because scour will decrease the value of H, which is the soil depth around the pile (as shown in Figure 4.10), so the triangle for calculating the lateral soil support will decrease from 1 to 2 after general scour and reduce to 3 after local scour.

In the absence of project-specific data, for an isolated pile, a local scour depth equal to $1.5D$ and an overburden reduction depth equal to $6D$ may be adopted, where D *is* the pile outside diameter.

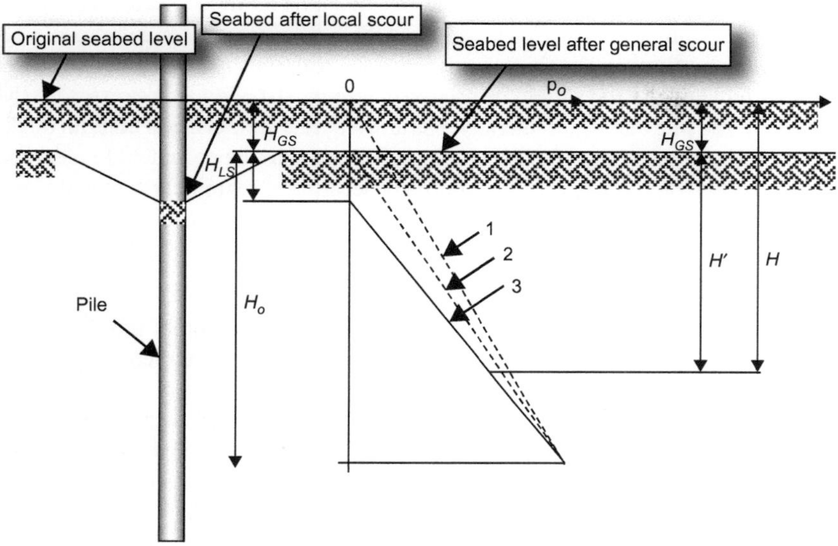

FIGURE 4.10 Relation between scour and lateral soil support.

Reduction in lateral soil support is due to:

1. a lower ultimate lateral pressure caused by decreased vertical effective stress p_o, and
2. a decreased initial modulus of subgrade reaction (E_S).

There is no generally accepted method to allow for scour in the *p-y* curves for offshore piles. There is a method for evaluating p_o and E_S as a function of scour depths. In this method, general scour reduces the p_o profile uniformly with depth, whereas local scour reduces *p* linearly with depth to a certain depth below the base of the scour pit.

Subgrade modulus reaction values (E_S) may be computed assuming the general scour condition only. Other methods, based upon local practice and previous experience in the same location, may be used instead.

Another area of concern in installing piles for offshore structures is the adequacy of existing hammers to produce the required pile penetration to support the applied load on the platform. The wave theory is based on the fact that each hammer blow produces a stress wave that moves along the length of the pile at the speed of sound. The entire length of the pile is not stressed simultaneously, in contrast to conventional dynamic formulas.

4.8 PILE WALL THICKNESS

The wall thickness of the pile may vary along its length and may be controlled at a particular point by any one of several loading conditions or requirements.

The allowable pile stresses should be the same as those permitted by the American Institute of Steel Construction (AISC) specification for a compact hot-rolled section. A rational analysis considering the restraints placed upon the pile by the structure and the soil should be used to determine the allowable stresses for the portion of the pile that is not laterally restrained by the soil. General column buckling of the portion of the pile below the seabed need not be considered unless the pile is believed to be laterally unsupported because of extremely low soil shear strengths, large computed lateral deflections or some other reason.

4.8.1 Design Pile Stresses

The pile wall thickness near to the seabed, and possibly at other points, is normally controlled by the combined axial load and bending moment that results from the design loading applied to the platform.

The bending moment curve on the pile may be calculated with soil reactions determined taking into consideration possible soil removal by scour. It may be assumed that the axial load is removed from the pile by the soil at a rate equal to the ultimate soil-pile unit skin friction divided by the appropriate pile safety factor, as specified in Table 4.11. When lateral deflections associated with cyclic loads at or near the seabed are relatively large, exceeding y_c for soft clay, consideration should be given to reducing or neglecting the soil-pile adhesion by friction through this zone.

4.8.2 Stresses Due to Hammer Effect

Each pile head on which a pile hammer will be hammered should be checked for stresses due to impact load and the weight of the auxiliary equipment. These loads may be the limiting factors in establishing the maximum length of add-on sections. This is particularly true in cases where piling will be driven or drilled on a batter, which is common in fixed offshore platforms, as shown in Chapter 3. The most frequent effects include: static bending, axial loads and arresting lateral loads generated during initial hammer placement.

Experience indicates that reasonable protection from failure of the pile wall due to the above loads is provided if the static stresses are calculated as follows:

1. The pile should be considered as a freestanding column with a minimum effective buckling length factor equal to 2.1 and a minimum reduction factor C_m of 1.0.
2. Bending moments and axial loads should be calculated using the full weight of the pile hammer, cap, and leads acting through the center of gravity of their combined masses, as well as the weight of the pile and taking into consideration the pile batter eccentricities.

The bending moment so determined should not be less than that corresponding to a load equal to 2% of the combined weight of the hammer, cap and leads applied at the pile head and perpendicular to its centerline.

Allowable stresses in the pile should be calculated with the allowable stress design by AISC. Note that the one-third increase in allowable strength should not be considered.

More attention should be paid to the stresses that occur in the freestanding pile section during driving, as shown in Figures 4.11 and 4.12. Generally, stresses are checked based on the conservative criterion that the sum of the stresses due to the dynamic stresses caused by the impact from the hammer and the stresses due to axial load and bending (the static stresses) should not exceed the minimum yield stress of the steel.

The engineering office should clearly define in the calculation notes, drawings and specification the required pile hammers that will be used.

A method of analysis based on wave propagation theory should be used to determine the dynamic stresses.

In general, it may be assumed that column buckling will not occur as a result of the dynamic portion of the driving stresses. The dynamic stresses should be less than 80% to 90% of yield strength, depending on specific circumstances,

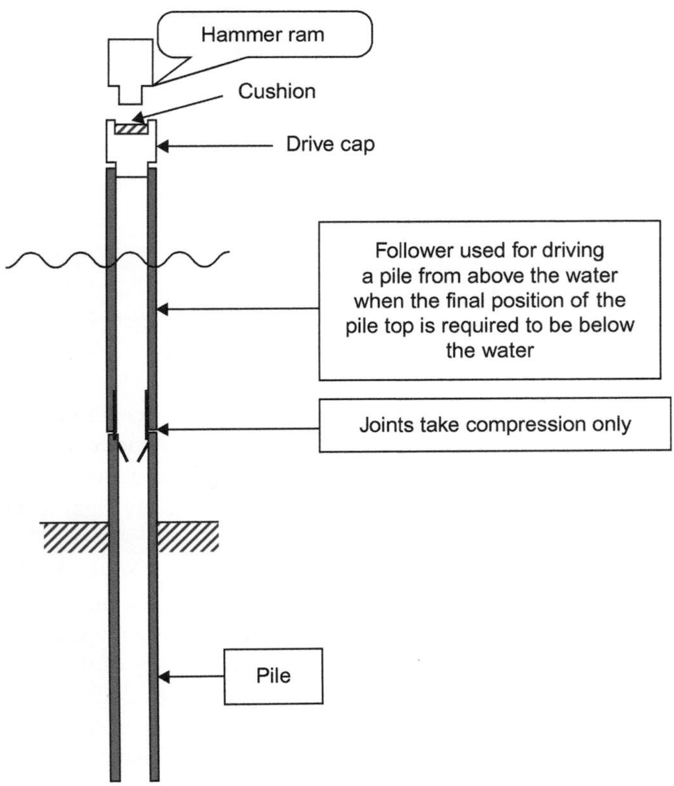

FIGURE 4.11 Typical arrangement for pile driving.

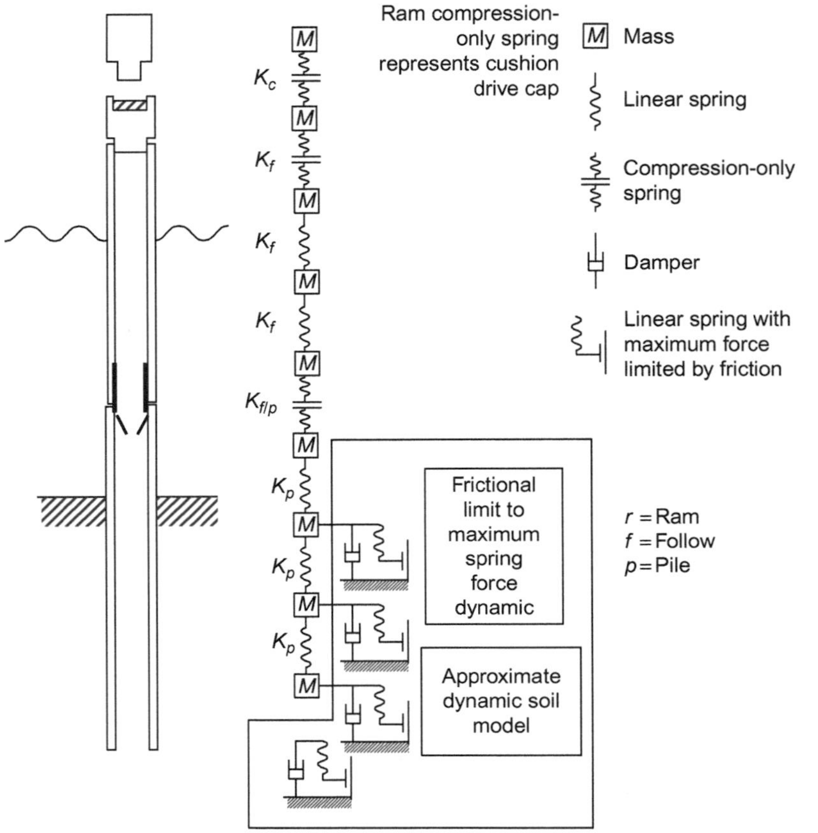

FIGURE 4.12 Dynamic analysis model of pile driving.

such as the location of the maximum stresses down the length of the pile, the number of blows, previous experience with the pile-hammer combination and the confidence level in the analyses. In most cases, special considerations apply to avoid damage to the appurtenances when there are significant driving stresses may be transmitted into the structure.

The static stress during driving may be taken to be the stress resulting from the weight of the pile above the point of calculation plus the pile-hammer components actually supported by the pile during the hammer blows, including any bending stresses resulting therefrom. It is the responsibility of the contractor to control the hydraulic hammers so that the driving energy does not exceed the rated energy, because usually only a limited deviation is considered in the engineering study. It is important to calculate carefully the static stresses present due to the hydraulic hammers, because there are possible variations in driving configurations, for example, in installing and driving the piles without lateral restraint, while at the same time the pile is exposed to wave and wind forces during driving.

In the past, several case histories were reported that describe some of the unusual characteristics of foundations on carbonate soils and their often poor performance. It has been shown from numerous pile load tests that piles driven into weakly cemented and compressible carbonate soils mobilize only a fraction of the capacity, as low as 15%, predicted by conventional design methods for siliceous material of the type generally encountered in the Gulf of Mexico. On the other hand, dense, strongly cemented carbonate deposits can provide higher capacity values than the equation.

The energy is determined primarily by the mass of the ram and its impact velocity:

$$E = \{1/2\} m V^2 \tag{4.50}$$

Note that only 60% to 70% of the energy is typically transferred to the drive cap from the ram, as shown in Figure 4.16. It is obvious that the greater the energy, the greater will be the penetration; on the other hand, the greater will be the risk of damaging the pile. The maximum stress in the stress wave is largely determined by the velocity of the ram. It is worth mentioning that, for easy driving conditions, long duration and a low-stress waveform are the best, and they could be achieved by a heavier and slower ram and a soft cushion.

Using the finite difference method, the ram is represented by a concentrated mass, and the required information about the ram is available from the hammer manufacturer. The efficiency of the hammer depends on the conditions as well as the operating procedures, as shown in Table 4.18 for various hammer types.

4.8.3 Minimum Wall Thickness

API RP2A defines that the minimum wall thickness of the pile, based on the diameter to thickness *(D/t)* ratio of the entire length of a pile, should be small, to avoid a local buckling when the stresses reach the yield strength of the pile material. It is very important to consider the installation process

TABLE 4.18 Efficiency for Different Hammer Types

Hammer	Efficiency
Single-acting steam or air	0.75–0.85
Double-acting steam or air	0.70–0.80
Diesel	0.85–1.0
Hydraulic	0.85–0.95

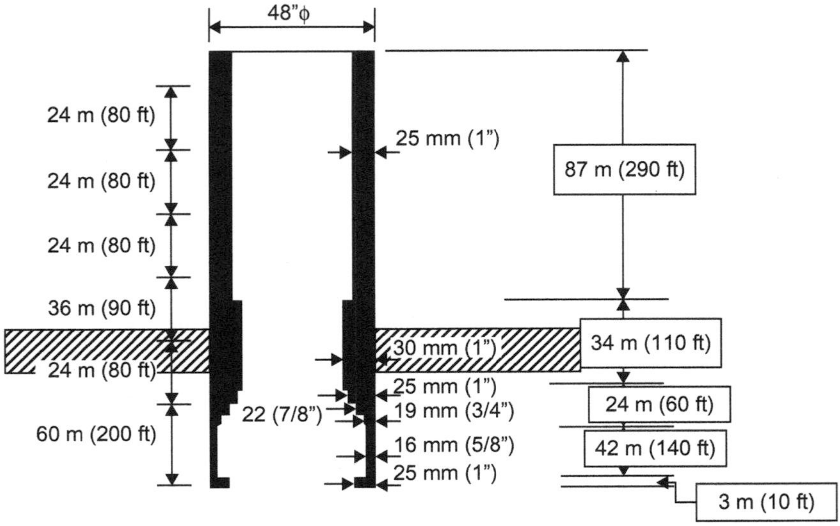

FIGURE 4.13 Example of design and assembly for offshore pile

when choosing the pile thickness and also the service life of the piling, because there should be a corrosion allowance to cover the corrosion effect during the platform's lifetime.

For piles that are to be installed by driving, where sustained hard driving of 820 blows per meter with the largest size hammer is anticipated, the minimum piling wall thickness should be more than

$$t = 6.35 + D/100 \tag{4.51}$$

where t = wall thickness (in inches or mm) and D = diameter (in inches or mm).

The requirement for a smaller D/t ratio when hard driving is expected may be relaxed when it can be shown by past experience or by detailed analysis that the pile will not be damaged during its installation. A typical example of pile thickness at different depths is shown in Figure 4.13.

4.8.4 Driving Shoe and Head

The driving shoe and head are usually the responsibility of the contractor, with approval from the engineering office. The purpose of the driving shoes is to assist piles in penetrating hard layers and to reduce driving resistances. Note that different design considerations apply for each use.

If the soil report indicates that there is a hard layer, the driving shoe should be designed to prevent high driving stresses at and above the transition point between the normal and the thickened sections at the pile tip.

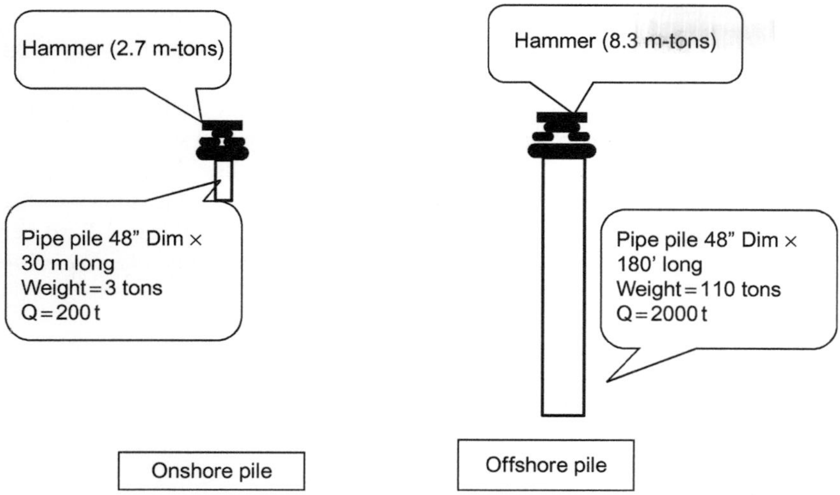

FIGURE 4.14 Comparison of pile driving onshore and offshore.

On the other hand, the design of the pile shoe should be checked to guarantee that the shoe will not reduce the end-bearing capacity of the soil plug below the value assumed in the design. Because external shoes tend to reduce the skin friction along the length of pile above them, external shoes are not used.

The installation contractor is responsible for designing the driving head at the top of the pile, and it should be designed to ensure that it is fully compatible with the proposed installation procedures and equipment.

In an example comparing pile driving onshore with offshore pile driving (as shown in Figure 4.18), the ultimate capacity of the pile onshore is 200 tons and the hammer needs to produce 2.7 tons, while for the offshore pile, the capacity can reach 2000 tons with three times higher energy, for about 8.3 tons with the same pile diameter (48 inches).

4.8.5 Pile Section Lengths

In defining the pile section lengths, consideration should be given to:

1. The capability of the lifting equipment to raise, lower and stab the sections
2. The capability of the lifting equipment to place the pile-driving hammer on the sections to be driven
3. The possibility of a large amount of downward pile movement after the penetration of a jacket leg closure
4. The amount of stress that will develop in the pile section while it is being lifted

5. The pile wall thickness and material properties at field welds
6. Avoiding interference with the planned concurrent driving of neighboring piles
7. The type of soil in which the pile tip is positioned. In addition, static and dynamic stresses due to the hammer weight and operation should be considered. Each pile section on which driving is required should contain a cutoff allowance to permit the removal of material damaged by the impact of the pile-driving hammer. The normal allowance is 2 to 5 ft (0.5 to 1.5 m) per section. This cutoff allowance should be made at a conveniently accessible elevation.

4.9 PILE DRIVABILITY ANALYSIS

The pile drivability analysis has three stages:

1. Evaluation of soil resistance to driving
2. Wave equation analysis
3. Estimate of blowcount versus pile penetration

The procedures used to evaluate the soil resistance to driving are empirical and have been developed from the back-analysis of pile-driving records. Their use is therefore limited to pile drivability assessment by wave equation analysis and they are not intended to provide an estimate of the ultimate axial capacity of foundation piles.

In the present case, as is often true, some of the parameters required for the wave equation analysis (step 2) depend on the maximum achievable penetration, which is calculated in step 3. Thus, to ensure consistency between the steps, an iterative analysis has been carried out.

4.9.1 Evaluation of Soil Resistance to Driving (SRD)

Different procedures are used for evaluating SRD in cohesionless versus cohesive soils. They are discussed below. As is the case for static pile capacity analysis, the components of shaft resistance and end bearing in SRD are evaluated separately, then combined to give total driving resistance (Toolan and Fox, 1977).

The variability of the soil conditions across the site and some anticipated variation in hammer performance are likely to influence the apparent driving resistance. Furthermore, the driving resistance during continuous driving is known to be considerably lower than when driving is restarted after an interruption long enough to allow soil set-up. To account for these factors, upper-bound and lower-bound SRD profiles have been formulated for a given design soil profile, based on the recommendations of Stevens et al. (1982).

4.9.2 Unit Shaft Resistance and Unit End Bearing for Uncemented Materials

In cohesive soil, the unit skin friction has been assessed based on the method proposed by Semple and Gemeinhardt (1981). This method was developed from back-analysis of pile installations in normally consolidated to heavily overconsolidated clays from many areas.

The unit shaft resistance component of SRD is derived using the API RP2A (1984) procedure for static pile capacity, modified by a pile capacity factor F_p, which is a function of the overconsolidation ratio (OCR), as follows:

$$F_p = 0.5(\text{OCR})^{0.3} \tag{4.52}$$

The unit end-bearing component of SRD is taken as $9S_u$.

The OCR is defined as the ratio of the maximum past effective consolidation stress and the present effective overburden stress. OCR is a function of the undrained shear strength ratio (S_u/p'), which is equal to 0.22 in normally consolidated clay with a shear stress angle equal to 26°. In general, OCR should be obtained by CPT.

In cohesionless (granular) soil, shaft resistance and end-bearing components of SRD can be derived using the API (1984) procedure for static pile capacity together with the soil parameters specified by Stevens et al. (1982) for the particular soil type. A limiting skin friction of 15 kPa has been taken for the calcareous and carbonate sand layers.

4.9.3 Upper- and Lower-Bound SRD

Based on Stevens et al. (1982), four cases have been assessed to obtain upper- and lower-bound SRD values in uncemented and weakly cemented soil layers:

1. SRD for lower-bound, coring pile is the outside shaft friction in addition to the inside shaft friction as half the outside. End bearing is unit end bearing multiplied by steel annulus area.
2. SRD for upper-bound, coring pile is similar to the previous case but with the full shaft friction on the inside of the pile.
3. SRD for lower-bound, plugged pile is the outside shaft friction in addition to the end bearing multiplied by full cross-section area.
4. SRD for upper-bound, plugged pile for cohesionless soil layers is with outside shaft friction from case 3 increased by 30% and end bearing from case 3 increased by 50%. For cohesive layers, outside shaft friction is as in case 3, and end bearing from case 3 is increased by 67%.

For the sandstone and limestone layers, plugged conditions are unrealistic, and only the coring cases are considered.

SRD calculations have been made to cover the range in unit values set out above. Sample results are shown in Figure 4.15. As shown in the figure, a range in SRD is provided, reflecting the various combinations in unit values given above. The lower bounds are likely to be applicable to the minimum resistances

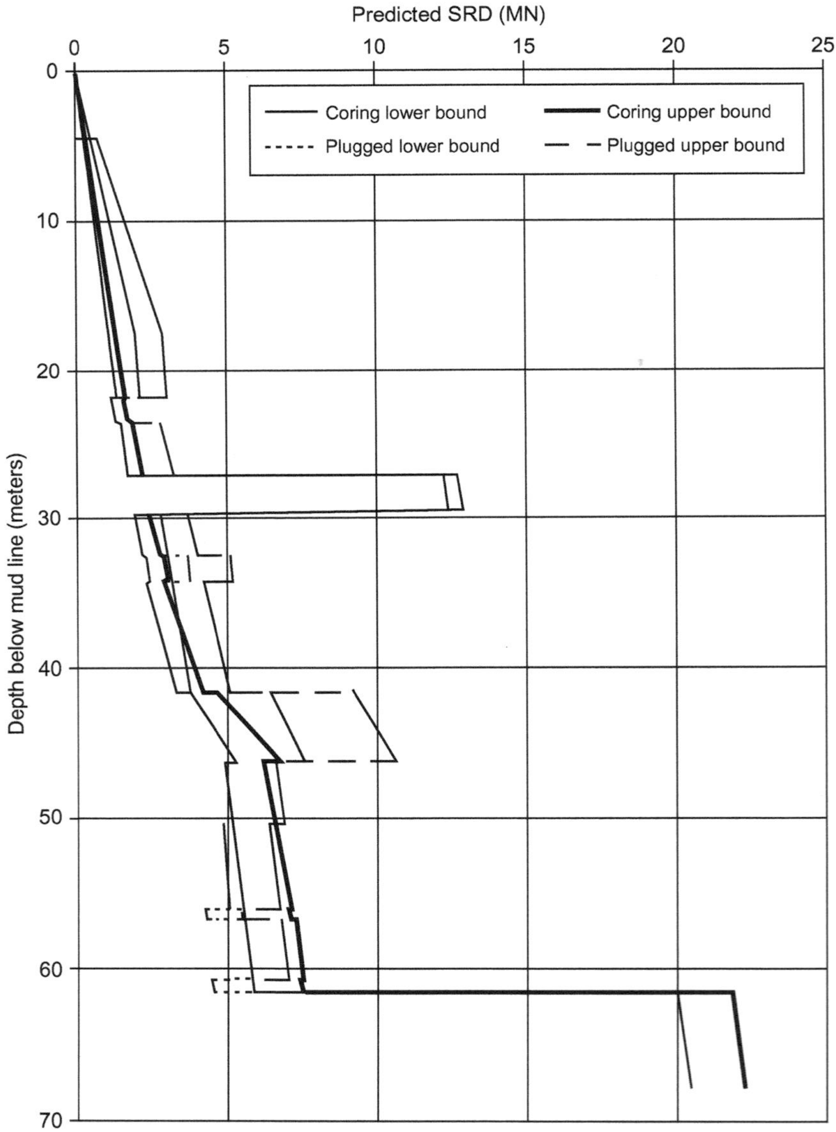

FIGURE 4.15 Example of soil resistance to driving (SRD) for 30-inch pipe pile with $1\frac{1}{4}$-inch thickness.

during continuous driving, with upper bounds indicative of local variations in soil conditions and resistances expected immediately on restarting a drive (i.e., soil set-up condition).

4.9.4 Results of Wave Equation Analyses

Blowcount versus SRD curves have been developed for the hammers listed in Table 4.20, using a software program.

The input parameters for the wave equation analyses differed for each hammer, pile and SRD bound, primarily in terms of the percentage of SRD that was due to skin friction. This has an effect on the amount of energy lost through damping, because of differences in damping values for skin friction and end bearing. The following damping values were used:

Skin damping for purely cohesive soil, 0.65 s/m
Skin damping for cohesionless soil, 0.15 s/m
Tip damping, all soils, 0.50 s/m

Skin damping for a combination of soil types was assessed by taking account of the fraction of total skin resistance due to each type and linearly interpolating between the purely cohesive and purely cohesionless values. Quakes of 0.1 inch were taken in all calculations.

Results are shown in Figure 4.16. There are two curves for each hammer analyzed, corresponding to input parameters associated with the relevant lower- and upper-bound SRD cases.

4.9.5 Results of Drivability Calculations

By combining the SRD versus depth curves and the wave equation results, one obtains predicted blowcount versus depth. Results are shown in Figure 4.17 and are summarized in Table 4.20.

According to a case study in the Red Sea, refusal will occur at 27.2 m penetration if the Delmag 080 hammer is used. For the Menck MRBS 3000, high blowcounts will be experienced during penetration and refusal may occur should, for example, the thickness or strength of cemented material at 27.2 m increase significantly from that identified in the borehole. For the Menck MRBS 3900, and indeed for all the hammers analyzed, refusal is indicated at 61.5 m penetration.

4.9.6 Recommendations for Pile Installation

The lower-bound curve of blowcounts versus depth in Figure 4.21 applies if there is no interruption during driving. The upper-bound curves represent estimates of effects of delays during driving. To ensure an efficient offshore pile-driving operation, it is recommended that delays during driving be avoided if possible. Attention should be paid to ensuring that the blowcount does not become excessive and that no pile-tip damage occurs, for all hammers analyzed.

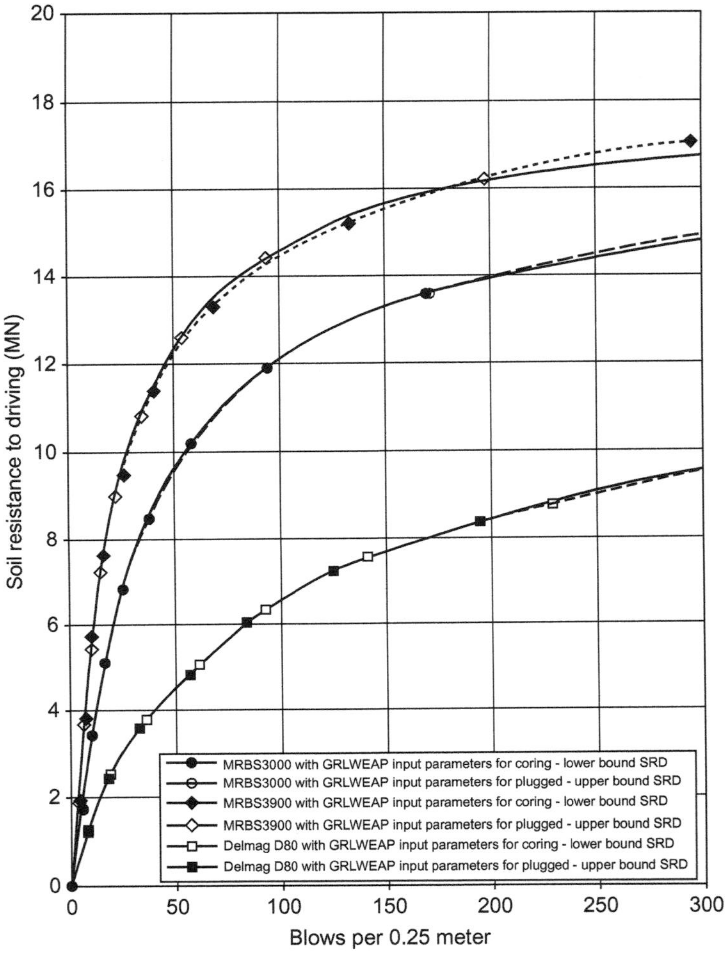

FIGURE 4.16 Blowcount versus SRD curves for three hammer types for 30-inch pipe pile with $1\frac{1}{4}$-inch wall thickness.

Depending on the pile-driving plant finally selected, it may be advisable to have equipment readily available during piling operations in case refusal or particularly high blowcounts occur at 27.2 m below the seabed. If refusal occurs, an assessment may be needed of the effect on ultimate pile capacity and therefore on target penetration.

The pile analysis usually assumes uniform wall thickness over the entire pile length. Should the wall thicknesses of the piles finally selected differ from this, then the drivability analyses should be repeated to assess the effect on blowcounts and achievable penetration. It is recommended that consideration be

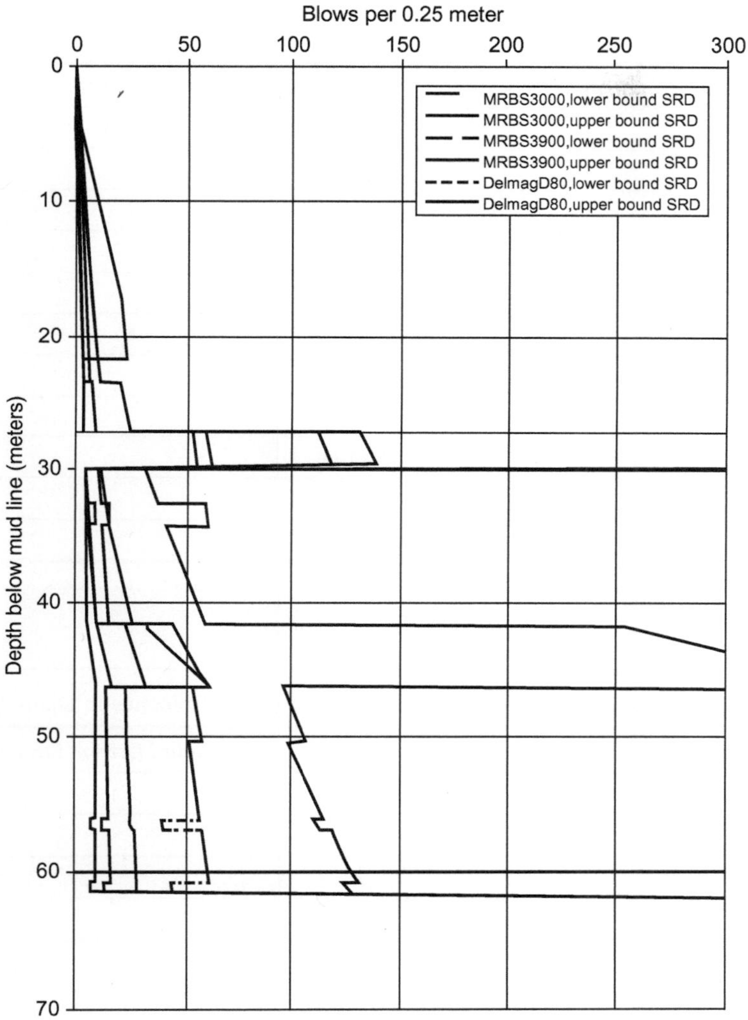

FIGURE 4.17 Blowcount versus depth for three types of hammers and 30-inch pile with $1\frac{1}{4}$-inch wall thickness.

given to incorporation of a pile shoe to prevent undue distress to the pile during driving through the cemented layers.

The shoe should be externally flush, should have an appropriate outer bevel profile at the tip, and should have an increased wall thickness over a suitable length. As with an overall increase in pile wall thickness requiring some additional drivability assessment, the incorporation of a pile shoe will also need to be with regard to its effect on pile-driving behavior.

TABLE 4.19 Minimum Pile Wall Thickness

Pile Diameter, in mm (inches)	Nominal Wall Thickness, in mm (inches)
610 (24)	13(1/2)
762 (30)	14(9/16)
914 (36)	16(5/8)
1067 (42)	17(11/16)
1219 (48)	19(3/4)
1524 (60)	22(7/8)
1829 (72)	25(1)
2134 (84)	28(11/8)
2438 (96)	31(11/4)
2743 (108)	34(13/8)
3048 (120)	37(11/2)

TABLE 4.20 Summary of Drivability Analysis for Pile 30 Inches in Diameter

Hammer	Achievable Penetration	Maximum Driving Stresses
Delmag D-80	27.2	202
MRBS 3000	27.2	220
MRBS 3900	61.5	252

Achievable penetration is based on refusal criteria of 15 blows/0.25 m.

4.10 SOIL INVESTIGATION REPORT

After all soil investigation tests on site and in the laboratory are completed, the following curves, which assist in pile design, should be presented in the soil investigation report: the curve for the relationship between the depth below the seabed and the unit skin friction (Figure 4.18) and the curve for the relationship between the pile depth below the seabed and the unit of end bearing (Figure 4.19).

If the pile diameter is not identified, the study should include the pile capacity in compression for different pile diameters, as presented in Figure 4.21, and the pile capacity in tension for different pile diameters, as shown in Figure 4.22.

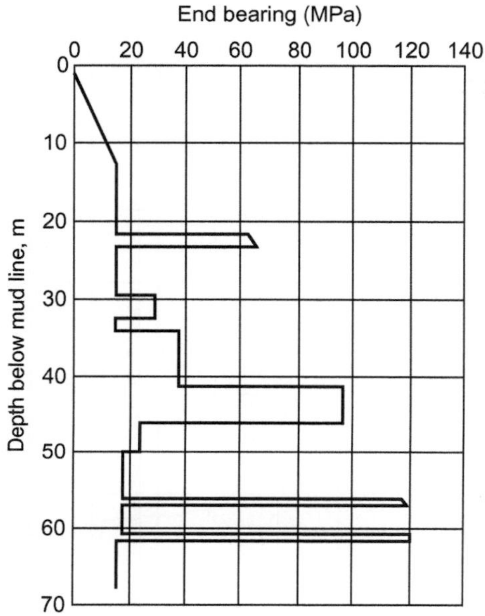

FIGURE 4.18 Relation between the shaft friction capacity and the depth below the seabed.

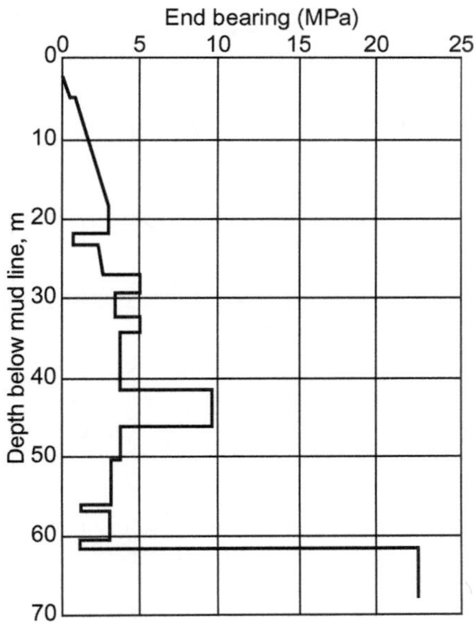

FIGURE 4.19 Relation between the end-bearing capacity and the depth below the seabed.

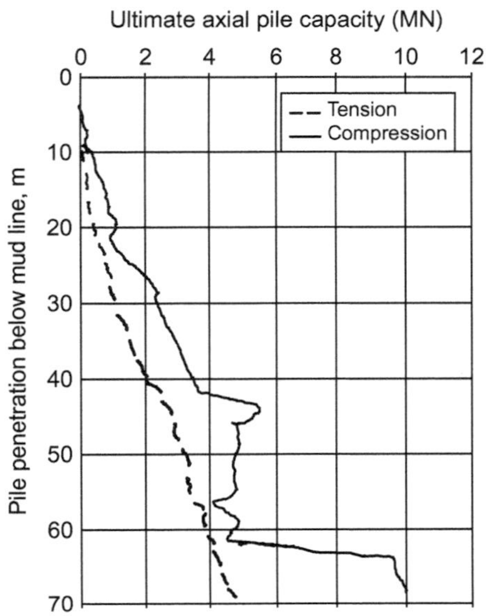

FIGURE 4.20 Relation between the ultimate axial pile capacity in tension and compression.

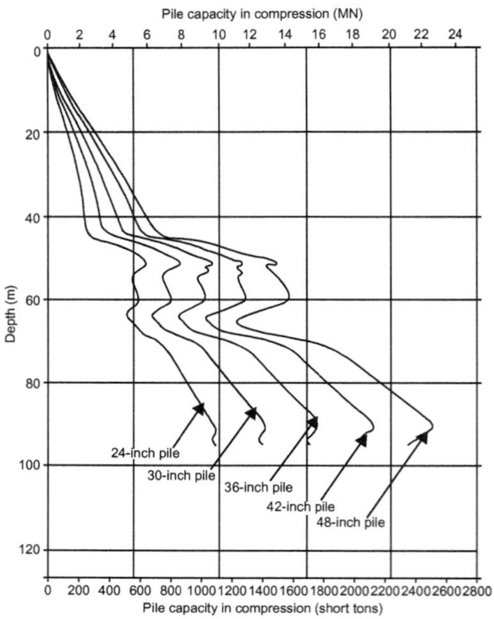

FIGURE 4.21 Pile capacity in compression with depth for different pile diameters.

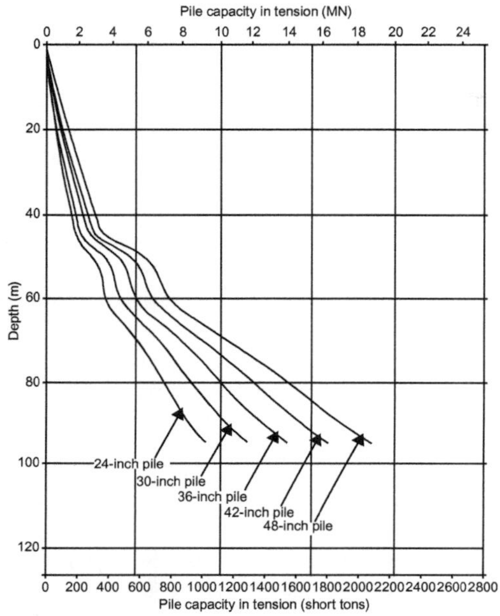

FIGURE 4.22 Pile capacity in tension for different pile diameters.

BIBLIOGRAPHY

American Society for Testing Materials (ASTM), 1968. Book of ASTM standards; with Related Materials, Part 23 Book of ASTM Standards; with Related Materials. American Society for Testing and Materials, University of California. Standards, Part II, USA.

American Society for Testing Materials (ASTM), 1972. Underwater Soil Sampling, Testing, and Construction Control. ASTM Special Technical Publication SOl, ASTM International, ASTM International, 100 Barr Harbor Drive, West Conshohocken, PA 19428-2959, United States.

American Society for Testing Materials (ASTM). 1996. Standard Test Method for Performing Electronic Friction Cone and Piezocone Penetration Testing of Soils, ASTM D 5778-95 (Reapproved 2000), ASTM Standards on Disc Volume 04.09: Soil and Rock (II): D 5714 – Latest, 100 Barr Harbor Drive, West Conshohocken, PA 19428-2959, United States.

American Society for Testing Materials (ASTM), 2004. Standard method of deep quasi-static cone and friction-cone penetration tests of soil. ASTM Standard D 3441, ASTM International, 100 Barr Harbor Drive, West Conshohocken, PA 19428-2959, United States.

American Society for Testing Materials (ASTM), ASTM D-5778 Standard Test Method for Performing Electronic Friction Cone and Piezocone Penetration Testing of Soils, 100 Barr Harbor Drive, West Conshohocken, PA 19428-2959, United States.

Angus, N.M., Moore, R.L., 1982. Scour repair methods in the southern North Sea. In: Proceedings of the 14th Offshore Technology Conference, OTC 4410, Offshore Technology Conference, Houston, TX, pp. 385–395.

API RP2A, 1984. Recommended Practice for Planning, Designing, and Constructing Fixed Offshore Platforms, 15th ed. American Petroleum Institute, Washington, DC.

API RP2A, 2007. Recommended Practice for Planning, Designing, and Constructing Fixed Offshore Platforms, 20th ed. American Petroleum Institute, Washington, DC.

Arup, O., et al., 1986. Research on the behavior of piles as anchors for buoyant structures—Summary report. Offshore Technology Report OTH 86 215, Department of Energy, London, England.

ASTM, 2004. Standard method of deep quasi-static cone and friction-cone penetration tests of soil. ASTM Standard D 3441, ASTM International, West Conshohocken, PA, p. 7.

Audibert, J.M.E., Dover, A.R., 1982. Pile load test: cyclic loads and varying load rates. J. Geotech. Eng. Div. ASCE, 108 (GT3), 501–505.

Bailey, E., Davis, G.L., Henderson, H.O., 1971. Design of an automatic marine corer. Third Annual Offshore Technology Conference. Preprint vol. 1, Paper No. 1365, pp. 397–416.

Baldi, G., Bellotti, R., Ghionna, N., Jamiolkowski, M., Pasqualini, E., 1986. Interpretation of CPTs and CPTUs, 2nd Part: Drained penetration of sands. In: Field Instrumentation and In-Situ Measurements: Proceedings of the 4th International Geotechnical Seminar, Singapore, November 25–27, 1986. Nanyang Technological Institute, Singapore, pp. 143–156.

Bea, R.G., June 1978. Engineering fixed offshore platforms to resist earthquakes. In: Proceeding ASCE Specialty Conference on Earthquake Engineering and Soil Dynamics, Pasadena, California, pp. 1357–1385.

Bea, R.G., 1980. Dynamic response of piles in offshore platforms. In: Proceeding of the Speciality Conference One on Dynamic Response of Pile Foundations—Analytical Aspects, ASCE, Geotechnical Engineering Division, October 30, 1980, pp. 80–109.

Bea, R.G., 1982. Soil strain rate effects on axial pile capacity. In: Proceedings of the 2nd International Conference on Numerical Methods in Offshore Piling, Austin, TX, April 29–30.

Bea, R.G., 1984. Dynamic response of marine foundations. In: Proceedings of the Ocean Structural Dynamics Symposium '84, Oregon State University, Corvallis, USA.

Bea, R.G., Audibert, J.M.E., 1979. Performance of dynamically loaded pile foundations. In: Proceedings of the Second International Conference on Behavior of Offshore Structures, BOSS '79, Imperial College, London, England.

Bea, R.G., Litton, R.W., Nour-Omid, S., Chang, J.Y., Vaish, A.K., 1984. A specialized design and research tool for the modeling of near-field pile-soil interactions. In: Proceedings of the Offshore Technology Conference, OTC 4806, Houston, TX, pp. 249–262.

Bea, R.G., Vahdani, S., Guttman, S.I., Meith, R.M., Paulson, S.F., 1986. Analysis of the performance of piles in silica sands and carbonate formations. In: Proceedings of the18th Annual Offshore Technology Conference, OTC 5145, Houston, TX.

Begemann, H.K.S., 1965. The friction jacket cone as an aid in determining the soil profile. In: Proceedings of the 6th ICSMFE, Montreal, Quebec, vol. I, pp. 17–20.

Bijaker, E.W., 1980. Physical causes of scour. Paper presented at a seminar on scour held by Society of Underwater Technology, London, England. Proceedings of a one day seminar 16 December 1980 Society for Underwater Technology London (1980).

Bogard, J.D., Matlock, H., 1990. Applications of model pile tests to axial pile design. In: Proceedings of the 22nd Annual Offshore Technology Conference, OTC 6376, Houston, TX.

Bogard, J.D., Matlock, H., Audibert, J.M.E., Bamford, S.R., 1985. Three years' experience with model pile segment tool tests. In: Proceedings of the Offshore Technology Conference, OTC 4848, Houston, TX.

Bond, A.J., Jardine, R.J., 1995. Shaft capacity of displacement piles in a high OCR clay. Geótechnique 45 (1), 3–23.

Briaud, J.L., Garland, E.E., Felio, G.Y., May 1984. Loading rate parameters for piles in clay. In: Proceedings of the Offshore Technology Conference, Houston, TX.

Chow, W.Y., Herbich, J.B., 1978. Scour around a group of piles. In: Proceedings of the 10th Annual Offshore Technology Conference, OTC 3308, Houston, TX, pp. 2243–2254.

Clarke, J. (Ed.), 1993. Large-Scale Pile Tests in Clay. Thomas Telford, London.

Clausen, C.J.F., Aas, P.M., Karlsrud, K., 2005. Bearing capacity of driven piles in sand, the NGI approach. In: Proceedings of the International Symposium on Frontiers in Offshore Geotechnics [ISFOG—05]. A.A. Balkema, Rotterdam, pp. 677–682.

Coyle, H.M., Reese, L.C., 1966. Load transfer for axially loaded piles in clay. ASCE. J. Soil. Mech. Found Div. 92 (1052), 1–26.

CUR Centre for Civil Engineering Research and Codes, 2001. Bearing capacity of steel pipe piles. CUR, Gouda, CUR-Report.

De Reister, J., 1971. Electric penetrometer for site investigations. ASCE. J. SMFE. Div. 97(SM-2), 457–472.

Desai, C.S., Holloway, D.M., 1972. Load-deformation analysis of deep pile foundation. In: Proceedings of the Symposium on Applications of the Finite Element Method in Geotechnical Engineering, U.S. Army Engineers Waterways Experiment Station, Vicksburg, Mississippi, pp. 629–656.

Doyle, E.H., McClelland, B., Ferguson, G.B., 1971. Wire-line vane probe for deep penetration measurements of ocean sediment strength. Third Annual Offshore Technology Conference. Preprint vol. 1, Paper No. 1327, pp. 21–32.

Dunnavant, T.W., Clukey, E.C., Murff, J.D., 1990. Effects of cyclic loading and pile flexibility on axial pile capacities in clay. In: Proceedings of the Offshore Technology Conference, OTC 6374, Houston, TX.

Focht, J.A., Jr., Kraft, L.M., Jr., 1986. Axial performance and capacity of piles. In: McClelland, B., Reifel, M.D. (Eds.), Planning and Design of Fixed Offshore Platforms. Van Nostrand Reinhold Company, New York.

Fugro, 1995. Final report, foundation design—Bridge piles, Jamuna Bridge, Bangladesh. Fugro Engineers BV Confidential Report No. K-2380/120 to HDEC, June 1995.

Fugro, 2004. Axial pile capacity design method for offshore driven piles in sand. Fugro Engineers BV Report No. P1003 to API, Issue 3, August 2004.

Gallagher, K.A., St. John, H.D., 1980. Field scale model studies of piles as anchorages for buoyant platforms. Paper EUR 135, European Offshore Petroleum Conference, London, England.

Georgiadis, M., 1983. Development of *p-y* curves for layered soils. In: Proceedings of the Conference on Geotechnical Practice in Offshore Engineering, Austin TX. ASCE, New York, pp. 536–545.

Hironka, M.C., Green, W.C., 1971. A remote controlled seafloor incremental corer. Third Annual Offshore Technology Conference, Preprint vol. 1, Paper No.1325, pp. 13–20.

Hvorslev, M.J., 1949. Subsurface exploration and sampling of soils for civil engineering purposes. Waterways Experiment Station, Vicksburg, Mississippi. (Reprinted by Engineering Foundation, New York, 1962.)

IRTP/ISSMGE International Society of Soil Mechanics and Geotechnical Engineering, 1999. International Reference Test Procedure for the Cone Penetration Test (CPT) and the Cone Penetration Test with Pore Pressure (CPTU). Report of the ISSMGE Technical Committee 16 on Ground Property Characterisation from In-situ Testing. In: Proceedings of the Twelfth European Conference on Soil Mechanics and Geotechnical Engineering, Proceedings of the XIIth ECSMGE. 2195-2222. Balkema, Amsterdam.

ISO (International Organization for Standardization), 2005. ISO 22476-1 (DIS 2005), Geotechnical Investigation and Testing—Field Testing—Electrical Cone and Piezocone Penetration Tests, International Standard ISO 22476-1.

Jamiolkowski, M., Ghionna, V.N., Lancellotta, R., Pasqualini, E., 1988. New correlations of penetration tests for design practice. In: De Ruiter, J. (Ed.), Penetration Testing 1988: Proceedings of the First International Symposium on Penetration Testing, ISOPT-1, vol. 1. A.A. Balkema, Rotterdam, pp. 263–296.

Jardine, R., Chow, F., Overy, R., Standing, J., 2005. ICP Design Methods for Driven Piles in Sands and Clays. Thomas Telford, London.

Karlsrud, K., Haugen, T., 1985. Behavior of piles in clay under cyclic axial loading—Results of field model tests. In: Proceedings of the 4th International Conference on Behavior of Offshore Structures, BOSS '85. Delft, The Netherlands.

Karlsrud, K., Nadim, F., Haugen, T., 1986. Pile in clay under cyclic axial loading field tests and computational modeling. Proceedings of the 3rd International Conference on Numerical Methods in Offshore Piling. Nantes, France.

Kolk, H.J., 2000. Deep foundations in calcareous sediments. In: Al-Shafei, K.A. (Ed.), Engineering for Calcareous Sediments: Proceedings of the Second International Conference on Engineering for Calcareous Sediments, Bahrain, February 21–24, 1999, vol. 2. A.A. Balkema, Rotterdam, pp. 313–344.

Kolk, H.J., Van Der Velde, E., 1996. A reliable method to determine friction capacity of piles driven into clays. In: Proceedings of the 28th Annual Offshore Technology Conference, OTC 7993, Houston, TX.

Kolk, H.J., Baaijens, A.E., Senders, M., 2005. Design criteria for pipe piles in silica sands. In: Gourvenec, S., Cassidy, M. (Eds.), Frontiers in Offshore Geotechnics ISFOG 2005: Proceedings of the First International Symposium on Frontiers in Offshore Geotechnics, University of Western Australia, Perth, Taylor & Francis, London, pp. 711–716.

Kraft, L.M., Lyons, C.G., 1974. State of the art: Ultimate axial capacity of grouted piles. In: Proceedings of the 6th Annual Offshore Technology Conference, OTC 2081, Houston, TX.

Kraft,, L.M., Jr., Cox, W.R., Verner, E.A., 1981. Pile load tests: Cyclic loads and varying load rates. ASCE J. Geotech. Eng. Div. 107 (GT11), January 1981.

Kraft, L.M., Focht, J.A., Amarasinghe, S.F., 1981. Friction capacity of piles driven into clay. ASCE J. Geotech. Eng. Div. 107 (GT11), 1521–1541.

Ladd, C.C., Foott, R., 1974. New design procedure for stability of soft clays. ASCE J. Geotech. Eng. Div. 100 (GT7), 763–786.

Lehane, B.M., Jardine, R.J., 1994. Displacement pile behavor in glacial clay. Cana. Geotech. J. 31 (1), 79–90.

Lehane, B.M., Randolph, M.F., 2002. Evaluation of a minimum base resistance for driven pipe piles in siliceous sand. ASCE J. Geotech. Geoenviron. Eng. 128 (3), 198–205.

Lehane, B.M., Schneider, J.A., Xu, X., 2005a. A review of design methods for offshore driven piles in siliceous sand. University of Western Australia Geomechanics Group, Report No. GEO: 05358, September, 2005.

Lehane, B.M., Schneider, J.A., Xu, X., 2005b. The UWA-05 method for prediction of axial capacity of driven piles in sand. In: Gourvenec, S., Cassidy, M. (Eds.), Frontiers in Offshore Geotechnics ISFOG 2005: Proceedings of the First International Symposium on Frontiers in Offshore Geotechnics, University of Western Australia, Perth, 19–21 September 2005. Taylor & Francis, London, pp. 683–689.

Lysmer, J., 1978. Analytical procedures in soil dynamics. Report No. UCB/EERC-78/29, presented at the ASCE Geotechnical Engineering Division Specialty Conference on Earthquake Engineering and Soil Dynamics, Pasadena, CA.

Matlock, H., 1970. Correlations for design of laterally loaded piles in soft clay. In: Proceedings of the 2nd Annual Offshore Technology Conference, OTC 1204, Houston, TX.

Matlock, H., Foo, S.H.C., 1979. Axial analysis of piles using a hysteretic and degrading soil model. In: Proceedings of the Conference on Numerical Methods in Offshore Piling, Institution of Civil Engineers, London, England.

McAnoy, R.P.L., Cashman, A.C., Purvis, O., 1982. Cyclic tensile testing of a pile in glacial till. In: Proceedings of the 2nd International Conference on Numerical Methods in Offshore Piling, Austin, TX.

McClelland, B., 1972. Techniques used in soil sampling at sea. Offshore Magazine 32 (3), 51–57.

McClelland, B., Ehlers, C.J., 1986. Offshore geotechnical site investigations. In: McClelland, B., Reifel, M.D. (Eds.), Planning and Design of Fixed Offshore Platforms, Van Nostrand Reinhold, New York.

Meyer, P.C., Holmquist, D.V., Matlock, H.T.L., 1975. Computer predictions of axially loaded piles with non-linear supports. Proceeding of the Offshore Technology Conference, OTC 2186, Houston, TX.

Meyerholf, G.G., 1956. Penetration tests and bearing capacity of cohesionless soils. J. ASCE, Vol. 82, No. SMI, January, pp. 1–19.

Miller, T.W., Murff, J.D., Kraft, L.M., 1978. Critical state soil mechanics model of soil consolidation stresses around a driven pile. In: Proceedings of the 10th Annual Offshore Technology Conference, OTC 3307, Houston, TX.

Murff, J.D., 1980. Pile capacity in a softening soil. In. J. Numer. Anal. Methods. Geomech. 4(2), 185–189.

Niedoroda, A.W., Dalton, C., Bea, R.G., 1981. The descriptive physics of scour in the ocean environment. In: Proceedings of the 13th Annual Offshore Technology Conference, Houston, TX, OTC 4145, pp. 297–304.

NORSOK Standard G-001 Rev. 2, October 2004. Marine soil investigations.

Novak, M., Sharnouby, B.E., 1983. Stiffness constants of single piles. J. Geotech. Eng., ASCE, 109 (7), 961–974.

Ocean Industry, 1970. New development in ocean bottom sampling. 5(10), 32.

Olson, R.E., Analysis of Pile Response Under Axial Loads, Report to API, USA, Geotechnical engineering report / Geotechnical Engineering Center, University of Texas at Austin, December 1984.

O'Neill, M.W., Hassan, K.M. 1994. Drilled shafts: Effects of construction on performance and design criteria. In: Proceedings of the International Conference on Design and Construction of Deep Foundations, 1. U.S. Federal Highway Administration (FHWA) pp. 137–187.

O'Neill, M.W., Murchison, J.M. 1983. An evaluation of *p-y* relationships in sands. Prepared for the American Petroleum Institute Report PRAC 82-41-1. University of Houston, Houston, TX.

O'Neill, M.W., Hawkins, R.A., Audibert, J.M.E., 2019. Installation of pile group in overconsolidated clay. ASCE J. Geotech. Eng. Div. 108(GT11), 1369–1386.

Peck, R.B., Hanson, W.E., Thornburn, T.H., 1974. Foundation Engineering. John Wiley & Sons, New York, p. 312.

Pelletier, J.H., Doyle, E.H., 1982. Tension capacity in silty clays—Beta pile test. In: Proceedings of the 2nd International Conference on Numerical Methods of Offshore Piling, Austin, TX.

Pelletier, J.H., Sgouros, G.E., 1987. Shear transfer behavior of a 30-inch pile in silty clay. In: Proceedings of the 19th Annual Offshore Technology Conference, OTC 5407, Houston, TX.

Pelletier, J.H., Murff, J.D., Young, A.C., 1995. Historical development and assessment of current API design methods for axially loaded piles. In: Proceedings of the 27th Annual Offshore Technology Conference, OTC 7157, Houston, TX.

Poulos, H.G., 1983. Cyclic axial pile response—Alternative analyses. In: Proceedings of the Conference on Geotechnical Practice in Offshore Engineering. Austin, TX, ASCE, New York, pp. 403–421.

Randolph, M.F., 1983. Design considerations for offshore piles. In: Proceedings of the Conference on Geotechnical Practice in Offshore Engineering, Austin, TX, ASCE, New York, pp. 422–439.

Randolph, M.F., Leong, E.C., Houlsby, G.T., 1991. One-dimensional analysis of soil plugs in pipe piles. Géotechnique 61 (4), 587–598.

Randolph, M.F., Murphy, B.S., 1985. Shaft capacity of driven piles in clay. In: Proceeding of the 17th Annual Offshore Technology Conference, OTC 4883, Houston, TX.

Randolph, M.F., Wroth, C.P., 1979. An analytical solution for the consolidation around a driven pile. Int. J. Numer. Anal. Methods in Geomech. 3 (3), 217–230.

Reese, L.C., Cox, W.R., 1975. Field testing and analysis of laterally loaded piles in stiff clay. In: Proceedings of the 5th Annual Offshore Technology Conference, OTC 2312, Houston, TX.

Roesset, J.M., Angelides, C., 1979. Dynamic stiffness of piles. In: Proceedings of the Conference on Numerical Methods in Offshore Piling. London, England, pp. 75–81.

Seed, H.B., De Alba, P., 1986. Use of SPT and CPT tests for evaluating the liquefaction resistance of sands. In: Proceedings of In Situ '86. ASCE, New York.

Semple, R.M., Gemeinhardt, J.P., 1981. Stress history approach to analysis of soil resistance to pile driving, vol. 1. Offshore Technology Conference, Houston, TX, pp. 165–172.

Semple, R.M., Rigden, W.J., 1984. Shaft capacity of driven pipe piles in clay. In: Proceedings of the Symposium on Analysis and Design of Pile Foundations, San Francisco, ASCE, New York.

Smith, E.A.L., 1962. Pile driving analysis by the wave equation, Part 1, Paper No. 3306. Transactions ASCE 127, pp. 1145–1193.

Stevens, R.S., Wiltsie, E.A, Turton, H., 1982. Evaluating pile drivability for hard clay, very dense sand and rock. In: Proceedings of the 14th Offshore Technology Conference, vol. 1. Houston, TX, pp. 465–482.

Thompson, G.W.L., Jardine, R.J., 1998. The applicability of the new Imperial College design method to calcareous sands. In: Proceedings of the Conference on Offshore Site Investigations and Foundation Behavior. Society for Underwater Technology, London, pp. 383–400.

Tomlinson, M.J., 1994. Pile Design and Construction Practice, fourth ed. E. and F.N. Spon, London.

Toolan, F.E., Fox, D.A., 1977. Geotechnical planning of piled foundations for offshore wellhead jacket centers. Proceedings of the Institute of Civil Engineers 62 (Part 1), 221–244.

Vermeiden, J., 1948. Improved sounding apparatus, as developed in Holland since 1936. In: Proceedings of the 2nd International Conference on Soil Mechanics, vol. 1. Rotterdam, pp. 280–287.

Vijayvergiya, V.N., 1977. Load movement characteristics of piles. In: Proceedings of the Ports '77 Conference, vol. II. ASCE, pp. 269–284.

Whitehouse, R., 1998. Scour at Marine Structures. Thomas Telford, London.

Wood, D.M., 1982. Laboratory investigations of the behavior of soils under cyclic loading: A review. In: Pande, G.N., Zienkiewicz, O.C. (Eds.), Soil Mechanics—Transient and Cyclic Loads. John Wiley & Sons, New York.

Youd, T.L., Idriss, I.M., 2001. Liquefaction resistance of soils: Summary report from the 1996 NCEER and 1998 NCEER/NSF workshops on evaluation of liquefaction resistance of soils. ASCE J. Geotech. Geoenviron. Eng. 127 (4), 297–313.

Chapter 5

Fabrication and Installation

5.1 INTRODUCTION

The construction of a fixed offshore platform is very specific. Therefore, the contractor responsible for fabrication and installation should be specialized and have a competent staff and reasonable facilities for the project. During this phase, the design of the platform should be checked against transportation, lifting and installation. There should be strong communication and interface management in the project between the engineering office and the fabrication and construction contractors. In some cases, the sea fastening or lifting will be done by the contractor but it should be checked and approved by the engineering company. Note that the launching and lifting of the platform component are designed by the engineering company but the input data for the analysis are delivered to the engineering company based on available barges, cranes and other equipment.

5.2 CONSTRUCTION PROCEDURE

The engineering firm performing the design of a jacket should take into consideration the lifting, launching or self-floating, which depend primarily on the available offshore installation equipment and the water depth. In general, the preference is to lift the jacket into place. The size of jackets has been increasing as offshore lifting capacity has grown. For jackets in shallow water, the height of the jacket is of the same order as the plan dimensions, so the erection is usually carried out vertically in the same direction as the final installation. Therefore, in this case the jacket may be lifted or skidded onto the barge.

Jackets designed for deeper water depths are usually erected on their side. Such jackets are loaded by skidding out onto a barge. Historically, most large jackets have been barge launched. This method of construction usually involves additional flotation tanks and extensive pipework and valving to enable the legs to be flooded for ballasting the jacket into the vertical position on site. This method of construction is currently applicable for jackets up to 25,000 tons. The key fundamental elements are the contract strategy, the quality plan and the requirements during construction until commissioning and start-up.

In considering the construction philosophy and contract strategy, the objectives of achieving quality requirements and efficiency are of fundamental importance.

Construction of the offshore structure jacket goes through a series of very distinct stages from the start of fabrication until the load-out. The stages begin with obtaining the steel sections, which are delivered from the manufacturer. The intiation of fabrication in some stages will be manual, with less quality control. Thus, efficiency decreases as progress is made through the operations. A third basic consideration is that risk increases with each progressive stage.

It is clear, therefore, that, as a general principle, as much work as possible should be undertaken in the earlier, more productive, higher-quality, less risky phases of the project.

Some of the principles that reduce the time and cost of construction are:

- Subdivision into as large components and modules as it is possible to fabricate and assemble.
- The fabrication of major components in the best location and under the best conditions applicable to each component.
- Delivery of critical materials by early planning of the flow of components to their assembly site.
- Providing adequate facilities and equipment for assembly, including such items as synchrolifts and heavy-lift cranes.
- Simplification of the configurations and standardization of details, grades and sizes. Avoidance of excessively tight tolerances.
- In general, structural systems should be selected that utilize skills and trades on a relatively continuous and uniform basis. Procedures that are overly sensitive to weather conditions should be avoided; processes that are weather-sensitive should be completed during shop fabrication (for example, application of the protective coating).

The quality-management system is a vital and integral component of all aspects of offshore fabrication because small mistakes can affect the integrity of the structure over time. Essentially, quality management involves ensuring that what is produced is what is needed. The requirements for documentation, hold point, audits, reviews and corrective actions are part of the quality-assurance process. They are crucial tools for controlling the project's execution and for providing verifiable evidence of the fabricator's competence.

Quality control, inspection and testing should be performed during all phases of construction to ensure that specified requirements are being met. The most effective quality scheme prevents the introduction of defective materials and workmanship into a structure, rather than finding problems after they occur.

5.3 ENGINEERING OF EXECUTION

Practically, the engineering firm that performed the design should follow the execution during each phase to ensure that the design requirements are fulfilled, and the firm should be available in case of a change request or for clarification. A general method of execution depends on the jacket design stage, because the shape of the jacket, its form and its properties require quite specific methods of load-out, offshore transportation and installation. Note that the construction phase is the contractor's responsibility, and there is considerable interfacing of engineering requirements. In the earlier phases (i.e., procurement through assembly and erection), the contractor, while being limited by design specification requirements, has freedom of choice with regard to the exact method of execution adopted. However, in all phases the contractor is required to demonstrate that the methods adopted are compatible with the specification requirements and do not affect the integrity of the structure.

Each phase of execution has its own specific engineering requirements, which are determined by the processes executed during that phase. The processes range from those that are largely repetitive early in execution to one-off activities in the latter phases. Accordingly, the engineering that supports procurement and shop fabrication is voluminous but repetitive (e.g., material take-offs, shop drawings, cutting plans, etc.). The assembly and erection phases are supported by a mix of repetitive engineering, as scaffolding, and specific studies for a limited series of activities. The volume of construction work for a large jacket is usually around 130,000–150,000 hours.

When designing larger components, consideration must be given to their subdivision into elements that will not distort when fabricated and that can be relatively easily assembled without welding/dimensional problems. For instance, joints are categorized as either complex or simple from the execution viewpoint, based on the number of separate fitting-welding nondestructive testing (NDT) cycles required during fabrication and the possibility of automatic welding between the node can and the tubular member during sub-assembly. The number of fitting or welding-NDT cycles depends on the existence of ring stiffeners and the number and disposition of stubs. For reasons relating to weld distortion and to allow automatic welding, it is almost essential that ring stiffeners be installed prior to fitting/welding of stubs. This adds an extra cycle to the fabrication of the node. Thus, ring stiffeners are best avoided. Where this is not possible, care should be taken to define them at an early stage on critical joints.

Node stubs can be classified as simple or overlapping. Overlapping stubs add at least one complete cycle to node fabrication and should therefore be avoided where possible. The minimum separation between the weld toes of adjacent simple stubs is typically specified as 50 mm. However, this distance is too small to allow simultaneous welding of adjacent stubs and a distance of 150 mm is more practical.

5.4 FABRICATION

The main element in fabrication and erection is the welding and its quality control. Welding procedures should be well prepared, by defining the steel grades, joint design, thickness range, welding process, welding consumables, welding parameters, principal welding position, preheating at working temperature and post-weld heat treatment.

The main factor in controlling the welding quality is selection of the welder. Ideally, the welder will have been continuously working for 6 months before the start of work. In addition, all welders should be qualified for the type of work assigned to them and certified accordingly.

After fabrication, non-destructive testing (NDT) should be performed. In most cases, NDT includes radiographic testing, such as gamma or x-ray, ultrasonic testing (UT) and magnetic particle (MP) testing. Both the weld itself and the heat-affected zone should have notch toughness properties equal to those specified for the members.

If you perform a macro-section cut through the weld, it should present a regular profile, with smooth transitions to the base material without significant undercuts or excessive reinforcement. The quality-control supervisor should not accept any cracks or cold lap as lack of fusion. The fracture mechanics toughness of heavy welded joints should be verified by crack-opening displacement tests.

The steel itself should be tested by mechanical testing, such as tensile tests, bend tests, Charpy V-notch tests and hardness tests.

The construction supervisor should postpone any welding when surfaces are humid or damp. In case of high wind speed, welding should be postponed unless there is an arrangement for suitable protection. In cold weather, heating of the enclosed space can be used to raise the temperature above the dew point, so that the groove for welding is dry at the time of welding, since preheating removes moisture.

Note that stress relief is normally not required for the range of wall thickness used in the jackets and piles of offshore structures in moderate environments, such as the Gulf of Mexico (GoM); on the other hand, it is frequently required for the thicker members of large deck structures and for the joints of the thicker-walled jackets of North Sea platforms.

Manual welding of all higher-strength steels and of normal-strength steel having a carbon equivalent greater than 0.41 should be carried out with low-hydrogen electrodes.

For special structural steels and for all repair welding, Det Norske Veritas (DNV) requires the use of extra-low-hydrogen electrodes. It is recommended (Gerwick, 2007) that all piling be welded with low-hydrogen electrodes in order to prevent fracture under impact.

Welding consumables should be kept in sealed moisture-proof containers at 20°C–30°C, but in any event at least 58°C above ambient temperature. Opened containers should be stored at 70°C–150°C, depending on type of electrode. When electrodes are withdrawn for use, they should be kept in heated containers and used within 2 hours. Consumables that have been contaminated by moisture, rust, oil, grease or dirt should be discarded. Surfaces to be welded should be free of mill scale, slag, rust, grease, oil and paint. Edges should have a smooth and uniform surface.

The fitter should be working before the start of welding. A check of the dimensions should be performed. Misalignment between parallel members should not exceed 10% of the thickness or 3 mm. If the thickness of abutting members differs by more than 3 mm, the thicker member should be tapered by grinding or machining to a slope of 1:4 or flatter, as shown on Figure 5.1. Certain completed welds that are critical for fatigue endurance may be required to be ground to a smooth curve. This also reduces the probability of brittle fracture.

Welds that are essentially perpendicular to the direction of applied fluctuating stresses in members important to the structural integrity are normally to be of the full-penetration type and, where possible, they should be welded from both sides. Intersecting members for which the welding details have not been specified in the design should be joined by complete-penetration groove welds.

The construction contractor should detail all lifting plates, padeyes and others that are subject to dynamic impact stress, so that welds are not perpendicular to the principal tension, taking into consideration that welds acting in shear are much less sensitive to cracking than welds in tension.

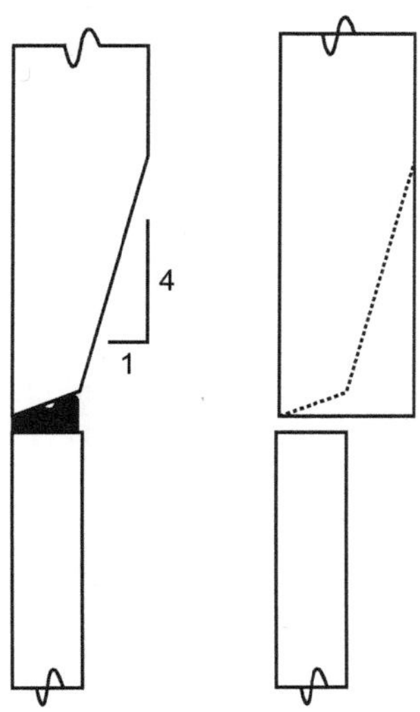

FIGURE 5.1 Grinding or machine tapering of the steel plate for welding penetration.

Full-penetration welds must be used and fillet welds avoided. However fillet welds for sealing purposes are required by DNV to have a leg length of at least 5 mm, whereas API RP2A requires only 3 mm. All temporary plates and fittings should be subjected to the same requirements for welding procedures and testing as the material of the member to which they are affixed.

Temporary cutouts should be of sufficient size to allow sound replacement. In order to reduce the stress concentration, the corner should be rounded. In general, if welds are perpendicular to the principal tension of a member subjected to dynamic impact, then great care must be taken to avoid undercutting.

If a defective weld is found, the defect should be rectified by grinding, machining or welding, as required. Welds with insufficient strength, ductility, or notch toughness should be completely removed prior to repair. If arc-air gouging is used to remove a defective weld, it should be followed by grinding. Whenever a discontinuity is removed, the gouged and ground area should be examined by MP testing or other suitable methods to verify complete removal. Repair welding should use extra-low-hydrogen electrodes and an appropriate preheating temperature, usually 25°C above the level used for production welding and at least 100°C.

All welds should be subjected to visual inspection and NDT as required by the project specifications for fabrication and construction process. As part of quality assurance, all destructive testing should be properly documented and identified so that the tested areas will be ready to be a reference for any audit during fabrication and construction and after completed installation of the structure.

Accurate cutting and beveling require more care and consequently take more time than welding and its quality-control checks.

It is useful to control cutting by the computerized technique, which ensures that all intersecting tubulars will fit properly. In many cases, the welding can then be carried out by semi-automatic welding equipment.

Because documentation is very important during the complicated construction procedure required for most offshore structures, the contractor should make a special effort to set up a quality-assurance system that will ensure proper records of all testing and that will be ready for any audit from a client or an internal audit.

Welding machines must be properly grounded to prevent underwater corrosion damage. Since welding machines are normally operated by direct current (DC), the discharge to ground may occur under water at piping penetrations or other similar points of concentration.

Fabrication of offshore steel structures should follow applicable provisions of codes, such as, for example, the AISC specification for the design, fabrication, and erection of structural steel for buildings.

The offshore tubular piles are fabricated with longitudinal seam welds and circumferential butt welds, according to the API-RP2A requirement, which states that spiral welded pipe cannot be recommended. There are always limitations on pile wall thickness and diameter. Generally, their selection is affected by their application to offshore structures and deepwater marine projects, and the available hammer capability that will be used. However, advances in the control and reliability of technology are continually being made. In China, for the Hangzhou Bay Bridge, spiral-welded piles 2.0 m in diameter and with a 28-mm wall thickness have been successfully installed.

The steel beams from rolled shapes or box girders can be spliced, but for cantilever beams, no splice should be located closer to the point of support than one-half the cantilevered length. For continuous beams, there should be no splice in the middle one-fourth of the span, or in the eighth of the span nearest a support, or over a support.

The most difficult joint is the X joint of two or more tubular members that connect in the same node. In most normal practice, the thicker member with the larger diameter should continue through the joint and the smaller member frame into it. In a number of recent large and important jackets, the intersection node is specially fabricated so that several or all intersecting members are continuous through the joint. In this case, the node is fabricated separately, so that it can be properly treated in the shop, and the members are joined to the node by simple full-penetration butt welds, as shown in Figures 5.2 and 5.3.

This same procedure has been employed quite effectively for jackets that have to be completed at remote areas, because the nodes are fabricated separately and shipped to the site. Then the main legs and braces are joined by butt welds. Cast steel nodes are being used increasingly, in order to eliminate the critical welding details.

Typical details for the proper bevel and weld for tubular members framing into or overlapping another member are shown in Figure 5.3. Grinding the external profile of the weld may be required in order to improve the fatigue endurance.

In most cases, the web-to-flange connection consists of continuous double-fillet welds. Welds should have a concave profile and a smooth transition into flange and web. The connection between flanges and plates for stiffening the flanges should be a full-penetration weld made from both sides.

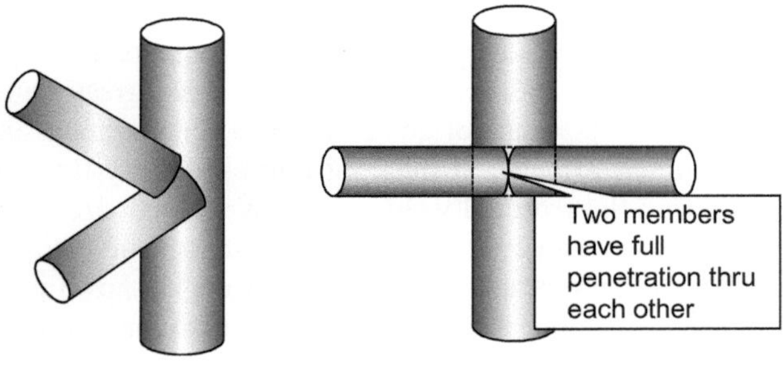

FIGURE 5.2 Fabrication of nodes.

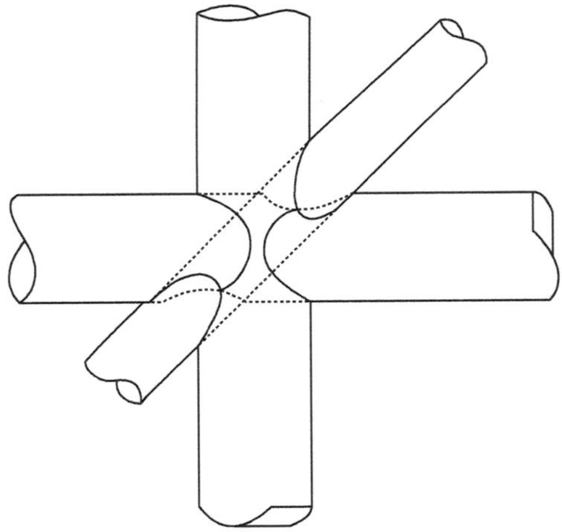

FIGURE 5.3 Intersection node with full sections carried through the joint.

Stiffener plate-to-web connections may be continuous double-fillet welds. Weld metal and heat-affected zone notch toughness should not be less than the minimum toughness requirements for the girder.

High-strength bolts are not usually used in connections for offshore structures but they may be used in temporary structures. They are especially suitable for field connections, where spray and wave-induced vibration make it difficult to obtain high-quality welds. They are also an effective means for making connections under cold conditions.

In accordance with the steel detailing and type of construction, the best method of ensuring adequate torque should be selected. In a large joint with multiple bolts, either the abutting plates should be pre-milled or shims should be employed to ensure a tight fit.

Fabrication specifications for offshore jackets are defined explicitly by the designer. In most cases, they are based on one or more of the well-known codes, with additional requirements dictated by the specific design, client standards and project specification and other considerations. In general, during fabrication and in all the construction phases, a great deal of attention is focused on the welding process and its quality control. In particular, greater attention has been focused on the importance of complete joint penetration groove welds, elimination of "notch effects" at the root and especially the cap of joint welds, and achieving the required weld profile. Welds that are critical for fatigue endurance may be required to be ground to a smooth curve, to prevent brittle failure. Typical welding details from API RP2A are as shown in Figure 5.4, which depicts tubular members framing into or overlapping another member with access from one side only. Figure 5.4 presents the welding procedure for a tubular joint based on the thickness of the brace, *t*, and the chord, *T*, noting

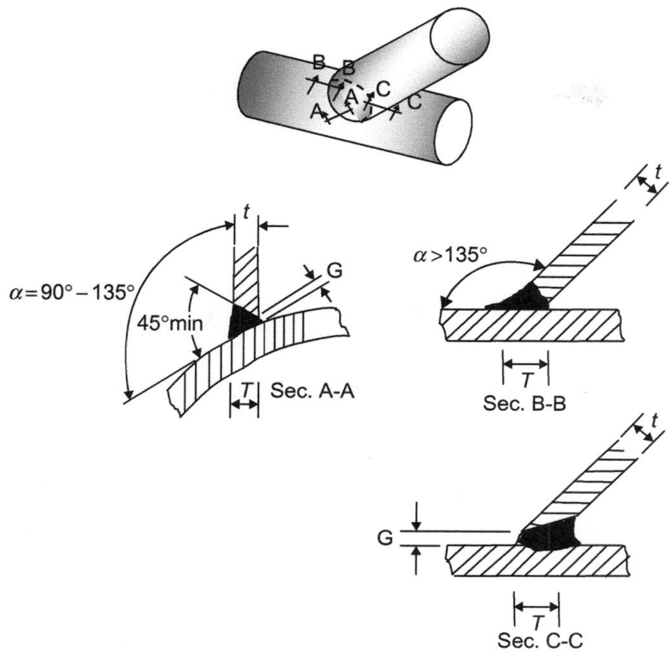

FIGURE 5.4 Welded tubular connection.

that α is the angle formed by the exterior surfaces of the brace and chord at any point on their joint line. However, a lot of emphasis is placed on designing stubs that can be welded from both sides.

Table 5.1 presents the relation between the groove angle and the opening distance. The relation between the angle of inclination and the minimum welding thickness is illustrated in Table 5.2.

Welds are usually subject to 100% visual, magnetic particle inspection (MPI) and ultrasonic test (UT) inspection. The weld acceptance criteria, as the maximum weld undercut length ($t/2$ or 10 mm), and maximum depth ($t/20$ or 0.25 mm), imply an exceptionally high quality of welding.

The location and orientation of circumferential and longitudinal welds during construction are based on minimizing interference and ensuring the minimum distance between circumferential welds. Special attention is required for items like pile sleeve shear plates, launch runners, mud mats and others where planned avoidance of weld interference is critical.

The weld testing should be applied to plates and fittings as the member to which they are being affixed. There is also an overriding necessity to ensure that such attachments are located at a safe distance from main structural welds in order to minimize the risk of defect propagation.

TABLE 5.1 Relation between Groove Angle and Opening Distance

Groove Angle	Opening Distance M (mm)
>90°	0–4.8
45°–90°	1.6–4.8
<45°	3.2–6.4

TABLE 5.2 Relation between Angle of Inclination (α) and Minimum Welding Thickness

α (degrees)	Min *S*
>135	<1.75 t
50–135	1.25 t
35–50	1.50 t
<35	1.75 t

In general, sub-assemblies are executed so that at least one of the two edges that will mate during subsequent assembly/erection has a cut-off allowance. This procedure provides flexibility, so that the sub-assemblies can be sent to the field with the cut-off allowance in place and cut to fit on location. Alternatively, they can be cut to exact dimensions during sub-assembly where the as-built dimension has already been determined.

5.4.1 Joint Fabrication

The main element of the structure is the joints, which are frequently geometrically complex. Accordingly, joint fabrication presents a challenge to the construction team, for welding and dimensional control.

On a complex jacket, the designer may specify the joint cans or the whole joint, including stubs and ring stiffeners, in a material with specified through-thickness properties. This requirement is specified by the designer to overcome the tearing or punching effects likely to be sustained by these elements during their design life.

The designer may also increase the thickness or reinforce the cans to withstand local stresses. Finally, in an effort to ensure that node welds contain minimal levels of residual stress due to fabrication, thermal stress relief or post-weld heat treatment (PWHT) of the heavier, more restrained welds may be prescribed.

API RP2A and ISO 19902 provide specific tolerances for final fabrication. The contractor must work within the allowable tolerances and must monitor weight-control requirements at each phase of construction. In general, joint fabrication tolerances are tight, as typical working points have an allowable ±6 mm from the drawings, stub angle allowable deviation is ±1 minute and all braces' allowable deviation is usually ±12 mm of the design dimension.

The typical fabrication process for a conventional joint, assuming that the joint can with or without ring stiffeners has already been fabricated, commences with profiling of stubs and terminates with UT inspection of the finished joint after PWHT.

5.4.2 Fabrication Based on ISO

ISO 19902 contains the fabrication requirements and tolerances for fixed steel offshore structures. All materials, welding, weld preparations and inspection should be in accordance with ISO. Welding and assembly sequences should be designed to minimize distortion.

Tubular Members and Joints

According to ISO, circumferential welds are not permitted within cones and node stubs unless specified on design or approved shop drawings.

The minimum spacing between circumferential welds in tubulars should be the smaller of 915 mm (36 inches) or 1 diameter of the tubular. Ring stiffeners in tubular members should be at least 100 mm (4 inches) from circumferential welds except that, where this is not possible, ISO recommends the welds to be overlapped by at least 10 mm (3/8 inch).

Rings used to stiffen cone-cylinder junctions should not be offset, and the welds should be overlapped by at least 10 mm (3/8 inch) to avoid coincident location of weld toes.

Figure 5.5 presents the longitudinal and circumferential welds in tubular joints, which should avoid the most critical areas of a joint.

The offset between longitudinal seams in adjacent cans in tubular joints should be maximized, taking into account connections at each end of each member, and in no case should be less than 90°, except that the location of a longitudinal seam may be adjusted by no more than 300 mm (12 inches) to avoid the prohibited locations.

All fabrication work should be performed based on safe practices, using equipment that complies with all applicable regulatory requirements, local standards, and any appropriate regional requirements.

Construction staff must be competent and qualified to perform their task. All fabrication, assembly and erection should be in accordance with approved procedures.

Unless detailed otherwise on the design drawings, the "default" layout rules and fabrication are based on API or ISO or another standard mentioned in the project specification.

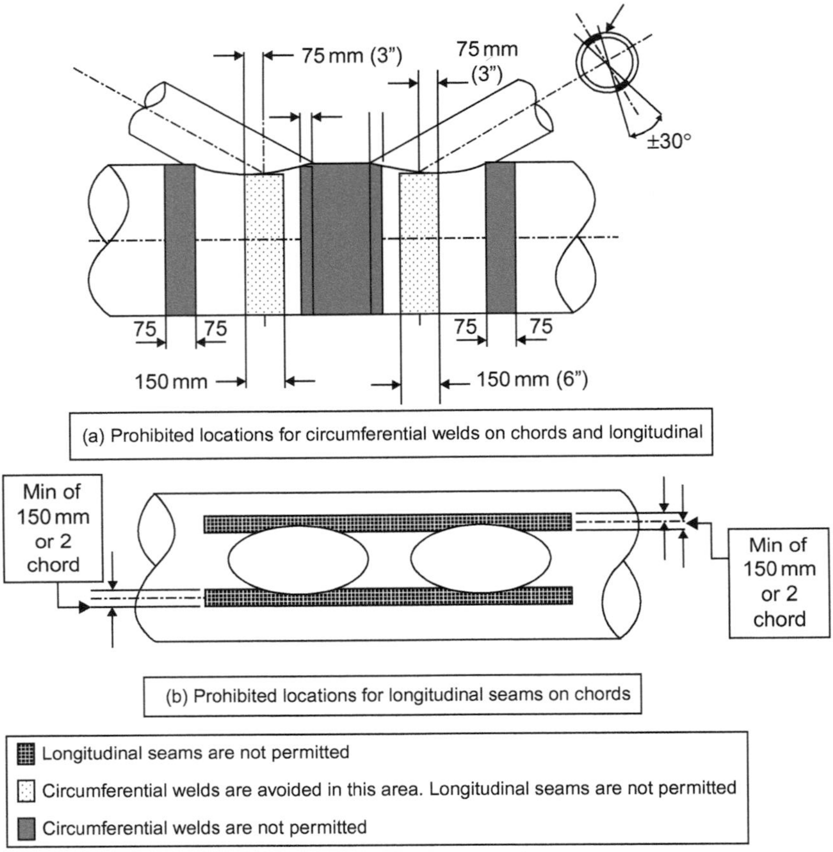

FIGURE 5.5 Longitudinal and circumferential welds.

Grouted Pile to Sleeve Connections

In the case of pouring a grout in the annulus between the pile and leg, the steel surfaces in contact on which interface shear is to be developed must be free of loose material, mill scale, grease and any other material that can reduce grout-to-steel friction.

Care should be taken during and after the installation of packers to prevent damage from handling, high temperatures and splatter from welding and debris. Any debris that could cause damage to the packers during the installation process should be removed. Measures such as dressing of the pile tip to a smooth profile and positioning of centralizers on the sleeve directly above the packers should be taken to minimize potential packer damage during pile stabbing.

For piles, the surface and followers should be untreated bare metal, free from mill varnish, oil and paint, except for any specified markings. The area of the surface used for markings should be the minimum required consistent with adequate identification.

Fabrication aids and temporary attachments are any members, such as erection braces, sling stops, scaffolding supports, padeyes, dogs, walkways, and others used for fabrication purposes. Material used for fabrication aids and temporary attachments should comply with the project specification.

Fabrication aids and temporary attachments used for lifts during fabrication should be designed and approved from the engineering office and should be shown on the shop drawings and detailed in the fabrication procedures.

The quality-control supervisor should monitor the welding procedure specifications and welder and tracker qualifications used to perform welding on fabrication.

When fabrication aids and temporary attachments occur in areas that become part of a completed weld area, within 300 mm (12 inches) of an intersecting weld, or in areas that require coating, the attachment should be removed by oxy-fuel cutting a minimum of 3 mm (1/8 inch) above the surface of the structural member and grinding the remainder flush with the surface of the member. If the ground areas lie close to or underneath a weld, the surface should be inspected using the magnetic particle test after grinding and before the permanent weld is executed.

Fabrication aids and temporary attachments that occur in areas that do not require coating and that are located more than 300 mm (12 inches) from an intersecting weld should be seal-welded and should be removed to within 6 mm (1/4 inch) of the surface of the member with sharp edges removed by grinding. Oxy-fuel cutting may be used for the bulk removal of a fabrication aid or temporary attachment, but it should not be used within 3 mm (1/8 inch) of the surface of the underlying component. In general, fabrication aids and temporary attachments should not be removed by hammering or oxy-fuel washing.

Heat Straightening

There is usually a heating procedure that should be prepared by the quality-control supervisor based on the project specifications so that straightening of members distorted by welding may be undertaken using localized heating. The maximum temperature of the heated areas should be checked with heat-indicating equipment.

Steel plate complying with the specification requirements may be rolled down to a diameter-to-thickness ratio of 20:1 with no additional treatment. If a diameter-to-thickness ratio of less than 20:1 is required, mechanical tests should demonstrate that the steel retains its mechanical properties. Post-forming heat treatments, or hot forming, can be necessary to ensure the properties are maintained.

The mechanical tests should be performed if the hot-rolled sections are heated to straightening, to guarantee that the mechanical properties are maintained after forming and welding.

The use of wedges or other techniques to assist rolling should be subject to the visual acceptability of the product and to acceptable mechanical test results on test specimens taken from regions of maximum strain.

According to ISO recommendation, any materials that require straightening before use, to achieve the specified tolerances, may be cold straightened if the strain does not exceed 5%. Similarly, tubular cans for members or joints may be re-rolled after welding to meet the dimensional tolerances if the strain during re-rolling does not exceed 5%.

Rat-holes, Penetrations and Cut-outs

Rat-holes, penetrations and other cut-outs should be avoided wherever possible. If the rat-holes are approved by the engineering office for critical needs, they should be marked on shop drawings and are subject to the following requirements:

- Permanent rat-holes should have a radius of 50 mm (2 inches) or twice the plate thickness, whichever is larger, and should be ground smooth. Fillet welds should be returned through the rat-holes.
- Temporary rat-holes should have a radius not less than twice the plate thickness and should be reinstated to an approved welding procedure. Note that filling with weld metal is not permitted.
- The radius of any approved cut-outs other than rat-holes should not be less than 100 mm (4 inches). It is preferred that the area be checked using UT before cutting to ensure that no cut is made within 300 mm (12 inches) of internal stiffeners. The original plate should be clearly marked and should be welded back into the cut-out.

Movement, Erection and Roll-up of Sub-assemblies

To ensure good-quality construction, sub-assembly and weld-up should be done at the ground level or under cover. When sub-assemblies are moved and installed into position in larger sub-assemblies or within the structure, care should be exercised to ensure that no members or joints within the sub-assemblies are overstressed or distorted.

Suitable supports may include: webbing slings; wire slings, when the components are protected from damage; and temporary lifting attachments welded to the sub-assemblies.

Sub-assemblies may be rolled into the correct orientation in the structure or to allow lifting at the correct orientation into the structure. Roll-up saddles should be designed and positioned to ensure that the structure is not overstressed during roll-up and should be well greased or otherwise lubricated.

Fabrication Tolerances

As per the contract, the contractor should provide qualified persons approved by the owner's representative and also should deliver good equipment and the necessary instruments to perform checks and to monitor and to control dimensions within the allowable tolerance. Tolerances should be checked at each stage in accordance with the fabrication procedures and the final survey should meet the defined tolerances. Fitters should be doing their own checking as the job progresses and their tape measures and straight edges do not need calibration. The final survey should be on complete sub-assemblies and on the completed structure after PWHT if applicable.

Based on ISO, personnel responsible for the final survey either should be qualified surveyors or should have had at least 5 years' experience of similar work. Instruments used should be in accurate adjustment and should have current, valid calibration certificates.

According to ISO, tolerances are guided by the following:

- Local tolerances should be controlled for structural components and sub-assemblies so that the accumulation of such tolerances does not affect the specified global tolerances.
- The specified tolerances should apply at all stages of fabrication and assembly.
- Allowance should be made for weld-gap tolerances and weld shrinkage in all component, sub-assembly and global tolerance calculations.
- Where tolerances have to be derived from a formula to be expressed in terms of a component dimension, such as wall thickness, the results should be taken to the nearest mm (0.04 inch).
- Tolerances should be based on theoretical setting-out points and centerlines of the structure referenced to permanent approved datum points, such as the coordinated survey stations, and should be corrected to a temperature of +20°C (68° F).
- Fabrication and yard assembly supports should be set to within ±5 mm ($\pm\frac{1}{4}$ inch) of the appropriate position shown on the approved workshop drawings. Where no such drawings exist, fabrication should be carried out from a level plane to within ±5 mm ($\pm\frac{1}{4}$ inch).
- The dimensional tolerance of launch-way centerlines should be within ±20 mm ($\pm\frac{3}{4}$ inch) of the theoretical position and should also be within ±6 mm ($\pm\frac{1}{4}$ inch) of its reference elevation. The variation in elevation between any two points on a launch-way should not exceed 3 mm ($\frac{1}{8}$ inch) within any 3 m (10 ft).

Leg-spacing Tolerance

The global (horizontal) tolerances for leg spacing at plan bracing levels are as shown on Figure 5.6 and are detailed below.

- The horizontal center-to-center distance between adjacent legs at the top of a structure where a deck or other structure is to be placed (stab-in nodes) should be within 10 mm ($\frac{3}{8}$ inch) tolerance from the construction drawings.
- The horizontal center-to-center distance between legs at other locations should be within 20 mm ($\frac{3}{4}$ inch) of the design values.

Vertical Level Tolerance

The global tolerances for vertical levels of plan bracing are as shown on Figure 5.7 and are detailed below.

- The elevation of plan bracing levels should be within ±13 mm (±$\frac{1}{2}$ inch) of the construction drawings.

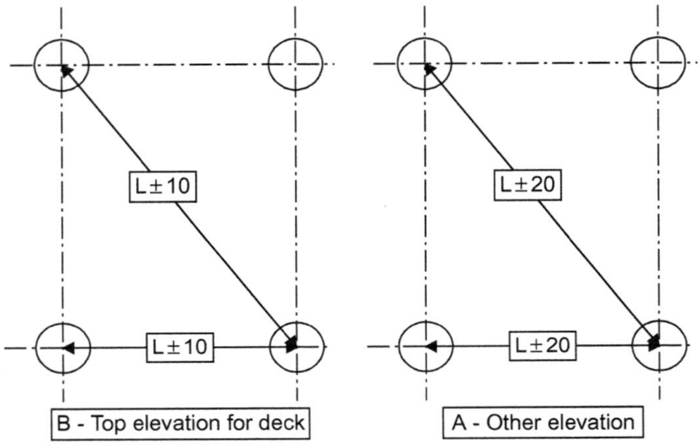

FIGURE 5.6 Horizontal tolerance.

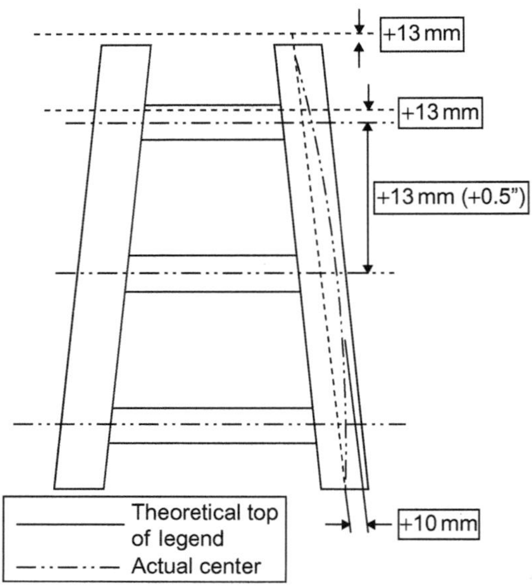

FIGURE 5.7 Vertical tolerance (according to ISO).

- The vertical level of braces within a horizontal plane should be within ±13 mm (±$\frac{1}{2}$ inch) of the construction drawings.
- The vertical distance between plan bracing elevations should be within ±13 mm (±$\frac{1}{2}$ inch) of the construction drawings.

Tubular Member Tolerance

For tubular members with thicknesses of 50 mm (2 inches) or less, the difference between the major and minor outside diameters (the out-of-roundness) at any point of a tubular member should not exceed the smaller of 1% of the diameter or 6 mm ($\frac{1}{4}$ inch).

For tubular members with thicknesses of more than 50 mm (2 inches), the difference between the major and minor outside diameters at any point of a tubular member should not exceed 12.5% of the wall thickness.

For tubular members with a diameter of 1200 mm or more, and a thickness of 100 mm or less, the difference between the major and minor outside diameters at any point of a tubular member may increase to 13 mm ($\frac{1}{2}$ inch), provided the tolerance on the circumference is less than 6 mm ($\frac{1}{4}$ inch).

The difference between the actual and nominal outside circumferences at any point of a tubular member should not exceed the smaller of 1% of the nominal circumference or 13 mm ($\frac{1}{2}$ inch).

Figure 5.8 presents the tolerances for the straightness of tubular members and beams, based on ISO, which are:

- Straight to be within ±10 mm (±$\frac{3}{8}$ inch) for length ≤12 m
- Straight to be within ±13 mm (±$\frac{1}{2}$ inch) for length >12 m.

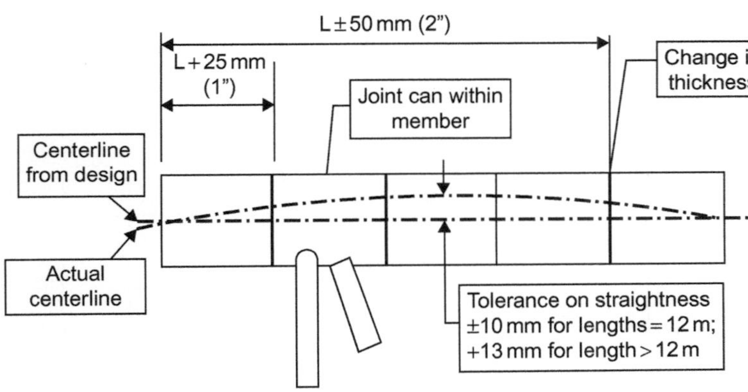

FIGURE 5.8 Tolerances for positioning of cans within members and straightness of members.

In addition, the tolerances for the straightness of tubular members and beams when assembled into the structure are:

- Between the ends of the node stubs, braces should be straight to within 0.12% of the length
- Between the ends of the node chords, braces should be straight to within 0.10% of the length.

The tolerances for the location of cans with different wall thicknesses within tubular members are as shown in Figure 5.8 and as detailed below:

- For joint cans, within 25 mm (1 inch) of the construction drawings.
- For other changes of wall thickness, within 50 mm (2 inches) of the construction drawings.

Figure 5.9 illustrates the tolerances for the mismatch at circumferential welds and longitudinal seams in tubular members.

- For double-sided welds, the smaller of 10% of the thicker tubular or 6 mm ($\frac{1}{4}$ inch)
- For single-sided welds, the smaller of 10% of the thicker tubular or 3 mm ($\frac{1}{8}$ inch)

Where these tolerances are exceeded, the transitions should be made by weld profiling to a maximum slope of 1:4.

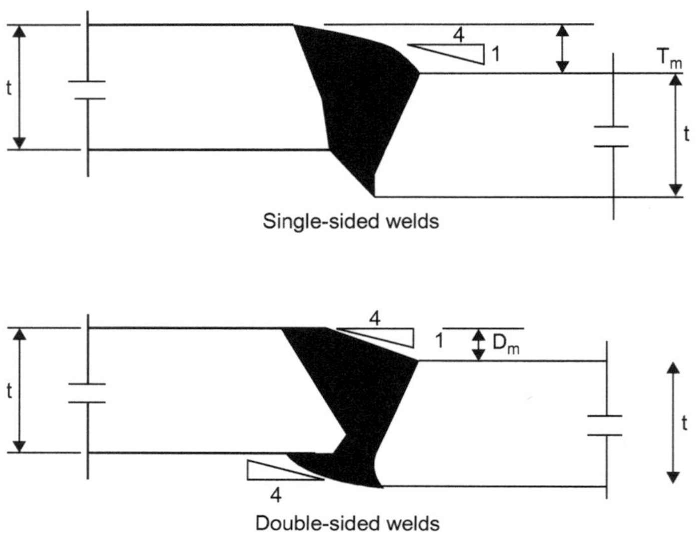

FIGURE 5.9 Joint mismatch tolerances.

Tolerance in Leg Alignment and Straightness

In addition to the tolerances for all tubular members, the tolerances for the alignment and straightness of legs are as shown in Figure 5.10 and are detailed below.

- For structures with more than 4 legs, at each planned bracing level the legs should be aligned to within 10 mm ($\frac{3}{8}$ inch) of a straight line.
- Between horizontal bracings, legs should be straight to within 10 mm ($\frac{3}{8}$ inch).
- Between nodes, legs should be straight to within 10 mm ($\frac{3}{8}$ inch).

Tubular Joint Tolerances

For tubular joint tolerances, a best-fit work point should be determined, taking into account all the design and as-built dimensions of the complete tubular joint. The best-fit work point should be within 15 mm ($\frac{5}{8}$ inch) of the design position.

The alignment of a brace-stub best-fit centerline or, for point-to-point construction, the brace centerline, should be within 13 mm ($\frac{1}{2}$ inch) of the design work point (WP), as in Figure 5.11.

The lengths of cans within the chord of a tubular joint should be within the range of design length +10 mm to design length −5 mm, and the location of the circumferential weld between chord cans should be within 5 mm of design location (Figure 5.12).

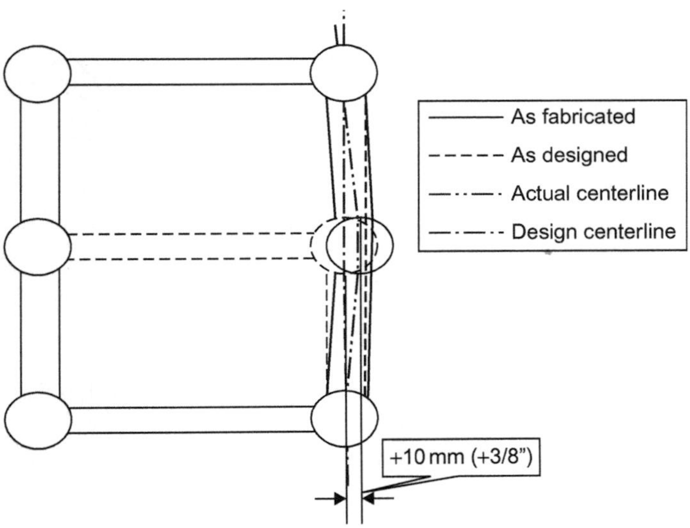

FIGURE 5.10 Allowable tolerance in the leg alignment.

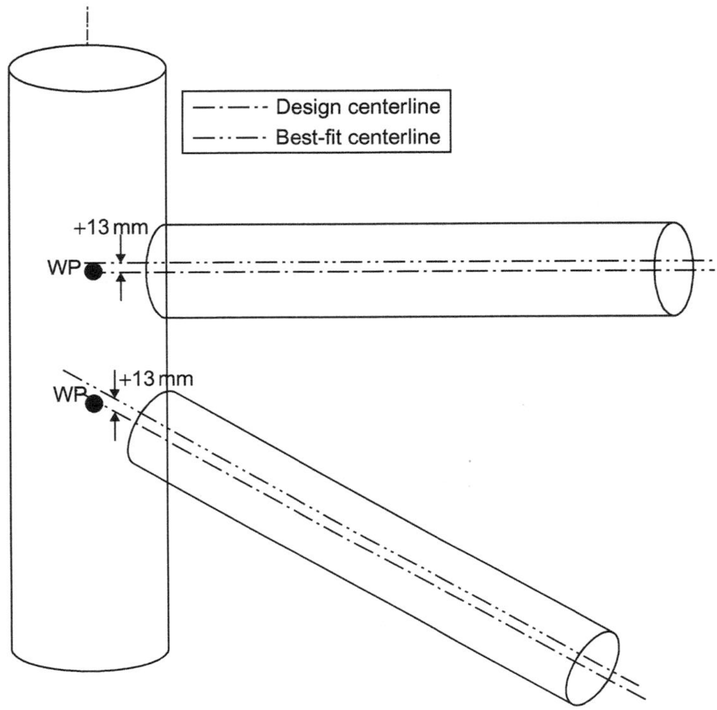

FIGURE 5.11 Allowable tolerance on brace stubs at tubular joints.

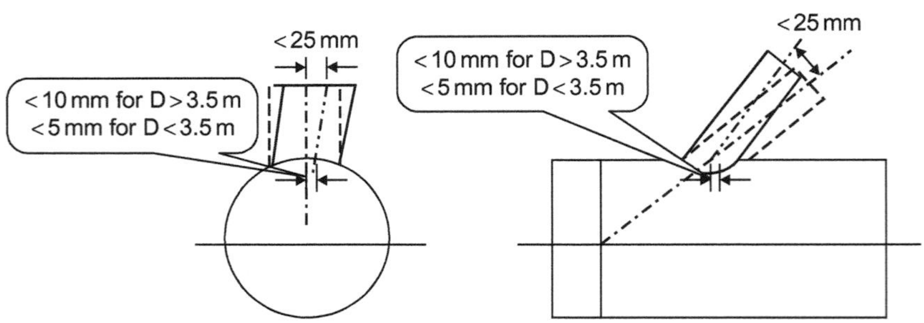

FIGURE 5.12 Allowable tolerance for tubular joints (according to ISO).

The lengths of the brace stub of a tubular joint should be within the range of design length +50 mm to design length −5 mm.

The tolerances on the positioning and alignment of stubs on tubular joints are (see Figure 5.11).

- The centerline of the brace at the intersection with the chord wall should be within 5 mm of the construction drawing where chord diameter <3.5 m.

- The centerline of the brace at the intersection with the chord wall should be within 10 mm of the design position where chord diameter >3.5 m.
- The angular orientation of the brace stub should be within 10 ft of the design orientation.
- The position of the centerline of the brace end of the stub should be within 5 mm of the design position.

Where plates carrying in-plane faces are arranged to form a cruciform structure, the misalignment should not exceed 50% of the thickness of the thinnest non-continuous member or 10 mm, whichever is smaller.

Stiffener Tolerances

The stiffener tolerances will be:

- At or within 150 mm of a conical transition, to within 3 mm ($\frac{1}{8}$ inch) of the construction drawing
- In launch legs, to within 3 mm ($\frac{1}{8}$ inch) of the construction drawing
- In tubular joints other than at conical transitions of launch leg nodes, to within 5 mm ($\frac{1}{4}$ inch) of the design position
- At all other locations, to within 10 mm ($\frac{3}{8}$ inch) of the construction drawing.

The tolerances for ring-stiffener cross-section, where the web of the stiffener is perpendicular to the centerline of the tubular, should be within 2.5% of the web height, as shown in Figure 5.13. On the other hand, the flange of the

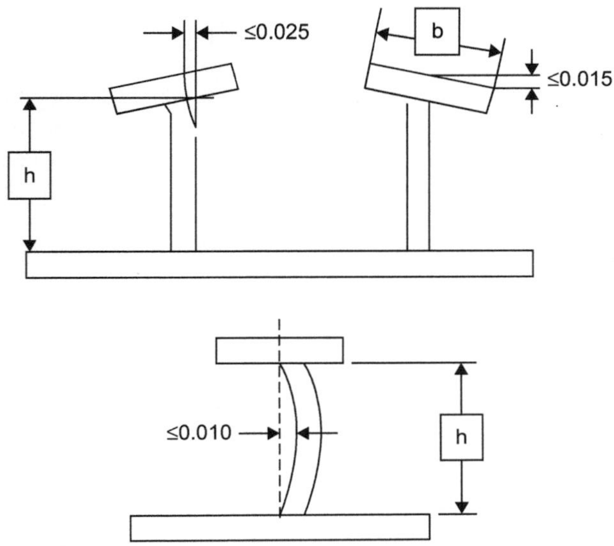

FIGURE 5.13 Tolerances for ring-stiffener cross-section (according to ISO).

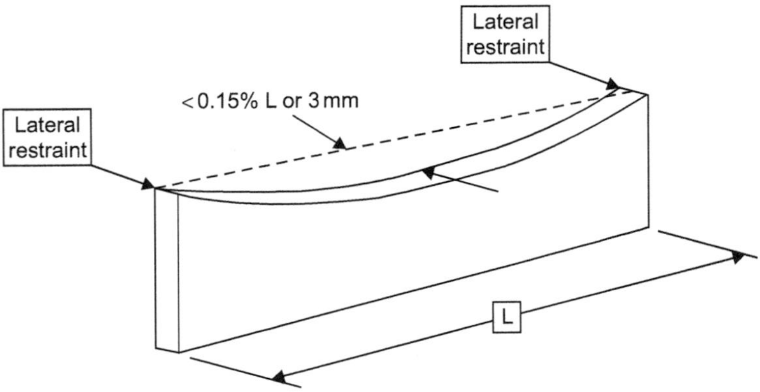

FIGURE 5.14 Tolerance for ring-stiffener straightness.

stiffener should be parallel to the centerline of the tubular to within 1.5% of the flange width and the web of the stiffener should be flat over its height to within 1.0% of the web height.

Out-of-straightness for longitudinal or diaphragm stiffeners in tubular members should be limited to 0.15% of the length of the stiffener (L) or 3 mm, whichever is larger, as shown in Figure 5.14.

Conductor Guides and Pile Tolerances

The tolerances for conductors are the same as for pile guides, sleeves and appurtenance supports and should be related to a best-fit line through the centers of the guides, sleeves and supports, as shown in Figure 5.15. The tolerance for the center of each guide/sleeve and the best-fit line should be less than 10 mm ($\frac{3}{8}$ inch) at the top guide and where the vertical spacing between guides is less than 12 m (40 ft). The tolerance of the center of each guide and the construction drawing line should be less than 13 mm ($\frac{1}{2}$ inch) in other cases. For pile sleeves, the tolerance should be checked at the mid-height of each set of centralizers.

The tolerance of the center of each appurtenance support and the best-fit line should be less than 10 mm ($\frac{3}{8}$ inch) at the top and less than ±25 mm (±1 inch) elsewhere.

In addition to the tolerances for other dimensions of tubular members, the tolerances of the straightness of a pile should be:

- In any 3 m (10 ft) length, piles should be straight to within 3 mm ($\frac{1}{8}$ inch).
- In any 12 m (40 ft) length, piles should be straight to within 10 mm ($\frac{3}{8}$ inch).
- In any length over 12 m (40 ft), piles should be straight to within 0.1% of the length considered.

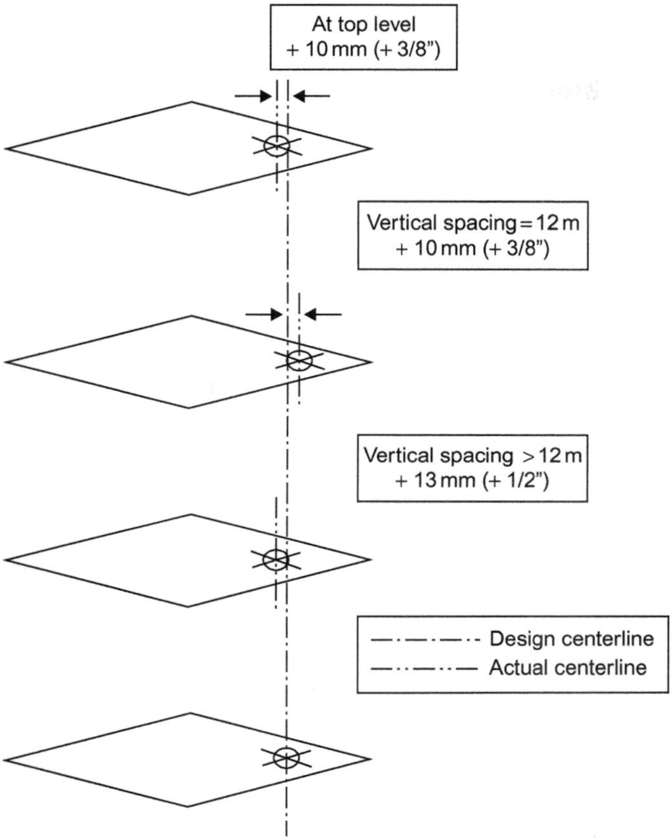

FIGURE 5.15 Tolerance for conductor guide alignment.

The design of anodes is discussed in detail in Chapter 6, but, during construction, the tolerances for the locations of anodes should be less than 300 mm (12 inches) and less than 10° circumferentially of the design position.

Dimensional Control

Quality control (QC) should be within the overall quality plan that was established before the start of fabrication. The owner should verify that all persons and equipment required for QC are available.

Of all the QC areas that require attention, dimensional control is emphasized in the code and specifications, especially for offshore structures. However, it is clear that attention must be paid to the dimensions that have structural significance, such as the straightness of elements, ovality of tubular members, eccentricities at node joints, and the like. It is also clear that, on a jacket, the global alignment/verticality of items such as pile sleeves, conductor guides, launch runners and others are also important. Finally, dimensional control of items that are intended for "mating" or "removal" offshore (e.g., piles/pile sleeves, jacket top, buoyancy tank or supports, etc.) is vital to the efficiency of offshore installation. Therefore, there are many aspects where attention to dimensional control is justified even if the overall design might occasionally benefit if the designer did not always require that everything fit so tightly.

Surveys should be performed using survey techniques and technology that enable survey accuracy that is better than the specified tolerances.

The methods and equipment used should enable verification of the survey accuracy. All instruments used should be in accurate permanent adjustment, should have current, valid control certificates and should be subject to a program of periodic checking.

The principal reason for requiring such accurate dimensional control of joints and tubular members during fabrication is not the structural consequences of out-of-tolerance components, but because the parts may not fit together in the yard. In the case of the tubular steel jacket, the theoretical allowable tolerances for node stub eccentricity are suitable from the structural analysis point of view, while the actual tolerances are very tight, so they will be less than the theoretical tolerances because of the fitting of the components together during the later phases of construction.

There are approved construction drawings, so during fabrication any measurement verification must be within the allowable tolerances, which should be based on ISO, API, AISC or another standard cited in the project specification document.

Dimensional control of node fabrication, in particular, involves potentially intricate calculations in the shop. However, the most successful systems simply involve the inclusion on the shop drawings of several additional "checking" measurements and the correct marking of the node can, stub generators and offsets.

All jacket and topside structures should be checked for loads applied during fabrication. The loads should be based on the proposed fabrication methodology. Consideration should be given to structure support points used for weighing and load-out. Site wind loads should be included as part of the load condition. These checks should be carried out during detailed design, not basic design.

5.5 JACKET ASSEMBLY AND ERECTION

The jacket assembly consists of sub-assembly parts such as the joints, tubular members, boat landing and others. The sub-assembly can be considered an intermediate stage between standard shop fabrication for joints, tubular members, and other sub-assembly elements and the whole assembly or erection. The assembly of the structure jacket is shown in Figure 5.16.

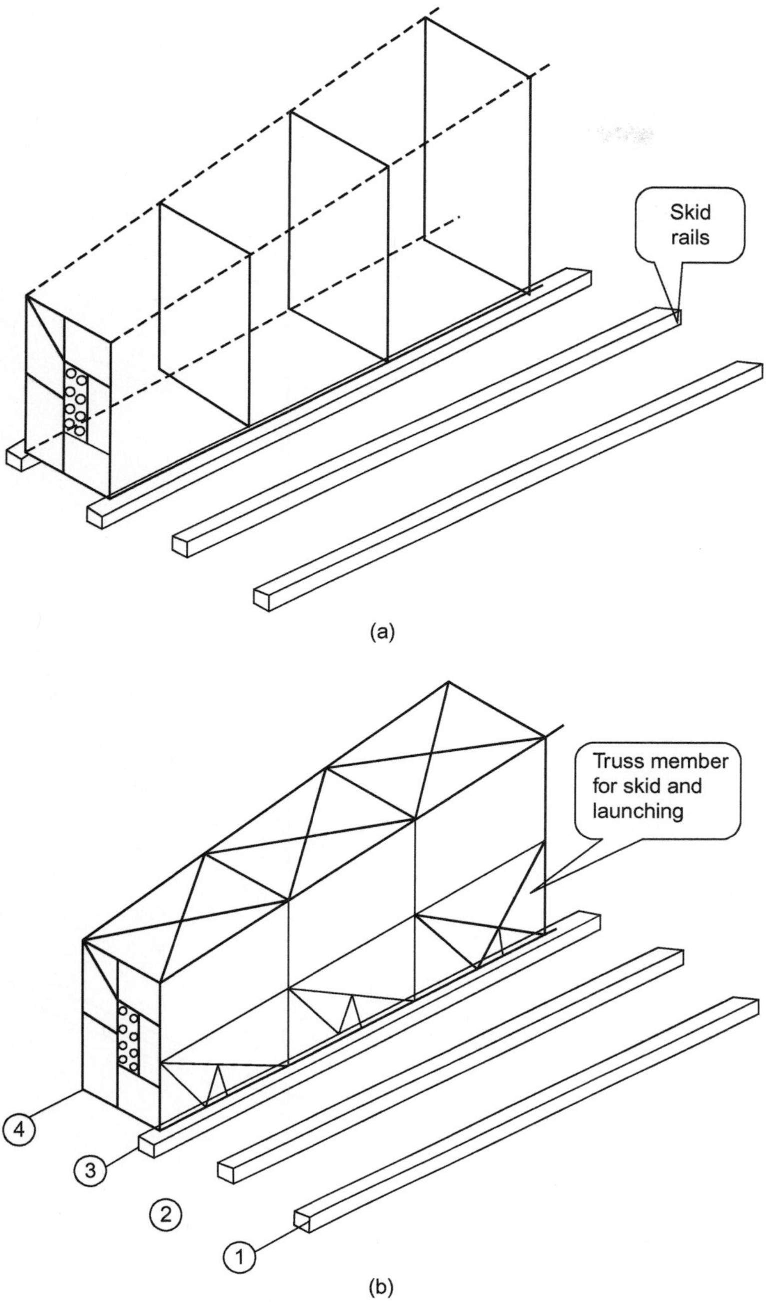

FIGURE 5.16 (*Continued*).

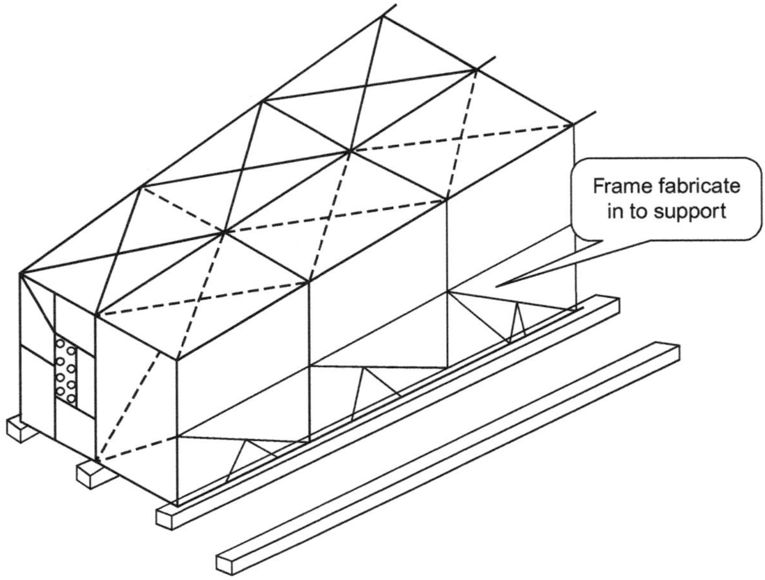

(c)

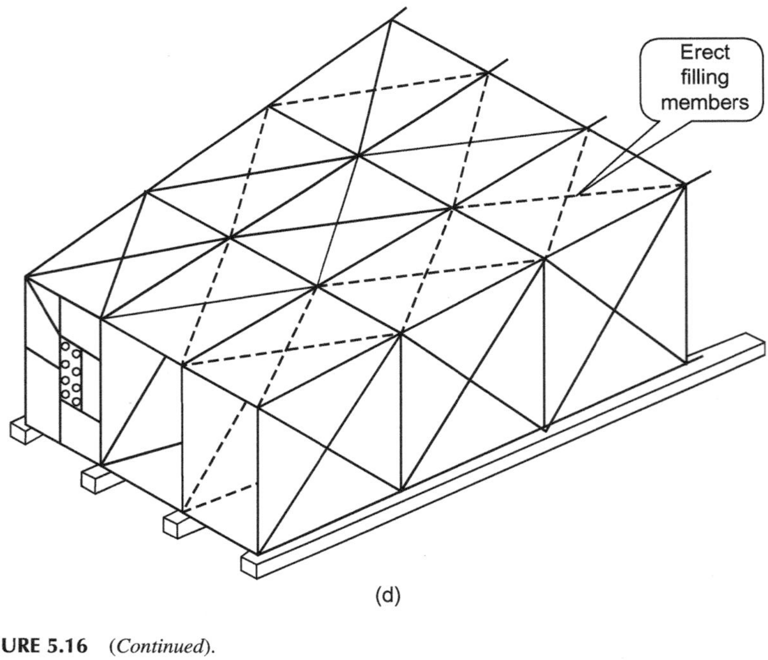

(d)

FIGURE 5.16 *(Continued).*

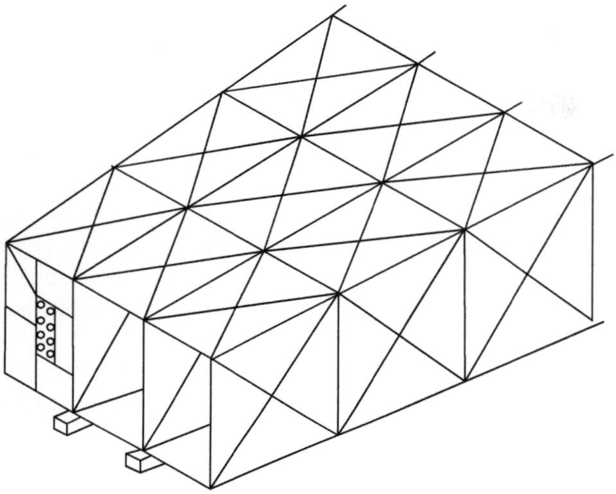

(e)

FIGURE 5.16 (a) Steps in erection of plan frames; (b) Erection of bay; (c) Frame roll-up; (d) Roll-up of the last frame and fill in between; (e) Jacket ready for roll-out and launching.

The sub-assembly should be managed so that the maximum number of welds are done in the shop; shops have the highest weld quality because many node and tubular welds can be double-sided and/or automatic when performed in the fabrication shop.

When defining sub-assemblies, the principal factors to be considered are:

- Dimensions (e.g., size and weight), which are the main factors governing the method of transportation
- Welding sequence (a difficult welding sequence can induce stresses during sub-assembly welding/erection and could cause distortion of a member)
- Construction difficulties (for example, short, large-diameter infillings are difficult to erect vertically and are best included in sub-assemblies, if possible).

Shop-fabricated sub-assemblies and loose items are incorporated into assemblies that constitute the major lifts of the erection sequence. Usually, for a large jacket, the assemblies typically consist of four types:

- Jacket levels incorporating conductor guide frames
- Top frames
- Jacket rows, such as bents or partial bents
- Pile-sleeve clusters

The assembly and erection phases are based on the following objectives:

- Increase the opportunity for access around the jacket during execution and for ground assembly
- Decrease the number or erection of joints in the main structural elements, such as jacket legs, launch runners, rows and levels. Align critical areas, such as conductor guides, pile sleeves and launch runners.
- Sub-assemble principal structural elements of the jacket, such as jacket legs, rows and levels. Sub-assemble, and where possible pre-test, systems such as grouting and ballasting. Include the maximum quantity of secondary items, such as anodes, risers, J-tubes and caissons. Coat or paint required areas prior to erection.
- Where possible, reduce the use of temporary structures, such as scaffolding, walkways and lifting aids that require subsequent removal, and pre-install such aids where they are necessary.

Usually, the assembly of a jacket frame, often having a base spread of 50 m or more, places severe demands on field layout and survey and on temporary support and adjustment bracing (see Figure 5.17). Such large dimensions mean that thermal changes can be significant. Temperature differences may be as great as 30°C between dawn and afternoon and as much as 15°C between various parts of the structure, resulting in several centimeters of distortion.

FIGURE 5.17 Frame jacket located horizontally.

Furthermore, high temperatures tend to induce residual stresses in the structure. Because of the difficulty associated with thermal distortion, it is normal to "correct" all measurements to a standard temperature, 20ºC.

For joints, elastic deflections are also a source of difficulty in maintaining tolerances. Foundation settlement or displacements under the skid beams and temporary erection skids must be carefully calculated and monitored.

In the construction plan, each assembly should be completed before the lifting process is started, which requires definition of the location, orientation and face-up or face-down placement of each assembly to be matched with the lifting procedure.

Central coordinates for each assembly are usually shown on the layout drawings. The central coordinates are then used as local benchmarks for erection of the assembly.

Note that the QC check of dimensions should be performed before and after welding to verify the measurements by check-through measurement. During sub-assembly, fabrication should aim for the positive allowable tolerance to compensate for the shrinkage that will happen due to welding. Perhaps the most fundamental rule in fitting is avoidance of applying excess force on the member during fitting prior to welding or to force stresses on unwelded members through the welding sequence, since such conditions cannot be foreseen by the designer and will account for more stress than was considered in the design calculation.

The sequence of events that applies to all types of assembly is:

- Preparation of assembly support and staging
- Rough setting of assembly main structure and position tacking. Dimensional control of assembly main structure.
- Infilling of secondary structure and position tacking. Dimensional control of assembly and secondary structure.
- Pre-weld inspection. Weld-out of structure subject to continuous inspection and according to approved sequence.
- Installation of appurtenances, such as anodes, supports, walkways, risers, J-tubes and caissons, of grouting and ballasting, and of scaffolding, lifting aids, erection guides and temporary attachments.
- Overall NDT, dimensional control
- Blasting and painting or touch up. Removal of temporary assembly supports and staging.
- Preparation for transport, lift and erection.

In this phase, assembled, sub-assembled and fabricated structures, together with loose items, are incorporated into the final structure according to the sequence of erection shown in Figure 5.16.

Jacket frames are typically laid out flat and then lifted upward by more than one crawler crane. Coordinating such a rigging and lifting operation requires thoroughly developed three-dimensional layouts, firm and level foundations for the cranes and very competent operators.

Structural analysis should be performed to check the structure's member stresses during the erection process for a given assembly; this is usually done using a computer model with all relevant structural characteristics. The assembly is analyzed for the load cases that correspond approximately to the support conditions of the assembly, with assumed locations of the cranes, bogies, saddles and others when the panel is being transported and when it is in horizontal and vertical positions. Structural analysis for lift and transport identifies the worst cases from the structural stress perspective. These cases are then analyzed to determine the maximum stresses and displacements. The calculations should show that global and local stresses are within allowable limits according to API/AISC codes.

The sequence for erection of all major components is:

- Technical appraisal of lift methods. Calculations for crane configuration, rigging accessories.
- Preparation of cranes for lift, preparation for rigging and transport assembly to lift location. Roll-up into position with scaffolding and staging in position, if possible.
- Preparation of fixing system and wind bracing, which is usually done by means of guy wires and turnbuckles. Weld-out at least sufficient to allow crane release.
- Crane release. Removal of rigging and temporary attachments.

Jacket structural completion is followed by a short phase during which all the jacket systems, both permanent and those required during installation, are completed and rendered functional.

Figure 5.17 presents the fabrication of one face of the jacket frame on the ground; the erection of the other side is shown in Figure 5.18.

The erection of additional steel framing to strengthen the jacket structure during launching is presented in Figure 5.19.

After construction of the first frame, erection of temporary and permanent horizontal bracing is started, as shown in Figure 5.20, at the same time that construction of the diagonal bracing is started, as shown in Figure 5.21.

Figure 5.22 illustrates the erection of the last horizontal level for the jacket, whereas Figure 5.23 presents fabrication of the conductor guide with the horizontal frame.

The last face on the jacket is fabricated on the ground, as shown in Figure 5.24, where one can see the exterior conductor guide, and then this face is lifted, as shown in Figure 5.25, to finalize construction of the jacket.

The fabrication of the topside is shown in Figure 5.26.

The accessories, such as the flare tip, riser guard and boat landing, are fabricated on site and are attached to the jacket after its installation.

(a)

(b)

FIGURE 5.18 (a), (b) Erection of the side frame.

FIGURE 5.19 Erection of strength framing for the launching process.

FIGURE 5.20 Start of erection of the horizontal bracing.

FIGURE 5.21 Erection of X bracing with the horizontal bracing.

FIGURE 5.22 Erection of the last horizontal level for the jacket.

FIGURE 5.23 Fabrication of the conductor guide with the horizontal frame.

FIGURE 5.24 Fabrication of the last face frame for the jacket.

FIGURE 5.25 Erection of the last face of the jacket.

FIGURE 5.26 Fabrication of the topside.

5.6 WEIGHT CONTROL

The weight-control report serves to monitor or remove weight and ensures that at all stages throughout the life of a project, weight assessments are up to date and traceable through back-up documentation.

The objectives of weight control are to properly report and to record all appropriate weight and center-of-gravity data associated with the platform in a suitable form and data are included for the topside, support frame, modules, helideck and bridges as appropriate. Data are also included for the steel space frame with all attachments and for piles of the sub-structure throughout the life of the project.

From the beginning of front-end engineering design (FEED) up to final platform construction, the weights and centers of gravity of topside and sub-structures are critical factors upon which major decisions are made. Such decisions typically cover the selection of construction pad at the fabrication yard; the means for, and route of, load-out; the need or otherwise for civil works; the selection of transportation barge, with associated ballasting and quay-clearance requirements; the selection of crane barge and preferred lifting arrangement to facilitate installation; the direction of topside inventory changes during the platform operating phase to accommodate production changes; and the selection of the optimal solution for platform abandonment should it be necessary at the end of the platform's useful life.

Notwithstanding the above, the progress of initial topside, sub-structure and foundation design is contingent upon the timely provision of reliable topside weight and center-of-gravity data.

5.6.1 Weight Calculation

The topside weights, in addition to drilling loads and other temporary operating loads, include essentially permanent contributions from all associated engineering disciplines, including structural, mechanical, electrical, instrument, communication, HVAC, safety and architectural. It is useful to categorize the weight elements associated with all disciplines other than structural and architectural as either main equipment or bulk materials.

Therefore, topside weights may be deemed to comprise the following six elements:

1. Structural steel
   This typically includes modules, decks and all main framing, deck plating and stiffening, all equipment supports integrated into the deck structure, and major facility supports, such as pipe racks, walkways, service platforms, stairs, ladders, handrails, crane pedestal and bridge supports.
   In general, structural steel consists of three categories:
   1. Primary steel
   2. Secondary steel
   3. Temporary steel items, typically including sea fastenings, bumpers, guides, lifting and rigging gear (i.e., slings, shackles, spreader beams, etc.) and secondary padeyes

2. Architectural
This typically includes firewalls, ceilings, doors, windows, flooring and floor finishes, furniture, partitions, kitchen fixtures, fittings and utensils, appliances, toilet fixtures and accessories, acoustics, insulation, windshield cladding, weather louvers, heat shields and others in the living quarters modules and emergency shelter.
3. Main equipment
This typically includes all tagged mechanical, electrical and instrument equipment and should include the skid weight, drip pan and all other items shown in the vendor package identification.
4. Bulk materials
These typically include all mechanical, piping and associated fittings and supports, electrical, communication, HVAC and safety items that have not been listed under equipment.
Piping bulk should typically include all process and utility piping, piping valves, pipe supports, trace heating, insulation, protective coatings and all associated accessories that have not been included in the equipment skids. Electrical bulk should include all electrical cables, cable trays, ladders, MCI, supports, light fixtures and junction boxes. Instrumentation bulk should include all instrument cable and all instrument piping and associated supports (such as control valves), instrument workshop equipment, all untagged control room auxiliary equipment, and all instrument equipment not listed under equipment.
HVAC bulk should typically include all ducts, duct supports, insulation, flow dampers, grills and the like for ventilation and air conditioning. Safety bulk should typically include fire monitors, hose reels, deluge valves, fire and gas detectors, halon and CO_2 systems, portable fire extinguishers, life buoys, life jackets, life rafts, survival suits and firemen's equipment.
5. Live loads
Live loads are the temporary operating loads on the topside. They typically include bulk stores, laydown area loads, crane loads, helicopter, emergency shelter, luggage, consumable and personnel effects and all the loads defined in Chapter 2.
6. Drilling loads
Although drilling loads from the required drilling equipment and tools are essentially temporary operating loads, they are treated separately and are not covered by the live load allowance.

Weight Engineering Procedures

At the start of the project and as soon as the preliminary equipment layout drawings are established, each discipline should use the weight and center-of-gravity control form to furnish all the initial weight and center-of-gravity information to the structural engineer.

Throughout the duration of the project, the lead structural engineer is responsible for gathering and recording the weight and center-of-gravity information from all disciplines and for generating the weight report. He is also responsible for reporting the status of the weight and center of gravity for the platform components, such as topside modules, decks, bridges and others, as well as for the sub-structure, to each discipline and to the project manager/engineer on an agreed-upon regular basis.

On the other hand, each discipline, such as electrical, piping, mechanical and others, is responsible for furnishing all the necessary weight and center-of-gravity information for equipment and any items associated with their discipline to the lead structural engineer. Each discipline is also responsible for updating such information at previously agreed-upon regular intervals at least monthly or when deemed necessary by the lead structural engineer.

Note that realistic weights and weight allowances can be obtained from vendors by specifying in the request for quotation (RFQ) packages that vendors guarantee, within agreed-upon limits, their quoted weights and center-of-gravity data when submitting their quotations.

In general, the lead structural engineer is usually responsible for delivering the weight-control report and should draw management's attention to any undesirable weight trends or problems and should suggest corrective actions as appropriate.

Classification of Weight Accuracy

The three basic classifications of weight accuracy are as follows. The weight allowances proposed for each project stage are shown in Table 5.3.

- Conceptual: This is based on initial estimates, possibly obtained from past projects. At the end of conceptual design, the structural weight estimates should be based on the preliminary structure design by weight take-off.
- Detailed: This is based on weight take-off and vendor information.
- Fabrication: This is based on approved drawings and final take-off.

Weight information for all items on the platform should be recorded in a manner consistent with the following definition.

Functional weight conditions are:

- *Dry weight* is a single-item weight or the weight of a collection of items characterized by its dead weight alone. This condition should typically exclude any operating fluids, spares, maintenance, tools, packing and temporary transportation materials. However, the recorded dry weight of equipment delivered with lubricants or coolants or the like and pre-installed should be deemed to include such additional weights.
- *Operating weight* is the weight of all equipment and bulks, containing all relevant fluids and supply weights, under normal operating conditions.

TABLE 5.3 Weight Allowances by Project Stage

Project Stage	Available Data	Allowance for Topside (%)	Allowance for Sub-structure (%)
Conceptual design	Historic volumetric	15	10
	Vendor catalog	15	
	Vendor data/quotation	12	
	Calculated from material take-off (MTO)	10	
	Historical weight for similar equipment	5	
	Design change allowance	5	5
	Fabrication change allowance	5	5
Detailed design	Vendor catalog	15	5
	Vendor data/quotation	10	
	Calculated MTO from approved for construction drawings	5	
	Design change allowance	5	5
	Fabrication change allowance	5	5
Fabrication	Vendor data/quotation	5	3
	Design change allowance	0	0
	Fabrication change allowance	5	2

- *Hydrotest* generally occurs during topside hook-up and commissioning and includes weight contributions from all permanent and temporary facilities together with all weight elements associated with hydrotesting.
- *Load-out* is the condition that exists during the activity of loading out the platform (topside, sub-structure) from the fabrication facility into the transportation barge.
- *Lift weight* exists during lifting of the structure. The lift weight includes the weight of the structure with the weight of all equipment and bulks actually being lifted. It should also include: all temporary lifting accessories, such as lifting slings, shackles and support frames; tie-down and support beams; and any shipped-loose items temporarily placed on the structure during lifting, but excluding hook-up spools, infill steel or other items that are to be lifted separately.

Allowances and Contingencies

The allowances and contingencies of weight are usually based on the engineering firm's experience. The contingency allowances are as follow, where the definition of allowance and contingency is obviously stated (for terminology, see Table 5.4).

1. ITEM ACCURACY ALLOWANCE (I_A)
   This allowance represents the anticipated weight growth of each individual item, or collection of items, resulting from inaccuracies inherent in the method of establishing the base weight. The value of the I_A is dependent on the degree of definition of each individual item or collection of items and the level of confidence in the weight estimate at any particular time. Three categories apply:
   1. Conceptual design allowance (C_A)
   2. Detailed design allowance (D_A)
   3. Fabrication design allowance (F_A)

TABLE 5.4 Offshore Lifting Terminology

Term	Definition
Allowance	An amount, expressed in terms of a percentage of base weight, which experience of past projects has shown to be consumed during the various phases of project execution.
Barge	The floating vessel, normally nonpropelled, on which the structure is transported.
Base weight	The base weight of any individual item or collection of items is specified to a functional weight condition and is, at the time of estimating, the best available estimate of weight for that item or collection of items exclusive of all allowances and contingencies.
Bending reduction factor	The factor by which the breaking load of a rope or cable is reduced to take account of the reduction in strength caused by bending around a shackle, trunnion or crane hook.
Breaking load	The load at which a rope or sling will break. The breaking load for a sling takes into account the termination efficiency factor.
Cable-laid sling	A cable made up of 6 ropes laid over a core rope, with suitable terminations at each end.
Certificate of approval	The formal document issued by the inspection company when, in its judgment and opinion, all reasonable checks, preparations and precautions have been taken, and an operation may proceed.
Consequence factor	A factor to ensure that main structural members have an increased factor of safety related to the consequence of their failure.

TABLE 5.4 Offshore Lifting Terminology—cont'd

Term	Definition
Contingency	An amount (in tons) to accommodate future changes in topside functional requirements instigated by management or by unexpected production changes.
Crane vessel	The vessel, ship or barge on which lifting equipment is mounted. It is considered to include a crane barge, crane ship, derrick barge, floating shear-legs, heavy lift vessel and semi-submersible crane vessel (SSCV).
Determinate lift	A lift where the slinging arrangement is such that the sling loads are statically determinate and are not significantly affected by minor differences in sling length or elasticity.
Dynamic amplification factor	The factor by which the gross lift weight is multiplied, to account for dynamic loads and impacts during the lifting operation.
Grommet	A grommet is comprised of a single length of unit rope laid 6 times over a core to form an endless loop.
Factored weight	The factored weight of any individual item is characterized by its base weight multiplied by the product of all relevant allowances.
Gross weight	The calculated weight of the structure to be lifted, including contingencies, or the weighed weight, including weighing allowance.
Hook load	The hook load is the summation of the lift weight and the rigging weight.
Indeterminate lift	Any lift where the sling loads are not statically determinate.
Lift point	The connection between the rigging and the structure to be lifted. May include padear, padeye or trunnion.
Lift weight	The lift weight is the gross weight in addition to the allowance for dynamic effects.
Load-out	The transfer of topside or jacket from land onto a barge by horizontal movement or by lifting.
Load-out, lifted	A load-out performed by a crane.
Minimum required breaking load	The minimum allowable value of breaking load for a particular lifting operation.
Net weight	The calculated or weighed weight of a structure, with no contingency or weighing allowance.
Padear	A lift point consisting of a central member, which may be tubular or a flat plate form, with horizontal trunnions around which a sling or grommet may be passed.

(*Continued*)

TABLE 5.4 Offshore Lifting Terminology—cont'd

Term	Definition
Padeye	A lift point consisting essentially of a plate, reinforced by cheek plates if necessary, with a hole through which a shackle may be connected.
Rigging	The slings, shackles and other devices, including spreaders, used to connect the structure to be lifted to the crane.
Rigging weight	The total weight of rigging, including slings, shackles and spreaders.
Rope	The unit rope from which a cable-laid sling or grommet may be constructed, made from either 6 or 8 strands around a steel core.
Safe working load (SWL)	The safe working load of a sling, shackle or lift point is the maximum load that the sling may raise, lower or suspend under specific service conditions.
Sea fastenings	The system used to attach a structure to a barge or vessel for transportation.
Skew load factor (SKL)	The factor by which the load on any lift point or pair of lift points is multiplied to account for sling mismatch in a statically indeterminate lift.
Sling breaking load	The breaking load of a sling, being the calculated breaking load reduced by the termination efficiency factor or bending reduction factor, as appropriate.
Sling eye	A loop at each end of a sling, usually formed by an eye splice or mechanical termination.
Splice	That length of sling where the rope is connected back into itself by tucking the tails of the unit ropes back through the main body of the rope, after forming the sling eye.
Spreader bar (frame)	A spreader bar or frame is a structure designed to resist the compression forces induced by angled slings, by altering the line of action of the force on a lift point into a vertical plane.
Termination efficiency factor	The factor by which the breaking load of a wire or cable is multiplied to take account of the reduction of breaking load caused by a splice or other end termination.
Trunnion	A lift point consisting of a horizontal tubular cantilever, around which a sling or grommet may be passed. An upending trunnion is used to rotate a structure from horizontal to vertical, or vice versa, and the trunnion forms a bearing around which the sling, grommet or another structure will rotate.

The item accuracy I_A appropriate to the beginning of each design phase is stipulated in Table 5.3. The average level of I_A should reduce with time through each design phase as a function of design maturity.

2. DESIGN CHANGE ALLOWANCE (D_A)
   This allowance provides for design changes during the detailed design phase. These design changes are a normal part of the design activity and represent the optimization of the design in satisfying the preferred approach defined during the conceptual design of the project.
3. FABRICATION CHANGE ALLOWANCE (F_A)
   This allowance provides for changes to design during the fabrication phase. These changes are a normal part of the fabrication activity and represent the optimization of the design in satisfying previously unidentified constraints arising during the fabrication phase. Examples are: steel section substitutes; pipework, cable work or ductwork re-routing; over-rolling of plates; weld metal; and paint. The F_A is applied on a modular basis and has the same value for each functional weight condition.
   The factored weight of an item or collection of items is obtained as:

$$\text{Base Weight of Item} \times (1 + I_A) \times (1 + D_A) \times (1 + F_A)$$

Management Contingency (MC)

Management contingency provides for initially unidentified changes in topside functional requirements that should be checked by the management during the conceptual design and then realistically spread over appropriate regions of the topsides for recording purposes. This contingency may optionally be agreed to be zero. At the end of conceptual design, the management contingency either will have been consumed and hence be represented by identifiable weight elements or will have been added to the then-existing level of operating contingency.

Operating Contingency (OC)

During the operation phase, usually there are many changes due to changes in the mode of operation. The OC provides built-in reserve topside capacity to facilitate these operational changes in functional requirements. The value of the OC, in tons, is to be agreed upon at the outset for conceptual regions of the topsides for recording purposes. This contingency may optionally be agreed to be zero.

The not-to-exceed weight of the platform (e.g., topside or sub-structure) is obtained as Factored Weights + MC + OC.

5.7 LOADS FROM TRANSPORTATION, LAUNCH AND LIFTING OPERATIONS

The topside structure and jacket component are subjected to critical loadings during construction operations. Some jacket members and joints may be subjected to high bending and punching shear loads while braces and bents are assembled into a jacket in the fabrication yard. Analysis of such assembly loading conditions requires sequential simulation of jacket geometry and loads and knowledge of the jacket assembly plan and procedures.

During transportation of the jacket to the site on the barge, the jacket and tie-down braces, their connections and the transportation barge are subjected to significant dynamic accelerations and inclined self-weight loads. These motions and resulting dynamic loads must be simulated in incremental loading sequences to determine the highest stressed components. Some bracing may be needed only for the jacket transportation phase, and some of these braces may have to be removed before the jacket is installed on site, to reduce in-place wave loads.

During its launch to the sea, the jacket will be subjected to significant inertia and drag loadings. In general, the most critical loading occurs as the jacket starts tilting around the launch beam and rapidly descending to the sea. At this position, the tilting beams exert high concentrated loads on the stiff bracing levels. These require a launch bracing system specially designed to distribute and reduce the launch forces. As the jacket hits the water plane and rapidly descends into the sea, the leading jacket braces may experience high drag and inertia forces.

It is worth mentioning that another critical case of loading is crane lift of the deck or jacket from the transportation barge. In such lifting operations, deck and jacket members and connections may be loaded in directions that differ from their in-place loading directions. Additionally, lifting slings that are redundant or shorter/longer than planned may result in loads that are substantially different from those calculated for idealized conditions: in the case of a four-sling lift, if one sling is shorter than planned, three instead of four slings may carry the entire deck loads. An unplanned load distribution may also be caused by a center of gravity that is at a location somewhat different than calculated. Furthermore, lifting padeyes and lugs are components with high consequences of failure. A single padeye failure may result in the loss of the entire deck and jacket and the crane. Such critical components and their connections to the structures lifted must be designed for higher safety factors. Safety factors of 4 or more against ultimate capacity are commonly used for padeyes, their connections to the structure, and the associated lifting gear.

5.8 LIFTING PROCEDURE AND CALCULATIONS

Lifting forces are functions of the weight of the structural component being lifted, the number and location of lifting eyes used for the lift, the angle between each sling and the vertical axis and the conditions under which the lift is performed.

All members and lifting point connections for the lifted component must be designed for the forces resulting from static equilibrium of the lifted weight and the sling tensions. Moreover, API RP2A recommends that, in order to compensate for any side movements, lifting eyes and the connections to the supporting structural members should be designed for the combined action of the static

FIGURE 5.27 Transport of the topside to the barge.

sling load and a horizontal force equal to 5% of this load, applied perpendicular to the padeye at the center of the pinhole. All these design forces are applied as static loads if the lifts are performed in the fabrication yard.

On the other hand, if the lifting derrick or the structure to be lifted is on a floating vessel, then dynamic load factors should be applied to the static lifting forces.

In particular, for lifts made offshore, API RP2A recommends two minimum values of dynamic load factors, which are 2.0 and 1.35. The factor 2.0 is to be considered in designing the padeyes as well as all members and their end connections framing the joint where the padeye is attached, while the factor 1.35 is to be used in designing all other members transmitting lifting forces.

For load-out at sheltered locations, the corresponding minimum load factors for the two groups of structural components become, according to API RP2A, 1.5 and 1.15, respectively.

Figure 5.27 shows lifting of the deck to start pullout onto the barge.

Lifting terminology is summarized in Table 5.4.

5.8.1 Lifting Calculations

For any lifting requirement, the calculations carried out should include the following allowances, factors and loads or equivalent weight contingency factors, as already discussed.

Weight control should be performed by means of a well-defined, documented system, in accordance with current good practice, such as ISO Draft International Standard ISO/DIS 19901–5—"Petroleum and natural gas industries—specific requirements for offshore structures—Part 5: Weight control during engineering and construction."

Where a limiting design sea state is derived by calculation or model tests, the limiting operational sea state should not exceed (0.7 × the limiting design sea state).

Calculated Weight

- Class A weight control will be needed if the project is weight- or center-of-gravity–sensitive for lifting and marine operations, or has many contractors to interface
- Class B weight control applies to projects where the focus on weight and center of gravity is less critical for lifting and marine operations
- Class C weight control applies to projects where the requirements for weight and center-of-gravity data are not critical

Unless it can be shown that a particular structure and specific lift operation are not weight- or center-of-gravity–sensitive, then Class A weight control will be needed. If the weight estimate is for likelihood accuracy 50%, then a reserve of not less than 5% should be applied. The extremes of the center-of-gravity envelope should be used.

A reserve of not less than 3% should be applied to the final weighed weight.

$$\text{Gross weight} = \text{calculated or weighed weight} \times \text{reserve} \tag{5.1}$$

Unless operation-specific calculations show otherwise, for lifts by a single vessel, the dynamic amplification factors (DAF) should be applied as in Table 5.5. Alternatively, the DAF may be derived from a suitable calculation or model test.

TABLE 5.5 DAF in Different Locations

			Onshore	
Gross Weight, W (tons)	Offshore	Inshore	Moving	Static
$W \leq 100$	1.30	1.15	1.15	1.0
$100 < W \leq 1000$	1.20	1.10	1.10	1.0
$1000 < W \leq 2500$	1.15	1.05	1.05	1.0
$W > 2500$	1.10	1.05	1.05	1.0

Where the lift is from a barge or vessel alongside the crane, vessel or barge, motions must be taken into account, as well as the crane boom-tip motions.

$$\text{Lift weight} = \text{gross weight} \times \text{DAF} \tag{5.2}$$

For offshore lifts by two or more vessels, the lift weight, as computed above, should be multiplied by a further DAF of 1.1.

Hook Load

The hook load is calculated in case of loading on a padeye or the structure, and the lift weight as defined before should be used. Loads in slings and the total loading on the crane should be based on hook load, where:

$$\text{Hook load} = \text{lift weight} + \text{rigging weight} \tag{5.3}$$

Rigging weight includes all items between the padeyes and the crane hook, including slings, shackles and spreaders, as appropriate.

Note that the definition of padeye here is taken to include any type of lift point, including padear, trunnion or other types.

Skew Load Factor (SKL)

For indeterminate four-sling lifts using matched slings, a skew load factor (SKL) of 1.25 should be applied to each diagonally opposite pair of lift points in turn. For determinate lifts, the SKL may be taken to be 1.0.

$$\text{Vertical padeye load} = \text{padeye resolved lift weight} \times \text{SKL} \tag{5.4}$$

where the padeye resolved lift weight is the vertical load at each padeye, taking into account lift weight and center of gravity only. Where the allowable center-of-gravity position is specified as a cruciform or other geometric shape, then the most conservative center-of-gravity position within the allowable area should be taken, until the position can be determined with confidence.

Resolved Padeye Load

The resolved padeye load is the vertical padeye load divided by the sine of the sling angle:

$$\text{Resolved padeye load} = (\text{vertical padeye load})/\sin\theta \tag{5.5}$$

where θ is the sling angle between the sling and the horizontal plane. Provided the lift-point is correctly oriented with the sling direction, then a horizontal force equal to 5% of the resolved padeye load should be applied, acting through the centerline and along the axis of the pinhole or trunnion. If the lift point is not correctly oriented with the sling direction, then the computed force acting along the axis of the pinhole or trunnion plus 5% of the resolved padeye load should be applied.

Sling Force

The sling force is the vertical padeye load plus the sling weight (per sling) divided by the sine of the sling angle:

$$\text{Sling force} = (\text{vertical padeye load} + \text{sling weight})/\sin\theta \qquad (5.6)$$

The minimum safety factor for sling or grommet-breaking load after resolution of the load based on center-of-gravity position and sling angle should be not less than 2.25.

Crane Lift Factors

For a two-crane lift, the resolved load at each crane should be multiplied by the following (according to DNV):

$$\begin{aligned} \text{Center of gravity factor} &= 1.03 \\ \text{Tilt factor} &= 1.03 \\ \text{Crane resolved lift weight} &= (\text{statically resolved lift weight into each crane}) \\ &\quad \times (\text{center-of-gravity factor}) \times (\text{tilt factor}) \end{aligned} \qquad (5.7)$$

For a two-crane lift with two slings to each hook, the load resolved to each padeye should be multiplied by a yaw factor:

$$\begin{aligned} &\text{Yaw factor} = 1.05 \\ &\text{Padeye resolved lift weight} = \\ &(\text{crane resolved lift weight, resolved to each padeye}) \times (\text{yaw factor}) \end{aligned} \qquad (5.8)$$

Two-crane lifts with other rigging arrangements will require special consideration.

Part Sling Factor

Where a two-part sling passes over, around or through a shackle, trunnion, padear or crane hook, other than at a termination, the total sling force should be distributed into each part in the ratio 45:55.

$$\text{Sling load} = \text{sling force} \times 0.55 \text{ (for two-part slings)}$$

Termination Efficiency Factor

The breaking load of a sling ending in an eye splice should be assumed to be the calculated rope breaking load multiplied by a factor as follows: for hand splices, 0.75; for resin sockets, 1.00; for swage fittings, "Superloop," 1.00.

$$\begin{aligned} \text{Sling breaking load} &= \text{rope breaking load} \\ &\quad \times \text{termination efficiency factor} \end{aligned} \qquad (5.9)$$

Other methods of termination will require special consideration.

Bending Efficiency Factor

Where any rope is bent round a shackle, trunnion, padear or crane hook, the breaking load should be assumed to be the calculated rope breaking load multiplied by a bending efficiency factor

$$\text{Bending efficiency factor} = 1 - 0.5/(P_d/r_d)0.5 \qquad (5.10)$$

where P_d is the pin diameter and r_d is the rope diameter.

This results in the bending efficiency factors shown in Table 5.6.

$$\text{Sling breaking load} = \text{rope breaking load} \times \text{bending efficiency factor} \qquad (5.11)$$

Grommets

Grommets require special consideration, to ensure that the rope breaking load and bending efficiency have been correctly taken into account.

The core of a grommet should be discounted when computing breaking load. The breaking load of each part of a grommet is therefore usually taken as 6 times the unit rope breaking load, with a factor to account for the spinning losses in cabling. This factor is normally taken as 0.85.

$$\text{Grommet breaking load (each part)} = 0.85 \times 6 \times \text{breaking load of unit rope} \qquad (5.12)$$

Typically, a grommet will be used with one end over the crane hook and the other end connected to a padeye by a shackle. The bending efficiency factors at each end may differ, and the more severe value should be taken. Bending efficiency is derived as before, where rope diameter is the single-part grommet diameter. The total breaking load of the grommet used in this manner is: 2 × (single-part grommet breaking load) × (more severe bending efficiency factor).

Shackle Safety Factors

The minimum shackle breaking load, where this can be reliable determined, should the largest value of the minimum required sling breaking load as calculated before from Equation (5.11), or

$$\text{Sling force from Equation (5.6)/DAF}$$

so that the suitable required shackle size can then be defined.

TABLE 5.6 Bending Efficiency Factor

P_d/r_d	0.8	0.9	1.0	1.5	2.0	3.0	4.0	5.0
Factor	0.44	0.47	0.50	0.59	0.65	0.71	0.75	0.78

Where the shackle is at the lower end of the rigging, the weight of the rigging components above the shackle (including DAF and taking account of sling angle) may be deducted from the sling force.

Consequence Factors

The consequence factors shown in Table 5.7 should be applied to the structure, including the lift points and their attachments into the structure. These consequence factors are applied based on the calculated lift point loads after consideration of all the factors. If a limit state analysis is used, then the additional factors should also be applied. A lifting calculations flow chart, with the various factors and their application, is illustrated in Figure 5.28.

5.8.2 Lifting Structural Calculations

Structural calculations, based on the load factors discussed above, should include adequate load cases to justify the structure. For example, for an indeterminate, four-point lift, the following load cases should normally be considered:

a. Base case, using lift weight, resolved to the lift points, but with no skew load factor
b. Lift weight, with skew load factor applied to one diagonal
c. Lift weight, with skew load factor applied to the other diagonal

In all cases, the correct sling angle and point of action, and any offset or torsional loading imposed by the slings, should be considered.

The overall structure should be analyzed for loadings. The primary supporting members should be analyzed using the most severe loading, with a consequence factor of 1.15 applied, as in Table 5.7

An analysis of the lift points and attachments to the structure should be performed, using the most severe load resulting, and a consequence factor of 1.35. The 5% side load should also be applied, as should any torsional load resulting from the 45:55 two-part sling loading, if applicable.

Where the lift point forms a structural node, then the calculations should also include the loads imposed by the members framing into it.

TABLE 5.7 Consequence Factors

Lift points, including spreaders	1.35
Attachments of lift points to structure	1.35
Members directly supporting or framing into the lift points	1.15
Other structural members	1.00

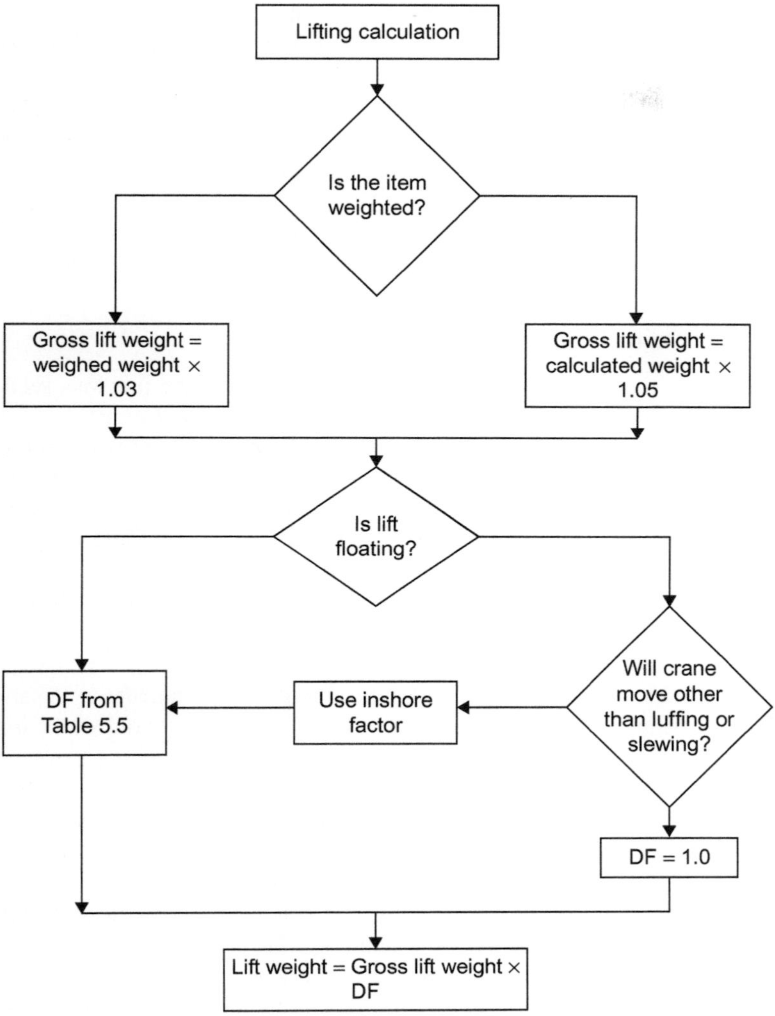

FIGURE 5.28 Procedure for calculating lifting weight.

Spreader bars or frames, if used, should be similarly treated, with load cases as above. A consequence factor of 1.35 should be applied to lift points, and a factor of 1.15 should be applied to members directly supporting the lift points, as presented in Table 5.7.

Stress levels should be within those permitted by the latest edition of a recognized and applicable offshore structure code. The loads should be treated as a normal serviceability level functional load with associated load and resistance safety factors. In the case of a working stress code, the one-third increase in strength by taking environmental loadings into consideration should not be allowed; similarly, for a load resistance factor design code (LRFD)/partial factor code, the load factor would be greater than that used for ultimate conditions.

Limit state analysis or a load resistance factor design approach may be applied according to a recognized code, provided that the total load factor is not less than the product of all the required factors, multiplied by a further factor of 1.30.

The material reduction factor should be: for elastic design of steel structures, not less than 1.15; and for plastic design of steel structures, not less than 1.30.

5.8.3 Lift Point Design

In addition to the structural requirements as described in the previous section, the following should be taken into account in the lift point design.

Adequate clearance is required between cheek plates, or inside trunnion keeper plates, to allow for ovalization under load. In general, the width available for the sling should be not less than (1.25 D + 25 mm), where D is nominal sling diameter. However, the practical aspects of the rigging and de-rigging operations may demand a greater clearance than this.

In general, for fabricated lift points, the direction of loading should be in line with the plate rolling direction. Lift point drawings should show the rolling direction.

Through-thickness loading of lift points and their attachments to the structure should be avoided if possible. If such loading cannot be avoided, the material used should be documented to be free of laminations, with a recognized through-thickness designation.

Pinholes should be bored or reamed and should be designed to suit the shackle proposed. Adequate spacer plates should be provided to centralize shackles.

Cast padears should be designed taking into account the geometrical considerations, the stress analysis process and the manufacturing process and quality control.

The extent of NDT should be submitted for review. Where repeated use is to be made of a lift point, a procedure should be presented for re-inspection after each lift.

5.8.4 Clearances

The clearance around the lifting object and crane vessel should be studied in the lifting procedure. The required clearances will depend on the nature of the lift, the proposed limiting weather conditions, the arrangement of bumpers and guides and the size and motion characteristics of the crane vessel and the transport barge.

Subject to the above, for offshore lifts, the following clearances should normally be maintained at each stage of the operation.

Clearances around Lifted Object

The clearance between any part of the lifted object (including spreaders and lift points) and crane boom is not less than 3 m.

The vertical clearance between the underside of the lifted object and any other previously installed structure, except in the immediate vicinity of the proposed landing area, is not less than 3 m.

The distance between the lifted object and other structures on the same transport barge should not be less than 3 m.

The horizontal clearance between the lifted object and any other previously installed structure, unless purpose-built guides or bumpers are fitted, is more than or equal to 3 m. The 3 m clearance also is reasonable between traveling block and fixed block at maximum load elevation.

Clearances around Crane Vessel

Denton (2009) recommended that, when mooring the crane vessel adjacent to an existing platform, clearances should be 3 m between any part of the crane vessel and the platform and 10 m between any anchor line and the platform.

Where the crane vessel is dynamically positioned, a 5-m nominal clearance is used between any part of the crane vessel and the platform.

The clearance between the crane vessel and seabed, after taking account of tidal conditions, vessel motions, increased draft and changed heel or trim during the lift, is 3 m.

The clearances around mooring lines and anchors stated below are given as guidelines to good practice. The specific requirements and clearances should be defined for each project and operation, taking into account particular circumstances, such as:

- water depth
- proximity of subsea assets
- survey accuracy
- the control ability of the anchor-handling vessel
- seabed conditions
- estimated anchor drag during embedment
- the probable weather conditions during anchor installation

Operators and contractors may have their own requirements, which may differ from those stated below, and these should govern if they are more conservative.

Clearances should take into account the possible working and stand-off positions of the crane vessel, and the moorings should never be laid in such a way that they could be in contact with any subsea asset. This may be relaxed when the subsea asset is a trenched pipeline, provided it can be demonstrated that the mooring will not cause frictional damage or abrasion.

In any case, moorings should not be run over the top of a subsea completion or wellhead. Whenever an anchor is run out over a pipeline, flow line or umbilical, the anchor should be securely stowed on the deck of the anchor-handling vessel. In circumstances where either gravity anchors or closed stem tugs are used, and anchors cannot be stowed on deck, the anchors should be double secured through the additional use of a safety strap or similar method.

The vertical clearance between any anchor line and any subsea asset should be not less than 20 m in water depths exceeding 40 m, and 50% of water depth in depths of less than 40 m.

Clearance between any mooring line and any structure other than a subsea asset should be more than 10 m.

When an anchor is placed on the same side of a subsea asset as the crane vessel, it should be placed more than 100 m from the subsea asset.

When the subsea asset lies between the anchor and the crane vessel, the final anchor position should be more than 200 m from the subsea asset.

During lifting operations, crossed mooring situations should be avoided wherever practical. Where crossed moorings cannot be avoided, the separation between active catenaries should be not less than 30 m in water depths exceeding 100 m, and 30% of water depth in water depths less than 100 m.

If any of the clearances are impractical because of the mooring configuration or seabed layout, a risk assessment should be carried out and special precautions taken as necessary.

Figure 5.28 presents the procedure for calculating the lifting weight; the procedure for calculating the loads on the padeye is illustrated in Figure 5.29. After obtaining the padeye loads, the structural members and padeye are required to be checked through a procedure presented in the flowchart in Figure 5.30. Figure 5.31 presents the procedure for checking the rigging facilities. The barge operator should have a chart that defines the crane's capability to lift the structure to or from the barge; a sample chart is shown in Figure 5.32.

Figure 5.32 presents the lifting capacity for the crane on a barge, which is the relation between boom radius, hook height and the lifting capacity. This chart should be included in the lifting procedure, which is usually delivered from the construction company and reviewed by the engineering firm and the company representative.

5.8.5 Lifting Calculation Report

Calculations should be presented for the structure to be lifted, demonstrating its capacity to withstand, without overstress, the loads imposed by the lift operation, with the load and safety factors, and the load cases.

The calculation package for lifting should contain the following minimal requirements:

a. Plans, elevations and sections showing main structural members
b. The structural model. This should account for the proposed lifting geometry, including any offset of the lift points.

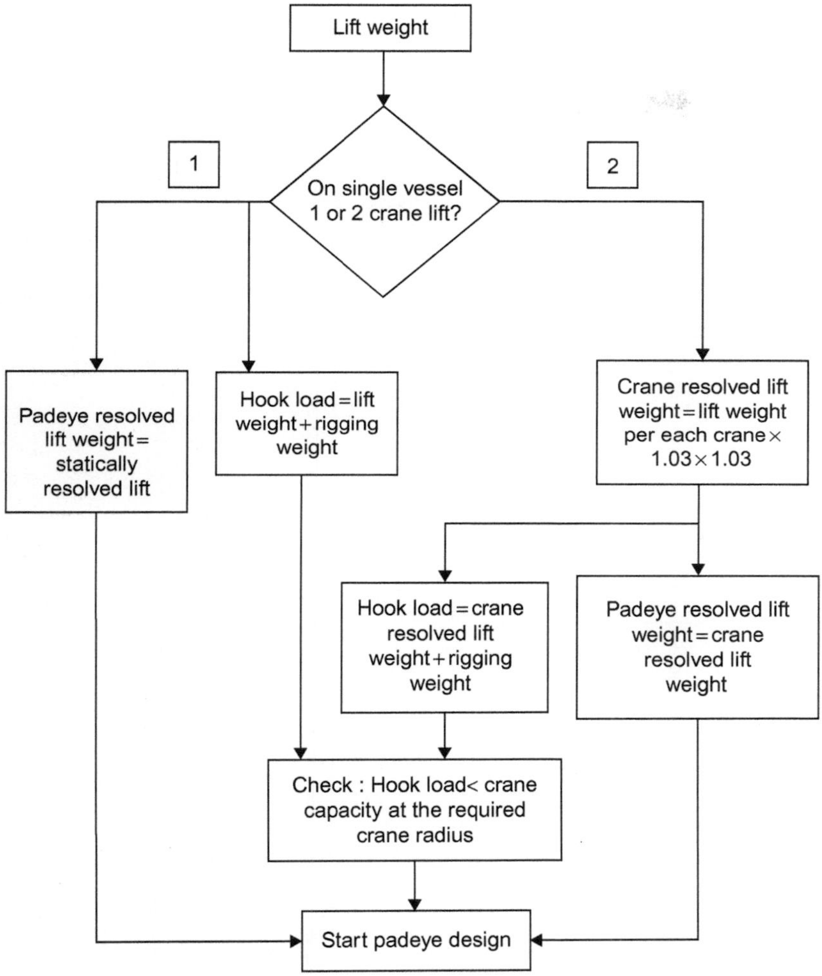

FIGURE 5.29 Procedure for calculation of loads on the padeye.

c. The weight and center of gravity

d. The steel grades and properties

e. The load cases imposed

f. The codes used

g. A tabulation of member unity checks, or a statement that unity checks are less than 0.8

h. Justification, or proposal for redesign, for any members with a unity check in excess of 1.0.

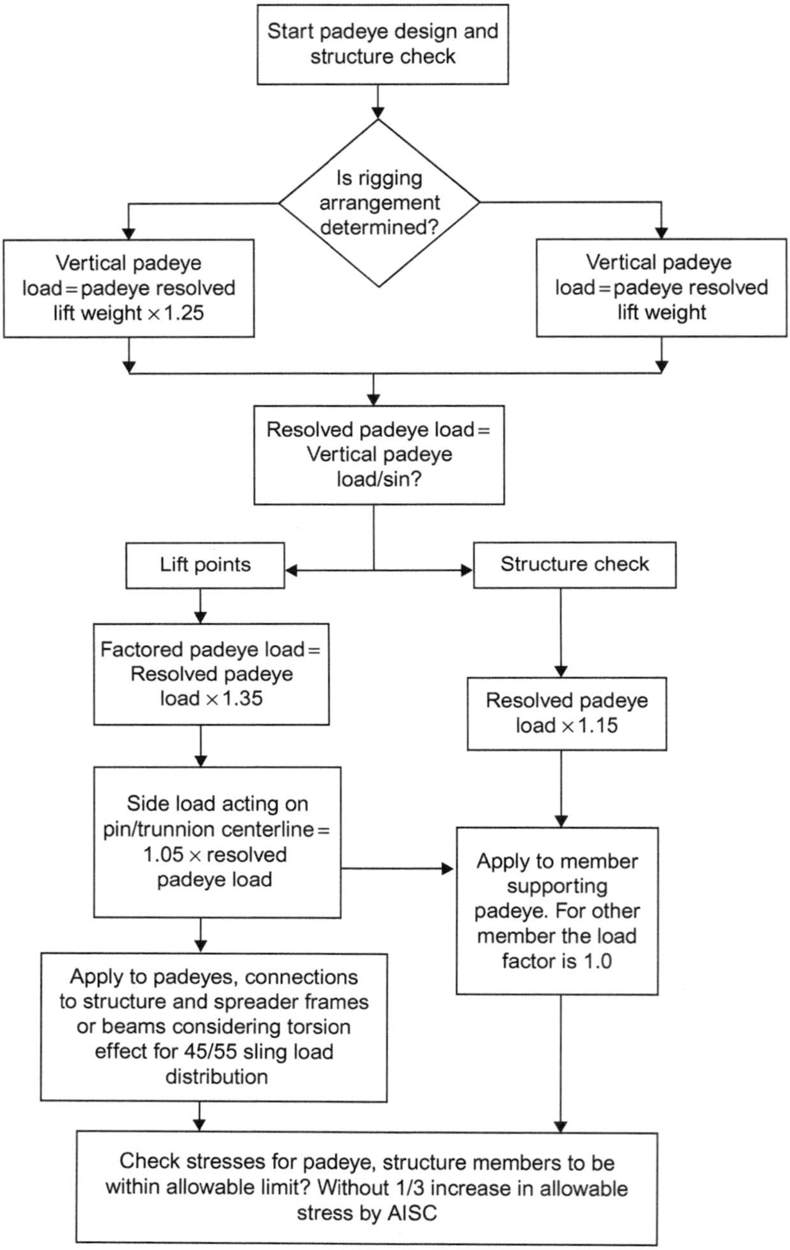

FIGURE 5.30 Check of the structural members and padeye.

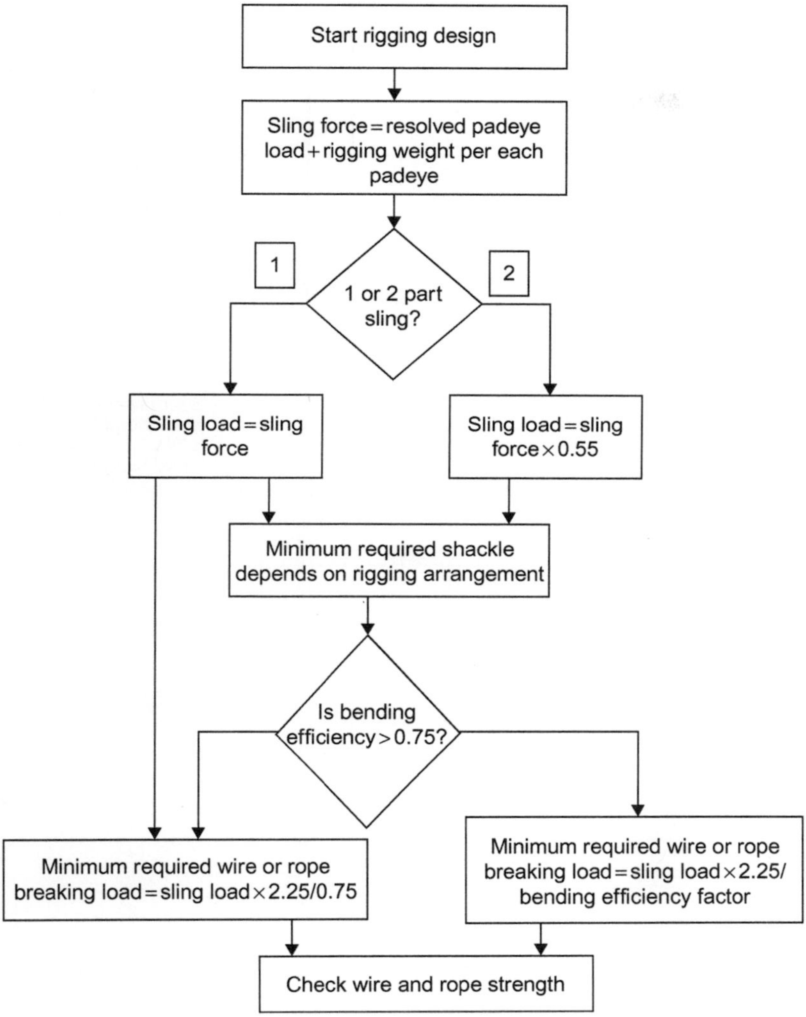

FIGURE 5.31 Check of rigging facilities.

A similar analysis should be presented for spreader bars, beams and frames. An analysis or equivalent justification should be presented for all lift points, including padeyes, padears and trunnions, to demonstrate that each lift point and its attachment into the structure is adequate for the loads and factors set out.

A proposal should be presented showing:

1. The proposed rigging geometry, including dimensions of the structure, center-of-gravity position, lift points, crane hook, sling lengths and angles, as well as shackle dimensions and "lost" length around hooks and trunnions.

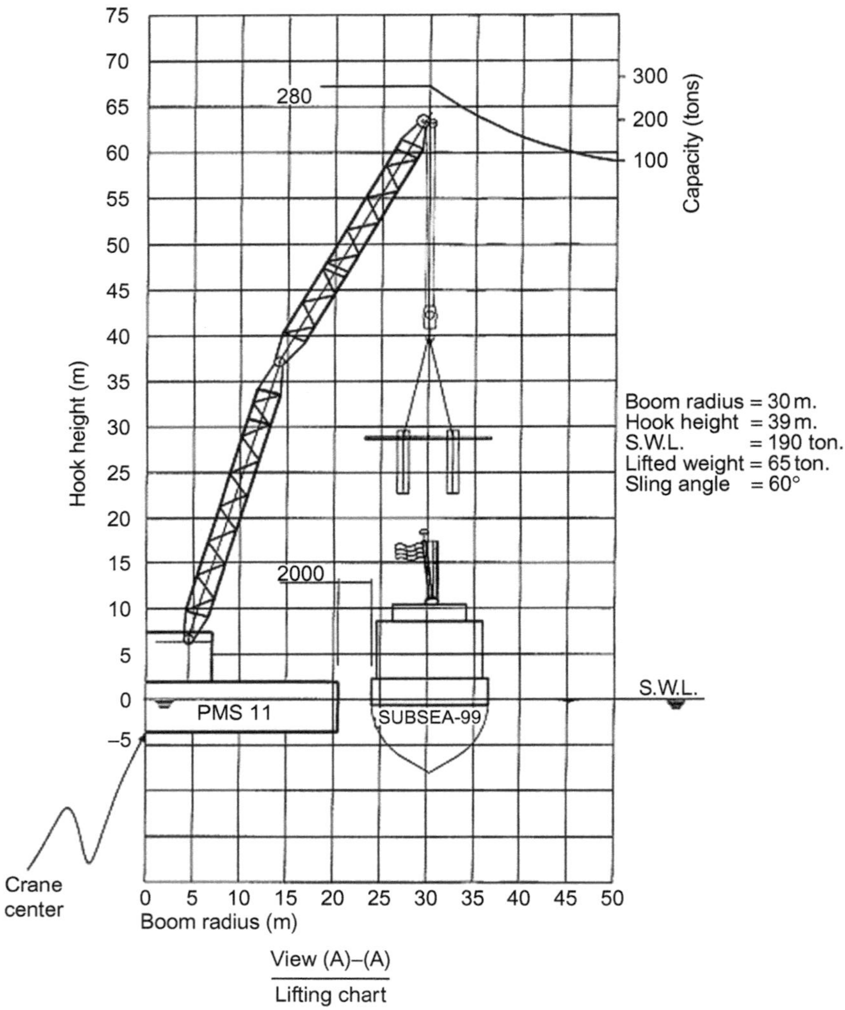

FIGURE 5.32 Lifting capacity of the crane.

2. A computation of the sling and shackle loads and required breaking loads, taking into account the safety factors, as described
3. A list of actual slings and shackles proposed, tabulating
   - Position on structure
   - Sling/shackle identification number
   - Sling length and construction diameter minimum breaking load for slings, and safe working load and minimum breaking load for shackles
   - Direction of lay
4. Copies of inspection/test certificates for all rigging components.

Shackles manufactured should deliver a test certificate that should be no more than 5 years old, and, if they are not new, a report of an inspection by a competent person since the last lift.

For each sling and shackle, a detailed record of all previous lifts should be presented, including the date and calculated load for each lift.

The Crane Vessel

The crane chart information should be submitted to the construction and fabrication contractor. The information should include, as appropriate:

- Vessel general arrangement drawings and specification
- Details of registry and class
- Mooring system and anchors
- Operating and survival drafts
- Crane specification and operating curves
- Details of any ballasting operations required during the lift.

The mooring arrangement for the operation and stand-off position should be submitted. This should include the lengths and specifications of all mooring wires and anchors, and a mooring plan showing adequate horizontal clearances on all platforms, pipelines and any other seabed obstructions. An elevation of the catenary for each mooring line, for upper and lower tension limits, should demonstrate adequate vertical clearance over pipelines.

Table 5.8 presents a checklist for verifying the topside lifting analysis, as a quality assurance measure for the engineering process in this phase.

TABLE 5.8 Checklist for Topside Lift Analysis

Items	Check Point	Check (Yes/No)
Computer Model		
	It is assumed that the model is checked for dimensions, elevations, member group and section properties in the in-service analysis and is upgraded to suit the current analysis.	
1	Check the input data and focus on the member group properties for slings, such as elastic modulus, E, shear modulus, G, and density.	
2	Member end releases for slings	
3	Member end offsets for slings at padeye locations	
4	Member properties: Kx, Ky, Lx and Ly for changed support conditions	

(Continued)

TABLE 5.8 Checklist for Topside Lift Analysis—cont'd

Items	Check Point	Check (Yes/No)
5	Support conditions for hook point(s)	
6	Adequacy of analytical springs	
7	Modeling of sling and hook for 75%–25% sling load distribution (sling mismatch)	
Loads		
	It is assumed that the load calculations are verified in the in-service analysis and only relevant load cases are picked for the current analysis. Remove future loads, post-installed items, operating weights and live loads	
1	Load contingencies	
2	DAF, skew load factors, consequence factors	
3	Proportionate distribution of sling loads (75%–25%)	
4	Calculations for center-of-gravity shift	
5	Check load combinations, include rigging loads	
Analysis Results		
1	Enclose sea-state summary	
2	Enclose member check report: review overstressed members	
3	Enclose joint check summary: review overstressed joints (check $F_y = 2/3F_u$ for chords of high-strength members)	
4	Enclose spring reactions equal to zero for correct hook location in the model	
5	Enclose sling forces equal to zero moment and shear; there are only axial forces	
6	Enclose deflection plots, member unity check ratio plots	
7	Weight-control report extract enclosed and the weight comparison carried out	
Padeye Design		
1	Appropriate selection of padeye plates, sling and shackle size	
2	Padeye design force as per the latest lift analysis	
3	Padeye stresses checked at the attachment to the main structure	
4	Stresses in the member-supporting padeye	

TABLE 5.8 Checklist for Topside Lift Analysis—cont'd

Items	Check Point	Check (Yes/No)
Miscellaneous		
1	Design of spreader beam or frame	
2	Design of padeyes on the spreader beam or frame, if they exist	
3	Lift drawing enclosed and reviewed and the derrick barge main block and auxiliary block load-carrying capacity checked. Check the requirement for load shedding if the hook load is greater than the load-carrying capacity of the derrick barge.	

TABLE 5.9 Checklist for Jacket/Topside Transportation Analysis

S/N	Items to Check	Check (Yes/No)
	Computer Model	
	It is assumed that the model is checked for dimensions, elevations, member group and section properties in the in-place/load-out analysis and is upgraded to suit the current analysis.	
1	Member end releases for sea fastening	
2	Sea fastening material yield strength	
3	Boundary conditions for dead load case	
4	Boundary conditions for tow load case	
	Loads	
	It is assumed that the load calculations are verified in the in-service as load-out analysis and only relevant load cases are picked for the current analysis.	
5	Remove future loads, operating weights and live loads, add rigging loads	
6	Load contingencies in static and tow analysis	
7	Load combinations in static analysis	
8	Center-of-rotation data	
9	Roll and pitch direction representation	
10	Consideration of self-weight during inertia load generation.	

(Continued)

TABLE 5.9 Checklist for Jacket/Topside Transportation Analysis—cont'd

S/N	Items to Check	Check (Yes/No)
11	Roll and pitch acceleration data	
12	Coefficients for the lateral load components due to combinations	
13	Primary and secondary load case identification	
14	Load combinations	
15	Chord strength reduction in the joint-can input file. (Check $F_y = 2/3\ F_u$ for chords of high-strength members.)	
16	Allowable stress modifiers	
	Analysis Results	
17	Enclose load case summary for dead load case	
18	Enclose reaction summary from combined analysis	
19	Cross-check reaction summary values from basic motion equations	
20	Enclose member check report: review overstressed members	
21	Enclose joint check summary: review overstressed joints	
22	Enclose model plots: joint/group/section names, Kx, Ky, Lx, Ly and Fy	
23	Enclose deflection plots, member unity check ratio plots	
	Sea-Fastening Design	
24	Check for adequacy of base plate connections for sea fastenings (weld size less than the barge deck plate thickness, gusset connections for uplift forces preferred)	
25	Check for adequacy of doubler plate connection (weld, doubler plate) on the topside members	
	Barge Deck-Strength Check	
26	Check for barge deck transverse girder adequacy	
27	Check for adequacy of weld between barge deck plate/girder for uplift	
28	Check for buckling strength of bulkheads for sea-fastening reactions	
29	Check for combined vertical/lateral loads on end bulkheads	
30	Check for combined vertical/lateral loads on internal bulkheads	
31	Check for barge flexibility effect of sea-fastening forces	

5.9 LOAD-OUT PROCESS

After the erection of the jacket and the topside, the load-out process is started; this process generates load-out forces on the structure as it moves from the fabrication yard to the barge.

If the load-out is carried out by direct lift, then, unless the lifting arrangement is different from that to be used for installation, lifting forces need not be computed, because lifting in the open sea creates a more severe loading condition that requires higher dynamic load factors. If load-out is done by skidding the structure onto the barge, a number of static loading conditions must be considered, with the jacket structure supported on its side. In load-out of the jacket, the loading conditions affect the structure in the different positions of the jacket during the load-out phases, as shown in Figure 5.33, from movement of the barge due to tidal fluctuations, marine traffic or change of draft, and from possible support settlements. Since movement of the jacket is slow, all loading conditions can be taken as static. Typical values of friction coefficients for calculation of skidding forces are:

- steel-on-steel without lubrication, 0.25
- steel-on-steel with lubrication, 0.15
- steel-on-Teflon, 0.10
- Teflon-on-Teflon, 0.08

All structures should be checked for the loads applied during load-out. The contractor should define explicitly the proposed method of load-out (i.e., skidded, trolleyed or lifted). The following should be considered:

1. Dry loads only should be used, together with weights for lifting gear, sea fastenings and others. The loads should be based on the weight-control report.
2. For skidded or trolleyed load-out:
   - A horizontal load of 15% of the vertical reaction on one skid rail should be applied.
   - Total loss of one vertical support should not affect the structure jacket if it is being supported by the remaining supports only.
   - Supports should be assumed to be hinged.
   - Wind loads for a return period of 1 year should be included with this load condition in the structure analysis check.
   - In most cases, the total support points are about 6–15 supports for the load-out.

The direction of the pull-out force is shown by the arrows in Figure 5.34.

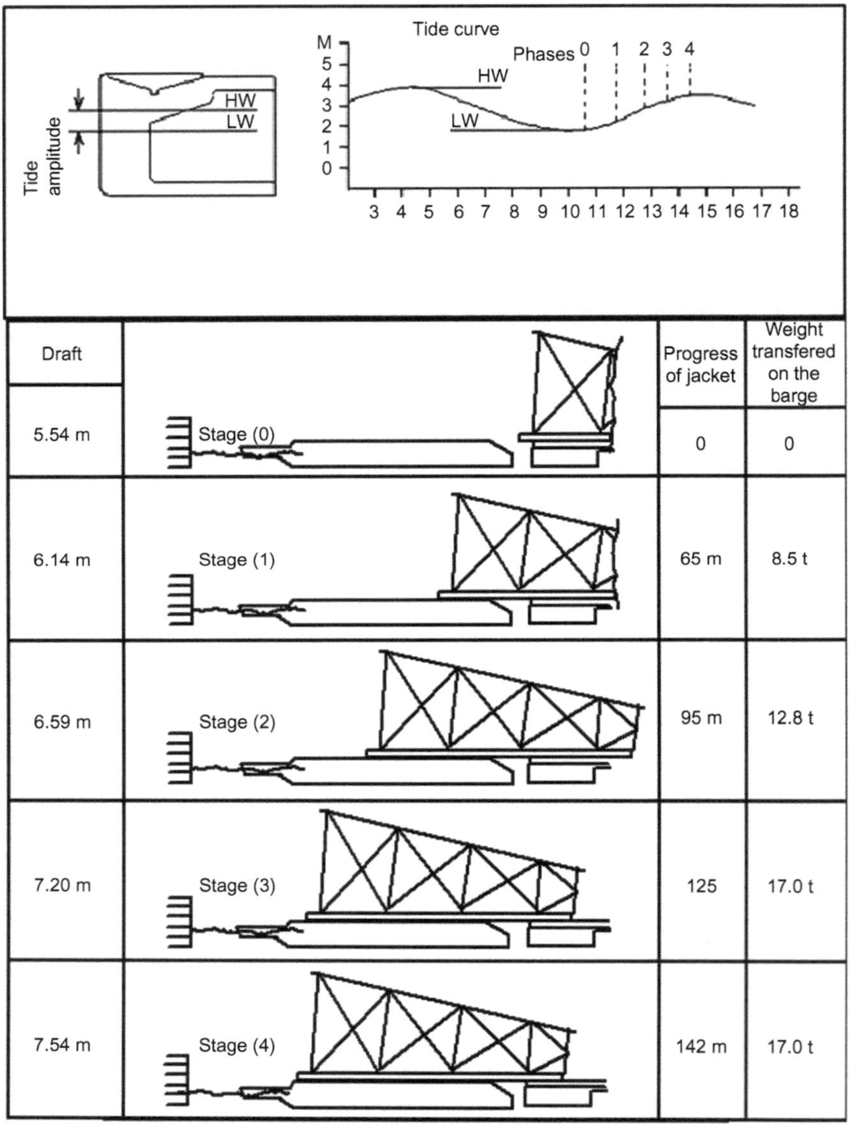

FIGURE 5.33 Jacket load-out stages.

(a)

(b)

FIGURE 5.34 (a) Load-out of the topside. (b) Moving the topside onto the barge.

5.10 TRANSPORTATION PROCESS

During transportation of the deck or jacket structure from the fabrication yard to its offshore location, the forces that will affect the structure depend upon the structure's weight and geometry and the support conditions supplied by the barge or by buoyancy, as well as on the environmental conditions that prevail during transportation.

A large role in the offshore structure project, during construction as well as operation, is played by the vessels that transfer people and equipment to the platform and that perform construction. The different types of vessels are described next.

5.10.1 Supply Boats

A supply boat is a boat having a large, open bay astern, as wide and as long as feasible, to enable the boat to deliver and supply all kinds of cargo. The "well" or open bay should be long enough to accommodate pipes that are nominally 12 m in length but may run 2 m or more longer. So a 15–20-m well is common. Supply boats are constantly increasing in displacement and capacity; a vessel with a 1000-ton displacement was considered a large boat until recent North Sea needs, sea states, and distances led to increases to 1500-, 2500-, and even 3500-ton displacements.

While supply boats are designed primarily to transport personnel with equipment, they must have maneuvering ability for close-in work alongside. They also need reinforced gunwales and heavy fenders to absorb the impact of contact with other vessels.

5.10.2 Anchor-handling Boats

Anchor-handling boats are specially designed to pick and move anchors, even in rough seas. Therefore, they are short, highly maneuverable vessels. The stern of an anchor-handling boat is open and armored so that wire or buoys can be dragged in as required. There is a winch at the forward end of the well so that, by means of a line, a wire line pendant or buoy can be quickly dragged on board. In addition, hydraulic-assist equipment is available.

5.10.3 Towboats

The main function of the towboat is to push the barge. There are many types of towboats. The large, ocean-going, long-distance towboat is capable of operating for around 30 days without refueling. It is designed to move to any part of the world to carry out a major towing job. It may be up to 80 m or more in length and carry a crew of 15 to 25. Boats of this type can run light at speeds of 12–15 knots. Harbor and other inland towboats are smaller and more maneuverable.

FIGURE 5.35 Tow boat.

Bollard pull is the force exerted by the towboat running full ahead while secured by a long line to a stationary bollard; that is, the boat is making no headway through the water. The relationship between indicated horse power (IHP) and bollard pull is such that a boat with around 10,000 IHP can exert 100–140 tons of static bollard pull. The effective bollard pull falls off as the speed through the water increases. The largest tugs have static bollard pulls of over 300 tons.

For safe and efficient operation, towboat length should be equal to or more than 11 times the expected maximum significant wave height. In major storms, the boat may have to cut loose and subsequently recover its tow after the storm has passed.

5.10.4 Towing

The first principle of towing is that the attachments to the structure or barge must always be sufficiently strong that they do not fail or damage the structure under a force that breaks the towline. In selecting the tow wire, consider that the actual breaking strength of wire rope is typically 10%–15% greater than the guaranteed minimum breaking strength.

Note that actual breakage usually occurs under a dynamic load rather than a static load. It is important that, during overload, the structure or vessel being towed remains undamaged. A usual requirement is that the ultimate capacity of any towline attachment to the unit be at least four times the static bollard pull and at least 1.25 times the breaking strength of the towline from the largest tug to be used on that attachment. In case of emergency, for towing ahead, one spare attachment point should be available with the required fitting.

A second principle of towing is that the towing force must be able to be resisted through a significant range of horizontal and vertical angles, thus imparting shear and bending, as well as tension, on the towing attachment.

A typical arrangement of a single boat towing with a bridle is shown in Figure 5.36. Often, a high-impact overload happens due to high-tension force, so the weak pendant, as shown in the figure, will be break. In this case, the shackle is pulled back on deck by a fiber rope pendant, a new pendant is fitted, and the towline is reconnected. While passing through restricted waters and during final positioning, the towline may be shortened to permit better control. In some cases, due to the configuration of a structure such as a caisson or TLP structure, movement of the structure will be accomplished by a boat that pushes the structure, using a fender in the boat and attached to the structure.

The moment of inertia of a towed structure, especially a large one such as an offshore caisson, is tremendous. It tends to keep moving ahead long after pull has ceased. A constant concern for boats towing in congested traffic conditions or in ice is that if the boat is stopped, the towed vessel or structure may overrun it.

Furthermore, the inertia of the towed structure makes it difficult to slow down or change direction. In a narrow channel, therefore, additional boats may be used alongside and astern.

Figure 5.37 presents the types of motion that affect floating structures.

In order to minimize the associated risks and to secure safe transport from the fabrication yard to the platform site, it is important to plan the operation carefully, following the recommendations in API RP2A (2007), including consideration of:

- Previous experience along the tow route
- Exposure time and reliability of predicted "weather windows" during transportation

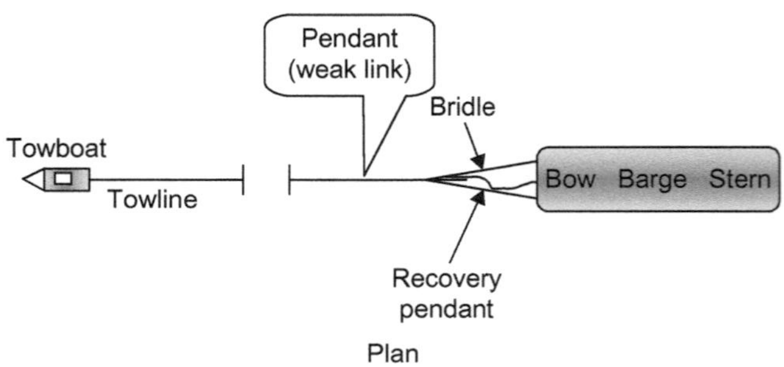

FIGURE 5.36 Typical towing arrangement for a barge.

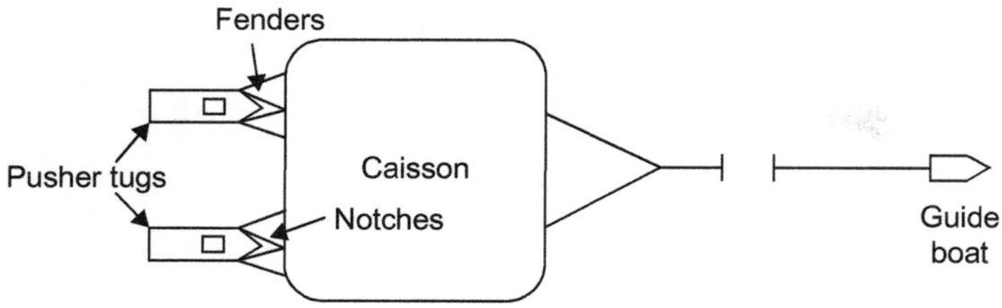

FIGURE 5.37 Types of motions.

- Accessibility of safe havens
- Seasonal weather systems
- Appropriate return period for determining design wind, wave and current conditions, taking into account characteristics of the tow, such as size, structure, sensitivity and cost.

Transportation forces are generated by the motion of the tow, the structure and the supporting barge. Loads will result from the design winds, waves and currents. If the structure is self-floating, the loads can be calculated directly. According to API RP2A, towing analyses must be based on the results of model basin tests or appropriate analytical methods and must consider wind and wave directions parallel, perpendicular and at 45° to the tow axis. Inertial loads may be computed from a rigid body analysis of the tow and by combining roll and pitch with heave motions, when the size of the tow, magnitude of the sea state and experience make such assumptions reasonable. For open-sea conditions, the following may be considered typical design values:

Single-amplitude roll: 20°
Single-amplitude pitch: 10°
Period of roll or pitch: 10 s
Heave acceleration: 0.2 g

When transporting a large jacket by barge, stability against this large size is a primary design consideration because of the high center of gravity of the jacket (Figure 5.38). Moreover, the relative stiffness of jacket and barge may need to be taken into account, together with the wave forces that could result during a heavy roll motion of the tow, when structural analyses are carried out for designing the tie-down braces and the jacket members affected by the induced loads. There are special computer programs (or a module for the software used for structure analysis) available to compute the transportation loads in the structure-barge system and the resulting stresses for any specified environmental condition.

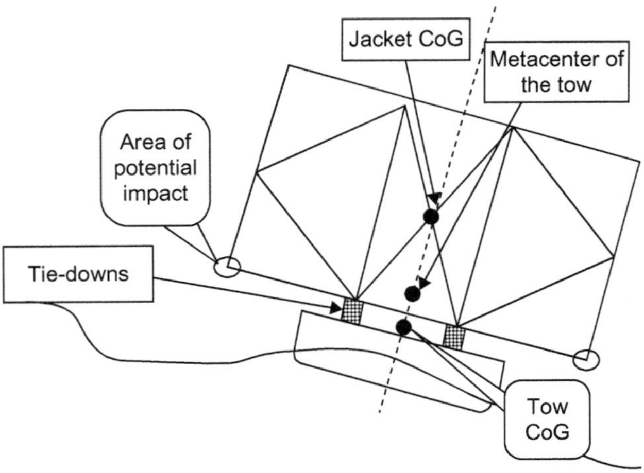

FIGURE 5.38 Center of gravity for a barge launching a jacket.

According to Gerwick (2007), when towing out in the open sea, boats lengthen their towlines to offset the wide range of loads in the lines caused by the waves and swells.

If the towed structure is considered a deep-draft vessel (as, for example, some of the offshore platforms in the North Sea have drawn from 110 to 120 m), the structure will be towed by a deep-draft vessel, and the towline, if attached to the structure below water near the center of rotation, can have a steep inclination, which may pull the stern of the boat down into the water.

Required channel widths in sheltered areas are usually twice the beam, but this must be considered in relation to the environmental conditions and navigational accuracy. According to Gerwick (2007), in exposed areas the channel width depends on currents and navigational accuracy and thus may vary from about 600–1500 m in the relatively short distance of 12 km. Unlike a towed ship or barge having a draft of around 8 to 10 m and a width from 30 to 40 m, an offshore structure like a deep-water caisson may have a draft of over 100 m and a width of 100–150 m. Therefore, it is not enough to plot only the position of the area of crossing; the extremities must also be considered. For a restricted area, the current speed and direction should be surveyed on the surface and at a reasonable depth.

If a structure is being towed by two boats between two islands, the location of the electronic and visual stations to achieve a safe crossing will be set on the islands.

Towed vessels and shallow-draft structures may have an actual draft greater than their mean draft. This may be due to trim, squat, list or wind heel, or it may be due to the lower density of fresh water discharging from a river into the adjacent sea.

The usual requirement for the distance clearance between maximum static draft and minimum water depth should not be less than 2 m or 10% of the maximum static draft, whichever is lesser, plus an allowance for motion. The maximum static draft should be the actual measured draft at the deepest point, with allowance for errors in measurement, initial trim, and water density change. The motion allowance should include the maximum increase in draft due to towline pull, wind heel, roll, pitch and heave. These values can best be determined by model tests. In practice, most fixed offshore platform structures will be governed by the 2-m minimum clearance.

The dynamic accelerations of the towed structure should generally be limited to 0.2 g, to minimize forces acting on the tie-downs and to minimize adverse effects on personnel.

The towing horsepower selected should be sufficient to hold the towed structure against significant wave heights, 40-knot sustained wind speed, and current speed for the region of work. Obviously, these arbitrary parameters have to be adjusted to the region involved.

The metacentric height (GM) is the distance between the center of gravity of a ship and its metacenter (Figure 5.39). (When a ship is heeled, the center of buoyancy of the ship moves laterally. The point at which a vertical line through the heeled center of buoyancy crosses the line through the original, vertical center of buoyancy is the metacenter, as shown in Figure 5.38.) The GM is used to calculate and to ensure the stability of a ship and it must be determined before the ship proceeds to sea. The GM must equal or exceed the minimum required GM for that ship for the duration of the forthcoming voyage.

The GM should have a positive value, usually 1–2 m for a large offshore structure. The maximum inclination of the towed structure under conditions

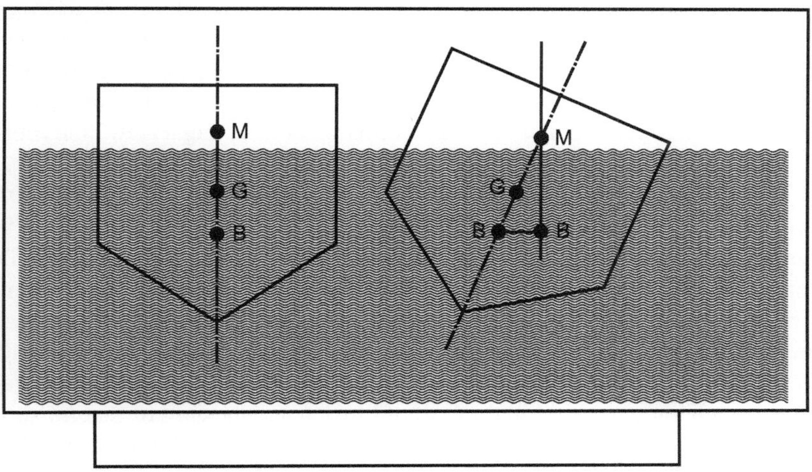

FIGURE 5.39 Metacentric height.

of wind at 60 km/h and full towline pull is not to exceed 5°. The maximum inclination of the towed structure under the 10-year storm for the season involved, with no towline pull, is not to exceed 5°. The static inclination under half the total towline pull, in still water, is not to exceed 2°.

5.10.5 Drilling Vessels

A drilling vessel is usually available during drilling activity for exploration or for normal drilling of wells during operation. In some cases, the same rig can be used in installing the platform, especially a platform of smaller size. Drilling vessels are large offshore vessels, fully equipped, including appropriate mooring gear and heavy-lift equipment for direct vertical pulls on the drill string, and they have a central moon pool, which is called an open well, that provides direct and partially protected access to the sea below with minimal wave action at the interface.

Drilling vessels can be worked in greater depths, so they have been used for many offshore construction tasks, from setting subsea templates to pipeline repair and seafloor modifications.

In some cases, the offshore drilling vessel may be a semisubmersible, or it may be a large ship hull, especially configured to minimize roll, but nevertheless, such a ship-shaped vessel does have inherently more roll response than a semisubmersible. The drilling derrick is equipped with a large hoist that can reach 500 tons or greater direct lift.

5.10.6 Crew Boats

Any company with a fleet of offshore structure platforms has daily need for crew boats to transfer staff and operators to the platforms from onshore. Crew boats are also used for small construction tasks or minor modification on the platform to transfer the team of workers with their tools. Crew boats must work during any sea conditions that permit work to be carried out in a reasonable and practical manner. The rule of thumb for selecting a crew boat is that the required horsepower be proportional to the square of the velocity.

In general, the governing factor for crew boat safety is the metacentric height (GM), as discussed above, so accelerations should be minimized by adopting a boat with as low a GM as is consistent with safety. A high GM means a quick roll response and physical discomfort for the passengers. A boat may get into pitch resonance with head-on or nearly head-on seas. This may be modified by changing the speed or the heading, or both. If the boat's length exceeds the wavelength, pitch response is reduced; however, this is usually only practical in the Gulf of Mexico and the Gulf of Suez or similar conditions, and it is not practical in the Pacific or North Atlantic, due to their longer waves.

5.10.7 Barges

A barge is considered a floating workshop (Figure 5.40). The offshore construction barge must be long enough to have minimal pitch and surge response to the waves in which it normally works, wide enough in beam to have minimum roll, and deep enough to have adequate bending strength against hog, sag, and torsion, as well as adequate freeboard. The plate of the deck should be continuous enough to enable it to resist the membrane compression, tension, and torsion introduced by wave loading. Side plates must carry high shear, so the sides usually have a stiffener to resist against buckling.

Impact loadings can come from wave slam on the bow, from ice, and from boats and other barges hitting against the sides. Unequal loads may be incurred in bending of the bottom hull plates during intentional or accidental grounding and of the deck plates due to cargo loads. Typical offshore barges run from 80 to 160 m in length, and the width should be 1/3–1/5 the length, while the depth will typically run about 1/15 of the length. From a practical point of view, barges with this depth have been found to give a reasonably balanced structural performance under wave loadings. Barges subjected to minimal wave loadings and required for operations in shallow water may have depths as low as 1/20 of the length.

Offshore barges typically have natural periods of roll of 5–7 s. Unfortunately, this is also sometimes coincides with the period of wind and waves; hence resonant response does occur. Fortunately, damping is very high, so that while motion in a sea will be significant, it reaches a situation of dynamic stability. Consideration must be given to the need to temporarily weld padeyes to the deck in order to secure cargo for sea. These padeyes must distribute their load into the hull. They will be subjected to fatigue and to impact loads in both tension and shear; therefore, a better design has special doubler plates fixed over the internal bulkheads so that padeyes may be attached along them. For welding, low-hydrogen electrodes should be used. Alternatively, posts may be installed, running through the deck to be welded in shear to the internal bulkheads.

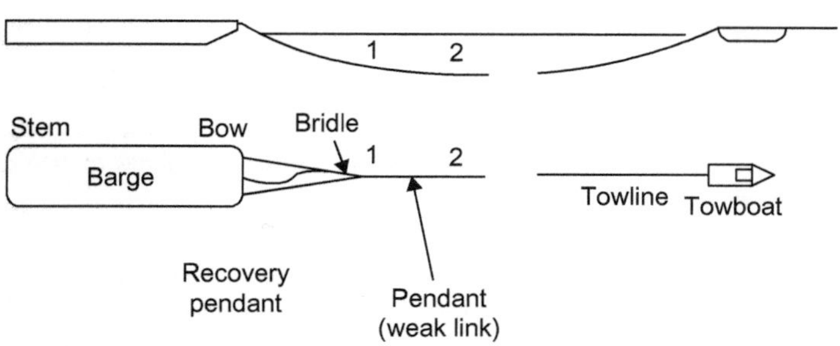

FIGURE 5.40 Material barge.

When heavy loads are skidded on or off a barge, they punish the deck edge and side because of the concentrated loading. Skid beams are often arranged to partially distribute the load to interior bulkheads.

Thus, sea fastenings are designed to resist the static and dynamic forces developed under any combination of the six fundamental barge motions (roll, pitch, heave, yaw, sway or surge). The dynamic component is due to the inertial forces that develop due to acceleration as the direction of motion changes.

Roll accelerations are directly proportional to the transverse stiffness of the barge, which is measured by its metacentric height (GM) (see Figure 5.39). Since a barge typically has a large GM, roll accelerations are severe. Conversely, if high cargo, such as the topsides or jacket, cause the GM to be low, the period and amplitude of roll and the static force resulting from the load are greater, but the dynamic component may be less.

The loads that apply to the fastenings are mainly from waves, which are cyclic, so sea fastenings tend to work loose as the wire rope stretches and wedges and blocking fall out. Under repeated loads, fatigue may occur, especially at welds. Welds made at sea may be especially vulnerable because the surfaces may be wet or cold. Using low-hydrogen electrodes in welding will help in this case. Chains are the preferred method for securing the transportation of jacket and topside in the sea, since chain does not stretch.

The effect of the accelerations is to increase the lateral loading exerted by the cargo due to the inclination of the barge by a factor of two or more. Flexing of the barge can also have a significant effect on support forces and the sea fastenings. Therefore, deeper, and hence stiffer, barges will experience a smaller range of loads than shallow, less stiff barges.

For decks or jackets, which are valuable cargo, sufficient freeboard should be provided to ensure stability, even if one side compartment or end compartment of the barge has been flooded, which, in most cases, means the submergence of the hull to the deck line, plus an arbitrary load of 3 m of water on deck.

Proposals are often made to build a structure on a barge and then to submerge the barge by ballasting and to float the new structure off.

5.10.8 Crane Barges

A crane barge is an offshore barge that has a sheer-legs crane or fully rotating crane. A sheer-legs crane can pick loads and luff but not swing. The sheer legs consist of an A-frame made up of two heavy, tubular members or trussed columns held back by heavy stays to the bow.

A sheer-legs barge is maneuvered by deck engines, tugs, or mounted outboard engine propellers. The crane barge positions its stern at the side of the material barge, picks the load, and then moves as necessary to set the load in exact position. Modern torque-converter deck engines and propellers with variable pitch allow a high degree of accuracy in positioning, on the order of 50 mm. One of the advantages of a sheer-legs crane barge over a fully revolving derrick barge is that the load is always picked over the stern end, hence preventing list from the swing of the crane.

The deck engines of a sheer-legs crane barge must be adequate to control the barge's motion in yaw, sway, and surge to a very close tolerance despite the state of the sea. The sheer-legs crane barges have a capacity of around 3000 tons.

A crane barge was successfully used to put jackets into pre-installed frames with a tolerance of only 50 mm.

5.10.9 Offshore Derrick Barges (Fully Revolving)

Fully revolving derrick barges are the traditional barge for use in constructing a new platform or performing a major modification on the platform. Derrick barges are very costly but they are capable of doing a huge amount of work. As with sheer-legs crane barges, they are fitted with deck engines and have full mooring capability. Figure 5.41 shows a derrick barge.

A typical marine derrick barge has a capacity in the range of 50 to 400 tons, whereas an offshore derrick barge has a crane capacity of 500–1500 tons. To handle ever-larger modules and deck sections, capacities have been rapidly increased in recent years, with the latest offshore derrick barges having two cranes, rated at 6500 tons each, with a total capacity of 13,000 tons.

The derrick barge represents an optimization of opposing demands. Structural and naval architectural considerations require the derrick to be located forward of the stern a distance 20%–25% of the length, that is, at the one-quarter or one-fifth point. The barge should be wide enough that, as the crane swings, there is adequate distribution of the structural load. On the other hand, the

FIGURE 5.41 Derrick barge.

effective reach of the crane and its load capacity are diminished by the distance from the boom seat to the stern or side of the barge.

The main advantages of the derrick barge is its close control of positioning, allowing it to be able quickly to reach any point in three-dimensional space, and its ability to be oriented in the most favorable direction to minimize boom tip displacements and accelerations

Note that a load suspended from a boom tip is a pendulum, and, while the load line length is usually too long for direct resonance, the load may tend to get dynamic amplification from lower-frequency energy. Therefore, the practical solution is to raise or lower the load quickly through those positions that develop an amplified response.

5.10.10 Jack-up Construction Barges

In cases of high wind speed and higher wave height, in turbulent sea areas, or with breaking waves, such as in shoal or coastal water, the jack-up barge is the suitable vessel for use in construction.

The jack-up barge is more expensive than the other construction barges, but its main advantage is that it incurs no stand-by time, because its crew can work all the time.

The typical sequence starts with the barge moving to the site with its legs raised. Upon arrival at the site, it is moored with a spread mooring. Construction jack-ups can operate only in relatively shallow water, 30–60 m, with 150 m as an extreme, so the use of a taut mooring is practicable, although up to 300 m may be adopted for drilling rigs.

With waves and swells usually of less than 1 m, the barge's legs are lowered to the sea floor and are allowed to penetrate under their own weight. In some soils, penetration can be aided by jetting and vibration. Using the jacks on one leg at a time, the bearing capacity of the soil should be checked. With all legs well embedded, the barge is jacked up clear of the water. This is the most critical phase, since wave slap on the underside of the barge may cause impact loads on the jacks and may shift the barge laterally, bending the legs. To cushion the impact, special hydraulic cushioning may be connected to nitrogen-filled cylinders; alternatively, neoprene cushioning may be employed. Once well clear, the barge is raised up to its working height. Then the legs may be cut loose, one at a time, and a pile hammer used to gain even greater penetration. A jack-up barge is shown in Figure 5.42.

Since uneven settlements may take place as a result of time, operations and wave energy input into the legs, the jacks have to be periodically reactivated to equalize the load on each jack. This is especially necessary during the first few days at a site. The mooring lines are reattached, slack. The barge is then jacked down until it is afloat. Once again, the critical period is when the waves are hitting the underside. The mooring lines are tightened. Then the legs are jacked free, one at a time.

(a)

(b)

FIGURE 5.42 Jack-up barge: (a) near the platform; (b) configuration.

Local eddies formed around the legs due to currents will lead to scour and will reduce the lateral capacity.

Steel mats are usually built onto the bottom of the legs, so that when the legs are jacked down they take their temporary support from the sea floor. A short stub pipe sleeve may penetrate below the mat to provide shear resistance against sliding. The main leg is jacked down through the sleeve. Since jack-up performance is so highly dependent on sea-floor soils, it is essential that a thorough geotechnical evaluation, including at least one boring, be made at each site. Of particular concern are layered soils, in which a leg may gain temporary support but then suddenly break through.

A general rule of thumb is to plot the previous leg positions (if known) and to space the new leg locations 4–5 diameters away. In clay soils, where jack-ups have previously worked around the site, holes will have been left that now may be partially empty or filled with loose sediments. If a leg is seated adjacent to such a hole, it may kick over into it, losing both vertical and lateral support and bending the leg. Of course, another advantage of mat-supported jack-up legs is that the mats can span local anomalies. Furthermore, jack-up rigs are usually equipped with two sets of six or eight legs.

One disadvantage of the jack-up occurs during the transfer of loads from barges or supply boats. In this case, the jack-up concept again becomes weather sensitive, because the barges must not be allowed to contact the legs or they may damage them.

In general, jack-ups provide a fixed platform, free from motion in response to the seas.

5.11 TRANSPORTATION LOADS

All structures should be checked for the inertia loads applied during sea transportation. Consideration should be given to the support points used for sea fastening. The following should be considered:

- Structure self-weight
- Equipment and bulk self-weight
- Transportation inertia loads
- Roll, 20°; Period, 10 s
- Pitch, 10°; Period, 10 s
- Heave: ±0.2 g
- Center of rotation is 60% above barge keel at longitudinal midship of the transport barge
- The transportation inertia loads should be combined as roll ± heave and pitch ± heave
- Wind loads for a return period of 10 years (1 minute mean) should be included
- The support points should reflect the support points adopted during load-out

The sea fastenings fix the jacket or the topside (the module) to the barge that transports it from the fabrication yard to its offshore location. Transportation is performed aboard a flat-top barge or, if possible, on the deck of the crane vessel.

The module must be fixed to the barge in order to withstand barge motions in rough seas. The sea fastenings are determined by the positions of the framing in the module as well as by the "hard points" of the barge.

As shown in Figure 5.43, the jacket rests vertically on the barge if it is not tall. Figure 5.44 presents the jacket being transported lying horizontally on the barge. A structure analysis, as discussed previously, will be run again, taking into consideration the fixation points and the movement of the barge. This phase requires cooperation between the installation company and the engineering firm that performed the design. After the engineering office receives the data from the installation company, it runs the structure analysis, and if it is determined that some member will be damaged, it will be changed in this stage. It is advantageous if cooperation between the installation company and the engineering company begins early, to avoid the need for any change in structure configuration.

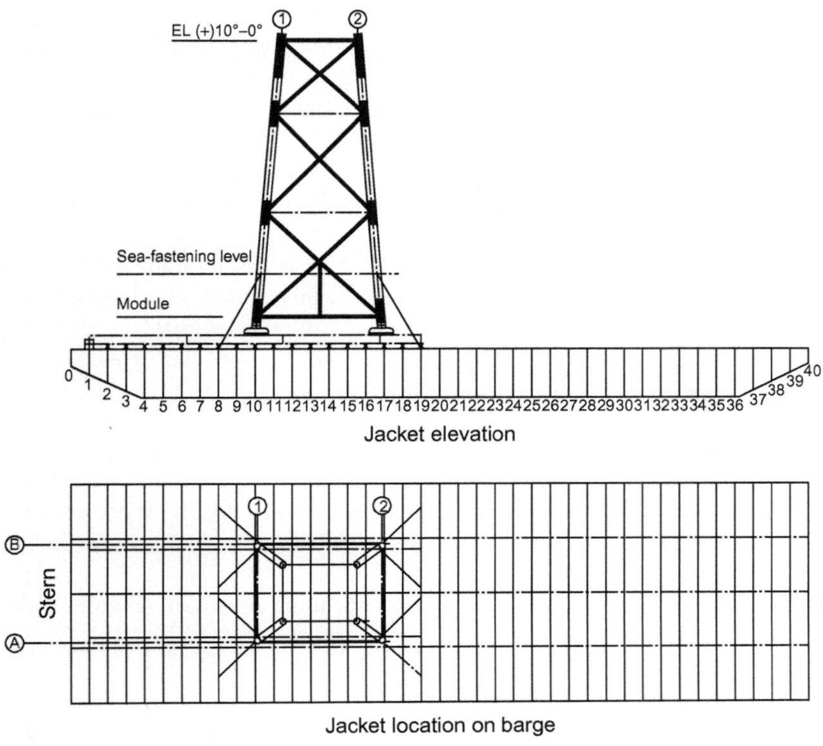

FIGURE 5.43 Jacket sea fastening during transportation.

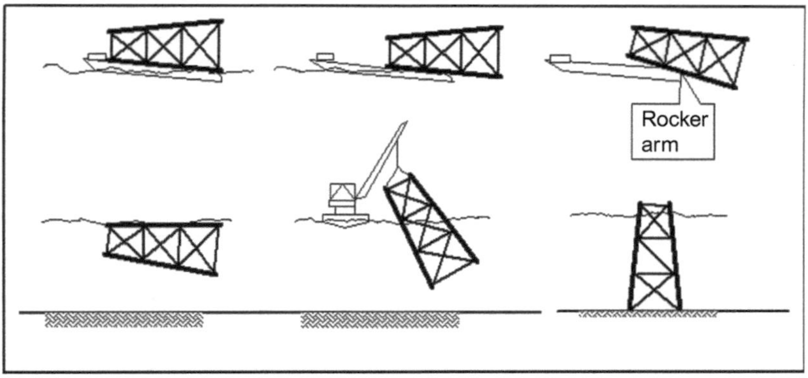

FIGURE 5.44 Launching and installing a jacket.

Skew load factor (SKL) accounts for sling fabrication tolerance or any other inaccuracy in the sling length. SKL should be calculated based on the DNV recommendations. In the absence of exact information, SKL should be set to 1.25 for a typical indeterminate four-point single-hook lift.

As an alternative to the SKL, the lift weight (hook weight) might be distributed in a 75%–25% split between each pair of slings in turn. All structural members, padeyes, shackles and rigging components should be designed or checked for both load distributions.

For padeye design, an additional lateral force equivalent to 5% of the sling force should be applied at the eye of the padeye, in conjunction with the design sling load.

The criterion of 75%–25% split or SKL of 1.25 is based on the variation in fabrication tolerance ±0.25%. If, for any reason, this cannot be achieved, then the SKL must be modified.

5.12 LAUNCHING AND UPENDING FORCES

Launching is the most critical process in platform construction, because in launching, the jacket is affected by different stresses during transfer from the barge to the sea and during the subsequent upending into its proper vertical position to rest on the seabed. A schematic view of these operations is shown in Figure 5.44.

There are six stages in a launching and upending operation (Figures 5.45 to 5.48):

- Jacket is put in a stable position on the barge
- Jacket slides along the skid beams
- Jacket rotates on the rocker arms
- Jacket rotates and slides simultaneously

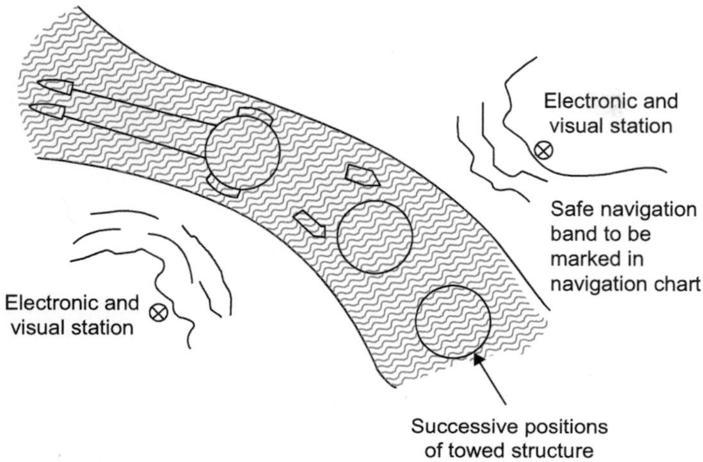

FIGURE 5.45 Start of launching the jacket.

FIGURE 5.46 Installation of the deck on the pile.

FIGURE 5.47 Crane barge near the platform.

FIGURE 5.48 Lifting the jacket for installation.

- Jacket detaches completely and comes to its floating equilibrium position
- Jacket is upended by a combination of controlled flooding and simultaneous lifting by a derrick barge.

The loads induced, static as well as dynamic, can be evaluated by appropriate analyses, which also consider the action of wind, waves and currents expected during the operation.

To start the launch, the barge must be ballasted to an appropriate draft and trim angle and subsequently the jacket must be pulled toward the stern by a crane. Sliding of the jacket starts as soon as the downward force from gravity and the crane pull exceeds the friction force. As the jacket slides, its weight is supported on the two legs that are part of the launch trusses. The support length keeps decreasing and reaches a minimum, equal to the length of the rocker beams, when rotation starts. It is generally at this instant that the most severe launching forces develop in reaction to the weight of the jacket. During the last two stages, variable hydrostatic forces arise that have to be considered at all members affected. The engineering firm should have done buoyancy calculations for every stage of the operation to ensure fully controlled, stable motion. Computer programs are available commercially to perform the stress analyses required for launching and upending and also to present the whole operation graphically.

The typical launch barge is very large and strongly built, is long and wide, and is subdivided internally into numerous ballast compartments since it must support a progressively moving jacket weighing thousands of tons. Heavy runner beams or skid beams extend the length of the barge, as shown in Figure 5.45. These girders distribute the jacket's load to the barge structure. The stern end of the barge, over which the jacket will rotate and slide into the water, requires special construction.

First, for a short period of time the stern will have to support the full weight of the jacket.

Second, since this reaction force has to be transmitted into the jacket, it must distribute the reaction over as long a length as feasible to avoid a point reaction. The jacket will be sliding on its specially reinforced runners shown in Figure 5.45; even so, they need a distributed rather than a point reaction. Hence, the stern of the barge is fitted with a rocker section that rotates with the jacket as it slides off.

For load-out of the jacket at the fabrication yard, the usual method is to ground the launch barge at the appropriate depth so that the barge deck is at the same elevation as the yard. Therefore, the jacket can be skidded out onto the barge with no change in relative elevations. When the load-out is performed with the barge afloat, then ballast must be rapidly adjusted to maintain the relative elevation at the barge deck as the load of the jacket comes on. Step-by-step adjustments under computer control are used to adjust deck elevation and trim. A launch barge is also fitted with heavy cranes on the bow to pull the jacket onto the barge and later, by re-rigging through sheaves on the stern, to pull the jacket off the barge during the launching process. The base of a deep-water jacket may reach to 60 m wide, so it will overhang the sides of the barge significantly.

The topside and jacket structures should be structurally checked against the effect of impact loads during the float over for installation and the loads should be calculated as a parameter in the motion analysis by the naval architect. The analysis should consider the selected barge, the stiffness of the structures and leg mating units and the full range of installation scenarios from initial docking through to final load transfer. Vertical loads should be based on the dry weight of the structure obtained from the weight-control report.

5.13 INSTALLATION AND PILE HANDLING

After launch of the jacket and lifting of the jacket by the crane barge, as shown in Figure 5.48, the next step is to start driving the piles into the legs.

There are various ways of providing lifting points for positioning pile sections. Padeyes are welded in the fabrication yard; their design should take into consideration the changes in load direction during lifting. Padeyes are then carefully cut before lowering the next pile section.

Figure 5.49 presents configurations for internal and external stabbing guides.

Figure 5.50 presents the sequence of the pile installation process.

Different solutions for connecting pile segments back-to-back are welding (shielded metal arc welding (SMAW) or flux-cored) or segments held temporarily by internal or external stabbing guides (see Figure 5.49). Welding time depends upon the pile wall thickness [because it takes 3 hr for a 1-inch thickness (25.4 mm) and 16 hr for a 3-inch thickness (76.2 mm) (typical)], the number and qualification of the welders, the environmental conditions, and the mechanical connectors, breech block (twisting method) and lug type (hydraulic method).

In summary, pile installation includes:

- Lifting from the barge deck
- Positioning over pile by booming out or any other method depending on the weather condition

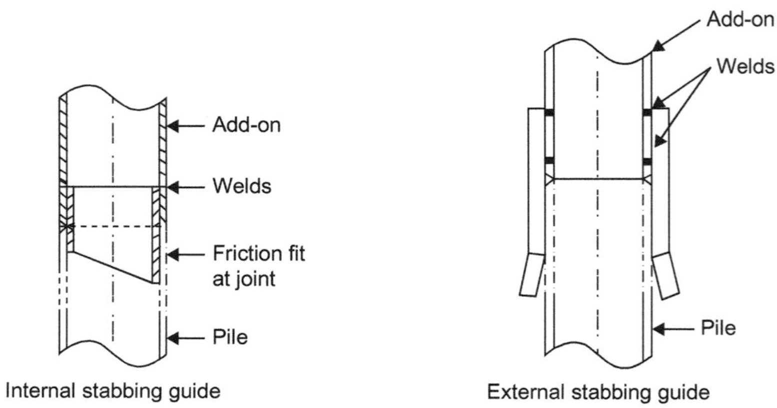

FIGURE 5.49 Configurations for the different types of stabbing guide.

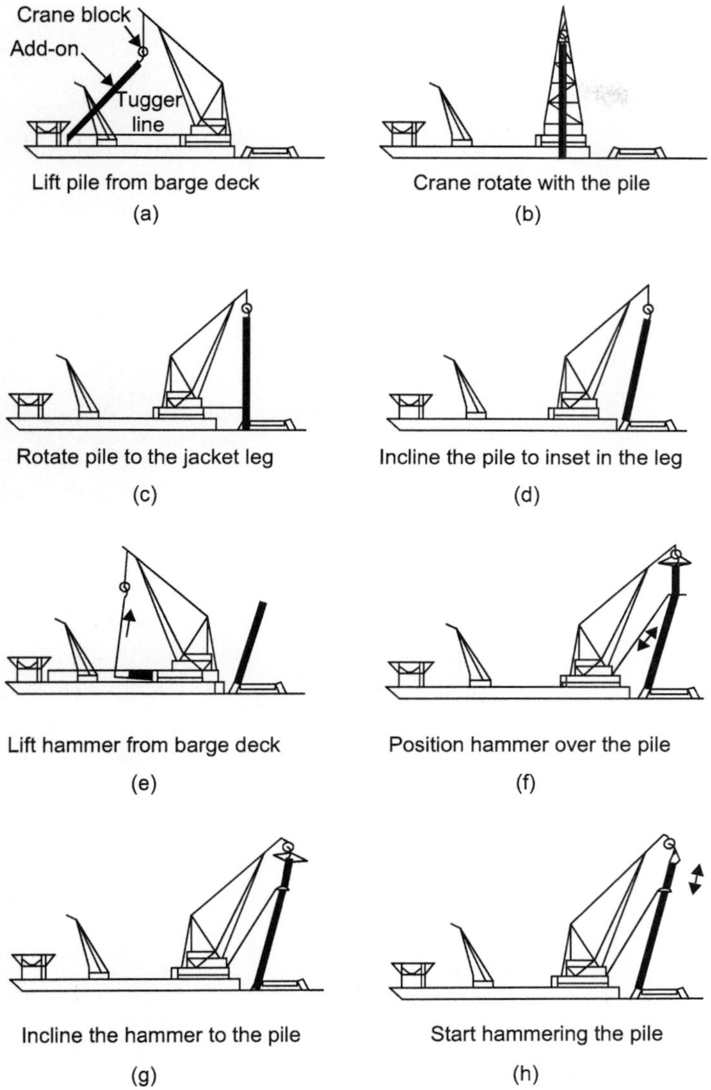

FIGURE 5.50 Pile installation process.

- Alignment of the pile cap
- Lowering leads after hammer positioning

Each step should be designed to prevent bending or buckling failure during installation and in-place conditions.

Pile lifting is as shown in Figure 5.51. One can see there are white markers along the pile length to define the distance of penetration of the pile exactly.

(a) (b) (c)

FIGURE 5.51 (a) Pile lifting. (b) Insertion of pile into previous one. (c) Pile lifting to insert.

Some penetration under the self-weight of the pile is normal. For soft soil conditions, particular measures are taken to avoid an uncontrolled run.

Piles are then driven or drilled until pile refusal. Pile refusal is defined as the minimum rate of penetration beyond which further advancement of the pile is no longer achievable because of the time required and the possible damage to the pile or to the hammer. A widely accepted rate for defining refusal is 300 blows/ft (980 blows/m).

Figure 5.51 shows the insertion of one pile into another one.

The shims are inserted at the top of the pile within the annulus between the pile and jacket leg and welded afterward. This metal-to-metal connection is achieved by a hydraulic swaging tool lowered inside the pile and expanding it into machined grooves provided in the sleeves at two or three elevations.

This type of connection is most popular for subsea templates. It offers immediate strength and the possibility to re-enter the connection should swaging prove incomplete.

Grout is extensively used to "cement" the annulus between pile leg and jacket sleeve. An annular gap of 50–100 mm is usually selected. The grout should flow from the bottom up.

The mix is generally cement plus water. Fly ash may be used to replace part of the cement in order to reduce heat of hydration. Silica fume may be added to

(a) (b)

FIGURE 5.52 (a) Resting the pile on the jacket. (b) Pile padeye for lifting.

promote thixotropic behavior, to increase strength, and to reduce bleed. Admixtures may be used to provide water reduction, retardation, and expansion characteristics. It is important that trial batches be made to ensure that the grout has the proper flow characteristics as well as strength. Flow rate should be kept low to avoid entrapment of voids. Grout should be overflowed to ensure that the initial mixture of cement and seawater is cleared. Pressures should be carefully controlled to prevent forcing the grout out from under the jacket sleeve; usually this exit is restricted by a grout retainer, but many times the grout retainer will have been damaged during pile driving. Therefore, a second grouting entry pipe is often provided, to permit the first grout to set and form a plug; then the main grouting is carried out through the upper entry port.

The final installation of the platform is presented in Figure 5.53.

FIGURE 5.53 Final installation.

BIBLIOGRAPHY

API RP2A, 2007. Recommended Practice for Planning, Designing, and Constructing Fixed Offshore Platforms, 20th ed. American Petroleum Institute, Washington, DC.

2009. Guidelines for Lifting Operations by Floating Crane Vessels, Noble Denton document 0027/ND Rev3. Noble Denton, London.

Gerwick, B.C., 2007. Construction of Marine Offshore Structures. CRC Press, Boca Raton, USA.

International Organization for Standardization (ISO), 2004. Petroleum and Natural Gas Industries—Offshore Structures—Part 2: Fixed Steel Structures, ISO/DIS 19902:2004. USA.

Chapter 6

Corrosion Protection

6.1 INTRODUCTION

Cathodic protection of the offshore structure is the responsibility of the offshore structure engineer. This chapter presents cathodic protection (CP) design methodology and what information the CP designer needs from the structural engineer as well as the sensitivity of the data provided for the output. Because most of the structure is affected by the cathodic protection, it is important to know the appropriate time and the methods available to assist in anode retrofitting of projects worldwide, especially due to maturation of existing offshore steel platforms.

Corrosion is a complex concept. Therefore, the aim here is to describe corrosion simply, in a form emphasizing only those aspects that are important in understanding corrosion of offshore structures.

In general, a key factor in corrosion is the environmental condition surrounding the structure. The definition and characteristics of this variable can be complex. Practically, it is important to keep in mind that the environment changes with time and conditions. It is also important to realize that the environment that actually affects a metal is the micro-environment at the local surface of the metal. It is indeed the reactivity of the local environment that determines the real corrosion damage. Thus, an experiment that investigates only nominal environmental conditions without consideration of local effects, such as flow, pH cells, deposits, and galvanic effects, is useless for lifetime prediction.

Corrosion is usually identified by the appearance of rust on the steel surface. The chemical reactions driving the corrosion process are caused by chloride attack. When corrosion of steel occurs, electrons in water cause the anodic reaction.

A steel offshore structure is exposed to saline water during its whole lifetime. The effect of the water on the integrity of the materials is thus important. Since steels are the metallic materials most commonly exposed to water, aqueous corrosion is discussed here, with a special focus on the reactions of iron (Fe) with water (H_2O). Metal ions go into solution at anodic areas in an amount chemically equivalent to the reaction at cathodic areas, as shown in Figure 6.1. In corrosion of steel, the following reaction usually takes place at anodic areas:

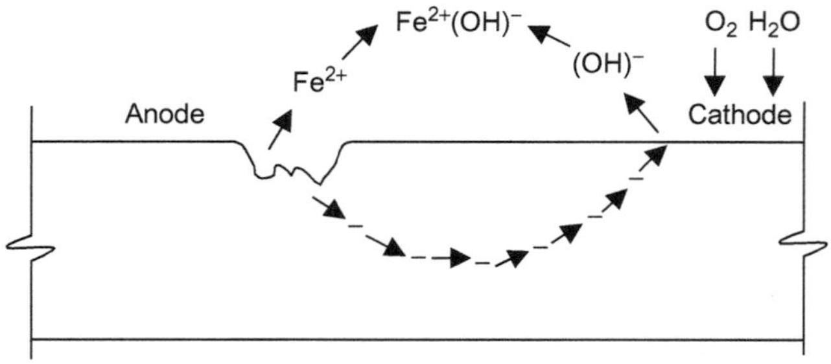

FIGURE 6.1 Corrosion process on the steel surface.

Anodic Reaction

$$Fe \rightarrow Fe^{2+} + 2e^{-} \tag{6.1}$$

The electrons from the anodic reaction will accumulate on other parts of the steel because another reaction uses the electrons with oxygen and water: this is called the cathodic reaction.

Cathodic Reaction

$$2e^{-} + H_2O + \frac{1}{2}O_2 \rightarrow 2OH^{-} \tag{6.2}$$

Equation (6.2) shows that hydroxide ions (OH^-) are present in a cathodic reaction. The OH^- increase the alkalinity and reduce slightly the effect of carbonates or chlorides. It is important to keep in mind that water and oxygen are the main reasons for corrosion.

As shown in the above equations and Figure 6.1, anodic and cathodic reactions are the first step in corrosion because the OH^- will react with ferrous iron (Fe^{2+}), as shown in Equation (6.3). This reaction produces ferrous hydroxide, which will react, as shown in Equation (6.4), with oxygen and water and produce ferric hydroxide, $Fe(OH)_3$, and the last component, which is the hydrate ferric oxide (rust). The chemical term for rust, $Fe_2O_3 \bullet H_2O$, is shown in Equation (6.5). This chemical reaction is shown graphically in Figure 6.1.

$$Fe^{2+} + 2OH^{-} \rightarrow Fe(OH)_2 \tag{6.3}$$

$$4Fe(OH)_2 + O_2 + 2H_2O \rightarrow 4Fe(OH)_3 \tag{6.4}$$

$$2Fe(OH)_3 \rightarrow Fe_2O_3 \bullet H_2O + 2H_2O \tag{6.5}$$

Saturated $Fe(OH)_3$ is nearly neutral in pH. A magnetic hydrous ferrous ferrite, $Fe_3O_4 \bullet nH_2O$, often forms a black intermediate layer between hydrous Fe_2O_3 and FeO. Hence rust films normally consist of three layers of iron oxides in different states of oxidation.

6.1.1 Corrosion in Seawater

The fixed offshore structure platforms in oil and gas projects are exposed to seawater all the time.

The corrosion problems for steel structures in seawater have been well studied over many years, but despite published information on materials behavior in seawater, failures still occur. The concentration of dissolved materials in the sea varies greatly with location and time as seawater is diluted by rivers, rain or melting ice or it is concentrated by evaporation. The most important properties of seawater are:

- Remarkably constant ratios of the concentrations of the major constituents globally
- High salt concentration, mainly sodium chloride
- High electrical conductivity
- Relatively high and constant pH
- Solubility for gases, of which oxygen and carbon dioxide in particular are of importance in the context of corrosion
- The presence of organic compounds
- The existence of biological life, to be further distinguished as microfouling (e.g., bacteria, slime) and macrofouling (e.g., seaweed, mussels, barnacles, and many kinds of animals or fish).

Some of these factors are interrelated and depend on physical, chemical and biological variables, such as depth, temperature, intensity of light and the availability of nutrients. The main numerical specification of seawater is its salinity.

Salinity was defined, in 1902, as the total amount of solid material (in grams) contained in one kilogram of seawater when all halides have been replaced by the equivalent of chloride, when all the carbonate is converted to oxide, and when all organic matter is completely oxidized. The 1902 definition was translated into Equation (6.6), where the salinity (*S*) and chlorinity (*Cl*) are expressed in parts per thousand.

$$S(\%) = 0.03 + 1.805Cl \tag{6.6}$$

The fact that Equation (6.6) gives a salinity of 0.03% for zero chlorinity was a cause for concern, and a program led by UNESCO helped to determine a more precise relation between chlorinity and salinity. The definition of 1969 produced by the UNESCO study is:

$$S(\%) = 1.80655Cl(\%) \tag{6.7}$$

The definitions of 1902 and 1969 give identical results at a salinity of 35% and do not differ significantly for most applications. The definition of salinity was reviewed again when techniques to determine salinity from measurements of conductivity, temperature, and pressure were developed. Since 1978, the Practical Salinity Scale defines salinity in terms of a conductivity.

The practical salinity symbol S for seawater is defined in terms of the ratio K of the electrical conductivity of a seawater sample at 15°C and the pressure of one standard atmosphere, to that of a potassium chloride (KCl) solution, in which the mass fraction of KCl is 0.0324356, at the same temperature and pressure. The K value exactly equal to 1 corresponds, by definition, to a practical salinity equal to 35.

The corresponding formula is given in Equation (6.8).

$$S = 0.0080 - 0.1692K^{0.5} + 25.3853K + 14.0941K^{1.5} - 7.0261K^{2} + 2.7081K^{2.5} \tag{6.8}$$

Note that in this definition parts per thousand (‰) is no longer used, but the old value of 35‰ corresponds to the new value of 35. Since the introduction of this practical definition, salinity of seawater is usually determined by measuring its electrical conductivity and generally falls within 32–35‰.

As discussed above, when corrosion occurs, the anodic reaction rate is exactly equal to the cathodic reaction rate. It was not mentioned before, but it is true that, in environments of good conductivity (such as in seawater or seabed mud), the corroding metal displays a single potential that lies between potential at cathode E_c and potential at anode E_a. In Figure 6.2(a), this condition is met where the anodic and cathodic curves cross. The potential at the

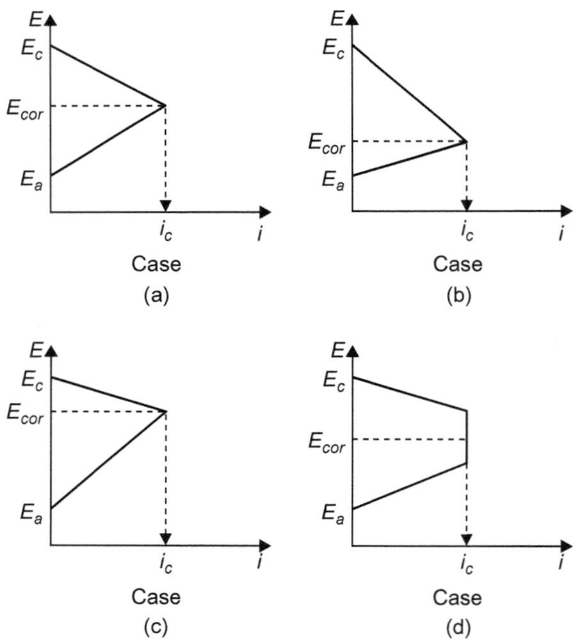

FIGURE 6.2 Evans diagrams.

crossover point is referred to as the *corrosion potential*, E_{cor}. It is the single potential exerted by a corroding metal referred to above. The current I_{cor} is referred to as the *corrosion current*, and it is an electrical representation of the corrosion rate. In reality, a corroding metal does not take up potential E_a or E_c but spontaneously moves to E_{cor}.

While the shape of the individual E–log I curves may vary, depending on environmental conditions, the manner in which the diagrams (*polarization diagrams*) are interpreted, in terms of E_{cor} and I_{cor}, remains the same.

Figure 6.2(a) presents an Evans diagram of the polarization curves for separate anodic and cathodic reactions intersecting at a point where the mean anodic and cathodic current densities are equal and presents the corrosion rate in terms of a mean corrosion current density (I_{cor}).

However, there is always some difference between the electrode potentials developed at anodic and cathodic sites on the metal surface. It may be the amount is significant in ohmic drop (iR drop) under conditions where corrosion macro-cells are formed when the anodic and cathodic areas are separated by a medium of high electrolytic resistance, so the Evans diagram is modified as shown in Figure 6.2(d).

Therefore, the mean corrosion rate I_{cor} is reduced, and the corrosion potential varies with the location between the limits of anode E_{cor} and cathode E_{cor}, with the positions of local anodes being indicated by the region of low corrosion potential. In the absence of significant ohmic drops, the mean corrosion rate (I_{cor}) depends on the magnitude of the difference between the reversible potentials of the anodic and cathodic reactions and on the average slopes of the anodic and cathodic polarization curves. If the anodic reaction is steeply polarized, as in the presence of a passive film, then I_{cor} is small and E_{cor} assumes a value that is close to the reversible potential of the cathodic reaction, as shown in Figure 6.2(c). On the other hand, if the cathodic reaction is steeply polarized, as in limited oxygen availability, the situation is as shown in Figure 6.2(b), with I_{cor} again small but the corrosion potential close to the reversible potential of the anodic reaction.

6.1.2 Corrosion of Steel in Seawater

The corrosion of steel in seawater, as well as in seabed mud, can be adequately represented by Equation (6.1), although the process normally proceeds to the precipitation of ferric hydroxide.

On clean steel in seawater, the anodic process occurs with greater facility than the cathodic process. Consequently, the corrosion reaction can proceed no faster than the rate of cathodic oxygen reduction. The latter usually proves to be controlled by the rate of arrival of oxygen at the metal surface, which in turn is controlled by the linear water flow rate and the dissolved oxygen concentration in the bulk seawater.

This may be represented on a polarization diagram, as shown in Figure 6.3. At first, the cathodic kinetics increase (get faster) as the potential becomes more

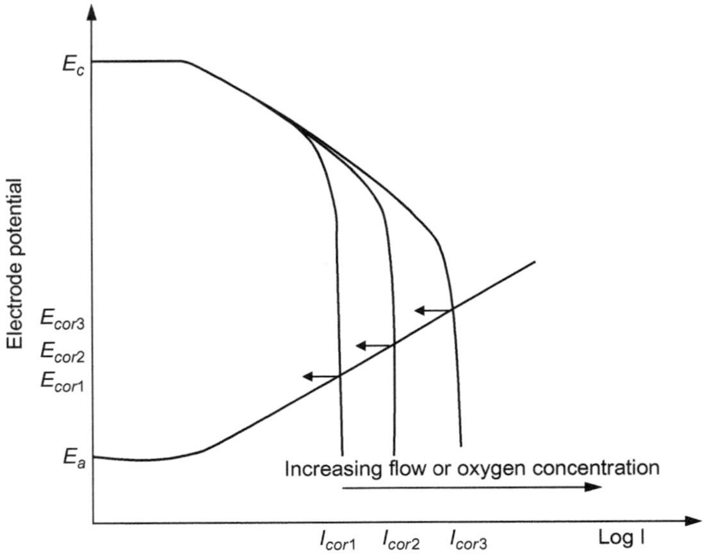

FIGURE 6.3 Polarization diagram with increasing oxygen concentration.

negative from E_c. This has the effect of depleting the oxygen immediately adjacent to the metal surface, thus rendering the reaction more difficult. Ultimately, a point is reached where the surface concentration of oxygen has fallen to zero and oxygen can then be reduced only as and when it reaches the surface. Further lowering of the potential cannot increase the cathodic reaction rate, because the kinetics are now governed by potential-independent diffusion processes, and a plateau, or limiting, current is observed. Figure 6.3 shows that the corrosion rate is then equal to this limiting current. The limiting current can be increased by increasing the oxygen flux, either by raising the bulk oxygen concentration (the concentration gradient gets steeper) or increasing the flow rate (the oxygen-depleted layer gets thinner). Both serve to increase the corrosion rate, as shown in Figure 6.3.

To a first approximation, it may be stated that the rate of corrosion of clean steel in aerated seawater under turbulent flow conditions is directly proportional to the bulk oxygen concentration and the linear velocity. Fick's first law of diffusion and the Chilton-Colbourn analogy can be used to calculate the precise effect of oxygen concentration and Reynolds number (flow rate) on corrosion. Accordingly, Ashworth (1994) estimated the maximum corrosion rates of clean steel in North Sea water at 7°C, as shown in Table 6.1.

In practice, corrosion products and marine fouling build up on steel as it corrodes in seawater, and they generally produce lower corrosion rates.

Rowlands (1994) suggests that the corrosion rate of fully immersed steel is fairly rapid in the first few months of exposure but falls progressively with time.

TABLE 6.1 Estimated Maximum Corrosion Rates of Clean Steel in North Sea Water at 7°C

Linear Flow Rate (m/s)	O_2 Concentration 6 ppm	7 ppm	8 ppm	9 ppm	10 ppm
0	0.080	0.094	0.107	0.120	0.134
0.3	0.091	0.107	0.123	0.138	0.154
0.4	0.096	0.111	0.128	0.144	0.160
0.6	0.104	0.121	0.138	0.156	0.174
1	0.120	0.140	0.160	0.179	0.199
2	0.160	0.187	0.213	0.240	0.266
4	0.240	0.280	0.320	0.360	0.400

A value of 0.13 mm/year may be taken as reasonably representative in any part of the world. However, pits may grow at 3 to 10 times that rate.

Many marine structures, particularly those in shallow waters, are simultaneously exposed to a number of discrete corrosive environments: the marine atmosphere, the splash zone, the tidal zone, the fully submerged zone and the mud zone. Commonly, the corrosion rate data are represented schematically, a practice that derives largely from the work of Humble (1949) as shown in Figure 6.6. Peak corrosion rates are found immediately below the mean low tide zone and in the splash zone. The typical mean corrosion rate in the splash zone, in case of quiet sea conditions, is estimated as 0.25 to 0.75 mm/year.

As shown in Figure 6.3, the corrosion rate is controlled by sluggish cathodic kinetics: in this case, the rate of arrival of oxygen at the surface and the effect of increasing oxygen availability.

The peak corrosion rate is often attributed to galvanic action between steel in contact with oxygen-rich surface waters, which is the cathodic area, and steel at somewhat greater depth exposed to waters of lesser oxygen content, which is the anodic area. It is difficult to conceive that the change in oxygen concentration with depth is sufficiently great to cause such an effect, and it may be that other factors come into play. It is essential to gather the following information before starting the design of the CP system. In fact, this list of required data can be considered a checklist for collecting information from the owner.

I. Structure design data
- Design life for the cathodic protection system and the structure
- Construction drawings with full details and dimensions

- General arrangement drawings showing the structure's relationship to the seabed, lowest astronomical tide level (LAT), mean water level, and maximum operational conditions
- Extent of use of, and application of, protective coating
- Availability of electrical power
- Proposed construction schedule
- Structure fabrication methods and fabrication on site
- Any weight limitations/constraints on the installed CP system
- Safety requirements
- Constraints and limitations on the installation and in-service maintenance and monitoring of the CP system
- High-strength steels or other metals used in the structure that may be subject to a reduction in mechanical properties when under cathodic protection

II. Offshore site location data

- Water depth, oxygen content, velocity, turbulence, temperature, resistivity, tidal effect and suspended soil
- Chemical composition of water
- Pollutants, depolarizing bacteria or marine borers present in the water or seabed
- Geological nature of the seabed and the probability that scour will occur
- Adjacent facilities, including pipelines, and details of their CP system
- Susceptibility to stratification of the water and the resultant effect on resistivity, temperature and oxygen content
- Performance history of previous or existing CP systems in the same environment
- Protective current density requirements to achieve the applicable protection criteria, obtained from site surveys or reliable documentary sources
- Susceptibility to adherent marine fouling, including type, rate of growth and variation with water depth

6.1.3 Choice of System Type

There are three types of cathodic protection system, each of which, when correctly designed, installed and operated, can effectively protect a fixed offshore steel structure for its design life. They are:

Sacrificial: anodes cast from reactive metals, usually zinc or aluminum alloys, as they are more electronegative than the structures requiring protection and require no external source of power
Impressed current: anodes manufactured from materials that are essentially inert and powered by an external source of direct current
Hybrid: a mixture of sacrificial anodes and externally powered impressed current anodes.

TABLE 6.2 Comparison of Different CP Systems

	Sacrificial Anode	Impressed Current	Hybrid
Advantages	Simple, reliable and free from in-service operator surveillance. System installation is simple. Permanent potential monitoring system not essential.	Flexible under widely varying operating conditions. Weight advantage for large-capacity, long-life systems.	Flexible under widely varying operating conditions. Weight advantage for large-capacity, long-life systems.
Disadvantages	Large weight penalty for large-capacity, long-life system. Response to varying operating conditions is limited. Hydrodynamic loadings can be high.	Relative complexity of system demands high level of detailed design expertise. System installation is complex, and it requires a power source. Perceived diver risk from electric shock. In-service operator surveillance required. Permanent potential monitoring system essential. Vulnerable to loss of power. It is not recommended for North Sea without full sacrificial back-up (i.e., part of a hybrid system).	Relative complexity of system demands high level of detailed design expertise. System installation is complex, and it requires a power source. Perceived diver risk from electric shock. In-service operator surveillance required. Permanent potential monitoring system essential.

The principal technical advantages and disadvantages of sacrificial, impressed current and hybrid systems are summarized in Table 6.2.

The use of the term *impressed current system* can be misleading, because for most offshore applications, an impressed current system is used in combination with a small number of sacrificial anodes, forming a hybrid system. Sacrificial anodes are provided in hybrid systems to ensure that adequate polarization of the critical nodes is maintained at all times, even if the power supply to the impressed current anodes fails or is switched off temporarily (e.g., to permit manual underwater inspection or cleaning of the structure by divers). Some early impressed current systems were provided with inadequate sacrificial anode back-up and significant corrosion damage was reported in times of unplanned (and planned) impressed current shut-downs.

The same considerations apply equally to jacket structures, with one important addition, namely, that a power source to drive the impressed current system is generally not available until the topside power generation equipment is installed and commissioned. On large deepwater jackets in the North Sea, this may be a year or more after installation of the jacket; protection for the interim period is provided by high-current, short-life sacrificial anodes.

It is strongly suggested that designers contemplating an impressed current system for North Sea applications provide full sacrificial back-up, with sacrificial anodes that provide full protection for a minimum of 2 years, plus an allowance for periods of possible impressed current system shut-down during subsea surveys and maintenance throughout the design life.

The obvious technical attractions of sacrificial systems, as illustrated in Table 6.3, make them the type most often chosen for offshore structures. Also, for many offshore structures, sacrificial systems are the most economical option for the owner when taking into account both capital expenditure and the running costs over the design life of the structure. However, generalizing about economic advantages and disadvantages can be misleading, because they differ widely for each type and size of structure and according to the design constraints imposed by the environmental conditions prevailing at different offshore locations. For this reason, economic advantages/disadvantages are not included in Table 6.2.

The hybrid system is used in Murchison and Hutton platforms due to the heavy weight constraints. The CP designers of the Murchison and Hutton platforms in the North Sea carried out detailed assessments of alternative sacrificial and impressed current designs. The assessments revealed that use of the sacrificial anode is essential but that the heavy weight can be reduced by using impressed current systems as the primary means of protection on both platforms, despite their vastly different structural configurations, as Murchison is a deepwater conventional jacket but Hutton was the world's first tension-leg platform.

Thus, a main advantage of using hybrid systems over sacrifical systems is weight-saving on Murchison and Hutton platforms. In the case of Hutton, the installed weight of the primary impressed current system, plus supplementary sacrificial anodes located close to the main node joints, was approximately 60 tons. On the other hand, an equivalent totally sacrificial system would have weighed around 250 tons. From a practical point of view, the most economical solution for buoyant structures, such as tension-leg platforms, is the impressed current system, which is different from the solution for the fixed offshore structure platforms.

The relatively simple geometry and large, flat surfaces of buoyant structures are ideally suited for protection to be provided by a small number of high-current, low-voltage, flush-mounted anodes. Cables to reference electrodes and anodes can be easily and economically routed through ballast tanks and human access pathways in the pontoons and columns of the hull.

TABLE 6.3 Comparison of Impressed Current CP and Sacrificial Anode CP Systems

	Impressed Current CP Systems	Sacrificial Anode CP Systems
Power source and connections	External continuous power source required. Can be inadvertently misconnected, resulting in reversed direct current (DC) polarity, which will accelerate corrosion. Isolation of anodes from the structure is essential. Fewer connections required, but more complex.	Independent of any power source. Straightforward welded or bolted connections to structure. Cannot be wrongly connected and connections are cathodically protected.
Control	Simple controls: manual or automatic. Automatic will maintain potentials within preset limits but requires additional fixed reference electrodes. Monitoring required at regular intervals.	A tendency for current to be self-adjusting. Dependent on initial design: if not adequate, will require additional anodes.
Anodes	Usually lighter and fewer numbers. May affect other structures in close proximity to anodes and should be assessed for any interaction.	Relatively heavy and large numbers. Bulk of material may restrict water flow and induce turbulence and drag. Less likely to cause interaction with neighboring structures because output is low. Lifespan varies with conditions, so replacement may be required at different times.
Damage	Anodes are lighter in construction and therefore less resistant to damage. Loss of an anode can be more critical to the effectiveness of the system due to high current output for each anode.	Anodes are robust and not susceptible to mechanical damage. Where a system has a large number of anodes, the loss of a few anodes has little overall effect on the system.
Maintenance	Equipment designed for long life but regular checks on electrical equipment in service are required.	Generally maintenance-free. Renewals usually required at regular intervals unless designed for life of structure.
Environment	May cause overprotection, resulting in coating disbondment or hydrogen-induced cracking of high-tensile steels. Less restricted by high-resistivity conditions.	Not practical for use in high-resistivity conditions. Limited current availability.

(*Continued*)

TABLE 6.3 Comparison of Impressed Current CP and Sacrificial Anode CP Systems—cont'd

	Impressed Current CP Systems	Sacrificial Anode CP Systems
Installation	Requires a high level of detailed design and installation expertise.	Straightforward. Often bulky and large numbers involved.
Hazards	Diver risk from electrical shock; system needs to be switched off when diver is near anodes.	Magnesium anodes can be used in potable tanks but never in areas containing hydrocarbons. Aluminum and zinc anodes must never be used in potable water tanks.
Cost	Generally, the initial cost is high but service cost is normally low.	Initial cost depends on design life but is relatively low; periodic renewals are necessary and cumulative costs are high.

It is worth mentioning that the impressed current systems may also be cost-competitive for traditional fixed offshore jacket structures with simple geometry, located in relatively shallow water depth. In this case the reference electrode and anode cables will be installed in substantial conduits and routed along the outside of the structural tubular members. But impressed current systems are less likely to be cost-competitive on large jackets of complex geometry located in harsh environments. In case of complex node geometries in the structure jacket, it will be difficult to allow large-capacity anodes to protect all surfaces adequately, because of shielding effects. Moreover, in a harsh environment, it will be difficult and expensive to route anode and reference electrode cables inside the structural tubular members in order to ensure their mechanical safety.

As a general rule, increasing anode operating temperatures causes a decrease in both anode ampere-hour capacity and driving potential; at temperatures exceeding 50°C, zinc alloys experience intergranular corrosion and they should not be used at low anode current densities, and the ampere-hour capacity of aluminum alloys tends to decrease significantly. In order to realize the performance claimed by anode manufacturers and thus to ensure the successful operation of the cathodic protection system, it is imperative that strict quality assurance and quality control of the anode manufacturing process should be achieved and maintained throughout production. (Quality assurance, quality control and tests are discussed below.) The requirements contained in DNV RPB, 401 (2005) are considered the minimum standards for offshore work, with supplementary requirements for specific project applications to be determined and specified by the designer.

6.1.4 Geometric Shape

Sacrificial anodes are generally cast in three basic geometric shapes: the long, slender, stand-off type; the flat-plate, flush-mounted type; and the bracelet type. In most cases, anode selection is performed by the owner, taking into account effects like sea current drag and interference with subsea interventions. If the anode type has not been specified by the owner, then the contractor selects the anode type, taking into account such factors as net anode mass to be installed and available space for location of anodes. The anode type further affects the anode utilization factor and the anode current output in relation to weight.

Typical examples of the first two basic geometries are shown in Figures 6.4 and 6.5.

The most common anode shape used for offshore structures is the long, slender type with a trapezoidal or circular cross-section with weight about 100 kg. The principal advantages of this anode geometry are high current output and good current distribution for a given mass, noting that a flush-mounted anode with the same net anode mass will have a lower anode current output and lower utilization factor. Another advantage is the simple fabrication and casting requirements.

In general, flat-plate anodes are the best solution to complex fabrications where space limitations prevent the use of larger stand-off anodes or if cathode current densities are low. Examples are heavily reinforced mud mats and large, flat, painted surfaces. The designer should determine if anode shapes can be more economically chosen from a manufacturer's standard units or whether, because of the large number required for a new structure, a preferred design could raise costs. Anode manufacturers offer a large variety of standard anode and insert core types, with the choice of steel inserts usually being among bar, tube or rod, in either straight lengths or prefabricated, weld-jointed shapes.

Stand-off and flush-mounted anodes may further be divided into "short" and "long," based on the length-to-width ratio, as shown in Figure 6.4 and

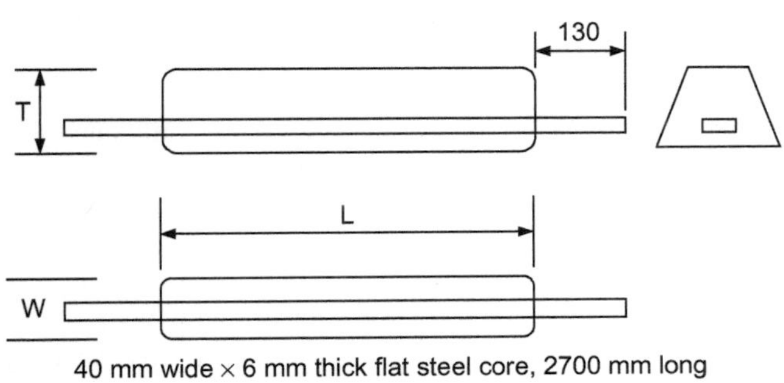

FIGURE 6.4 Flushed-mounted anode.

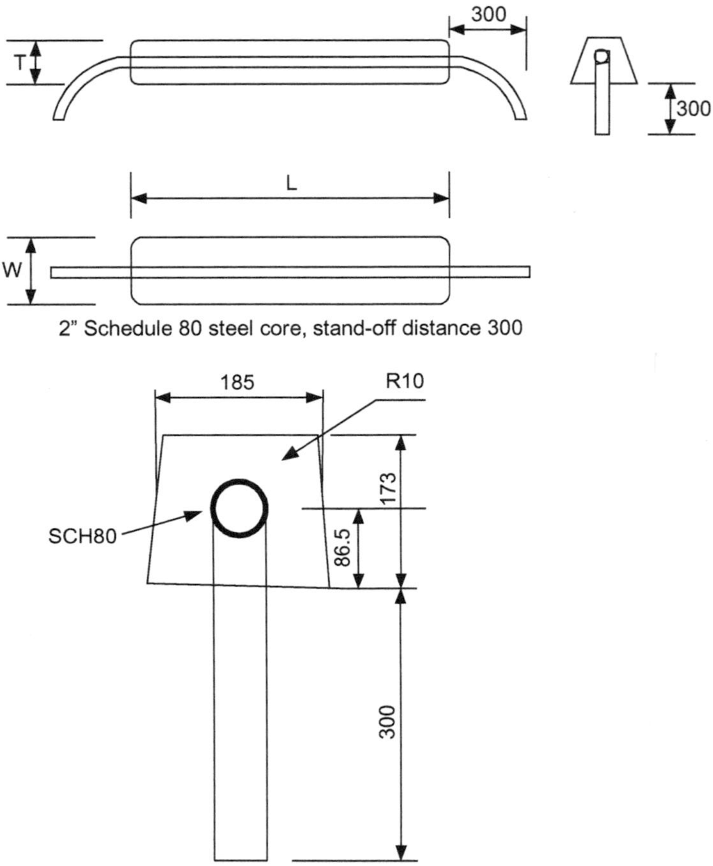

FIGURE 6.5 Stand-off anode.

Figure 6.5. The anode type selected is a main factor in anode resistance and utilization, as discussed below.

The slender stand-off type is typically cast on a tubular insert and is used for relatively large anodes (e.g., on platform substructures and subsea templates). The current output, I_a (A), in relation to net anode mass, M_a (kg), is high, as is the utilization factor (μ).

Modern stand-off anodes put the anode in a steel frame, called a sled, and connect it with a special clamp to the steel structure. This system is very easy to install and less costly in retrofitting an existing structure.

Stand-off anodes can be manufactured and obtained up to a net anode mass of several hundred kilograms. In surface waters, drag forces exerted by sea currents are significant.

Bracelet anodes formerly were used primarily for pipelines but now they are used on platform legs in the upper zone, combining a high current-output-to-weight ratio with low drag. All flush-mounted anodes should have a suitable coating system applied on the surface facing the protection object. This is to avoid build-up of anode corrosion products that could cause distortion and eventually fracture of the anode's fastening devices.

Type of anodes and any special requirements for anode fastening should be defined during the conceptual phase of CP design, taking into account forces exerted during installation as well as piling operation and the effect of wave forces during the structure's lifespan. For stand-off type anodes, special precautions may be necessary during anode design and distribution of anodes to avoid impeding subsea operations.

The insert should be structurally suitable for the anode weight and for the forces likely to be encountered during its lifetime, including impact, storm damage, wave action and, possibly, ice. The insert should normally be made from materials that can be welded to the structural steel. The typical grades of steel (BS4360 grades 40A, 43A or 50C, or API 5L grades B, X42 or X52) are usually used.

If anode inserts are fabricated by welding, the latter has to be in accordance with a recognized, quality-controlled standard. Inserts should be prepared by abrasive cleaning to a minimum standard of S BS1706 and follow NACE recommended practice. RP 0387 may be followed.

Aluminum-based anode steel insert specifications are similar to those for zinc, except that the surface must not be zinc coated nor galvanized after cleaning.

Bracelet anodes are the most commonly used type for protection of submarine pipelines, for which their wrap-around construction is ideally suited. They are rarely used on new offshore platform constructions because of their low current-output-to-mass ratio compared to long, slender anodes. However, bracelet anodes do lend themselves to retrofitting on existing structures and to supplementing or replacing the original failed, deficient or end-of-life cathodic protection systems.

6.2 COATINGS AND CORROSION PROTECTION OF STEEL STRUCTURES

The steel in offshore platforms is subject to different types of corrosion phenomena: atmospheric corrosion, splash zone corrosion, crevice corrosion, etc. (Recently, many steel structures in service in seawater have been corroded by the interaction of aerobic and anaerobic bacteria.) Typical rates of corrosion of uncoated steel in seawater are 0.15 mm/year in the splash zone and 0.07 mm/year in the submerged zone, with even more, up to 0.3 mm/year, in cold fast-running tides carrying silt or other abrasive sediments. Some studies for uncoated steel in seawater give rates of 0.127 mm/year. Corrosion rates in fresh water are about half of those in seawater.

The procedure for painting and coating of the steel members should be carried out as far as practicable in the shop, under strong supervision and with appropriate conditions of humidity and protection from the extremes of weather. The joint surfaces should, of course, be masked to permit welding. It is essential to know that field coating of the joints and touch-up of shop coats should be done only when the surfaces are dry and at the proper temperature. In some locations, portable tents or other protection will have to be provided and heaters or dehumidifiers, or both, may be required.

It is worth mentioning that coatings may delay initiation of corrosion by 10–20 years. It is extremely important that surface preparation be thorough and in accord with the specified requirements. The offshore environment will quickly degrade any coatings placed on damp steel or over mill scale or rust. Morning dew can quickly degrade a well-prepared surface. DNV rules require that the provisions for coating include the following, whcih can serve as a checklist:

- A description of general application conditions at the coating yard
- Method and equipment for surface preparation
- Ranges of temperature and relative humidity
- Application methods
- Time between surface preparation and first coat
- Minimum and maximum dry-film thickness of a single coat
- Number of coats
- Minimum total dry-film thickness
- Relevant drying characteristics
- Procedure for repair of damaged coating
- Methods of inspection (for example adhesion testing)

The relation between the coating breakdown as a percentage of area and the structure lifetime, which is considered as a factor in designing the CP system, is as shown in Table 6.4.

TABLE 6.4 Guide to Coating Breakdown Criteria for Cathodic Protection Design

	Coating Breakdown (% of Area)		
Year	Initial	Mean	Final
10	2	7	10
20	2	15	30
30	2	25	60
40	2	40	90

The manufacturer's specification should be followed clearly in surface preparation, and application of coating should be carried out when the surface temperature is more than the dew point or when the relative humidity of the air is below the limits recommended by the coating manufacturer. Special coatings are usually applied to steel that will be in the splash and atmospheric zones and to internal spaces that are exposed to seawater. In the case of sealed internal spaces permanently filled with seawater, corrosion inhibitors may be added to the water prior to sealing. The most effective coatings are organic coatings, rather than metallized zinc: vinyl mastic on urethane over zinc or zinc silicate in temperate zones, and phenolic over the zinc primer in Arctic and sub-Arctic conditions. The U.S. Corps of Engineers is currently providing corrosion protection to lock gates in West Virginia by shot blasting a single coat of zinc primer 0.625–0.1 mm thick, followed by two coats of zinc-rich vinyl immersion coating to 0.175 mm. Underwater and in the splash zone, they apply Copoxy Shop Primer, followed by a top coat of epoxy to 1.0 mm.

Sacrificial anodes or impressed current cathodic protection is normally used to protect offshore structures. Anodes must be carefully installed in accordance with the specifications to ensure that they cannot become dislodged during transport, launching, installation, pile driving and service. An adequate electrical connection between sacrificial anodes and the steel structure is essential. Impressed current is believed more effective because it is less likely to be shielded, but it requires continued monitoring and adjustment. If the electric current is shut down for a long time or frequently for any reason, corrosion will be accelerated. Coatings may be applied to members that will be underwater in service in order to minimize the requirements for cathodic protection, provided the coating has adequate resistance to cathodic disbondment. Zinc-based and aluminum-based alloys have been applied by thermal spray. Titanium-clad steel tubular piles were used on the Trans-Tokyo Bay Bridge. In the splash zone, additional protection may be provided by means of Monel wrap or copper nickel, austenitic stainless steel or carbon steel plate wrap.

Table 6.5 presents a guide to the minimum design current densities for cathodic protection in different locations.

For example, at least one offshore operator has successfully employed thermal sprayed aluminum coatings for critical high-strength steel components on a major North Sea structure, as an alternative to a conventional sacrificial anode system. Table 6.6 presents the positive and negative potential limits for cathodic protection for steel.

The benefit of using galvanic coating is that it provides an essentially uniform, protected potential of −800 to −900 mV Ag/AgCl over the entire coated surface area.

Figure 6.6 presents a sketch of the relative metal loss from an offshore steel structure member under different environmental conditions. It is obvious that the maximum corrosion rate occurs at the splash zone.

TABLE 6.5 Guide to the Minimum Design Current Densities for Cathodic Protection for Bare Steel

	Current Densities (mA/m^2)		
Location	**Initial**	**Mean**	**Final**
North Sea	180	90	120
Arabian Gulf	130	70	90
India	130	70	90
Gulf of Mexico	110	60	80
Indonesia	110	60	80

TABLE 6.6 Potential Limits for Cathodic Protection of Steel

	Positive Limit	Negative Limit
Aerated seawater	−800 mVAg/Ag	−1050 mV
	+250 mV Zn	0.0 mV
Anaerobic conditions	−900 mV Ag/AgCl	−1050 mV
	+150 mV Zn	0.0 mV

In general, corrosion of carbon steel in seawater is controlled by the availability of oxygen to the metal surface. Thus, under static conditions, carbon steel corrodes at between 100 and 200 mm/year, reflecting the oxygen level and temperature variations in different locations. As velocity causes a mass flow of oxygen to the surface, corrosion is very dependent on flow rate and can increase by a factor of 100 in moving from static or zero velocity to velocity as high as 40 m/s. Galvanizing confers only limited benefit under flow conditions, as corrosion of zinc also increases with velocity. For the thickness normally used in seawater piping, it will extend the life of the pipe for about 6 months. There is not much information available about the effect of temperature within the range normally encountered in seawater systems. It has been noted, at the LaQue Centre, that corrosion of carbon steel increases by approximately 50% between the winter (average temperature 7°C) and summer (27°C to 29°C) months. Although oxygen solubility tends to fall with a rise in temperature, the higher temperature tends to increase the corrosion rate.

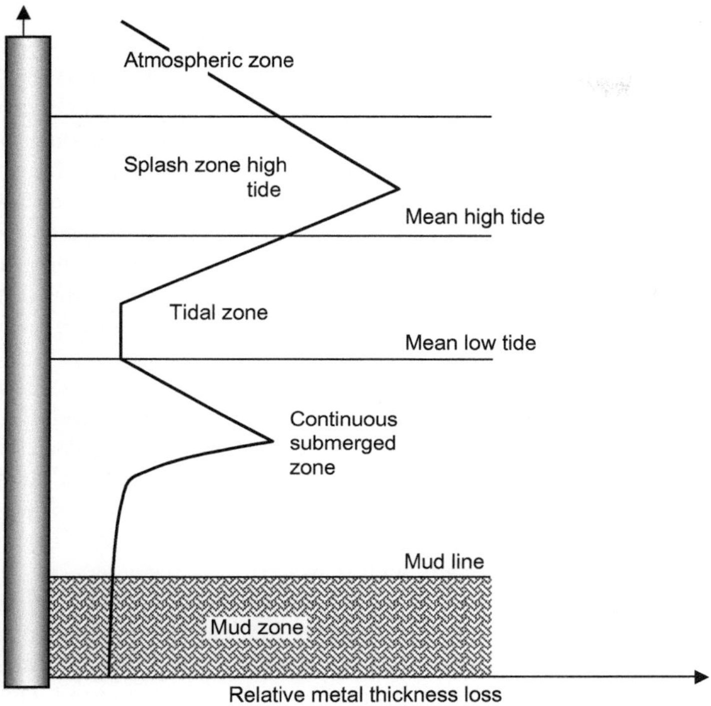

FIGURE 6.6 Corrosion profile of steel piling after 5 years of exposure (Humble, 1949).

6.3 CORROSION STRESSES DUE TO THE ATMOSPHERE, WATER AND SOIL

Atmospheric corrosion is a process that takes place in a film of moisture on the metal surface. The moisture film may be so thin that it is invisible to the naked eye. The corrosion rate is increased by an increase in relative humidity; by the occurrence of condensation, when the surface temperature is at or below the dew point; and by an increase in the amount of pollution in the atmosphere, because the corrosive pollutants can react with the steel and may form deposits on the surface. Experience has shown that significant corrosion is likely to take place if the relative humidity is above 80% and the temperature above 0°C.

It is worth mentioning that, when pollutants or hygroscopic salts are present, corrosion occurs at much lower humidity levels. The atmospheric humidity and air temperature in a particular region will depend on the climate conditions in that part of the world.

The location of the constituent element of a structure also influences corrosion. If the steel structures topside are exposed to the open air, climatic parameters such as rain and sunshine and pollutants in the form of gases or aerosols affect corrosion.

Special care should be taken when considering structures that are partly immersed in water or partly buried in soil, such as the offshore structure jacket. Corrosion under such conditions is often restricted to a small part of the structure, where the corrosion rate can be high. Exposure tests for estimating the corrosivity of water or soil environments are not recommended. The type of water—fresh, brackish or salt—has a significant influence on the corrosion of steel. Corrosivity is also influenced by the oxygen content of the water, the type and quantity of dissolved substances and the water temperature. For offshore structure jackets, animal or vegetable growth can accelerate corrosion.

Three different zones of immersion in water can be defined. First, the underwater zone is the area that is permanently exposed to water. Second, the intermediate zone is a fluctuating level zone, where the water level changes due to natural or artificial effects, thus giving rise to increased corrosion due to the combined impact of water and the atmosphere. Third, the splash zone is the area subject to wave and spray action, which can give rise to exceptionally high corrosion stresses, especially with seawater.

Corrosion in soil is dependent on the amount and the nature of mineral content of the soil and on the presenting organic matter, the water content and the oxygen content. The aeration of the soil is the main factor that influences its corrosivity. The oxygen content will vary and corrosion cells may be formed. Different types of soil and differences in soil parameters are not considered classification criteria in ISO 12944.

It is worth mentioning that, for the selection of a protective paint system, the special stresses to which a structure is subjected and the special situations in which a structure is located should be taken into consideration. Both the design and the use of the structure may lead to corrosion stresses not taken into consideration in the classification system.

6.3.1 Classification of Environments

For the purposes of ISO 12944, atmospheric environments are classified into six atmospheric corrosivity categories. To determine corrosivity categories, the exposure of standard specimens is strongly recommended. Table 6.7 defines the corrosivity categories in terms of the mass or thickness loss of standard specimens made of low-carbon steel and/or zinc after the first year of exposure. For details of standard specimens and the treatment of the specimens prior to and after exposure, see ISO 9226. Extrapolation of thickness losses to one year from shorter exposure times, as well as back-extrapolation from longer times, will not give reliable results and are therefore not permitted. The thickness losses obtained for steel and zinc specimens may sometimes fall into different categories. In such cases, the higher corrosivity category should be used. If it is

TABLE 6.7 Atmospheric Corrosivity Categories and Examples of Typical Environments

Corrosivity Category	Mass Loss per Unit Surface/Thickness Loss (After First Year of Exposure)				Typical Environments in a Temperate Climate	
	Low-Carbon Steel		Zinc			
	Mass Loss (g/m^2)	Thickness Loss (μm)	Mass Loss (g/m^2)	Thickness Loss (μm)	Exterior	Interior
C1 very low	≤10	≤1.3	≤0.7	≤0.1		Heated buildings with clean atmospheres, such as offices, shops, schools or hotels
C2 low	>10–200	>1.3 –25	>0.7–5	>0.1–0.7	Atmospheres with low levels of pollution, mostly rural areas	Unheated buildings where condensation may occur (e.g., depots, sports halls)
C3 medium	>200–400	>25–50	>5–15	>0.7–2.1	Urban and industrial atmospheres, moderate sulfur dioxide pollution; coastal area with low salinity	Production room with high humidity and some air pollution (e.g., food-processing plant)
C4 high	>400–650	>50–80	>15–30	>2.1–4.2	Industrial area and coastal area with low salinity	Chemical yards, swimming pools, coastal shipyards and boatyards
C5.I very high (industrial)	>650–1500	>80–200	>30–60	>4.2–8.4	Industrial areas with high humidity and aggressive atmosphere	Buildings or areas with almost permanent condensation and with high pollution
C5.M very high (marine)	>650–1500	>80–200	>30–60	>4.2–8.4	Coastal and offshore areas with high salinity	Buildings or areas with almost permanent condensation and with high pollution

TABLE 6.8 Corrosivity Categories for Water and Soil

Category	Environment	Examples of Environment and Structures
Im1	Fresh water	River installations, hydro-electric power plants
Im2	Sea or brackish water	Harbor areas with structures like sluice gates, locks, jetties and offshore structures
Im3	Soil	Buried tanks, steel piles and steel pipe

not possible to expose standard specimens in the actual environment of interest, the corrosivity category may be estimated by simply considering the examples of typical environments given in Table 6.7. The examples listed are a guideline that may be misleading. Only actual measurement of thickness loss will give the correct classification. Note that the corrosivity categories can also be estimated by considering the combined effect of the environmental factors: annual time of wetness, annual mean concentration of sulfur dioxide and annual mean deposition of chloride, as in ISO 9223.

Categories for Water and Soil

For structures immersed in water or buried in soil, corrosion is normally local and corrosivity categories are difficult to define. Table 6.8 lists three different environments and their designations. Note that, in many such environments, cathodic protection is involved.

Note that in Table 6.7 the loss values for the corrosivity categories are identical to those given in ISO 9223. In the coastal areas of hot humid zones, the mass or thickness losses can exceed the limits of the category; In the case of marine structures in category C5-M, special precautions must be taken when selecting their protective paint systems.

In most cases, corrosion behavior can be predicted from the type of climate. In a cold climate or a dry climate, the corrosion rate will be lower than in a hot temperate climate; it will be greatest in a hot, humid climate and in a marine climate, although considerable local differences can occur. The main concern is the time period during which the structure is exposed to high humidity (*time of wetness*—see Table 6.9).

6.3.2 Mechanical, Temperature and Combined Stresses

Erosion, which is caused by abrasive stresses, may occur in air due to particles, as in the case of sand entrained by wind. Surfaces that are subject to abrasion are considered to be exposed to moderate or severe mechanical stresses. On the other hand, in water, mechanical stresses may be produced by boulder movement, the abrasive action of sand, wave action and others.

TABLE 6.9 **Calculated Time of Wetness and Selected Characteristics of Various Climates, as in ISO 9223 (1992)**

Type of Climate	Mean Value of the Annual Extreme Values (°C)			Calculated Time of Wetness at RH >80% and Temperature >0°C (h/year)
	Low Temperature	High Temperature	Highest Temperature with RH >95%	
Extremely cold	−65	+32	+20	0–100
Cold	−50	+32	+20	150–2500
Cold temperate	−33	+34	+23	2500–4200
Warm temperate	−20	+35	+25	
Warm dry	−20	+40	+27	10–1600
Mild warm dry	−5	+40	+27	
Extremely warm dry	+3	+55	+28	
Warm damp	+5	+40	+31	4200–6000
Warm damp, equable	+13	+35	+33	

Mechanical stresses can be divided into three classes—weak, moderate and severe—and the criteria for each class are:

Weak: no, or very slight and intermittent, mechanical stresses, for example, due to light debris or small quantities of sand entrained in slow-moving water

Moderate: moderate mechanical stresses, due to solid debris, sand, gravel, shingle or ice entrained in moderate quantities in somewhat fast-flowing water, a strong current without entrained matter flowing past vertical surfaces, animal or vegetable growth and slight wave action

Severe: high mechanical stresses due to, for example, solid debris, sand, gravel, shingle or ice entrained in large quantities by fast-flowing water over horizontal or inclined surfaces and dense animal or vegetable growth, particularly if, for operational reasons, it must be removed mechanically from time to time.

The stresses will accumulate in medium or high temperatures. In most DNV standards, medium temperatures are defined as those between +60°C and +150°C, and high temperatures as those between +150°C and +400°C.

Temperatures of this magnitude occur only under special conditions during construction or operation. High temperatures are usually seen in fires, and medium temperatures don't really exist, except in some areas like the Arabian Gulf, where temperatures can reach 40 to 60°C, which is near the medium range.

When combined stresses affect steel, corrosion will increase. In general, corrosion develops more quickly on surfaces exposed simultaneously to mechanical and chemical stresses. Steel structures near salt water undergo additional mechanical stress due to impact, as in boat landings.

6.4 CATHODIC PROTECTION DESIGN CONSIDERATIONS

This section presents the method of cathodic protection that is primarily relevant to CP conceptual design, including the compatibility of CP with metallic materials and coatings. The discussion is intended as a guide for owners and their contractors in preparing for conceptual or detailed CP design.

The corrosion allowance should be taken into consideration during design and cannot be ignored.

6.4.1 Environmental Parameters

The major seawater parameters affecting CP design are not limited to the following:

- dissolved oxygen content in water
- sea currents
- temperature of the sea
- marine growth presence and its thickness
- salinity of the sea

In addition to the above, variations in seawater pH and carbonate content are also factors that affect the formation of calcareous layers associated with CP and thus affect the design of the CP system. In this case, more current is needed to achieve and to maintain the CP of bare metal surfaces. In seabed sediments, for the protection of the pile and the part of the offshore structure jacket embedded in the soil, the major parameters that affect the CP design are temperature, bacterial growth, salinity and sediment coarseness.

It is very important to know that all the above parameters depend on and vary with geographical location, depth and season. It is not feasible to give an exact relation between the seawater environmental parameters indicated above and cathodic current demands to achieve and to maintain the CP system. The main parameter for the CP design system for offshore structure platforms, default design current densities, i_c (A/m^2), are defined based on climatic region, which is related to mean seawater surface temperature and water depth.

If we know the ambient seawater temperature and salinity, we can define the specific seawater resistivity, ρ (ohm•m), which is used to calculate the anode resistance, R (ohm), a controlling factor for the current output from an anode.

6.4.2 Design Criteria

The cathodic protection system for the substructure should be designed in accordance with DNV RP B401-1993. The design criteria should also contain the conditions and assumptions that were taken into consideration.

Retrofit CP design may be limited to the platform structure, pile guides, piles and, where applicable, any conductors or appurtenances, including risers.

When the focus is on the platform, the design does not include pipelines or associated subsea structures that might be in close proximity and electrically connected to the platforms, other than making an allowance for some debris on the seabed.

The design life of the sacrificial anode cathodic protection (SACP) system should be defined based on the the owner's requirements, which take into consideration the lifetime of the platform.

Retrofit CP design uses conventional and modified CP criteria to maintain external structure potentials within the range of −800 to −1050 mV for an Ag/AgCl (seawater) reference electrode, throughout the remaining service life of the platform.

A seawater resistivity of 20 Ω/cm and an average seawater temperature of 22°C have been used in the design example discussed here.

In the case of platform jackets that were originally fitted with aluminum flush-fit, long, slender anodes more than 30 years previously, the anodes will have been either totally consumed or at end of their life. Because these anodes have a theoretical utilization factor of 85%, the anode retrofit design will not consider allowance for any remaining anode alloy.

In the case of retrofit, the designer should check the original coating system applied to the splash zone. For the purposes of CP retrofit design, an initial coating breakdown factor of 2% can be used for the first year, and a factor of 1% can be used for each year thereafter. This is slightly less than what DNV recommends but in some offshore operations it is accepted in practice. Assuming platforms originally installed 30 years earlier, the coating breakdown factors used in calculations for 25 years are an initial coating breakdown failure (CBF) of 35%, mean CBF of 48% and final CBF of 60%.

Based on DNV RP B401 (2005), the CP design document should recommend the following current densities to be used when designing offshore structure CP systems for wetted areas: 150 mA/m^2, 70 mA/m^2, 100 mA/m^2 for initial, mean and final, respectively. In buried areas, 20 mA/m^2 can be used for initial, mean and final.

The above current densities are recommended for new uncoated steel structures in tropical waters (>20°C) with a depth of 30 m or less. In practice, the above criteria are known to be conservative for retrofit designs when structures remain partially protected, and using these figures would result in a large mass of anode alloy and a substantial number of anode sleds.

DNV and other international standard organizations recognize that design parameters may be varied based on local conditions and experience gained in operation.

In the case of assessment of the original CP design for sacrificial anodes on offshore platforms, the assessment procedure should calculate the average current densities for the platforms based on potential levels over the last 25 years and compare them to total anode alloy weight loss over the same period. To obtain the most accurate current density figures, the total surface areas for each platform are added together, as are the number of anodes, to obtain the total mass of anode alloy consumed over 25 years.

For example, an assessment concluded, based on potential data, that the average current density had been 24 mA/m^2, whereas anode weight loss over the same period indicated an average current density of 23 mA/m^2. Based on the engineering office design experience, and according to the findings of the assessment of the offshore platform, the recommendation for wet areas was 35 mA/m^2, and for buried areas it was 10 mA/m^2 for initial, mean and final current density for platform retrofit CP designs.

6.4.3 Protective Potentials

A potential of −0.80 V relative to the Ag/AgCl/seawater reference electrode is generally accepted as the design protective potential *Ec* (V) for carbon and low-alloy steels. It has been argued that a design protective potential of −0.90 V should apply in anaerobic environments, including typical seawater sediments. However, it is advisable that the protective potential not be a variable in the design procedure.

For a correctly designed galvanic anode CP system, the protective potential for the main part of the design life will be in the range of −0.90 to −1.05 (V). Toward the end of the service life, the potential increases rapidly toward −0.80 (V), and eventually to even less negative values, referred to as *underprotection*.

The term *overprotection* is only applicable to protective potentials more negative than −1.15 (V). Such potentials do not apply for CP by galvanic anodes based on Al or Zn.

6.4.4 Negative Impact of CP on the Structure Jacket

Cathodic protection is accompanied by the formation of hydroxyl ions and hydrogen at the surface of the steel structure. This may cause disbonding of nonmetallic coatings by mechanisms that include chemical dissolution and electrochemical reduction processes at the metal-coating interface, possibly including build-up of hydrogen pressure at this interface. The process of coating deterioration is called *cathodic disbonding*.

Application of a surface preparation is required to achieve an optimum surface roughness, and some coating systems, such as those based on epoxy or polyurethane, have shown good resistance to cathodic disbonding by galvanic anode CP. For coating systems whose compatibility with galvanic anode CP is not well documented, the owner should be cautious and perform the necessary tests to ensure that the coating resists cathodic disbonding.

Testing of marine coatings' resistance to cathodic disbonding has been standardized, as in ASTM G8. Cathodic protection will cause formation of atomic hydrogen at the metal surface. Within the potential range for CP by aluminum- or zinc-based anodes (i.e., –0.80 to −1.10 V Ag/AgCl/seawater), the production of hydrogen increases exponentially toward the negative potential limit. The hydrogen atoms can either combine, forming hydrogen molecules, or become absorbed in the metal matrix. In the latter case, they may interact with the microstructure of components subject to high stresses, causing initiation and growth of hydrogen-related cracks, referred to as *hydrogen-induced stress cracking*.

Alloys like copper and aluminum are immune to hydrogen-induced stress cracking. High-strength titanium alloys should be avoided because they are very costly and are not readily available on the market.

Special techniques have been applied to confine the CP protective potential to a less negative range (e.g., –0.80 to –0.90 V), including the use of diodes and special anode alloys, but practical experience is limited. A major disadvantage of this approach is that the individual component or system needs to be electrically insulated from adjacent "normal" CP systems.

The explosive gas mixture of hydrogen and oxygen may be present in the cathodic protection systems in closed compartments without ventilation. However, the risk is moderate with Al- and Zn-based galvanic anodes.

It is very important to keep in mind that the CP application will assist in formation of a calcareous layer, consisting primarily of calcium carbonate, on bare steel surfaces. The thickness of the calcareous layer is typically on the order of a tenth of a millimeter, but thicker deposits may occur. The calcareous layer will reduce the current needed for maintenance of CP and is therefore beneficial. However, a calcareous layer may affect the connection of subsea electrical and hydraulic couplers with small tolerances. The application of an insulating layer of a thin-film coating, such as baked epoxy resin, is the main solution to this problem. An alternative solution is to electrically insulate the connectors from the CP system and to use seawater-resistant materials for all wetted parts. High-alloyed stainless steels, nickel-chromium-molybdenum alloys, titanium and certain copper-based alloys, such as nickel-aluminum bronze, have been used for this purpose.

The anodes may interfere with subsea operations and may increase drag forces imposed by flowing seawater, so these factors should be considered in the structure design.

6.4.5 Galvanic Anode Materials Performance

Aluminum and zinc are the main basic materials that are traditionally used in offshore structures. Note that the conceptual design in the front end engineering phase will report the generic type of anode material (i.e., aluminum- or zinc-based) that has been selected, and this information will be used in the next step of detailed engineering for the CP system. In general, aluminum-based anodes are traditionally used due to their higher electrochemical capacity. However, zinc-based anodes have sometimes been considered more reliable with respect to electrochemical performance for applications in marine sediments or internal compartments with high bacterial activity, which are both environments with anaerobic conditions.

Based on practical experience, ferritic and ferritic-pearlitic structural steels with specified minimum yield strength (SMYS) up to at least 500 MPa have proven compatibility with marine CP systems. Laboratory testing has revealed that hydrogen will induce stress cracking during extreme conditions of yielding. It is recommended that all welding be carried out according to a qualified procedure with 350 HV as an absolute upper limit. With a qualified maximum hardness in the range 300 to 350 HV, design measures should be implemented to avoid local yielding and to apply a reliable coating system as a barrier to CP-induced hydrogen absorption.

Welding of material that can change its formation should be followed by post-weld heat treatment to reduce the heat-affected zone hardness and residual stresses from welding.

Bolts made of martensitic steel heat-treated to SMYS up to 720 MPa, as per ASTM A193 grade B7 and ASTM A320 grade L7, have well-documented compatibility with CP. However, failures due to inadequate heat treatment have occurred, and, for critical applications, batch testing is recommended to verify a maximum hardness of 350 HV.

Design precautions should include measures to avoid local plastic yielding and use of coating systems qualified for resistance to disbonding by mechanical and physical or chemical effects.

In most design cases, anodes should have a closed-circuit potential of −1.10 V (or more negative) to copper/copper sulfate electrodes and should have a minimum efficiency of 80% (maximum consumption rate of 3.68 kg per ampere-year). Each foundry pour should be tested for chemical composition and closed-circuit potential. A suitable sample should be retained from each pour for performing a rate-of-consumption test. A minimum of three rate-of-consumption tests should be performed during the production run. In the event that any test indicates a consumption rate of more than the maximum limit mentioned above, tests should be performed on additional samples to confirm the adequacy of the anode composition.

The active anode material should be a proven aluminum-zinc-indium alloy suitable for long-term continuous service in seawater, saline mud or alternating seawater and saline mud environments.

TABLE 6.10 Performance Properties of the Alloy

Environment	Nominal Resistivity (Ohm/cm)	Anode Temperature (°C)	Minimum Alloy Capacity (A/hr/kg)	Working Potential (mV w.r.t.) Ag/AgCl/Seawater
Saline mud	100	40	1500	−1050
Saline mud	200	25	1865	−1050
Seawater	25	25	2000	−1070

Whoever will deliver the anode should propose an alloy offering the minimum electrochemical performance characteristics shown in Table 6.10, and the characteristics should be documented by long-term performance test reports.

Before starting the manufacturing process, the supplier should prove the ability of the proposed alloy to satisfy the above requirements, and minimum anode capacities and working potentials should be determined at operating anode current density, in the range of 0.25 A/m^2 to 1.0 A/m^2 for saline mud and 0.25 A/m^2 to 4.0 A/m^2 for free seawater.

The supplier should submit the full chemical analysis details and the specific gravity value of the proposed alloy to the client for approval and should demonstrate that the specification requirements can be satisfied at the compositional limits.

6.4.6 CP Design Parameters

Most of the design of the CP system depends simply on providing an electrical current to compensate for the loss of electrons through corrosion. So all the design parameters depend on providing a quantity of anode that will be distributed through out the structure to provide the required current under the environmental conditions that the structure is exposed to during its lifetime.

So the target is to select the quantity of anode that will be depleted after the structure's lifetime and that will provide the required current. The design concept and equations are available in guideline DNV B401, which is traditionally used by most designers, but it needs to be adapted according to the designer's experience with the geographic location of the structure.

All the parameters are usually defined by the designer of the CP system. However, sometimes certain or all CP design parameters have already been defined by the purchaser in a project document.

The design values recommended are consistently selected using a conservative approach; therefore, the values are expected to provide a service life that exceeds the design life of the CP system.

In the case of oil companies that have a fleet of fixed offshore structures, they may specify less, or (in certain cases) more, conservative design data, based on their own experience or other special considerations. Furthermore, the contractor, in addition to the owner, may propose use of alternative design data; however, any such data should then be accepted by the owner, preferably before the CP design work has started.

All electrochemical potentials associated with CP will be refered to the Ag/AgCl/seawater reference electrode. The potential of this reference electrode is virtually equivalent to that of the standard calomel electrode (SCE).

One change to note in the DNV RP 2004 revision is that the number of depth zones for design current densities has been extended from two to four, while the number of coating categories has been reduced from four to three.

Design Lifetime

The owner usually will specify the required lifespan for the CP system. The design life should take into account any period of time when the CP system will be active prior to operation of the platform.

It is also very important to take into consideration, when defining the CP lifespan, that the maintenance and repair of CP systems for fixed offshore structures are generally very costly and sometimes impractical. It is therefore normal practice to apply at least the same anode design life as the design life for the offshore structure platform. However, in certain circumstances, planned retrofitting of sacrificial anodes may be an economically viable alternative to the initial installation of very large anodes. This alternative should then be planned so that necessary provisions for retrofitting are made during the initial design and fabrication.

In general, the design life should be stated clearly on the basis of design and approved by the owner clearly before the design of the CP system is begun.

Current Densities for Design

The current density, I_c, is the cathodic protection current per unit surface area in A/m^2. There are initial and final design current densities, I_{ci} (initial) and I_{cf} (final), respectively, that give a measure of the anticipated cathodic current density demand to achieve cathodic protection of a bare metal surface within a reasonably short period of time. The main part of design is to calculate the required initial and final currents, which can define the number and sizing of anodes. During calculation, the effect of any coating on current demand should be considered by the application of the coating breakdown factor.

In case of a steel structure with some rust or mill scale, it is necessary to effect polarization of an initially bare metal surface. This is done by the cathodic current density that is called the initial design current.

In addition to that, the initial design cathodic current density is necessarily higher than the final design current density because the calcareous scale and possibly marine fouling layer developed during the initial phase reduce the subsequent current demand as the polarization resistance is reduced. A sufficient initial design current density enables rapid formation of protective calcareous scale and hence efficient polarization.

During the offshore steel structure's lifetime, it usually accumulates calcareous scale and marine growth, so the final design current density must be sufficient to repolarize a structure if such layers are partly damaged.

An appropriate final design current density and hence CP polarizing capacity will further ensure that the protection of the structure's jacket remains polarized to a potential of –0.95 to –1.05 V throughout the design life. In this potential range, the current density demand for maintenance of CP is lowest.

The initial and final current densities are the main parameters used to calculate the required number of anodes of a specific type to achieve a sufficient polarizing capacity. The CP system designer uses Ohm's law and assumes that the anode potential is in accordance with the design closed-circuit potential and the potential of the protection object is at the design protective potential for C-steel and low-alloy steel, i.e., –0.80 V.

The anode current decreases linearly when the cathode is polarized toward the closed-circuit anode potential, reducing the driving voltage for the galvanic cell. The total CP current for a CP unit, I_t, can be calculated from Equation (6.9):

$$I_t = \frac{E_c - E_a}{R} \tag{6.9}$$

where R (ohm) is the total anode resistance, E_c (V) is the global protection potential and E_a (V) is the actual anode (closed-circuit) potential.

The mean design current density, I_{cm} (A/m^2), is a measure of the anticipated cathodic current density once the CP system has attained its steady-state protection potential; this is typically 0.15 to 0.20 V more negative than the design protective potential.

The initial polarization period preceding the steady-state condition is normally short compared to the design life, and the time-weighted cathodic current density becomes very close to the steady-state cathodic current density.

The water depth and the geographical location are the main parameters affecting CP. Recommendations for initial/final and average (mean) design current densities, based on climatic regions and depth, are given in Table 6.11. These design current densities have been selected conservatively to account for harsh weather conditions, including waves and sea currents, but not erosive effects on calcareous layers by silt or ice. They further assume that the seawater at the surface is saturated with air (i.e., at 0.2 bar oxygen partial pressure).

Based on DNV B401, the data in Table 6.11 reflect the expected influence of seawater temperature and depth on the properties of a calcareous scale formed by cathodic protection and of the dissolved oxygen content. The properties of such layers are dependent on the seawater ambient temperature and, moreover, on certain depth-dependent parameters other than temperature. Note that oxygen is dissolved in the surface layer by dissolution from air and photosynthesis, so that the oxygen content at a great depth in a tropical region is likely to be substantially lower than in temperate or arctic surface waters of the same seawater temperature. Higher design current densities usually apply in the uppermost zone due to the wave forces and marine growth on degradation of calcareous scales and convective mass transfer of oxygen. Therefore, the design must account for the fact that, in certain areas, decomposition of organic material may reduce and ultimately consume all oxygen in the seawater. No such reduction in oxygen content is accounted for in Table 6.11.

In the case of bare steel surfaces buried in sediments, a design current density (initial/final and average) of 0.020 A/m^2 is recommended irrespective of geographical location and depth.

Bacterial activity may be the primary factor determining the CP current demand in the uppermost layer of seabed sediments. Further down into sediments, the current will be related to hydrogen evolution.

The design also has to take into consideration the effect of heat transfer due to convection in increasing the oxygen; this is usually done by an additional margin of CP current density.

The design current densities in Table 6.11 also apply for surfaces of any stainless steel or non-ferrous components of a CP system, including components in C-steel or low-alloy steel. For calculation of anode current output, a protective potential of −0.80 V then also applies to these materials.

Based on DNV, for aluminum components, or those coated with either aluminum or zinc, a design current density of 0.010 A/m^2 is recommended for initial and final currents, as well as mean values. For internally heated components, the design current density should be increased by 0.0002 A/m^2 for each °C that the metal/seawater is assumed to exceed 25°C.

The main goal of the CP design system is that the current density available to the structural steel surfaces will be sufficient, at any time during the design life of the protected offshore structure, to achieve the required potential range and maintain a calcareous deposit.

TABLE 6.11 Recommended Initial, Mean and Final Design Current Densities (A/m^2) as a Function of Depth, Climatic Region and Surface Water Temperature Based on DNV

Depth	Tropical (>20°C)			Sub-Tropical (12–20°C)			Temperate (7–11°C)			Arctic (<7°C)		
(m)	Initial	Mean	Final	Initial	Mean	Final	Initial	Mean	Final	Initial	Mean	Final
0–30	0.150	0.070	0.100	0.170	0.080	0.110	0.200	0.100	0.130	0.250	0.120	0.170
30–100	0.120	0.060	0.080	0.140	0.070	0.090	0.170	0.080	0.110	0.200	0.100	0.130
100–300	0.140	0.070	0.090	0.160	0.080	0.110	0.190	0.090	0.140	0.220	0.110	0.170
>300	0.180	0.090	0.130	0.200	0.100	0.150	0.220	0.110	0.170	0.220	0.110	0.170

TABLE 6.12 Recommended Constants *a* and *b* for Calculation of Paint Coating Breakdown Factors

Category	Description	a	b	
			Water Depth 0-30 m	Water Depth > 30 m
1	One layer of epoxy paint coating thickness >20 μm of dry film	0.10	0.10	0.05
2	One or more layers of marine paint coating by epoxy, polyurethane or vinyl base, with total nominal of the dry film thickness >250 μm	0.05	0.025	0.015
3	Two or more layers of marine paint coating by epoxy, polyurethane or vinyl base, with total nominal of the dry film thickness >350 μm	0.02	0.012	0.008

Maintenance of anode performance includes underwater inspection, as presented in Chapter 8, and should be documented for follow-up and monitoring of the CP system.

Coating Breakdown Factors

The use of nonmetallic coatings is very important, as they reduce the CP current demand for the protection of the structure, which will reduce anode weight correspondingly. For weight-sensitive structures with a long design life, the combination of a coating and CP is likely to give the most cost-effective corrosion control. If a very long design life is required by the owner, CP may be impractical unless combined with coatings.

If the CP of bare metal surfaces is known or expected to be high from the calculation, coating applications will be essential. In deep water, the formation of calcareous deposits may be slow. Coating should also be considered for surfaces that are partly shielded from CP by geometric effects.

In the case of large and complex offshore structure platforms, such as a multi-well production structure platform, the extensive use of coating is essential to reduce the overall current demand and to ensure adequate current distribution. To compensate for this, the design coating breakdown factors to be used for CP design (see Table 6.12) are deliberately selected conservatively to ensure that a sufficient total final current output capacity is installed. Consequently, calculations of the electrolytic voltage-drop away from the anodes using these coating breakdown factors and result in excessively high cathodic protection in terms of the estimated protection potential. This applies primarily to relatively long design lives when the calculated coating breakdown, and hence current demands and electrolytic voltage-drop, increase exponentially.

Use coatings is not a practical solution in the case of submerged parts of the offshore structures that need frequent inspections for fatigue cracks based on the underwater inspection plan, because this location is usually the critical welded joints of the jacket structure.

Metallic coatings on zinc or aluminum are compatible with galvanic anode CP. The organic coatings have no advantage in decreasing the current demand for CP. In addition, it is advisable to avoid zinc-rich primers because they are unsuitable for application with a CP system, due to their susceptibility to low electrical resistivity, which will require a high CP current demand.

The coating breakdown factor, f_c, represents the anticipated reduction in cathodic current density due to the application of an electrically insulating coating. When f_c is equal to 0, it means that the coating is 100% electrically insulating, thus decreasing the cathodic current density to zero. If f_c is equal to 1, the coating has no current-reducing properties.

The coating breakdown factor is a function of coating properties, operational parameters and time. As a simple engineering approach, f_c can be expressed as:

$$f_c = a + b \bullet t \tag{6.10}$$

where t is the coating lifetime in years, and a and b are constants that are dependent on coating properties and the environment (see Table 6.12).

It is worth mentioning that the effect of marine growth is highest in the upper 30 m, where wave forces may further contribute to coating degradation.

Another factor is periodic cleaning of marine growth in this zone.

Preferably, the owner will specify constants a and b for calculation of coating breakdown factors based on his own practical experience of specific coating systems in a particular environment. When the owner has not specified any such data, the default values in Table 6.12 should be used.

Once a and b are defined, mean and final coating breakdown factors, f_{cm} and f_{cf}, respectively, to be used for CP design purposes are to be calculated by introducing the CP design life, t_f (years):

$$f_{cm} = a + 0.5b\, t_f \tag{6.11}$$

$$f_{cf} = a + b \bullet t_f \tag{6.12}$$

For certain protection objects, with large uncoated surfaces, the initial coating breakdown factor, $f_{ci} = a$, may be applied to calculate the initial current demand to include coated surfaces.

If the calculated value exceeds 1, $f_c = 1$ is applied in the design. When the design life of the CP system exceeds the actual calculated life of the coating system, f_{cm} may be calculated as:

$$f_{cm} = 1 - \frac{(1-a)^2}{2bt_f} \tag{6.13}$$

To account for the effect of a coating system on coating breakdown factors, three coating categories have been defined for inclusion in Table 6.12.

The criteria for the three categories are:

Category 1: One layer of epoxy paint coating not less than 20 μm of dry film thickness
Category 2: One or more layers of marine paint coating by epoxy, polyurethane or vinyl base, with total nominal of the dry film thickness not less than 250 μm
Category 3: Two or more layers of marine paint coating by epoxy, polyurethane or vinyl base, with total nominal of the dry film thickness not less than 350 μm.

Note that category 1 includes the shop primer type of coating. It is assumed for categories 2 and 3 that the supplier will provide specific coating materials to be applied and will ensure its performance by documented qualification or by performing the required tests. It is very important to know that the assumption in design is that all the three categories of coating have been installed according to the manufacturer's recommendations and that surface preparation has included blast cleaning, in accordance with ISO 8501. The surface roughness should be controlled according to the manufacturer's recommendation. For any coatings applied without blast cleaning of surfaces, the coating breakdown factors f_{cm} and f_{cf} will be taken equal to 1, while the initial current demand may be calculated as for category 1.

Data about the performance of coatings on cathodically protected structures should be obtained, in particular for long service lives. The data in Table 6.12 should therefore be regarded as coarse but conservative engineering judgments. For any coating system not covered by the three coating categories above and with major potential effect on the overall current demand, the owner should specify or accept the applicable constants a and b.

The values of a and b for a depth of 30–100 m in Table 6.12 are applicable to calculations of current demands of flooded compartments and of closed compartments with free access to air.

It is worth mentioning that the constants in Table 6.12 do not account for significant damage to paint coatings during fabrication and installation. Therefore, if such damage is anticipated, the affected surface area is to be estimated and included in the design calculations as bare metal surface.

From technical experience of many owner companies, coating category 1 is not applicable for structural nodes, the surface areas of welds and up to 0.25 m adjacent to the welds, which are assumed to be bare and are included for the calculation of current requirements.

The average and final coating coverage and breakdown factors are determined for the coated surface in each zone.

Design Parameters for Galvanic Anode Material

The recommended compositional limits alloying and impurity elements for Al- and Zn-based anodes in Table 6.13 apply. The CP design parameters related to anode material performance are:

- design electrochemical capacity, ε (Ah/kg)
- design closed-circuit anode potential, E_{oa} (V)

The design electrochemical capacity, ε (Ah/kg), and design closed-circuit anode potential, E_{oa} (V), are used to calculate the design anode current output and the required net anode mass using Ohm's and Faraday's laws, respectively.

The design electrochemical capacity, ε (Ah /kg), in Table 6.13 is used for design unless otherwise specified or accepted by the owner. The data are applicable for seawater temperature up to 30°C as a yearly mean value.

Due to the relatively high anodic current densities that are utilized for anode testing, short-term laboratory examination of galvanic anode materials and anode electrochemical efficiency will typically result in values close to the

TABLE 6.13 Recommended Compositional Limits for Al-Based and Zn-Based Anode Materials

Alloying/Impurity Element	Zn-Based	Al-Based
Zn	NA*	2.5–5.75
Al	0.10–0.50	rem
In	NA*	0.015–0.040
Cd	≤0.07	≤0.002
Si	NA	≤0.12
Fe	≤0.005	≤0.09
Cu	≤0.005	≤0.003
Pb	≤0.006	NA

**NA is not available*

TABLE 6.14 Recommended Design Electrochemical Capacity and Design Closed-Circuit Potential for Anode Materials at Seawater Ambient Temperatures

Anode Material	Environment	Electrochemical Capacity (Ah/kg)	Closed-Circuit Potential (V)
Al-based	Seawater	2000	−1.05
	Sediments	1500	−0.95
Zn-based	Seawater	780	−1.00
	Sediments	700	−0.95

theoretical limit (e.g., ≥ 2,500 Ah/kg for Al-Zn-In material). Such data do not replace the recommended design values for electrochemical capacity. The use of electrochemical capacity greater than the default values based on DNV RP B401 (2005) in Table 6.14 should be justified by long-term testing. Even such testing will tend to result in slightly non-conservative values, as the testing time is still relatively short and the anodic current density is relatively high compared to the working conditions for real anodes. The design values for closed-circuit anode potential, E_{oa} (V), in Table 6.14 are used for design. The data are applicable for all seawater temperatures with a yearly average of max 30°C.

Anode Resistance

The anode resistance, R (ohm), is calculated using the following formulas, which are applicable to the actual anode shape. Calculations are performed for the initial anode dimensions and for the estimated dimensions when the anode has been consumed to its utilization factor.

For long, slender stand-off anodes, $L \geq 4r$:

$$R = \frac{\rho}{2\pi L}\left(\ln\frac{4L}{r} - 1\right) \tag{6.14}$$

where L is the anode length, ρ is the seawater resistivity (ohm/m) and r is the effective anode radius.

For short, slender stand-off anodes, $L < 4r$:

$$R = \frac{\rho}{2\pi L}\left[\ln\left\{\frac{2L}{r}\left(1 + \sqrt{1 + \left(\frac{r}{2L}\right)^2}\right)\right\} + \frac{r}{2L} - \sqrt{1 + \left(\frac{r}{2L}\right)^2}\right] \tag{6.15}$$

The equation is valid for anodes a minimum distance of 0.30 m from the protection object. For an anode-to-object distance less than 0.30 m but a minimum of 0.15 m, the same equation may be applied with a correction factor of 1.3.

For long, flush-mounted anodes with L greater than or equal to 4 times the width and thickness:

$$R = \frac{\rho}{2 \bullet S} \tag{6.16}$$

For noncylindrical anodes, $r = c/(2\pi)$, where c (in m) is the anode cross-sectional periphery. S is the arithmetic mean of anode length and width.

For short, flush-mounted, bracelet and other anode types:

$$R = \frac{0.315\rho}{\sqrt{A}} \tag{6.17}$$

where A is the anode surface area.

Resistivity of Seawater and Sediment

Seawater resistivity, ρ (ohm•m), influences the design of the CP system and this resistivity is affected by both the seawater salinity and temperature. Note that, in the open sea, salinity does not vary significantly and temperature is the main factor. On the other hand in the case of shore areas, particularly at river outlets and in enclosed bays, the salinity will vary significantly. It is recommended that the design of CP systems in such locations be based on resistivity measurements reflecting the annual mean value and the variation of resistivity with depth.

In comparison to seawater, the resistivity of marine sediments is higher by a factor ranging from about 2 for very soft clays to approximately 5 for sand. If resistance data for the sediment at the location are not available, the CP designer should use the highest factor for calculation of the resistance of any buried anodes.

In temperate regions with annual average surface water temperature of 7°C to 12°C, resistivities of 0.30 and 1.3 ohm·m are recommended as reasonably conservative estimates for the calculation of anode resistance in seawater and marine sediments, respectively, and independent of depth. If you want to use lower values for design, actual measurements should be documented, taking into account any seasonal variations in temperature.

The seawater data that will be used in the design should contain the structure's local annual average conditions versus depth. If the structure is exposed to seawater for which the temperature varies more than 5°C for a depth interval, that depth interval is split up into separate zones that cover no more than a 5°C interval each. The depth-averaged temperature is used for each zone created. Seawater resistivity (ρ) should be obtained from local seawater resistivity measurements corrected to average annual seawater temperature conditions.

Anode Utilization Factor for CP Design

The utilization factor of any particular anode design is greatly affected by the node shape and core insert configuration. The utilization factor represents the maximum volume of cast anode alloy that can be consumed before the anode can no longer deliver the current required, and it has to take into account the reduced size of the anode alloy and disbonding of anode alloy from the core at the end of the system's design life.

The anode utilization factor, u, is the fraction of anode material of an anode with a specific design that may be utilized for calculation of the net anode mass required to sustain protection throughout the design life of a CP system.

When an anode is consumed to its utilization factor, the polarizing capacity, which is determined by the anode current output, cannot be estimated due to loss of anode material support or rapid increase of anode resistance due to other factors.

The utilization factor is dependent on the anode design, particularly on its dimensions and the location of anode cores. Table 6.15 presents anode utilization factors, based on DNV RP B401, that are used for design calculations. According to BS, an anode utilization factor of 0.90 for the long, slender, stand-off type shall be considered in design.

Design Parameters for Current Drain

All items that are expected to become electrically connected to a CP system are considered in current drain calculations.

Offshore platforms often include temporary or permanent structural systems that are not considered to require CP but unfortunately will drain current from the CP system, such as the mooring systems for floating installations, or other

TABLE 6.15 Recommended Anode Utilization Factors for CP Design Calculations

Anode Type	Anode Utilization Factor
Long, slender stand-off, $L \geq 4r$	0.90
Short, slender stand-off, $L < 4r$	0.85
Long, flush-mounted $L \geq$ anode width $L \geq$ anode thickness	0.85
Short, flush-mounted, bracelet and other types	0.80

structural elements, such as piles and skirts, which do not need corrosion protection due to their increased wall thickness, which includes a corrosion allowance to account for the expected corrosion rates.

Calculations of current drain use the design current densities and coating breakdown factors for items requiring CP. Calculations of surface areas and current demands are carried out.

The design current densities and coating breakdown factors are important for calculation of current drains to structural elements that are not considered to need CP, but will be, or are expected to be, electrically connected to the protection object by the CP system being designed.

In case of buried steel structural surfaces, such as mud mats, skirts and piles, a current drain value of 0.020 A/m^2 is accounted for, based on the outer external surface area that is exposed to sediment. In some cases and based on the owner's specification for parts of steel skirts and piles to be buried in sediments, a current density (initial/final and average) of 0.025 A/m^2 is used. The current drain to open pipes covers an internal surface area equivalent to 10 times π times the diameter.

For open pile ends, the top internal surface is included for a distance of 5 times the diameter and is regarded as seawater exposed. Internal surfaces of piles filled with sediments do not have to be included.

Unless otherwise specified by the owner, a current drain of 5 A/well casing should be considered in CP design calculations.

Subsea wells that are cemented will have reduced current drain in comparison to platform wells, which are normally not cemented. The CP system design should consider the worst-case scenario in operations to have a conservative design. For example, subsea wells may account for significant current drain during installation or work-over and other interventions.

Current drain due to anchor chains is accounted for by 30 m of chain for systems with the mooring point topside only. On the other hand, if the mooring point is below the water sea level, the seawater-exposed section above this point is also included. A current drain for 30 m of chain is also included for CP of anchoring arrangements using chains.

6.4.7 Design Calculation for CP System

In design of a CP system for a fixed offshore platform, detailed design is normally preceded by conceptual design activity in the front end engineering phase. During the conceptual design phase, the type of anodes and fastening devices are selected, taking into account forces exerted on anodes during installation and operation. Moreover, any coating systems to be applied to specific areas or components should also be specified, as this is the main factor in calculating current demands for the CP system and also in calculating the required total net mass of anode material.

If no CP conceptual report has been prepared, then the basic concepts for detailed CP design are defined by the purchaser in some other reference document(s) to be included in an inquiry for CP detailed design.

The checklist of information and any optional requirements provided by the purchaser includes:

- conceptual CP system design report
- design life of CP system to be installed
- the project design basis, which are salinity and temperature as a function of depth for calculation of anode resistance, water depth and the level of seabed for platform substructures
- environmental and installation parameters affecting forces exerted on anodes
- structural drawings and information on coating systems as required for calculation, including components that may exert temporary or permanent current drain
- identification of any interfaces to electrically connected component systems with self-sufficient CP systems, such as pipelines and other connected facilities
- any specific requirement for CP design parameters to be applied, such as coating breakdown factors and current drain to wells
- any specific requirements for anode material and anode design

The purchaser ensures that valid revisions of drawings and specifications affecting calculation of current demand for CP and location of anodes are available to the contractor during the design work. All necessary information is provided to the contractor for calculation of the overall current demand (e.g., conductors for production platform substructures and production control system for subsea valve trees).

In the design of CP systems for the structure jacket, it is advisable to divide the areas into separate surface areas to be protected. The division into areas may be based on depth zones.

Calculation of Current Demand

To calculate the current demand, I_c (A), to provide adequate polarizing capacity and to maintain cathodic protection during the design life, the individual surface areas, A_c (m^2), of each CP unit are calculated and multiplied by the relevant design current:

$$I_c = A_c i \bullet f_c \tag{6.18}$$

Design current density, i_c (A/m^2), and the coating breakdown factor, f_c, if applicable, are then selected.

Calculation of the surface area of the jacket is complicated if done manually, so some simplification for complex joints is possible with limited accuracy. However these days there are many software programs available that can calculated it quickly with high accuracy. Also, any simplification that is performed should be tend toward the conservative. For major surface areas, such as the jacket, an accuracy of from −5% to +10% is adequate. For smaller components, the required accuracy may be lower, depending on whether or not a coating will be applied to such items and to the major surfaces.

Surface area calculations for each unit are documented in the CP design report. The contractor ensures that all items affecting CP current demand are included in the surface area calculations. The items may include various types of outfitting to be installed by different contractors.

For subsea production systems, production control equipment is typically manufactured from uncoated stainless steel, such as piping components, couplings, connectors, cable trays, and others. ROV override components are also often manufactured from stainless steel without a coating.

For items with major surfaces of uncoated metal, the CP current demands for both initial polarization and for polarization at the end of the design life, I_{ci} (A) and I_{cf} (A), respectively, are calculated, together with the mean current demand required to maintain cathodic protection throughout the design period, I_{cm} (A). For protection objects with current demand primarily associated with coated surfaces, the initial current demand can be deleted in the design calculations. For future reference, all calculated data are documented in the design report.

Calculation of Anode Mass

The total net anode mass, m_a (kg), required to maintain cathodic protection throughout the design life, t_f (years), is calculated from I_{cm} (A) for each unit of the mean current demand including any current drain:

$$M_a = \frac{8760 I_{cm} \bullet t_f}{u \bullet \varepsilon} \tag{6.19}$$

Anode Number Calculation

From the anode type selected, the number of anodes, N, anode dimensions and anode net mass are defined to meet the requirements for 1) initial/final current output, I_{ci}/I_{cf} (A) and 2) anode current capacity C_a (Ah), which relate to the CP current demand, I_c (A), of the protection object.

The preliminary sizing of anodes should be based on commercially available products, which requires liaison with potential anode vendors.

The individual anode current output, I_a (A), required to meet the current demand, I_c (A), is calculated from Ohm's law:

$$I_c = NI_a = \frac{N(E_{co} - E_{ao})}{R} = \frac{N\Delta E_o}{R} \tag{6.20}$$

where I_a (A) is the individual anode current output, E_{ao} (V) is the design closed-circuit potential of the anode material and R (ohm) is the anode resistance. The initial and final current output, I_{ai} and I_{af}, are calculated using the initial and final anode resistance, R_{ai} and R_{af}, respectively.

For calculation of anode resistance, E_{co} (V) is the design protective potential, which is −0.80 V. ΔE_o (V) is termed the *design driving voltage*.

Because the design driving voltage in Equation (6.20) is defined using the design protective potential for C-steel, the initial/final design current densities that define the anode current output capacity, and hence the driving voltage, refer to the required anode current output at this potential. Hence, the initial/final design current densities given in Table 6.11 are based on a protection potential of −0.80 V. This means that they are always used for calculations according to Equation (6.20) in combination with this potential, as well as if a more negative protection potential (e.g., ≤ -0.90 V) is aimed for.

The individual anode current capacity, C_a (A•h), is given by:

$$C_a = m_a \varepsilon u \tag{6.21}$$

where m_a (kg) is the net mass per anode. The total current capacity for a CP unit with N anodes thus becomes $N \bullet C_a$ (A•h).

Calculations are carried out to demonstrate that the following requirements are met:

$$C_{at} = NC_a \geq 8760 \bullet I_{cm} \bullet t_f \tag{6.22}$$

$$I_{ati} = NI_{ai} \geq I_{ci} \tag{6.23}$$

$$I_{atf} = NI_{af} \geq I_{cf} \tag{6.24}$$

C_{at} in Equation (6.22) is the total anode current capacity. I_{cm}, I_{ci} and I_{cf} in Equations (6.22), (6.23) and (6.24) are the current demands of a CP unit, including any current drain. 8760 is the number of hours per year. I_{ai} and I_{af}

in Equations (6.23) and (6.24) are the initial and final current output for the individual anodes.

If anodes of different sizes and hence different anode current capacities, C_a (Ah), and current output, I_a (A), are utilized for a CP unit, $N \bullet C_a$ and $N \bullet I_{ai}/N\{ts\} \bullet \{ts\} I_{af}$ will have to be calculated for each individual size and then added for calculation of the total anode current capacity (C_{at}) and total anode current output (I_{ati}/I_{atf}).

If the above criteria cannot be fulfilled for the anode dimensions and net mass initially selected, another anode size is selected and the calculations are repeated until the criteria are fulfilled.

Optimizing the requirements in Equations (6.22), (6.23) and (6.24) is an iterative process where a simple computer spreadsheet may be helpful. In general, if Equation (6.22) is fulfilled, but not Equation (6.23) and/or (6.24), a higher number of smaller anodes, or the same number of more elongated anodes, are to be used. On the other hand, if $N \bullet I_a$ in Equations (6.23) and (6.24) is much larger than I_{ci} and I_{cf}, fewer and/or more compact anodes may be applied.

Unless a high initial current capacity is deliberately aimed for, as with protection objects consisting primarily of uncoated metal surfaces, the anodes to be installed should have a similar anode current output (I_a) to net anode mass (m_a) ratio. It is worth mentioning that small anodes with high anode current output to net mass ratio will be more rapidly consumed than large anodes with a higher ratio, which could result in an insufficient total anode current capacity toward the end of the design life.

For anodes with the same anode resistance and hence the same anode current output, but a major difference in net anode mass due to different anode geometry, the anode with the lowest net anode mass will be consumed first. The anodes with the lowest anode resistance will be consumed first if anodes with the same net anode mass but with major differences in anode resistance and, therefore, anode current output are used.

From a practical point of view, the following factors are used:

Design protective potential (E_{oc}) = −0.800 V Ag/AgCl/seawater
Design closed circuit potential (E_{oa}) = −1.050 V Ag/AgCl/seawater

The anode current output is calculated for the initial and final life of the CP system. In the latter case, anodes of 10% remaining mass are used.

The number of anodes computed for the structure's initial and final currents, and the structure's current capacity, by zone, is as follows:

Anodes for initial current needs (N) = I_c (initial)/I_a (initial)
Anodes for final current needs (N) = I_c (final)/I_a (final)
Anodes for final current needs (N) = M/m [m = mass of anode material per anode]

The number of anodes provided within each zone is greater than the required number as calculated by zone, and also greater than the minimum number of anodes required to ensure adequate current capacity and proper current distribution.

The total number of anodes required is the sum of those needed for the three zones.

Anode Resistance Calculation

The anode resistance, R (ohm), is based on Equations (6.14) to (6.17), using the actual anode dimensions and specific resistivity of the surrounding environment. To calculate the initial anode resistance, R_i (ohm), the initial anode dimensions are inserted in the relevant anode resistance equations from Equations (6.14) to (6.17). The final anode resistance, R_f (ohm), is calculated based on the expected dimensions when the anode has been consumed to its utilization factor, u, as explained before.

At the end of its design life, t_f (years), after the anode has been consumed to its utilization factor, u, the remaining net anode mass, m_f (kg), is calculated from the following equation:

$$m_f = m_i \bullet (1 - u) \tag{6.25}$$

Note that the final volume of the anode to be used for calculation of R can be calculated from the remaining net anode mass, m_f (kg), the specific density of anode material, and the volume of insert materials. If you don't have data about details of anode inserts, to be on the conservative side, estimate the volume, or you can neglect the volume calculation.

For long and short slender stand-off anodes consumed to their utilization factor, a length reduction of 10% is assumed. In addition to that, assuming that the final anode shape is cylindrical, the final radius is calculated based on this length reduction, and the final anode mass.

For long, flush-mounted anodes, the final shape is assumed to be a semi-cylinder and the final length and radius are calculated as above.

For short, flush-mounted anodes, bracelet anodes and other shapes mounted flush with the jacket, the final exposed area is assumed to be equivalent to the initial area facing the surface to be protected.

The anode quantities should be calculated to ensure that the correct total net mass of anode material is provided based on mean current density requirements and to provide sufficient current to maintain polarization throughout the design life of the platform as defined by the owner, which is usually 25 years.

The total mass of sacrificial alloy required to protect each platform is calculated using the formula:

$$W = \frac{t_f E I_{cm}}{u} \tag{6.26}$$

where W is the total net weight of anode alloy in kg, t_f is the design life (years), E is the consumption rate of anode alloy (kg/year), I_{cm} is the mean current to protect the structure and u is the utilization factor.

The number of anodes is calculated once the total weight of anode material has been determined for each area to be protected, by dividing the total anode weight by the specified anode unit net weight, as in the following formula

$$\text{Number of anodes} = \text{Total weight/anode net weight}$$

The designer should ensure that the number of anodes for each area is capable of providing sufficient current throughout the remaining design life of the structure to achieve the target by maintaining polarization during the structure's lifetime. This can be achieved by calculating the initial, mean and final current output of each anode type used and comparing the results with the platform current demands based on surface areas and specified current densities.

The above equation is valid for anodes with a minimum stand-off distance of 300 mm from the structure. A correction factor of 1.3 is applied for anodes between 150 mm and 300 mm from the structure.

Anode Design Precautions

The contractor specifies in the CP design report the tentative dimensions and net mass for anodes to be used.

The design report should include the design of anode-fastening devices for anodes that will be affected by significant forces during installation and operation. There are special considerations for large anodes that will be installed on structural members subject to fatigue loads during pile-driving operations. Therefore, doubler or gusset plates may be used for large anodes.

To use the anode resistance formula in Table 6.16 for stand-off type anodes, the minimum distance from anode to protection object is 300 mm. However, for distances down to 150 mm, the formula can still be used by multiplying the anode resistance by a factor of 1.3.

Detailed anode design ensures that the utilization factor assumed during calculations of required anode net mass is met. Hence, it ensures that the anode inserts are still likely to support the remaining anode material when the anode has been consumed to its design utilization factor. In most cases, anode cores of stand-off anodes protrude through the end faces.

With the exception of stand-off anodes, a marine-grade paint coating (min 100 μm DFT) is specified for anode surfaces facing the protection object.

Anode Distribution on the Offshore Jacket

The calculated number of anodes, N, for a CP unit is distributed to provide a uniform current distribution, taking into account the current demand of individual members based on the different surface areas and coatings used. On fixed offshore platforms, special areas to be considered when distributing anodes are nodes, pile guides and conductors. The location of all individual anodes is shown on drawings.

TABLE 6.16 Steps in CP Design

Item	Calculation and Equations	Input/Output Data
Coating Breakdown		
Jacket HAT* to −3,000 and risers all depths uncoated at installation		0.350
Jacket −3000 to mud line uncoated at construction		0.35
Jacket coating damaged prior to placement		1.00
Jacket f_c (initial) total factor at placement		0.00
Conductor coating damaged during construction		1.00
Coating breakdown factor per RP		1.00
Average factor	f_c (average) = a + b • t_f /2	
	f_c (final) = a + b • t_f	
Average Breakdown Factors		
Jacket: HAT to −3.0	f_c (average) = a + b • t_f /2	0.475
Jacket: −3.0 to −30.0	Bare	1.0
Jacket: −30.0 plus		1.0
Conductors: HAT to −3.0		1.0
Conductors: −3.0 to −30.0		1.0
Conductors: −30.0 plus		1.0
Risers: HAT to −3.0		0.472
Risers: −3.0 to 30.0		0.472
Risers: −30.0 plus		0.472
Final Breakdown Factors		
Jacket: HAT to −3.0		0.6
Jacket: −3.0 to −30.0		1.0
Jacket: −30.0 plus		1.0
Conductors: HAT to −3.0		1.0

(Continued)

TABLE 6.16 Steps in CP Design—cont'd

Item	Calculation and Equations			Input/Output Data
Conductors: −3.0 to −30.0				1.0
Conductors: −30.0 plus				0.6
Risers: HAT to −3.0				0.6
Risers: −3.0 to 30.0				0.6
Risers: −30.0 plus				
Current demand jacket				
I_c (initial)				
Zone	A_c (m^2)	f_c (initial)	i_c (initial) (A/m^2)	I_c (A)
Jacket: HAT to −3.0	67.3563	0.350	0.035	0.825
Jacket: −3.0 to −30.0	692.6240	1.0	0.035	24.242
Jacket: −30.0 plus	0.00	1.0	0.035	0.00
I_c (average)				
Jacket: HAT to −3.0	67.3563	0.475	0.035	1.120
Jacket: −3.0 to −30.0	692.6240	1.0	0.035	24.242
Jacket: −30.0 plus	0.00	1.0	0.035	0.000
I_c (final)				
Jacket: HAT to −3.0	67.3563	0.60	0.035	1.414
Jacket: −3.0 to −30.0	692.6240	1.0	0.035	24.242
Jacket: −30.0 plus	0.00	1.0	0.035	0.000
Current demand conductors				
I_c (initial)				
Conductors: HAT to −3.0	30.099	1	0.035	1.053
Conductors: −3.0 to −30.0	151.736	1	0.035	5.311
Conductors: −30.0 plus	0.00	1	0.035	0.000
I_c (average)				
Conductors: HAT to −3.0	30.099	1	0.035	1.053
Conductors: −3.0 to −30.0	151.736	1	0.035	5.311
Conductors : −30.0 plus	0.00	1	0.035	0.000

TABLE 6.16 Steps in CP Design—cont'd

Item	Calculation and Equations			Input/Output Data
I_c (final)				
Conductors: HAT to −3.0	30.099	1	0.035	1.053
Conductors: −3.0 to −30.0	151.736	1	0.035	5.311
Conductors: −30.0 plus	0.00	1	0.035	0.000
Riser demand conductor				
I_c (initial)				
Risers: HAT to −3.0	11.124	0.35	0.035	0.136
Risers: −3.0 to 30.0	56.076	0.35	0.035	0.687
Risers: −30.0 plus	0.00	0.35	0.035	0.00
I_c (average)				
Risers: HAT to −3.0	11.124	0.475	0.035	0.185
Risers: −3.0 to 30.0	56.076	0.475	0.035	0.932
Risers: −30.0 plus	0.00	0.475	0.035	0.00
I_c (final)				
Risers: HAT to −3.0	11.124	0.60	0.035	0.234
Risers: −3.0 to 30.0	56.076	0.60	0.035	1.178
Risers: −30.0 plus	0.00	0.60	0.035	0.00
Pile penetration I_c (initial/final and average)	634.219	1.00	0.01	6.342
Well casing	4 × (2.5) A/well			10
Total drain current required				16.342
Current demand for jacket buried				
I_c (initial)	49.551	1.00	0.01	0.496
I_c (average)	49.551	1.00	0.01	0.496
I_c (final)	49.551	1.00	0.01	0.496
Total initial current, I_c (initial)				49.092
Total average current, I_c (average)				49.681
Total final current, I_c (final)				50.269

(Continued)

TABLE 6.16 Steps in CP Design—cont'd

Item	Calculation and Equations	Input/Output Data
	Mass (M) = I_c (average) * t_f * 8760/(u *e)	
Electrochemical anode efficiency		2.0
Utilization factor		0.90
Mass required		
I_c (Average)	49.681	
I_c (Average) * t_f * 8760/(u * e)	Total mass	6044 kg
Anode Resistance Calculations		
Anode initial physical dimensions		
Anode length		2.420 m
Anode width		0.187 m
Anode depth		0.170 m
Anode radius		0.114 m
Alloy density		2700 kg/m^3
Net anode volume		0.06 m^3
Anode mass		189.76 kg
Anode final physical parameters	Anode final mass m(final) = m(initial) * (1 – u)	18.976 kg
	Anode final length L(final) = L(initial) – 0.10 * u * L(initial)	2.202 m
Anode resistance, R (initial)		0.0453 Ω
Anode resistance, R (final)		0.0622 Ω
	Anode Output (I_a) = (E_c – E_a)/R	
Anode (Al) closed-circuit potential		–0.95 V
Protected cathode potential		–0.80 V
Anode driving potential		0.15 V
Anode output I_a (initial)		3.310 A
Anode output I_a (final)		2.412 A
Anode Number Calculation		
Total initial current/I_a (initial)		14.829
Total initial current/I_a (initial)		15.0

TABLE 6.16 Steps in CP Design—cont'd

Item	Calculation and Equations	Input/Output Data
Total final current/I_a (final)		20.845
Total final current/I_a (final)	Rounded up to integer	21.00
Anode number		**32**
Initial current available from anode number above		105.935 A
Total initial current required per calculation		49.092 A
Difference of initial current available and initial current required	Initial surplus	57 A
Final current available from anode number above		77.172 A
Total final current required per calculation		50.269 A
Difference of final current available and final current required	Final surplus	26.90 A
Mass Check		
Mass of anodes per distribution		6072 kg
Mass required for life		6044 kg
Difference of mass available and mass required	Mass surplus	28 kg
Average current from mass for design life		55 A
Anode Gross Weight Calculation		
Anode unit gross weight estimate		216 kg
Anode gross weight estimated on numbers above		6912 kg
Total gross weight		6912 kg

**HAT is the highest astronomical tide.*
I_c is the current demand for specified area
A_c is the individual area
f_c is the coating breakdown factor
i_c is the current density
t_f is the required lifetime per year
a, b as shown Table 6.12

From a practical point of view, the anodes used to provide cathodic protection of surfaces buried in sediments are located freely exposed to the sea.

To avoid interaction effects that reduce the current output, there should be enough spacing between the anodes. As far as practical, anode locations should be selected to cover the surface area of the steel structure but to be away from structural members that may reduce the current output. Note that the closest anode is problematic, especially in case of welding joints.

Shielding and interference effects become insignificant at a distance of about 0.5 m or more, except for very large anodes. Our design tends to be conservative, so that we can consider two adjacent anodes as one long anode, or as one wide anode, depending on their location in relation to each other, if it is expected the anodes will interfere with each other.

It is important to avoid locating anodes near welding or in areas with high fatigue loads. In addition, the location of anodes should take into account the restrictions from fabrication, installation and operation.

Sequential priority for anode placement is based upon the following considerations.

1. Anode distribution should normally begin with placement upon larger members, such as legs near nodes, and continue to minor members.
2. Placement should consider future added components, which are usually new conductors and risers as operations require.
3. Locations where attenuation effects will reduce good current distribution, such as well conductors and complex pile guides, will require a higher anode concentration.
4. Anodes are placed within a minimum distance of 150 mm of adjacent structural welds.
5. Anodes are placed within 100 mm of surfaces to be protected.
6. No anodes are to be located closer than 600 mm to tubular joints.

Proper anode placement should be performed as an iterative process, in which initial design drafts are adjusted to an optimum configuration, using the above guidelines. The placement process may result in conservative adjustments to anode dimensions in addition to the calculated total number to ensure adequate current distribution.

Anodes designed to protect the buried surface area are all installed on the bottom elevation of the structure. Figures 6.7 and 6.8 present examples of the distribution of anodes.

6.5 DESIGN EXAMPLE

Table 6.16 sets forth the steps of CP design for an offshore structure platform. The table can be useful in a spreadsheet for the design of a CP system.

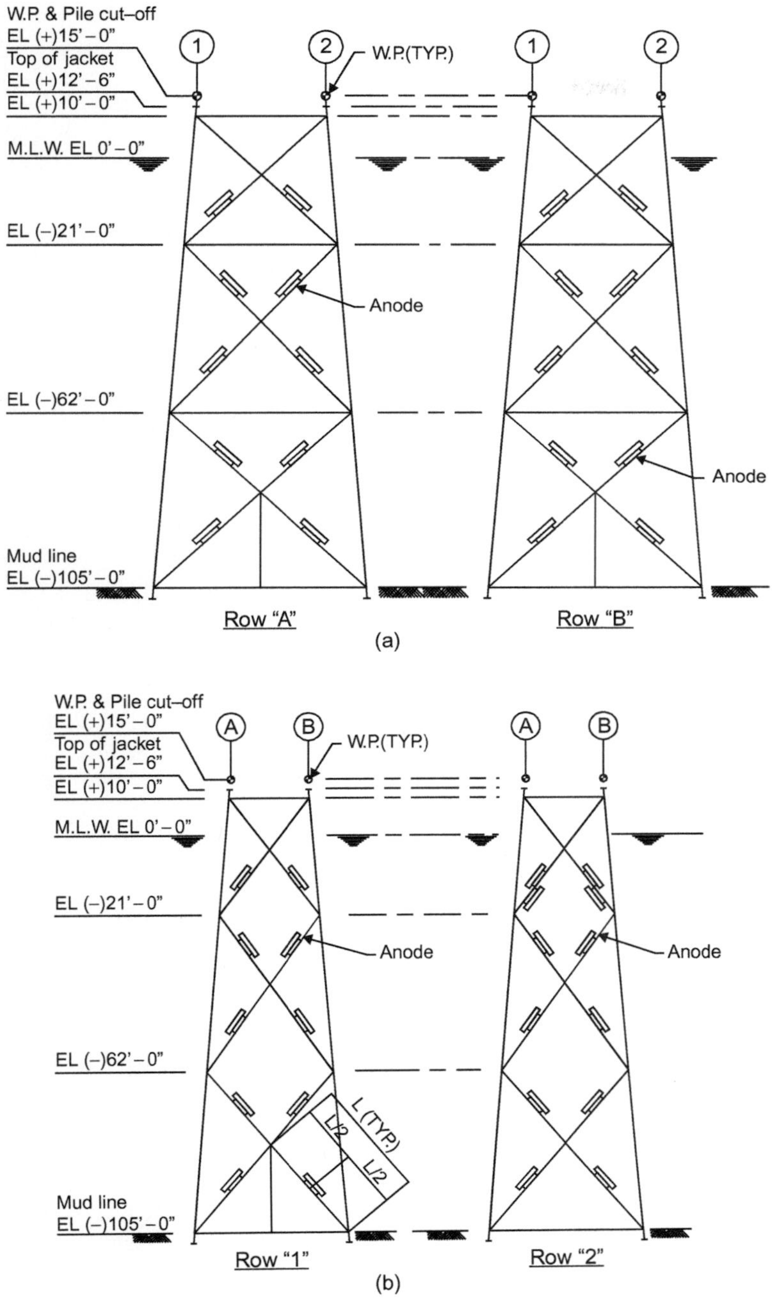

FIGURE 6.7 (a) Location of anodes on a 4-legged platform. (b) Location of anodes on a 4-legged platform for rows 1 and 2. (c) Location of anodes on a 4-legged platform in a different horizontal plan.

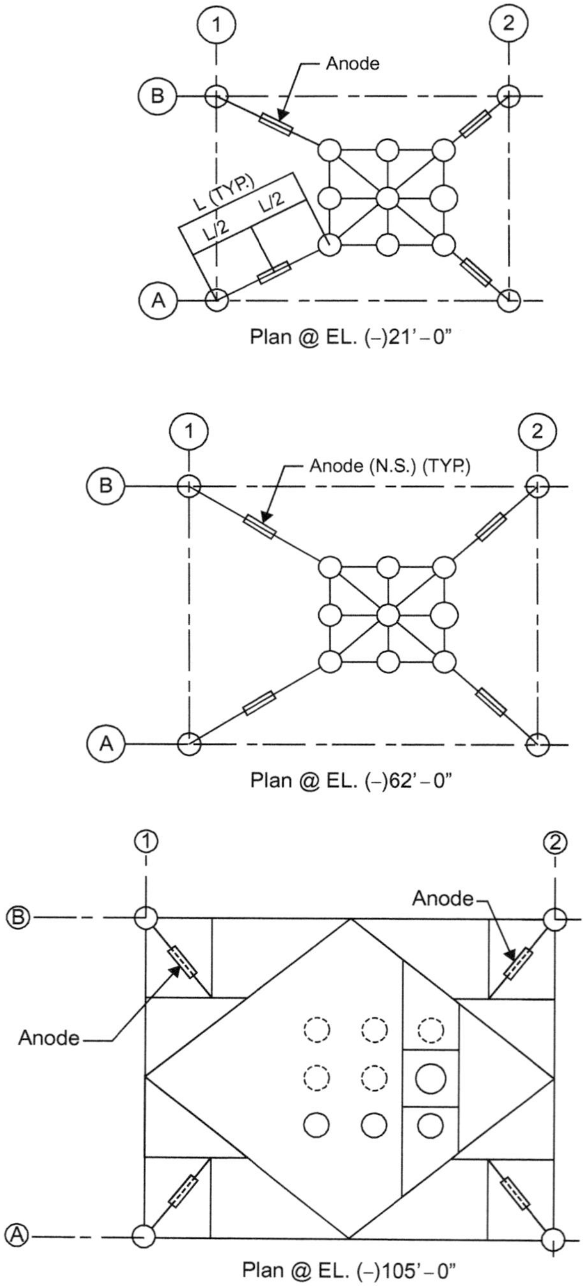

(c)

FIGURE 6.7 (*Continued*).

FIGURE 6.8 Locations of the anodes on the jacket during construction.

6.6 GENERAL DESIGN CONSIDERATIONS

Low-alloy steels with a minimum yield stress ranging from 275 to 450 MPa are specified for substructures for fixed offshore platforms. Special attention should be given to the use of higher-strength steels, particularly those with minimum yield stress above 400 MPa.

Sacrificial anodes are manufactured from an indium-activated aluminum alloy. The use of other aluminum activators has to be agreed to by the principal. Tin-activated aluminum alloys are not be used.

Mercury- or cadmium-activated aluminum alloys may be considered only if indium-activated aluminum ones are proven to be ineffective in providing protection. However, statutory regulations limiting the use of these alloys may apply.

The core material must be carbon steel and must comply with the specification requirements for tubular structure secondary members. The core diameter and wall thickness shall be agreed upon with the principal. Anode core attachment by welding is as specified by drawings and in conformance with the welding specification for the associated structure.

Anodes of the slender stand-off type should have a minimum stand-off, the closest distance between the aluminum alloy and the attached steel member, of 150 mm. Anodes should have a cylindrical or trapezoidal shape. The anode core should exit at the anode ends.

The sacrificial anode material should have a capacity greater than 3.6 kg/A/year (efficiency = 2420 Ah/kg). Anode material should have a potential in the range of −1.05 to −1.15 V versus Ag/AgCl/seawater. The number of anodes, their capacity and potential should be sufficient to meet both the polarization and current maintenance requirements for the steel over the design temperature range.

The closed-circuit anode potential (E_{oa}) used in design calculations for determination of indium-activated aluminum anode current output should be −1.05 V Ag/AgCl/seawater.

The objective of the design calculations is to compute by iteration the number of anodes and anode dimensions to verify the following:

- the initial current, final current, current capacity and current distribution requirements of the structure are fulfilled
- the lowest installed cost for the CP system, including cost of anode materials, molds, installation and coating systems (if used), is achieved.

To ensure adequate and efficient use of the anode material, the submerged structure should be divided into zones to be protected. Three zones are normally classified:

- Zone 1: This zone starts from mean sea level to just above the first horizontal member below it. The depth of this zone is either the trough depth of a 100-year extreme wave or 20 m, whichever is larger. The calcareous coatings of this zone are subjected to damage by storm waves.
- Zone 2: The zone below zone 1 and above sea bottom.
- Zone 3: The zone below the sea bottom mud line.

In zone 3, instead of calculating the surface area for well-casing strings extending beyond the end of the conductor, an allowance of 2.5–3 amperes per well should be provided. This allowance is provided in addition to the surface area contributions from other components, such as conductors and piles.

The external surface area of the buried part of piles and conductors is included in the calculations only up to a depth of 30 m into the seabed.

The current drain to open skirt piles, sumps and caissons should cover an internal surface area equivalent to 10 times π times their respective diameter.

The internal surfaces of all flooded structural members with unsealed flooding holes are included in the surface area calculations.

6.7 ANODE MANUFACTURE

Anodes are manufactured in accordance with the required specification and the supplier usually submits detailed anode and steelwork manufacturing drawings, complete with tolerances, as well as manufacturing and testing procedures for the client's approval.

The stand-off supports are formed from single lengths of carbon steel seamless tube or bar, bent using purpose-built bending equipment of proven performance to provide the minimum clearance specified between the centerline of the anode and the nearest jacket member. The tubular supports have their ends square to the anode axis and to each other within a tolerance of ±2 mm.

The bending method to be used for tubular inserts is proven to the purchaser's satisfaction to achieve consistent tolerances, with a maximum flattering (ovality) of 10% of the tube outer diameter. All production bends are checked for general dimensional tolerances and a minimum of 20% of bends are checked for flattering tolerance.

Inserts formed from flat steel bar are bent using reduced formers to produce bends of controlled radius. The supplier's proposed procedure is selected to ensure that the tolerances and acceptance criteria are fully complied with.

6.8 INSTALLATION OF ANODES

In the case of a new structure, the anodes are usually installed on the fixed offshore structure by welding in and by clamps. In the case of retrofitting, the anode supports the existing steel structure. When using the clamp type, electrical continuity is typically provided by a copper cable, attached to the anode support and the protection object by brazing, or by some special mechanical connection designed to ensure reliable electrical continuity. Electrical continuity cables may also be installed to provide electrical continuity to components of a CP unit without reliable electrical connection to anodes by welds, by metallic seals or by threaded couplings. The design requirements for such connections apply.

In the detailed design drawings of the CP system, it is important to consider the anode-fastening devices, which should be designed to be match the method of installation and operation of the offshore structure platform and any other special requirements.

Thermite welding is not recommended, but pin brazing may be considered.

The following information and any optional requirements (intended as a checklist) should be supplied by the purchaser:

- anode detailed drawings from CP design, or by manufacturer if completed
- the detailed drawing for CP design defining the location of individual anodes
- preparation requirements
- any special requirements for documentation
- any further amendments to, and deviations from, the order.

The contractor may specify that all work related to anode installation be described in an installation procedure specification. If applicable, this document should include, as a minimum, the following:

- materials specification and equipment to be used, including certificates and material data sheets
- receipt, handling and storage of anodes and other materials for anode installation requirement
- reference to welding and/or brazing procedure specifications and qualification of personnel carrying out welding/brazing
- anode fastening inspection and testing procedure.

All welding associated with anode installation should be qualified according to a recognized standard. Only qualified welders of brazing equipment may be used.

Installation of anodes is carried out according to the approved for construction drawings that will present locations of individual anodes and any other relevant specifications for fabrication of the protection object. To facilitate

FIGURE 6.9 Cow horn anode weighing 189 kg.

the installation, stand-off anodes to be mounted on structural members could be displaced laterally not more than one anode length and circumferentially a maximum of 30°.

For welding of anodes to structural members under high external loads, welded connections are placed at least 150 mm away from other welds, and a minimum of 600 mm away from structural nodes of jacket structures.

Installed anodes must be adequately protected during any subsequent coating work. Any spill of coating on anodes has to be removed. For coated objects, exposed anode cores are coated to the same standard.

Visual inspection of the welds and any brazed connections is the main inspection procedure for anode installation.

For welding to structural members, further nondestructive testing (NDT), such as radiographic or ultrasonic testing, may apply in accordance with the applicable fabrication specification.

Subsequent to completed anode installation, compliance with drawings for anode installation must be confirmed.

For brazed and mechanical connections for electrical continuity, measurements are carried out according to a documented procedure and with an instrument capable of verifying an electrical resistance of 0.1 ohm maximum.

Cow horn anodes are shown in Figure 6.9; a cow horn anode weighs 189 kg.

6.9 ALLOWABLE TOLERANCE FOR ANODE DIMENSIONS

In most cases, the anode tubular core shall be 80 mm, Schedule 80 pipe, extending for the full length of the anode casting, having a minimum stand-off height of 300 mm and being fabricated from ASTM 106 Grade B or similar material.

Welding is in accordance with AWS D1.1, Structural Welding Code. Steel anode cores manufactured from channel sections are not acceptable.

The anodes are located on the supporting structural members to minimize the impact of wave loads on the structure.

The weight of each anode should be within –0 to +5% surface defect tolerance.

Dimensional tolerances for the long, slender, stand-off anodes and flush-mounted anodes are:

- For anodes with tubular stand-off supports, the smaller of 2.5% of the anode's length or ±20 mm
- For flush-mounted anodes, the smaller of 2.5% of the anode's length or ±10 mm
- For all anodes, the smaller of 2.5% of the thickness, width or diameter or ±5 mm.

The straightness of each anode may not deviate by more than 2% from the longitudinal axis over the nominal anode length.

Bends on anode tubular supports are checked to ensure that flattering (ovality) of the pipe due to bending is within specified limits. Wrinkling of the inside of the bend is not acceptable.

The core should be centrally located within the casting length. A tolerance of ±5% on the nominal position of the core in the anode is permitted provided the stand-off distance meets the specified minimum anode/member clearance.

The stand-off support fixing centers (to the structure) must be within 20 mm of the nominal dimensions.

Each structural anode is weighed and should conform to the following weight tolerances:

- Net weight of individual anodes must be greater than 97% of the design net weight.
- Total anode net weight supplied may not be below the contract net weight.

6.9.1 Internal and External Inspection

Verification of anode weight and dimensions is carried out with the frequency and acceptance criteria specified in NACE RP 0387. The anode has to comply with tolerances in manufacturer's drawings, and a minimum of 10% of the anodes of a specific design are checked, based on the tolerance dimensions, as discussed before.

Note that, in any case, the anode surface should be generally free from defects affecting anode efficiency. Inspection for cracks and other surface irregularities is carried out on all anodes with the acceptance criteria as specified in NACE RP 0387; anodes exhibiting any of the following defects will be rejected.

- For zinc-based anodes, no cracks visible to the naked eye are acceptable.
- Cracks that are seen to penetrate into anode inserts are not acceptable.

- Shrinkage pits or cavities exceeding 10% of depth of anode thickness are not acceptable.
- Cracks exceeding 2 mm in width or greater than 150 mm in length and intersecting each other are not acceptable.
- Cracks exceeding 4 mm in width or 300 mm in length are not acceptable.
- Apparent slag or dirt inclusions are not allowed.
- All anodes must be free from slag, dross and any other inclusions having a particle size of 2 mm or greater.
- Surface irregularity on the top anode surface due to topping-up during solidification of the anode may not exceed ±5 mm of the nominal level. Topping-up may not exceed 0.5% of the total anode alloy volume and must be mechanically well bonded to the main body of the anode.
- Cold shuts or surface laps are not permitted.

Unless otherwise agreed, a minimum of two anodes of each size shall be subject to destructive testing to verify absence of internal defects and adequate location of anode inserts. The cutting procedure and acceptance criteria in NACE RP 0387 apply.

As a minimum, electrochemical testing is performed in first-day production tests for purchase orders exceeding 15,000 kg of net anode material and for each further 15,000 kg of production.

The anodes are transversely sectioned by single cuts at 25% and 50% of the nominal length. The cut faces, when examined visually without magnification, cannot contain more than:

- 0.25% of surface areas as gas holes or porosity, with individual pore size limited to 10 mm^2
- 20% of the insert perimeter with voids or porosity adjacent to the insert or any one face and no greater than an aggregate of 10% on all faces.

The cut faces are dimensionally checked to ensure compliance with the requirements of the core location. All anodes represented by a failed test are rejected.

BIBLIOGRAPHY

Ashworth, V., 1994. Corrosion. In: Shreir, L.L., Jarman, R.A., Burstein, G.T. (Eds.), Corrosion Control, 3rd ed, vol. 2. Butterworth-Heinemann, Oxford, p. 10.15.

ASTM A193/A193M-11 Standard Specification for Alloy-Steel and Stainless Steel Bolting for High Temperature or High Pressure Service and Other Special Purpose Applications, Book of standards Volume: 01.01. ASTM International, West Conshohocken, PA, USA.

ASTM A320/A320M-11 Standard Specification for Alloy-Steel and Stainless Steel Bolting for Low-Temperature Service. ASTM International, West Conshohocken, PA, USA.

British Standard BS EN 1706:1998 Aluminium and aluminum alloys. Castings. Chemical composition and mechanical properties.

British Standard BS4360:1990 Weldable structural steels. British Standards Institute, London.

Butler, G., Mercer, A.D., 1975. Corrosion of steel in potable waters. Nature 256, 719–720.

Det Norske Veritas (DNV), 1993. Recommended Practice, Cathodic Protection Design, DNV RP B401-1993.

Det Norske Veritas (DNV), 2005. Recommended Practice, Cathodic Protection Design, DNV RP B401-2005.

Humble, H.A., 1949. Corrosion, vol. 5, pp. 292–302. NACE, Houston, TX.

ISO 12944, 2007. Corrosion protection of steel structures by protective paint systems. International Organization for Standardization, Geneva, Switzerland.

ISO 9226, 1992. Determination of corrosion rate of standard specimens for the evaluation of corrosivity. International Organization for Standardization, Geneva, Switzerland.

Linder, B., August 1979. Magnetite anodes for impressed current cathodic protection. Materials Performance. Vol. 18: No. 8, p. 17. NACE International. Dallas, Texas.

UK Department of Energy. Design and operational guidance on cathodic protection of offshore structure subsea installation and pipeline, The Marine Mechnology Directorate Aberdeen, UK © MTD Ltd, 1990.

National Association of Corrosion Engineers (NACE) RP 0176-94 Corrosion control of steel fixed offshore platform associated with petroleum production, NACE, Houston, TX, 1994.

National Association of Corrosion Engineers (NACE) RP 0387-99 Metallurgical and inspection requirements for cast sacrificial anodes for offshore applications, NACE, Houston, TX, 1999.

Pourbaix, M., 1974. Atlas of Electrochemical Equilibria in Aqueous Solutions, NACE International, Houston, TX.

Rowlands, J.C., 1994. Sea Water Corrosion: Metal/Environment, 3rd ed. Butterworth-Heinemann, Oxford, pp. 2:60–2:71.

Smith, M.L., Weldon, C.P., 1997. Impressed current tensioned anode strings for offshore structures. NACE - CORROSION 97, Paper 97476. NACE International, Houston, TX.

Warne, M.A., August 1979. Precious metal anodes—Options for cathodic protection. Materials Performance. Vol. 18: No. 8. NACE International, Houston, TX.

Warne, M.A., Hayfield, P.C.S., March 1976. Platinized titanium anodes for use in cathodic protection. Materials Performance, Vol. 15: No. 3, Houston, TX.

Chapter 7

Assessment of Existing Structures and Repairs

7.1 INTRODUCTION

Worldwide, oil and gas offshore projects have the same problem: platforms are aging. The peak of oil exploration and development extended from the mid 1970s to the mid 1980s, especially in the Middle East, so platforms now have been aging for over 40 years. As stated in other chapters of this text, offshore structure platforms offer unique considerations in design, construction and applied loads.

When compared to traditional onshore structures, offshore structure platforms are very different. In this chapter, we focus on those differences. Because there are many steel structures, many researchers and many codes (i.e., ACI, BS, EC and Egyptian code, Indian code) worldwide, the methodology of assessment for traditional onshore structures can be very different from that for offshore structures. In addition, many steel structures have failed for different reasons, which provides us insight into evaluation and assessment of the structures.

Most of the offshore structure platform designs are performed according to API RP2A; however, API focuses its design on platforms for construction in the Gulf of Mexico (GoM), the area where construction of fixed offshore platforms was pioneered. (Around 1960, platform design was based on engineering offices' experience, and engineering firms proposed the API RP2A standard based on their experience. Since its initial development, API RP2A has undergone many revisions.) Because globally there are only a few thousand offshore platforms and the number of failed platforms is not great, there are only a few failed platforms (most of them due to fire) available for case studies, and there are a minimum number of engineering companies that have experience in realistic assessment of offshore platforms.

Before beginning the assessment process, it is important to highlight that structural analyses of old platforms were performed and the offshore structure has a special nature because the main forces affecting it are waves and wind, which change with climate change. This point is very critical and should be taken into consideration during the assessment process.

7.2 API RP2A: HISTORICAL BACKGROUND

The history of the API standard for offshore structures is a very important consideration in evaluating and assessing structures that have existed since the 1960s.

API was formed in 1919. API activity in the field of offshore structures was initiated with the formation of the Committee on Standardization of Offshore Structures. The committee's charge was to provide guidance that would avoid problems and that could serve as the basis for future regulation for offshore structures.

RP2A was the first publication of the committee, and when it was issued in 1969 it was only 16 pages long. API RP2A is now on its twentieth edition, and its alternative version, RP2A-LRFD, contains 224 pages. RP2A has always been recognized as an evolving document, and every edition has stated in the foreword that, as offshore knowledge continues to grow, recommended practice will be revised.

From the beginning, API RP2A was intended not as a complete specification, but as a supplement to other engineering codes and specifications. Furthermore, as more recent editions indicate, API RP2A is intended to supplement, rather than replace, individual engineering judgment. Development of API RP2A guidance is reviewed in the sections of this chapter for each of the main topics.

For systems considerations, until recently API RP2A has been primarily concerned with the design of every member separately. In the latest edition, a separate subsection is devoted to structural redundancy, and the guidance on framing patterns indicates consideration should be given to providing redundancy in the structure and the framing system; substitute load paths are to be preferred in design.

In the nineteenth edition (1991) of API RP2A, a section on the re-use of platforms was added, and in the twentieth edition (1993) a separate section on *minimum structures* was included. Minimum structures are defined as offshore structures that provide less redundancy than typical 4-legged template-type platforms or free-standing caisson platforms. API RP2A illustrated that the consequences of failure for minimum structures are usually lower than for conventional structures. Therefore, API suggests that minimum structures be checked by evaluating reserve strength and redundancy.

7.2.1 Environmental Loading Provisions

According to Oto (1999), one of the most important topics in API RP2A is the load effect due to environmental conditions; however, in early editions, the guidance on this topic was the most brief.

Morison's equation has always been recommended for evaluating horizontal wave force, and it has always been recognized that the appropriate drag and inertia coefficients depend in part on the wave theory being used. However, the API RP2A recommended values for drag and inertia coefficients have been revised over the years, with the most substantial change in the latest edition. It is important to know that the recommended value of the drag coefficient has changed from 0.5 in the early editions to 0.6, and it is now 1.05 for rough members.

In the first edition, no guidance was given for the most appropriate wave theory to use. Fifth-order Stokes wave theory and modified solitary wave theory were suggested by the third edition, and, as new wave theories became accepted, they were recognized in RP2A. The twentieth edition now contains detailed guidance on zones of applicability for wave theories, including stream functions, Stokes fifth-order and Airy (linear) wave theory.

API RP2A has always recommended that the environmental conditions data to which platforms will be designed should be selected by the owner. In 1986, the sixteenth edition suggested factors to be considered in the selection of the design conditions, including whether the platform is to be manned, the planned life of the platform and the cost of the platform, including estimated losses if the design criteria were to be exceeded. In the more recent editions, the intent is the same, although recognition of risk analysis has been added to justify the choice of the design event return period.

Prior to 1986 and the introduction of the sixteenth edition, RP2A did not state an explicit recurrence interval for the design conditions. Instead, it was suggested that the recurrence interval of the design event should be several times the planned life of the platform.

The Gulf of Mexico (GoM) has played a major part in offshore structures erected before 1970 as well as some unmanned satellite offshore platform structures. The regional guideline hydrodynamic parameters introduced in 1976 are believed to have been based on a 100-year recurrence period. A 100-year recurrence interval was also accepted and used for the reliability-based calibration of the LRFD version of RP2A, which began development in the late 1970s. However, not until the sixteenth edition (1986) did RP2A explicitly recommend that not less than the 100-year environmental design criteria should be considered where design events may occur while the platform is manned.

Morison's Equation

API RP2A suggests that Morison's equation (see Chapter 3) be used for evaluating the hydrodynamic loading on individual tubular members. Prior to the thirteenth edition (1982), the equation had only been written in terms of horizontal wave force; after this date, it was rewritten in terms of force normal to the axis of the member.

API RP2A always recognized that the appropriate drag and inertia coefficients (C_d and C_m) depend in part on the wave theory being used. In the first edition (1969), the recommended typical values for C_d ranged from 0.5 to 1.0, and for C_m from 1.3 to 1.7. By the third edition (1972), the recommended range for C_m was increased to 1.5–2.0.

In 1975, the sixth edition presented a revision to the recommended range of typical values for C_d; the lower value was increased so that the range became 0.6 to 1.0. These values remained unchanged until the sixteenth edition, which further revised the coefficients, and the range became 0.6 to 1.2 for C_d, and 1.3 to 2.0 for C_m.

However, when region-specific guidance was introduced in the seventh edition (1976), the recommended values for use in the GoM were 0.6 and 1.5 for C_d and C_m, respectively. In 1980, the eleventh edition was modified slightly to a constant drag coefficient of 0.6 and an inertia coefficient of 1.5 for members 1.8 m (6 ft) or less in diameter, increasing linearly to 2.0 for members 3 m (10 ft) in diameter and greater, and these values remained unchanged until the twentieth edition.

The twentieth edition (1993) provided the major revision to the environmental loading provisions, with significant changes in the recommended values of drag and inertia coefficients for use in conjunction with other factors (which are discussed below). For typical design situations, the following values are now recommended for unshielded tubular members: for smooth members, $C_d = 0.65$ and $C_m = 1.6$; for rough members, $C_d = 1.05$ and $C_m = 1.2$.

Wave Theories

The first edition of API RP2A gave no guidance for the most appropriate wave theory to use. Stokes fifth-order and modified solitary wave theory were suggested by the third edition, and as new wave theories became accepted, they were suggested in RP2A. The fifth edition (1974) illustrated Chappelear wave theory, and Dean's stream function and extended velocity potential wave theory were added in the seventh edition (1976).

The twentieth edition now contains detailed guidance on zones of applicability for wave theories, including stream functions of various order, Stokes fifth-order and Airy wave theory (Table 7.1).

Selection of Design Condition

As stated before, API RP2A has always recommended that the selection of environmental conditions to which platforms are designed should be made by the owner.

In the early editions, a classification system was suggested for platforms based on the class number, which was added to reflect the probability of the design condition being equaled or exceeded. Thus, a Class 100 platform would be designed to resist environmental loads that correspond to a 1% annual

TABLE 7.1 Member Wave Loading Parameters in Different API RP2A Editions

API RP2A Edition (Year)	Values of C_d	Values of C_m	Suggested Wave Theories
First and second (1969–1971)	0.5–1.0	1.3–1.7	No guidance
Third–fifth (1972–1974)	0.5–1.0	1.5–2.0	Stokes fifth-order and modified solitary; Chappelear added from fifth edition on
Sixth–fifteenth (1975–1985)	0.6–1.0	1.5–2.0	Stream function and extended velocity potential added from seventh edition on
Sixteenth–nineteenth (1986–1992)	0.6– 1.2	1.3–2.0	Stream function and extended velocity potential added from seventh edition on
Twentieth–(1993–2007)	Smooth/rough 0.65/1.05	Smooth/rough 1.6/1.2	Detailed guidance on zones of applicability

probability. Note that the class is defined in terms of platform environmental loading. Factors proposed for consideration in selecting the class of the structure, and hence the design condition, were:

Probability of personnel being on the platform
Prevention of possible pollution
Intended use of the platform
Planned life of the platform
Cost of the platform, giving consideration to both initial cost and estimated losses if the design criteria were exceeded.

In the first edition (1969), it was stated that economic risk evaluations for offshore platforms indicated that the optimum class number was several times the planned life in years. This design guide remained almost unchanged until the sixteenth edition (1986); before then, the only change was to remove the term class *number* in the fifth edition. In succeeding editions, the intent has remained the same, although explicit recognition of risk analysis has been added.

The definition of environmental loading for the extreme design condition in API RP2A (for use in U.S. waters) has always been based on a wave height of a given return period (usually 100 years) acting together with "associated" wind and "associated" current. However, in practice, no current was included in the design criteria for many of the platforms in the GoM. The regional guideline values have only recently been introduced in RP2A (twentieth edition).

The term "associated" has always been unclear. In the sixteenth edition (and subsequent editions) the following sentence was added without amplification: *Where there is sufficient knowledge of wave current joint probability, it may be used to advantage*. In the draft edition of RP2A-LRFD that was published in 1989, the definition for combined extreme wind, wave and current is discussed. The following three definitions are used:

1. 100-year wave + associated current and wind
2. 100-year platform load
3. 100-year wave + 100-year current + 100-year wind.

Offshore structures are affected by wave loading, and the definition preferred by API is to use the 100-year return period wave height with the *statistically expected* value of current and wind. It is noted that for structures whose extreme fluid loading is not dominated by waves, any "reasonable" combination of parameters leading to the 100-year return period load, such as base shear or overturning moment, may be used.

Using the 100-year return period wave height combined with the 100-year return period current speed and the 100-year return period wind speed is recognized as conservative, and for some structures, for example in areas of the GoM, too conservative.

Deck Clearance or Air Gap

The air gap is a very simple and very important factor that affects structure safety. Where waves strike a platform's deck, a large force affects the deck; however, it is only in the latest edition that general (nonregional) guidance is offered in RP2A.

The air gap was first discussed in RP2A in the seventh edition (1976), when regional guidance was introduced for the GoM. Then it was recommended that use of reference level wave heights for water depths greater than 200 feet (61 m) in the GoM *should result in a deck clearance elevation of at least 48 feet* (14.6 m) *above mean low water* (MLLW). This allowed for storm and astronomic tides, but it did not include an explicit safety margin.

When guidance for other regions in U.S. waters was introduced in the eleventh edition (1980), tabulated values of corresponding reference deck elevations were also included. The values included *an appropriate safety margin*, and were unchanged until 1986 when the sixteenth edition was introduced. Then, instead of providing values for minimum deck elevation, detailed guidance was provided for its evaluation; this included an explicit safety margin or air gap of at least 5 feet (1.5 m).

These recommendations still apply in the twentieth edition and the LRFD version; however, they are now considered part of the general provisions and not region-specific. In the latest editions it is also recommended that an additional air gap should be provided for any known or predicted long-term seafloor subsidence. For the GoM, explicit guideline values for deck height have been re-introduced: for deep water >500 feet (152.4 m), the recommended deck elevation remains 48 feet (14.6 m) above MLLW, but for shallower waters (depths of 100 feet or 30.5 m), it has been increased up to 53.5 feet (16.3 m).

The Latest Editions of RP2A WSD and LRFD

In general, the change to the loading guidance between subsequent editions of RP2A has been small. However, with the publication of the twentieth edition of RP2A in 1993 and the simultaneous introduction of the first edition of RP2A-LRFD, the environmental loading guidance has been extensively revised and updated. Several factors have been introduced or modified that are intended to produce more realistic estimates of platform hydrodynamic loading.

The changes include:

- Doppler effect of current on wave period, because a current in the wave direction tends to stretch the wavelength, making the wave less steep. Guidance is provided for the evaluation of an apparent wave period.
- Wave kinematics reduction factor, because "real" waves are not two-dimensional. The kinematics reduction factor is applied to the horizontal particle velocities and accelerations from two-dimensional wave theory in order to approximate the effects of directional spreading. For tropical storms, the factor given in RP2A is in the range 0.85 to 0.95, and for extra-tropical storms it is in the range 0.95 to 1.00. The commentary in API RP2A provides some further guidance on assessing the reduction factor and also discusses an irregularity factor to account for the nonsymmetric shape of real waves.
- Current blockage factor: current speed in the vicinity of the platform is reduced from the free stream value by "blockage," because the structure causes the incipient flow to diverge. The factor is in the range 0.7 to 1.0, depending on the "drag area" of the platform. The multiplying factor on current velocity is evaluated from:

$$1/(1 + (\text{hydrodynamic drag area}/4 \times \text{frontal area}))$$

- Conductor shielding factor: depending on the configuration and spacing of the well conductors, the wave force may be reduced due to hydrodynamic shielding. Guidance on the reduction factor to apply to the drag and inertia coefficients for the conductor array is given; the factor ranges from approximately 0.5 for very closely spaced conductors to 1.0 for spacing-to-diameter ratios greater than four.
- Current profile stretching: because the design current profile is specified only to the mean water level, the profile must be stretched to the local wave surface. "Nonlinear stretching" is preferred in RP2A, although linear stretching is considered to be a good approximation in some circumstances.

Together, these loading provisions provide a package that is intended to model platform hydrodynamic loading more realistically.

7.2.2 Regional Environmental Design Parameters

Some of the design parameters are the responsibility of the platform owner, as stated in API RP2A, and are usually requested by the engineering companies. RP2A did not contain any recommendations for regional design parameters until the seventh edition (1976), when guideline wave heights for the GoM were introduced. These values were referred to as the reference level, where it is really used as a guide only, along with guideline values for Morison coefficients, wave steepness (1/12) and other specific guidance (current was not mentioned). By the eleventh edition (1980), guideline wave parameters were provided for 10 areas in U.S. waters, covering offshore Alaska, California and the Atlantic coast. This has gradually been increased, and the twentieth edition covers 20 areas.

There are specialized companies and government authorities in most countries that can provide environmental data. The environmental data are obtained from mathematical hindcasting and predictive forecasting based on huge databases of environmental data (such as wind and wave measurements) and are supplemented with hurricane data taken from aircraft and free-floating buoys. It is understood that the peaks-over-threshold method is used to estimate significant wave height from 3-hour sea-state data. According to Forristall (1978), the short-term distribution of wave height within a sea state is assumed to be based on a 2-parameter Weibull distribution.

As illustrated in Table 7.2 for omnidirectional wave heights in the GoM, the guidelines were very uncertain to begin with, and, with time, the range of values has been narrowed.

API recognized that, based on specialist oceanographic and hydrodynamic advice, alternative metocean design parameters can be used; however, RP2A does not provide detail about a specific methodology to follow.

TABLE 7.2 Reference Wave Heights for the Gulf of Mexico

RP2A Edition (Year)	Reference Wave Height (m)	Range of Values (m)
First–sixth (1969–1975)	—	—
Seventh–fifteenth (1976–1984)	21.2	17.4–25.3
Sixteenth–nineteenth (1984–1991)	21.64	19.5–23.8
Twentieth (1993–)	20.73	Directional factors

7.2.3 Member Resistance Calculation

As discussed by OTO report index (1999), there were only been three occurrences of major changes to the member resistance equations in RP2A. The first was the introduction of member resistance formulations into RP2A in the sixth edition (1975). Before then, RP2A had recommended that the AISC provisions be used. Guidance for local buckling was introduced, along with provisions for hydrostatic pressure, and interaction equations for axial compression and bending stress and axial tension and hoop stress.

The second change was with the publication of the eleventh edition (1980). Equations were introduced for the assessment of allowable hoop stress, rather than using a design chart, and a formulation for the combined effects of axial compression, bending and hydrostatic pressure was introduced.

The third change occurred in 1987, with the introduction of the seventeenth edition, where a major change was made in the allowable bending stress, which was increased from 0.66 F_y to 0.75 F_y for members not susceptible to local buckling.

Finally, in 1993, the first edition of API RP2A-LRFD introduced some equations that had been modified. The modifications have not yet been introduced into the WSD version, and the member resistance formulations in the twentieth edition (introduced at the same time as LRFD) are identical to the those in the nineteenth edition.

7.2.4 Joint Strength Calculation

There have been ongoing changes in the recommendations for tubular joints due to continuous research and development. The first edition recommends that connections be designed to develop at least 50% of the effective strength of the members. However, in the first two editions of RP2A, no provisions for tubular joint strength were given except for a mention that the design and detailing of joints should be in accordance with good current practice. Not until the third edition (1972) was some specific guidance given for joint strength design. This was based on the punching shear concept, and it was very simplistic. The fourth edition (1972) introduced factors into the allowable stress formula to allow for the presence of load in the chord and the brace-to-chord diameter ratio.

However, not until late 1977, with the publication of the ninth edition, was allowance made in the allowable stress formulation for the joint configuration, such as T, X or K. A plastic reserve factor was also introduced to account for interaction between brace axial and bending stress; however, this term was later dropped in the fourteenth edition.

The fourteenth edition (1984) added a major change in joint design recommendations because new test data had proved that the previous API joint strength equations and guides were un-conservative. Therefore joint design based on the fourteenth edition is different than design based on the previous editions and earlier joint design is less conservative. In addition, the fourteenth edition provided a modified allowable stress formula and included a new and more realistic expression to account for the interaction effect of chord loads. The applied stress formula was also modified to be based on the nominal stress in the brace. The previous API formula is believed to have underestimated the calculated punching shear stress. Furthermore, an interaction equation was introduced to account for the combined effect of axial and bending stresses in the brace.

The fourteenth edition introduced the nominal load approach. The punching shear and the nominal load approaches are intended to give equivalent results. The nominal load approach was introduced because the punching shear approach does not always reflect the actual mode of failure.

Note that, according to OTO report index (1999), the fourteenth edition contained several printing errors, largely in the joint strength provisions, and it was withdrawn shortly after issue. In 1984, the fifteenth edition, which contained the corrections, was submitted.

The guidance has remained unchanged; however, in the twentieth edition (1993), further recommendations have been added on load transfer through the chord. Note that there is also a major change in joint strength calculation in API RP2A (2007).

7.2.5 Fatigue

In the early editions of RP2A, the guidance on fatigue is very simple and mentioned that, in the tubular joint design, consideration should be given to high stress, low cycle fatigue and brittle fracture, as well as due consideration should be given to the effects of salt water corrosion. Allowable design stresses for fatigue loading were suggested in AISC (1969), and by the fifth edition of RP2A, the fatigue provisions of AWS D1.1 (1972) were referenced. From the third edition, it was recommended that, based on a typical GoM storm history, and in lieu of a more rigorous analysis:

- Nominal brace stress due to design loading of a cyclic nature should not exceed 138 MPa at the joint.
- Simple joints are preferred for jacket leg joints.
- Complex joints should be detailed with smooth flowing lines.

In later editions, the above recommendations were modified slightly to form the basis for fatigue design. However, not until the eleventh edition (1980) were specific guidance on fatigue analysis and cumulative fatigue damage assessment provided, along with design curves of stress range versus predicted number of cycles (S-N curves). In 1986, in the seventeenth edition, API adjusted the fatigue guidance in light of new research on the effects of the weld profile and produced a lower fatigue design S-N curve. In addition, a simplified fatigue design approach was introduced to assess the allowable peak hot-spot stress as a function of water depth; the curves were calibrated for a typical GoM wave climate.

In 1993, the twentieth edition presented a modification to the peak hot-spot stress curves to account for the new wave force recipe. This edition also introduced a correction term to the S-N curves to account for the scale effect due to plate thickness.

7.2.6 Pile Foundation Design

The pile is very critical in assessing the existing platform because the structural analysis should include the geotechnical data. In some cases, geotechnical data for the nearest area are used as a guideline for performing the in-place analysis or the nonlinear analysis. Note that, for older platforms constructed from 1970 onward, the design depends on the practical wisdom and capability of the installation contractor. Pile foundation design guidance was initially limited in API RP2A. In the first edition (1969), safety factors were 1.5 and 2.0 for the extreme and operating conditions, respectively, and these have remained unchanged until the last edition.

Design guidance for the axial capacity of tubular piles in clay was initially based on engineering practice that had previously been followed for about 30 years in the GoM. This guidance was unchanged until the sixth edition (1975), when it was replaced by the so-called API method 2. This was a substantial change that led to a significant increase in design pile penetration. As stated in the OTO report, due to concerns raised by the projects, the previous method was reinstated in the seventh edition (1976) for highly plastic clays, such as those found in the GoM. API method 2 was categorized for use with other types of clay. The design guidance for clays remained almost unchanged until the seventeenth edition (1987), when it was completely changed and a new method was introduced. In the first two editions of RP2A, guideline bearing-capacity factors and soil friction angles for sands and silts were recommended for a limited range of soil types, along with limiting values. In the third edition, the limiting values were removed, and this guidance remained almost unchanged until the fifteenth edition was introduced in 1984. Following an extensive review of all the available test data, the guidance was changed extensively.

It is very important to highlight that one of the most significant changes was for piles under tension, where the earth pressure coefficient was increased from 0.5 to 1.0 for full displacement for plugged or closed-ended piles. Other changes were to expand the range of soil types covered by guideline parameters and to reintroduce limiting values on end-bearing and skin friction piles. The last edition of API RP2A (2007) presents more methodology for pile design using known CPT data, as discussed in detail in Chapter 4.

7.3 DEN/HSE GUIDANCE NOTES FOR FIXED OFFSHORE DESIGN

In 1974, the Den/HSE Guidance Notes were originally produced by the United Kingdom's Department of Energy (Den); they have since been re-issued by the Health and Safety Executive (HSE). The main purpose of the notes is to explain the Den/HSE view on procedures and requirements for offshore construction, installation and survey. For areas where no detailed guidance is given in the notes, other codes and design standards are referenced, including API and DNV.

New editions of the Guidance Notes have been published every five years or so, and the latest edition, the fourth, was published in 1990. However, amendments to the notes are issued as and when new information or guidance is available; seven amendments to the second edition and six amendments to the third edition have been issued. In addition, background documents are published on a frequent basis to supplement the guidance.

The notes contain guidance on a wide range of design considerations for offshore installations; some, like environmental loading, are covered in detail, while other areas are only mentioned.

In general, it is important to mention that HSE guidance for extreme environmental conditions usually matches with United Kingdom (UK) regulatory requirements, so it may be not useful in another global location. However, the guidance on environmental loading in the notes has evolved progressively to reflect new information and trends in design practice; more guidance and indicative values of wave height were introduced into the second edition in 1977. The guidance on environmental considerations is both descriptive and detailed, and the fourth edition in 1990 presents indicative values of most of the important environmental design parameters.

The latest guidance on environmental loading is based on widely used current practice, and for overall structural load it involves a package of interdependent assumptions that are considered to provide estimates of sufficient accuracy for design purposes.

7.3.1 Environmental Loading Provisions

The environmental loading sections in the Guidance Notes have evolved progressively to reflect new information and trends in design practice. In the latest edition, guidance is given on all types of environmental effects influencing platform loading, including: wind, waves, tides and currents, water depth, marine growth, air and sea temperature and snow and ice accretion.

In the 1st Edition of API RP2A (1969) much of the hydrodynamic guidance was discursive, but the 2nd Edition (1972) provided detailed guidance which was given together with indicative values of the main parameters. It is important to highlight that the guidance on environmental loading was virtually unchanged in the second edition (1972), but was completely changed and updated in the fourth edition. Therefore, in the assessment process it is necessary to define the year of design and the edition of the code that was followed.

The indicative values in the notes are intended for use only when site-specific measurements or other authoritative studies are not available. The data are generally presented as contours on maps covering the North Sea and northeast Atlantic. The contours are based on long-term measurements supplemented by mathematical modeling and do not take into account small-scale local effects.

7.3.2 Joint Strength Equations

Simple guidance on joint detailing was provided in the third edition (1972). When the fourth edition was published in 1990, detailed formulations for joint strength were introduced. These equations, produced following an extensive testing program and an assessment of worldwide test results, are based on the load approach, as opposed to punching shear approach. As in the contemporary API guidance, the equations are presented according to joint type and loading type, along with an interaction equation for combined load effects. However, in contrast to the API recommendations, the Den/HSE formulations are based on characteristic strength, which is the value at which not more than 5% of results in an infinite number of tests would fail. Also contrary to API practice, explicit safety factors are defined; 1.28 for extreme conditions, and 1.70 for operating conditions. The joint type classification system is similar to API's.

7.3.3 Fatigue

In the second edition (1971), there was an extensive guidance on fatigue analysis and assessment. This included detailed weld-type classification, a range of S-N curves which varies than AWS curves, and guidance on a number of factors that modify the standard S-N curves. These modification factors are used to account for unprotected joints in seawater and measures taken to improve the profile of welds. In the 1984 edition, included more guidance and recommendations about welding profile, and it introduced measures to account for the scale effect between the actual plate thickness and standard thickness for the S-N curves. The latter correction term has only recently been introduced into the API recommendations. In the fourth edition, Den/HSE introduced guidance on avoiding brittle fracture.

7.3.4 Foundations

Den/HSE includes some general guidance on factors to be considered when designing piled foundations, but for the site investigations guideline there is another document, which was published by HSE in 1986. The minimum safety factor for use with the extreme condition is recommended as 1.5. The 1984 edition suggested that the safety factor to be used should be chosen depending on the reliability of the soils data, load estimates, analytical methods and installation technique.

7.3.5 Definition of Design Condition

Prior to publication of the fourth edition, the extreme design condition in HSE guidance was based on a combination of the following minimum values:

- Maximum wave height, with average recurrence period of 50 years
- The maximum current
- The 1-minute mean wind speed, with an average recurrence period of 50 years.

In practice, these factors were assumed to be acting in the same direction. The Den/HSE notes also stated that any other combination of environmental factors that may cause greater stress either in the structure as a whole or in any element of the primary structure should be taken into account.

By 1990 and the introduction of the fourth edition, information on currents had improved somewhat and the definition of the minimum current for design was changed to current with an average 50-year return period.

In the fourth edition, the use of joint probability is recognized for assessing combinations of extreme parameters, but it is not believed to have been used in practice for North Sea platform design.

7.3.6 Currents

The term maximum current was not specified for the design condition in the second and third editions, other than it should take into account the tidal current, which is associated with surges, wind-generated currents and any currents due to other causes. In the North Sea in 1984, there was a shortage of measurement data in the database; in addition, the theory of near-surface currents and their variation with depth was not fully developed, and there were no maps giving information on extreme currents. However, a contour map of maximum tidal surface currents based on mean spring tides was presented; these data may have inadvertently been used for design in some cases.

In the fourth edition (1990), the guidance on currents was improved considerably, and the design event is now defined using the current with a minimum average return period of 50 years; this is evaluated using the mean spring tidal current and the 50-year return surge current. Contour maps are now presented for depth-averaged speed and direction of an average spring tide, and 50-year return period depth-averaged hourly-mean storm surge currents. Guidance on depth profiles is given, along with a discussion of other effects.

7.3.7 Wind

Note that the first edition presented the values for maximum 3-second gust speed and hourly-mean wind speed in the form of contour maps. The guidance on wind loading remained unchanged until the fourth edition, when it was extensively revised, and a new contour map of hourly-mean wind speed was introduced.

In this guideline, the average periods and a table of factors were provided to account for the effect of height above sea level.

7.3.8 Waves

In the second and third editions, contour maps of indicative 50-year wave heights and associated zero-crossing period were presented. The wave heights were defined as the most probable value of the height of the highest wave in a fully developed sea state lasting 12 hours. The maps were developed in 1977 by the Institute of Oceanographic Sciences, and the data were derived from instrumental field measurements and wind data.

In the fourth edition (1990), the definition of the design wave was fundamentally changed—instead of a 12-hour storm, the wave height was based on a 3-hour sea state. However, as a result of the simplified method used to derive the 1984 values, the overall effect on the values of the design wave largely cancelled out.

In the fourth edition, the indicative values of design wave height have been consistently evaluated and are generated from long-term data sets and mathematical models. However, it is recognized that the values are not definitive and that other values may be used where they can be justified.

In addition, in the fourth edition a table of factors is provided relating the 50-year significant wave height with individual wave height for a range of return periods from 50 years to 10,000 years. These factors are based on a sophisticated procedure that accounts for the fact that seas in the North Sea are not narrow-banded and allows for the possibility that the highest wave may not occur in the worst storm.

$$H_{\max,n}/H_s = kFq \tag{7.1}$$

where k is a factor defining the relationship between individual wave height and associated crest elevation c, $h = 2k\,c$. For HSE guidance purposes, it is recommended that $k = 0.9$. Note that q will be obtained by using the equation:

$$q = (1/2\log_e(n))^{1/2} + 1/16(1/2\log_e(n))^{-3/2} \tag{7.2}$$

It is worth mentioning that the product of k and F is very close to 1. Thus, the factors are very close to a commonly used approximation of the ratio between the most probable maximum height in a sea state and corresponding significant wave height, because F allows for the possibility of extreme waves occurring in less than extreme sea states. Wave height can be derived numerically from modeling of the total population of individual waves in a return period. In the North Sea, for a 50-year return period, F is around 1.12.

$$H_{\max,n}/H_s = (1/2\log_e(n))^{0.5} \tag{7.3}$$

Another change in the fourth edition introduced conversion factors to define a range for associated wave periods, rather than a contour map of values.

Design is intended to be based on the wave period within the recommended range producing the worst effect.

7.3.9 Deck Air Gap

In the early editions of the HSE code, no detailed guidance was given on air gap, other than where the second and third editions stated that the maximum crest elevation from the 50-year wave was required for calculating the clearance height of the superstructure.

However, the fourth edition has a subsection devoted to the topic. The notes state that the air gap should be based on an assessment of the probability of encountering extreme wave crests of return period greater than 50 years, but the air gap relative to the design extreme wave crest elevation should never be less than 15 m. Further guidance is given for structures with large-diameter members that diffract waves (D/L >1/5) and that may increase the maximum crest elevation.

7.3.10 Historical Review of Major North Sea Incidents

Knowledge of major offshore incidents serves a precautionary function in design. The Piper Alpha disaster had a major effect on the development of offshore design, particularly in the North Sea. The repercussions following any major incident inevitably lead to changes in design guidance and practice. Where fatalities occurred, as on Sea Gem, Alexander Kielland and Piper Alpha, the changes in practice have been far-reaching. Table 7.3 summarizes the major incidents.

TABLE 7.3 Major Incidents for Fixed Offshore Structures

Date	Platform	Type of Incident
1965	Sea Gem	Jack-up collapse due to brittle fracture; fatalities
1974	Auk A	Collision damage to platform
1976	Heather A	Jacket damaged by pile driving
1978	Transworld 58	Semi-sub fatigue damage to critical member
1980	Alexander Kielland	Semi-sub capsize because of loss of column resulting from fatigue damage; fatalities
1984	Ninian	Damage due to fabrication defect in primary structure
1984	Claymore	Damage due to fabrication defect in subsea brace
1988	Piper A	Platform collapse due to fire and explosion; 167 fatalities
2003	Various platforms	Platform damage and some failures in GoM due to Hurricane Lilli
2004	Temsah	Platform collapse and fire

7.4 HISTORICAL ASSESSMENT OF ENVIRONMENTAL LOADING DESIGN PRACTICE

Historically, and in present practice, environmental design loading is defined largely on the basis of owner company preference. Some companies take a conservative approach when choosing environmental design parameters and use 100-year events as the basis for the design condition, while others take a less onerous approach.

Until now, there has been no standard for environmental design parameters, the methods of design and the analysis procedure because the wave heights considered in design vary even in adjacent areas and with time. In August 2005, hurricane Katrina caused GoM wave heights that were the highest in 100 years. Because of this, the maximum wave height in designs was adjusted accordingly. In September 2005, hurricane Rita also caused record-setting wave heights, resulting in the designs, being adjusted again.

Different approaches are also taken in the choice of design current velocity and profile, drag coefficient and other factors.

7.4.1 Environmental Parameters for Structure Design

In practice, the extreme wave height and current used in the design of every fixed offshore platform in the North Sea (NS) have usually been obtained by using historical data from nearby sites.

While there is no standard procedure that is followed by all companies for establishing design NS wave heights, a common method is to fit a cumulative probability distribution for H_s to long-term field measurements, and the short-term distribution of wave heights in a sea state is then often assumed to follow a Rayleigh distribution. In some cases, the resulting wave heights have then been arbitrarily adjusted upward for design purposes.

Although the wave climate has not changed, the estimates of design environmental criteria have changed over the years. There are several reasons for this, and one main reason is that additional years of data measurement are available. Recording of NS wave measurements did not really begin until the early 1970s. Therefore, early platforms had to rely on design parameters extrapolated from a very limited number of years of data, and this affected the reliability of the platforms. Clearly, as time has gone on, more years of data have been accumulated, and the statistical uncertainty in deriving design parameters has been reduced. In addition, data have been recorded at more sites, and thus much of the uncertainty associated with interpolating values for new sites has been reduced.

For wave heights, the method of measurement and data recording and analysis have improved over the years, and the definition of significant wave height has been refined. Up until the mid 1970s, waves were usually measured by special equipment and the data were recorded for 20 minutes every 3 hours on paper chart recorders.

As discussed in Chapter 3, significant wave height was defined as the mean of the third highest waves. From the mid to late 1970s, analogue magnetic tapes were used more widely, and the data were processed to derive significant wave heights defined using the root mean square (rms) of the sea-state wave heights ($H=4\ H_{nns}$). In the early 1980s, the definition of significant wave height was based on the spectral moment ($H=4(\text{mo})^{05}$).

Methods of data reduction and statistical analysis have improved considerably from the manual methods of the early 1970s to the sophisticated computations undertaken today. In the early years, it was not unusual for only one or two distribution types to be considered and a best-fit line fitted by eye; now a wide range of distribution types are considered, a variety of data reduction methods are used, a number of sophisticated fitting methods are employed and significance or goodness-of-fit tests are undertaken.

Wave history is usually available from the government authority concerned with metocean data. In the UK, data recording environmental parameters for 25 years of winter storms (covering the period 1964–1989) are available on a 1-km grid covering the North Sea and the northeast Atlantic Ocean.

7.4.2 Fluid Loading Analysis

In the early years of North Sea development, only Airy and Stokes third- and fifth-order wave theories were available to designers for practical use. All theories were used for their simplicity, particularly for appurtenance design, but as design became more sophisticated, Stokes theories were used more widely, and today Stokes fifth-order wave theory is the most widely used theory for design.

Dean (1965) followed work primarily by Airy, and Stokes wave theories were shown to inaccurately predict wave particle kinematics for some combinations of water depth and wave period. The selection of the most suitable wave theory was improved following work by Le Mehaute (1976) and Barltrop et al. (1990). With advances in computing power, stream theory has become more popular, particularly in shallower water and for waves close to breaking. Stream function theory up to the eleventh order, as developed by Dean in 1965, or Chappelear velocity potential is now used.

All of the above are regular wave theories. Tromans et al. (1994) proved a theory based on random waves. A new wave theory has recently been developed by, and is being used by, the Shell company. It is said to model the irregular nature of real seas better.

Along with advances in wave theories, research and development have increased the speed of the computer software used to assess structural loading.

It is very important to emphasize that marine growth was not usually considered in design before the late 1970s. While the effects of marine growth on platform loading have usually been considered in the design of North Sea structures from at least the late 1970s, guidance on marine growth was included in RP2A only in 1993, with the introduction of the new loading provisions in the twentieth edition. Detailed guidance on marine growth is now given in the fourth edition of Den/HSE notes, but only descriptive guidance was given in the 1977 edition.

Due to increased knowledge and experience, the effect of breaking waves, wave slam and slap can now be assessed.

Techniques have improved for allowing for wave-current interaction and for stretching the current profile to the instantaneous wave surface. The so-called mass continuity method was widely used until the 1980s, but this assumption can be shown to be un-conservative. Until the mid 1980s, highly sheared current profiles were assumed for design; today, a more uniform slab profile is generally preferred for the extreme design conditions. In the twentieth edition of API RP2A, the Doppler effect of the current on wave period is included.

Another main area of change has been the hydrodynamic coefficients used in Morison's equation. Considerable research has been done over the years to quantify the coefficients. A variety of tests in wave flumes, involving multiple and inclined members, and full-scale test structures have been undertaken. The use of these data by operators in practice cannot be easily quantified.

The following Morison's coefficients are based on information from one major North Sea operator and may be considered indicative of general practice:

Up to 1971	$C_d = 0.5$	$C_m = 1.5$
1972–1980	$C_d = 0.6$	$C_m = 1.8$
From 1981	$C_d = 0.7$	$C_m = 2.0$

7.5 DEVELOPMENT OF API RP2A MEMBER RESISTANCE EQUATIONS

According to OTO report index (1999), every edition of API RP2A-WSD has included the following text (with minor modifications to the wording) referring to AISC for cases not covered by RP2A:

Unless otherwise recommended the platform should be designed so that all members are proportioned for basic allowable stresses specified by the AISC Specification for the design. Where the structural element or type of loading is not covered by this recommended practice or by AISC, a rational analysis should be used to determine the basic allowable stresses with factors of safety equal to those given by this recommended practice or by American Institute of Steel Construction AISC.

The earliest editions of RP2A did not contain any recommendations for allowable stresses and relied totally on the AISC or *rational analysis*. This reliance has been eroded over time, with more and more guidance being given in RP2A. The later editions of API RP2A-WSD warn that the AISC load and resistance factor design code is not recommended for design of offshore platforms.

It is worth mentioning that API RP2A-LRFD does not mention AISC in the main body of the code, but gives guidance on the use of AISC-LRFD in the commentary for nontubular members.

In every edition of API RP2A-WSD, the wording has been unchanged regarding increasing the allowable stresses for extreme environmental conditions:

The required section properties computed on this basis should not be less than required for design dead and live loads but in case of considering the effect of the lateral and vertical forces imposed by the design environmental conditions, the basic AISC allowable stresses may be increased by one-third.

7.6 ALLOWABLE STRESSES FOR CYLINDRICAL MEMBERS

7.6.1 Axial Tension

API RP2A did not contain a recommended expression for an axial tensile check until one was introduced in 1987 in the seventeenth edition. Until then, reliance was on the AISC. In 1969, as now, AISC recommended that the allowable tensile stress be given by the following equation, as presented in Chapter 3:

$$F_t = 0.6\,F_y \tag{7.3}$$

In the seventeenth edition, API RP2A-WSD adopted this recommendation explicitly.

7.6.2 Axial Compression

Unstiffened tubular members under axial compression are subject to the following three failure modes:

- material yield
- Euler column (overall) buckling
- local buckling.

In general, members of low *D/t* ratio are not subject to local buckling under axial compression. All editions of RP2A recommend that unstiffened tubular members be investigated for local buckling once the member *D/t* ratio exceeds a limiting value.

The first three editions of RP2A (that is, until 1972), recommended that local buckling be investigated for $D/t > E/12\,F_y$. For example, for $F_y = 345$ MPa, the limiting value is about 50.

The fourth through tenth editions of RP2A (that is, until 1979), recommended that local buckling be investigated for $D/t > 22{,}750/F_y$ (F_y in MPa). For example, for $F_y = 345$ MPa, the limiting value is about 66. In later editions, the limiting value is 60.

No further guidance on local buckling stress or allowable compressive stress was given in the first five editions of RP2A. AISC also did not contain any guidance on local buckling for tubular members until 1978, when the eighth edition of AISC was introduced, but AISC did contain formulae for overall buckling, as presented in Chapter 3 in Equations (3.81) and (3.82).

In 1975, in the sixth edition, API recommended that, where the limiting *D/t* ratio was exceeded, the allowable axial compression and bending stress for an unstiffened member should be determined by substituting a reduced yield stress F_{yr} for F_y in the appropriate AISC design formulae, where:

$$F_{yr} = \left[1 - \left(1 - \frac{22750/F_y}{D/t}\right)^2\right] F_y \tag{7.4}$$

In the eleventh edition (1980), where the *D/t* exceeded 60, expressions were provided for elastic and inelastic local buckling. The expressions were valid for *D/t* ratios up to 300, and for $t = 6$ mm. The elastic local buckling stress was calculated by:

$$F_{xe} = 0.6\,E\,t/D \tag{7.5}$$

In the twelfth edition, the constant 0.6 was replaced by the term $2C$, where C was recommended to be 0.3, thus giving the same result as in the eleventh edition.

$$F_{xc} = [1.64 - 0.23(D/t)1/4]F_y \tag{7.6}$$

It was then recommended that the minimum of $\{F_y$, F_{xe} and $F_{xc}\}$ should be substituted for F_y in the expressions for evaluating the allowable compressive stress.

Until the seventeenth edition (1987), API RP2A did not contain an expression for an allowable compressive stress; instead, the AISC was referenced. In the seventeenth edition, RP2A introduced the equation from AISC for the allowable compressive stress that is presented in Chapter 3 as Equations (3.81) and (3.82).

7.6.3 Bending

In the case of pure bending, the failure of a tubular member is usually precipitated by localized axisymmetric bulges on the compression side of the cylinder. As with local buckling for axial compression, the buckling behavior depends on the *D/t* ratio; at larger *D/t* ratios, both the moment and the rotational capacities of the tube decrease.

In the first five editions, RP2A did not contain any expressions for an allowable bending stress, and allowable stress for bending was treated very simply. AISC also did not contain any explicit guidance for tubular members until

1978; however, for noncompact members, the allowable bending stress was given as $F_b = 0.6\ F_y$.

In 1975, the sixth edition of API RP2A gave the allowable bending stress as (for F_y in MPa):

$$F_b = 0.66\,F_y \quad \text{for} \quad D/t \leq 22{,}750/F_y \tag{7.7a}$$

$$F_b = 0.66\,F_{yr} \quad \text{for} \quad D/t > 22{,}750/F_y \tag{7.7b}$$

where F_{yr} is the reduced yield stress as given in Equation (7.4).

In the eleventh edition (1980), F_{yr} was replaced by F_{xc}, and the limiting value was changed to 60, i.e., the high rotational capacity of the structure provides a ductile failure mode exhibiting very gradual load decay and the intermediate rotational capacity provides a semi-ductile failure mode.

$$F_b = 0.66\,F_y \quad \text{for} \quad D/t \leq 60 \tag{7.8a}$$

$$F_b = 0.66\,F_{xc} \quad \text{for} \quad D/t > 60 \tag{7.8b}$$

The seventeenth edition (1987) substantially changed the allowable bending stress from a maximum value of 0.66 F_y to 0.75 F_y, and allowance for local buckling was formulated explicitly in the expressions. Formulations for three regions are now provided in RP2A; they can be classified according to rotational capacity:

- high rotational capacity; ductile failure mode exhibiting very gradual load decay;
- intermediate rotational capacity; semi-ductile failure mode;
- low rotational capacity; little post-yield ductility, susceptible to local buckling and rapid load decay.

Because an unstiffened tubular member with a very high *D/t* ratio is unlikely to be practical in offshore structures, in the eighteenth edition the upper limit for *D/t* was changed to 300.

7.6.4 Shear

The sixth edition of API RP2A recommended for the first time that, for tubular members, the applied beam shear stress should be evaluated using one half of the gross cross-sectional area. However, not until the seventeenth edition was the allowable shear stress recommended by AISC introduced, as presented in Chapter 3. The seventeenth edition also introduced an expression for the applied torsional shear stress (for torsional shear calculation, see Chapter 3).

In the eleventh edition (1980), the equation for acting hoop stress from hydrostatic pressure was modified slightly by replacing the constant by ρ, where ρ is the density of seawater.

7.6.5 Hydrostatic Pressure

Unstiffened tubular members under hydrostatic pressure are subject to local buckling of the shell wall anywhere between restraints. The effect of pressure on the tube is magnified by any initial geometric imperfection or out-of-roundness. For closed-end tubular members, such as braces, hydrostatic pressure also imposes an axial compressive stress of 0.5 fh, some of which is taken by the structure and some of which passes into the member; the treatment of this force has always been a little unclear in RP2A, although the more recent editions are more definitive.

No provisions were included for hydrostatic pressure until the sixth edition (1975). Then the acting hoop stress from hydrostatic pressure was given as in Equation (3.55).

The allowable hoop stress was evaluated using a design chart. The L/D ratio and *D/t* ratio were used to derive the critical hoop strain, and the ultimate hoop stress was read from the chart using F_{ys}, which, in the absence of interaction with axial stress, was taken as F_y. For combined axial and hydrostatic stresses, F_{ys} was derived from the chart for use later. The allowable hoop stress was obtained from the ultimate value by dividing by an appropriate safety factor; suggested safety factors were given as:

1.25–1.5 Where the one-third increase in allowable stresses is appropriate, as when considering interaction with storm loads.

1.67– 2.0 Where the basic allowable stress would be used, as pressures which will definitely be encountered during the installation or life of the structure.

In the eleventh edition (1980), the equation for acting hoop stress from hydrostatic pressure was modified slightly by replacing the constant by $\gamma/2$, where γ is the density of seawater. Also, at this time API introduced a new method for evaluating the allowable hoop stress. The method is based on a critical hoop buckling stress derived according to whether the collapse mode is elastic or inelastic. The allowable hoop stress is the critical hoop stress divided by the safety factor, which was hardened in the eleventh edition to become 1.5 for the extreme condition and 2.0 for other conditions.

In the eleventh edition (1980), the critical hoop buckling stress was defined as:

$$F_{hc} = F_{he} \quad \text{for} \quad F_{he} \leq 0.667\,F_y \tag{7.9}$$

$$F_{hc} = 2.53\,F_y/(2.29 + (F_y/F_{he})) \quad \text{for} \quad 0.667\,F_y < F_{he} < 4.2\,F_y \tag{7.10}$$

$$F_{hc} = F_y \quad \text{for} \quad F_{he} \geq 4.2\,F_y \tag{7.11}$$

Elastic buckling is described as in Chapter 3 about critical. The formulations to evaluate F_{hc} for inelastic buckling are presented by the following equations, which were modified in the LRFD version issued in 1993.

$$F_{hc} = 0.45\,F_y + 0.18F_{he} \quad \text{for} \quad 0.55F_y < F_{he}$$
$$F_{hc} = 1.31\,F_y/(1.15 + F_y/F_{he}) \quad \text{for} \quad 1.6F_y < F_{he} \le 6.2F_y$$
$$F_{hc} = F_y \quad \text{for} \quad F_{he} > 6.2F_y$$

7.6.6 Combined Axial Tension and Bending

Prior to 1987, RP2A did not contain any provisions for combined axial tension and bending; however, AISC contained the following equation:

$$\frac{f_a}{0.6F_y} + \frac{f_{bx}}{F_{bx}} + \frac{f_{by}}{F_{by}} \le 1.0 \tag{7.12}$$

In 1987, the seventeenth edition introduced an interaction equation for combined axial tension and bending for the first time. The equation was simply a modification of the AISC equation to allow for the circular cross-sectional shape:

$$\frac{f_a}{0.6F_y} + \frac{\sqrt{f_{bx}^2 + f_{by}^2}}{F_b} \le 1.0 \tag{7.13}$$

The interaction equation (Equation 7.13) has been modified in the LRFD version.

7.6.7 Combined Axial Compression and Bending

API RP2A did not contain any provisions for combined axial compression and bending load effects until the sixth edition was introduced in 1975. The eighth edition of AISC contained the following two interaction equations, one to verify member stability and the other for plasticity. Where the axial component was small (≤ 0.15) an alternative equation was suggested. Buckling check:

$$\frac{f_a}{F_a} + \frac{C_{mx} f_{bx}}{(1 - f_a/F'_{ex})F_{bx}} + \frac{C_{my} f_{by}}{(1 - f_a/F'_{ey})F_{by}} \le 1.0 \tag{7.14}$$

and Strength check:

$$\frac{f_a}{0.6F_y} + \frac{f_{bx}}{F_{bx}} + \frac{f_{by}}{F_{by}} \le 1.0 \tag{7.15}$$

where

$$F_e = \frac{12\pi^2 E}{23(Kl/r)^2}$$

The reduction factors C_{mx} and C_{my} depend on the support conditions of the member, end moments, and whether transverse loading is applied: the factors take a value between 0.4 and 1.0.

In the case of circular section, such as pipe, these interaction equations for circular cross-sections were adopted as presented in Chapter 3, Equations (3.94) and (3.95).

When $f_a/F_a \leq 0.15$, two equations for check of buckling and strength will be used, as discussed in Chapter 3, Equation (3.86).

Guidance was given on the evaluation of the reduction factor C_m for different types of jacket and deck components, and it has remained unchanged. However, in the eleventh edition (1980), allowance was made for having different C_m and F_e values about the x and y axes, as in Equation (3.87).

The interaction Equations (3.86) and (3.87) have been modified in the LRFD version issued in 1993.

7.6.8 Combined Axial Tension and Hydrostatic Pressure

Combined axial tension and hydrostatic pressure was originally called *tension and collapse interaction* when it was introduced in the sixth edition. The terminology is now slightly misleading since the formulation also applies to bending. In the sixth edition, the following interaction equation was introduced:

$$(F_{ys}/F_y)^2 + (F_{ys}/F_y)A + A^2 \leq 1.0 \tag{7.16}$$

where

$$A = \frac{f_a + f_b - 0.5 f_h}{0.6 F_y} \tag{7.17}$$

F_{ys} was read from the design chart. It was stated that, under cyclic loads, tension and collapse interaction need not be investigated where both of the following apply: SF ≥ 2 for collapse alone and $f_a + f_b \leq 138$ Mpa.

The eleventh edition introduced an interaction equation based on the maximum strain energy theory for biaxial loading by Beltrami and Haigh, as in Equations (3.98), (3.99) and (3.100). The interaction equation has been modified in the API RP2A-LRFD introduced in 1993.

In API RP2A, the safety factors, SF, are now tabulated as in Table 7.4; however, it was not until the seventeenth edition that a separate safety factor for bending was introduced.

7.6.9 Combined Axial Compression and Hydrostatic Pressure

Combined axial compression and hydrostatic pressure was originally called compression and collapse. Again, when interaction was introduced in the

TABLE 7.4 API RP2A Safety Factors for Use with Hydrostatic Interaction, from Seventeenth Edition (1987)

	Loading			
Design Condition	Axial Tension	Bending	Axial Compression	Hoop Compression
1. Where the basic allowable stresses would be used (e.g., pressures that will definitely be encountered during the installation or life of the structure)	1.67	F_y/F_b	1.67–2.0	2.0
2. Where the one-third increase in allowable stresses is appropriate (e.g., when considering interaction with storm loads)	1.25	$F_y/1.33F_b$	1.25–1.5	1.5

The value used should not be less than the AISC safety factor for column buckling under axial compression.

eleventh edition, the terminology became slightly misleading, since bending is included in the equations. In 1980, three criteria had to be satisfied:

$$\frac{f_x - 0.5F_{ha}}{F_{aa} - 0.5F_{ha}} + \left(\frac{f_h}{F_{ha}}\right)^2 \leq 1.0 \tag{7.18}$$

$$\frac{f_x}{F_{xc}} SF_x \leq 1.0 \tag{7.19}$$

$$\frac{f_h}{F_{hc}} SF_h \leq 1.0 \tag{7.20}$$

where

$$f_x = f_a + f_b + (0.5f_h) > 0.5F_{ha} \tag{7.21}$$

In the seventeenth edition, the change in the safety factor for bending was allowed for and the second criterion was replaced by:

$$\frac{f_x + (0.5f_h)}{F_{xc}} SF_x + \frac{f_b}{F_y} SF_b \leq 1.0 \tag{7.22}$$

In addition, the seventeenth edition recognized that if $f_b > f_a + 0.5f_h$, the combined axial tension, bending and hydrostatic pressure criteria should also be satisfied.

7.6.10 AISC Historical Background

According to Brockenbrough (2003), the welding specifications based on AISC have been changed as shown in Table 7.5.

7.6.11 Pile Design Historical Background

Pile design depends on the configuration of the platform, the water depth and the construction method on site at the time of execution.

Because water depth has the most effect on pile design, Figure 7.1 presents the platform numbers in the North Sea before 1994, and the figure shows how the number of platforms with a water depth over 100 m (300 ft) increased rapidly after 1974.

From Tables 7.2 and 7.3, the effect of the limiting friction and end-bearing values can be seen at around 35 m for both the first and fifteenth editions for API RP2A. For compressive capacity, there is very little difference between the ultimate capacities predicted by the first edition and those predicted by the fifteenth edition; the only real change for dense sand is in the value of the lateral

TABLE 7.5 Historical AISC Allowable Stresses (N/mm^2) for Weld in ASD

Year	Steels and Welding Materials	Fillet Weld Shear	Tension	Compression
1934	A7/A9 steel	78	89.6	103.4
1939	A7/A9 steel	78	89.6	124.1
1946	A7/A9 steel: 60XX electrodes	93.8	137.9	137.9
1961, 1963	All steels: 60XX electrodes or subarc grade SAW-1. A7 and A373 steels: 70XX or subarc grade SAW-2. A36, A42, A441 steels: 70XX or subarc grade SAW-2.	93.8 93.8 108.9	Same as member all cases	Same as member all cases
1969	All steels and weld processes[1]	$0.3\ F_{uw}$	Same as member all cases	Same as member all cases
1989	No significant changes	$0.3\ F_{uw}$	Same as member all cases	Same as member all cases

Note that, for butt weld, the shear strength is 93 N/mm^2
Electrodes and matching base metals are defined. Allowable shear stress is 0.30 times nominal tensile strength of weld metal. 0.30 F_{uw} in supplement 3, 1974, permitted weld metal with a strength level equal to or less than matching base metal, except for tension members.

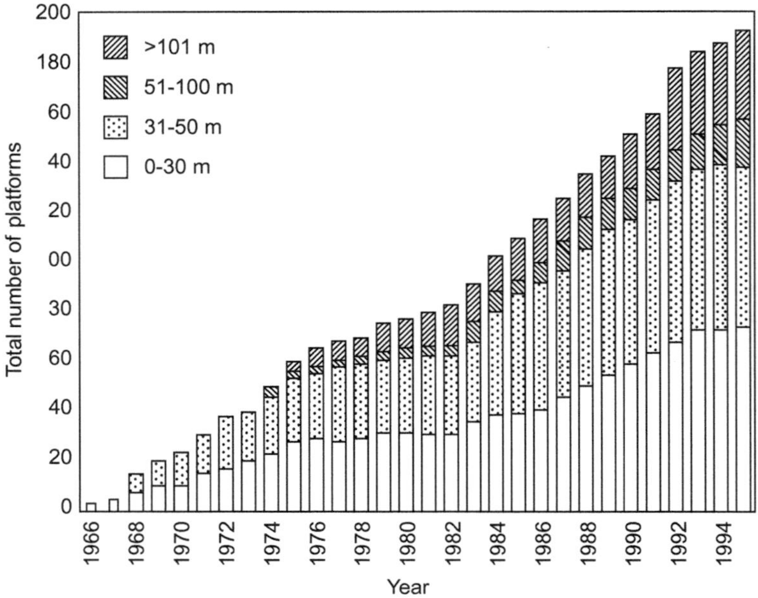

FIGURE 7.1 Bar chart of development with water depth.

earth pressure coefficient, K, which changed from 0.7 to 0.8. For the third to tenth editions (1972 to 1984), there is a reduction in ultimate compressive capacity of up to 30% for a given pile length if K is taken as 0.5.

The ultimate tensile capacity predicted by the recommendations in the first edition is almost identical to that in the third edition. However, the ultimate tensile capacity has been increased by up to 50% for the same pile length according to the fifteenth edition. In terms of pile length for a given pile load, the fifteenth edition leads to a reduction of around 7 m for this example.

Table 7.6 presents the ultimate capacities for a pile with a diameter of 1 m and a depth of 30 m in homogeneous dense sand.

Table 7.7 presents the ultimate capacity for 30 m penetration in homogenous normally consolidated clay.

This table and Figure 7.4 illustrate the significant reduction in ultimate pile capacity for guidance in the sixth edition, particularly at deeper penetrations. The seventeenth edition guidance has led to a reduction in capacity of around 25% from the guidance introduced in the seventh edition for a given pile length. In terms of pile length for a given capacity, the seventeenth edition leads to an increase of around 9 m for this example.

Note that there is no longer a difference between the pile ultimate tensile and compressive capacity for different editions in normally consolidated clay.

TABLE 7.6 Change in Ultimate Capacity for a 30-m Penetration in Dense Sand

RP2A Edition (year)	Ultimate Compressive Capacity (MN)	Ultimate Tensile Capacity (MN)
First–second (1969–1971)	11.5	3.8
Third–fourteenth (1972–1984)	9.0	3.8
Fifteenth–twentieth (1984–)	12.4	5.7

TABLE 7.7 Change in Ultimate Capacity for a 30-m Penetration in Clay Soil

RP2A Edition (year)	Ultimate Compressive Capacity (MN)	Ultimate Tensile Capacity (MN)
First–fifth (1969–1974)	5.06	4.64
Sixth (1975)	3.53	3.10
Seventh–sixteenth (1976–1987)	6.08	5.65
Seventeenth–twentieth (1988–)	4.58	4.16

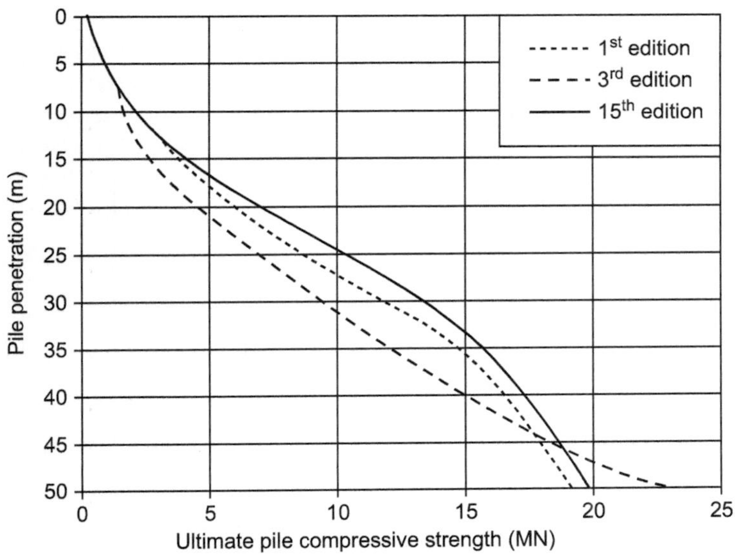

FIGURE 7.2 Changes in pile ultimate compressive capacity calculation for a dense sand, per API edition.

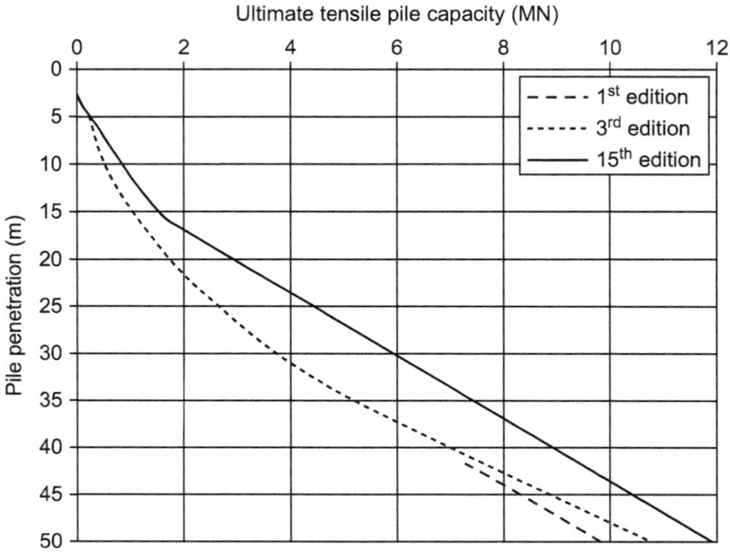

FIGURE 7.3 Changes in pile ultimate tensile capacity calculation for a dense sand, per API edition.

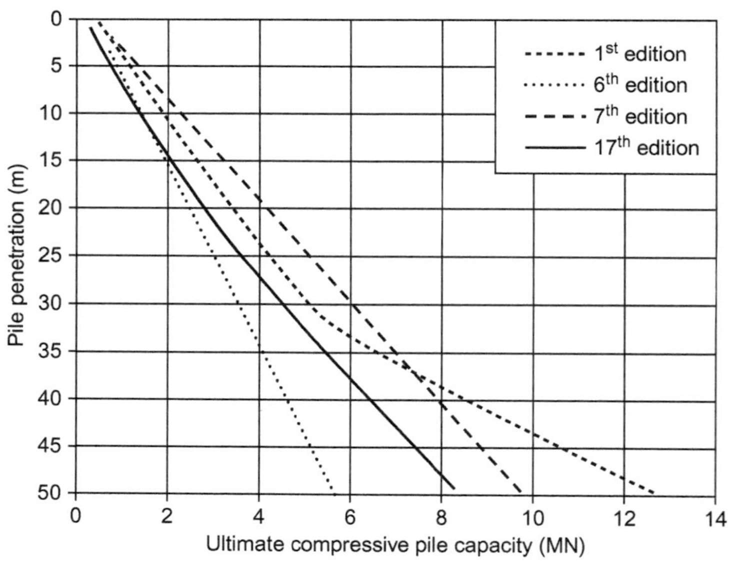

FIGURE 7.4 Changes in pile ultimate compressive capacity calculation for a normally consolidated clay, per API edition.

7.6.12 Effects of Changes in Tubular Member Design

The load calculation underwent a minor change in the new revision of the design code; however, the most changes were made in the resistance formula special to the strength of tubular joints, due to the continuous improvement in the software for finite element analysis.

The average reliability index evaluated for the resistance formulations in each RP2A edition is presented in Table 7.8. The table illustrates clearly that there has been very little change in the average reliability index over the years as a result of changes in the equation of resistance. Based on OTO report index (1999), the average reliability index for the twentieth edition, which is applicable from the seventeenth edition, is 2.25, and the averages for the various stress combinations vary from 2.12 for tension and bending to 2.85 for compression, bending and hydrostatic pressure.

By obtaining and evaluating the reliabilities for the early editions, it is found that the reliability has been influenced by components designed with hydrostatic pressure, which was not included in the early editions. It is important to know that in the North Sea, structures were not installed in waters over 50 m deep until the mid 1970s, and thus hydrostatic pressure was not a critical design factor until then.

The average reliability index with the hydrostatic pressure components for the early editions is shown in Table 7.9. The average reliability index due to changes in the resistance equations has remained virtually constant since

TABLE 7.8 Effect of Change in Resistance Formulations on the Average Reliability Index, Based on HSE

Date of Code	Average Reliability Index
1969–1975	2.15
1975–1978	2.3
1978–1994	2.25

TABLE 7.9 Effect of Change in Resistance Formulations and Hydrostatic Component Database on the Average Reliability Index

Date of Code	Average Reliability Index
1969–1981	2.25
1981–1987	2.32
1987–1991	2.25

TABLE 7.10 Effect of Change in Morison Coefficients on the Average Reliability Index, Based on HSE

Date of Code	Average Reliability Index
1970–1972	1.45
1972–1981	1.8
1981–1994	2.25

RP2A was first introduced. The maximum difference in the average reliability is between a reliability index of 2.25 for the twentieth edition and a value of 2.32 for the eleventh edition.

To illustrate the importance of environmental loading, Table 7.10 shows the effect on average reliability index of changes in the Morison coefficients. The table shows that the platform reliability for the platform design in 1970 was 1.45, but it changed to 2.25 for design after 1981, so there is an increase of about 55%. For the purposes of the table, the designs have been based on the latest resistance formulations in the twentieth edition, and all that have been changed are the drag and inertia coefficients.

In general, component reliability is most sensitive to environmental loading, particularly extreme wave height.

7.7 FAILURE DUE TO FIRE

The deck of an offshore structure platform in the oil and gas industry may be exposed to fire. So in most cases, after a fire, an assessment of the deck is requested to determine if the deck can be used or if it needs strengthening. There are many studies of the reduction on steel structures exposed to fire. The analysis of catenary action in steel beams under fire conditions is based on the field test done by Yingzhi Yin from Manchester University in 2003.

The large deflection behavior of steel beams at elevated temperatures is what was studied.

The parametric study investigated uniform temperature effect in lateral and axial restraint, in axial restraint only and without any restraint axially or laterally. According to European code EC3, steel strength will soften progressively from 100°C to 200°C. Note that only 23% of ambient-temperature strength remains at 700°C.

The EC1 (ISO 834) standard fire curve is shown in Figure 7.5, which presents a different EC1 time-temperature curve.

Fire resistance times are based on standard fire tests based on ASTME84-05, not on survival in real fires. Based on fire load and compartment properties (<100 m^2) is presented. The differences between hydrocarbon fire and external fire are shown in Figure 7.6.

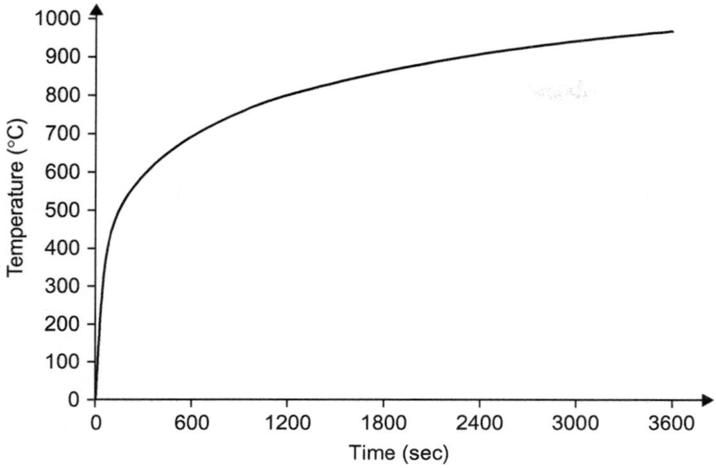

FIGURE 7.5 Standard fire test based on ISO 834.

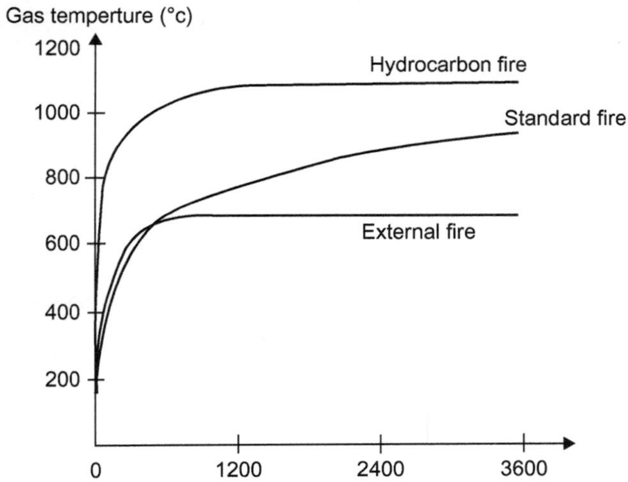

FIGURE 7.6 Type of fire and the maximum temperature.

In fire testing, usually the load is kept constant, fire temperature is increased using the standard fire curve and then the maximum deflection criterion for fire resistance of beams is measured. Then it is possible to define the load capacity criterion for fire resistance of columns.

Steel stress-strain curves at high temperatures are presented in Figure 7.7. Note that the elastic modulus at 600°C is reduced by about 70% and the yield strength at 600°C is reduced by over 50%.

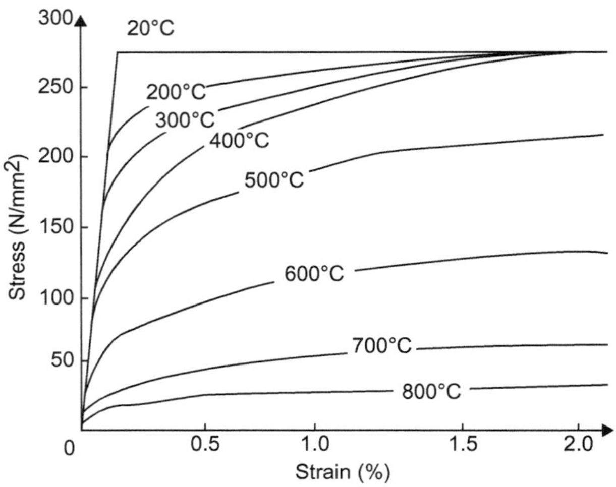

FIGURE 7.7 Stress-strain relationship at different temperatures.

Figure 7.8 shows that steel softens progressively from 100°C to 200°C and above. Only 23% of ambient-temperature strength remains at 700°C. At 800°C, strength reduced to 11% and at 900°C to 6%. In general, steel melts at about 1500°C.

Degradation of steel strength and stiffness is presented in Figure 7.8. Strength and stiffness reductions are very similar for S235, S275 and S355 structural steels and hot-rolled reinforcing bars.

7.7.1 Degree of Utilization

Based on European code EC3 for steel, the degree of utilization can be determined from the following equation:

$$\mu_0 = \frac{E_{f.d}}{R_{f.d.0}} \tag{7.23}$$

where $E_{fi.d}$ is the design loading of a member in fire and $R_{fi.d.O}$ is as a proportion of its resistance at ambient temperature but including the material safety factor for the fire limit state.

A degree of utilization can be used if there is no chance of overall or lateral-torsional buckling, η_f calculated as proportion of design loading at ambient temperature

$$\mu_0 = \eta_f \left(\frac{\gamma_{M.f}}{\gamma_{MI}} \right) \tag{7.24}$$

where η_f is the reduction factor for the design load level for the fire situation.

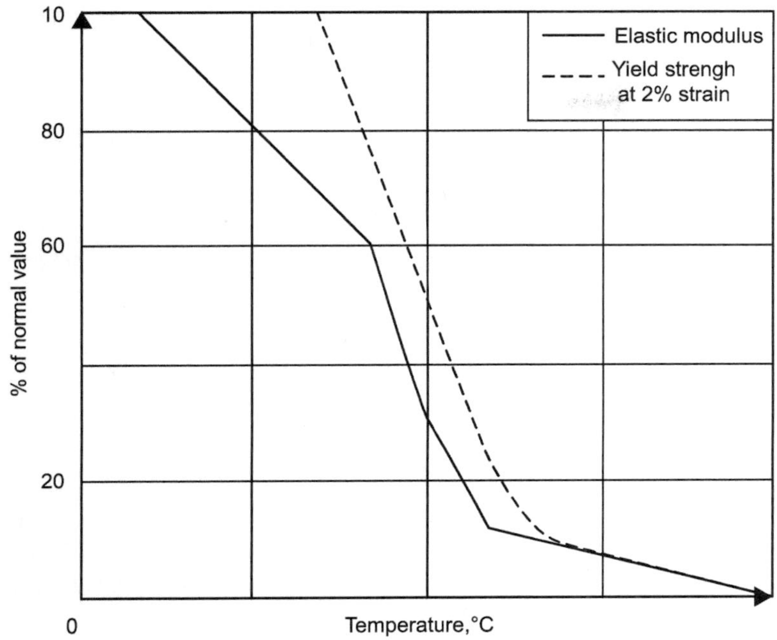

FIGURE 7.8 Percentage reduction in yield strength and elastic modulus by increasing temperature.

Based on EC3, for steel, material partial safety factors $\gamma_{M1} = 1.1$ and $\gamma_{M.fi} = 1$

$$\eta_f = \frac{E_f.d.t}{E_d}$$

It is relative to ambient temperature at the design load.

$$\eta_f = \frac{\gamma_{GA} G + \psi_{1.1} Q_{k.1}}{\gamma_G G_k + \gamma_{Q.1} Q_{k.1}} \tag{7.25}$$

where $Q_{k.1}$ and G_k are the characteristic values of the variable and permanent loads, respectively.

Note that the partial safety factor for permanent load γ_G is equal to1.35, and the partial safety factor for variable load $\gamma_{Q.1}$ is equal to 1.5 in the fire limit state, where $\gamma_{G.A} = 1$ for permanent loads and accidental design situation and $\psi_{1.1} = 0.5$ for combination factor, variable loads and offices.

7.7.2 Tension Member Design by EC3

Design Loading: $N_d = 247.95$ kN

Try IPE 100: $(100 \times 55 \times 8$ kg/m$)$

Design resistance: $N_{.Rd} = A_t f_y / \gamma_{MO}$

$= 1030 \times 0.275/[1.1] = 257.5$ kN > 247.95

Design loading in fire: $N_{f.d} = \eta_f N_d$

Combination factor $\psi_{1.1} = 0.5$

$G_{k.1}/Q_k = 2.0$

Load reduction factor $\eta_f = 0.46$

$N_{f.d} = 0.46 \times 247.95 = 114$ kN

Design resistance at 20°C, using fire safety factors:

$N_{f.20.Rd} = k_{y.20}\ N_{Rd}\ (\gamma_{M.1}/\gamma_{M.f})$

Strength reduction factor $k_{y.20} = 1.0$ at temperature 20°C (see Table 7.11)

$$N_{f.20.Rd} = 1.0 \times 257.5 \times ([1.1]/[1.0]) = 283.25 \text{ kN}$$

Critical temperature: Degree of utilization $\mu_0 = N_{f.d}/N_{f.20.Rd}$
$= 114/283.25 = 0.40$

Therefore, the critical temperature $\theta_c = 619$°C (see Table 7.12)

Note that, from Table 7.11, the intermediate value of the linear interpolation can be used.

The relative thermal elongation of steel $\Delta l/l$ should be determined from the following:

For 20°C $\leq \theta_a <$ 750°C

$$\Delta l/l = 1.2 \times 10^{-5}\theta_a + 0.4 \times 10^{-8}\theta_a^2 - 2.416 \times 10^{-4}$$

For 750°C $\leq \theta_a \leq$ 860°C

$$\Delta l/l = 1.1 \times 10^{-2}$$

For 860°C $< \theta_a \leq$1200°C

$$\Delta l/l = 2 \times 10^{-5}\theta_a - 6.2 \times 10^{-3}$$

where l is the length at temperature 20°C, Δl is the temperature-induced elongation and θ_a is the steel temperature in degrees C.

7.7.3 Unrestrained Beams

In the load resistance domain, lateral-torsional buckling capacity at compression flange maximum temperature $\theta_{a.com}$ is presented by the following equation:

$$M_{b.f.t.Rd} = W_{pl.y} K_{y.\theta.com} f_y \left(\frac{\chi_{LT.f}}{1.2}\right) \frac{1}{\gamma_{M.f}}$$

where reduced yield strength of compression flange $= k_{y.q.com}\, f_y$ at $\theta_{a.com}$ and reduction factor $\chi_{LT.fi}$ for flexural buckling is based on normalized slenderness:

$$\bar{\lambda}_{LT.\theta.com} = \bar{\lambda}_{LT} \sqrt{k_{y.\theta.com}/k_{E.\theta.com}}$$

TABLE 7.11 Reduction Factors for Stress-Strain Relationship of Carbon Steel at Elevated Temperatures

	Reduction Factors at Temperature θ Relative to the Value of f_y or E_a at 20°C		
Steel Temperature, θ	Reduction Factor (Relative to f_y) for Effective Yield Strength $k_{y,\theta} = f_{y,\theta}/f_y$	Reduction Factor (Relative to f_y) for Proportional Limit $k_{p,\theta} = f_{p,\theta}/f_y$	Reduction Factor (Relative to E_a) for the Slope of the Linear Elastic Range $k_{E,\theta} = E_{a,\theta}/E_a$
20°C	1.0	1.0	1.0
100°C	1.0	1.0	1.0
200°C	1.0	0.807	0.900
300°C	1.0	0.613	0.80
400°C	1.0	0.42	0.700
500°C	0.780	0.360	0.600
600°C	0.470	0.180	0.310
700°C	0.230	0.075	0.130
800°C	0.110	0.050	0.090
900°C	0.06	0.0375	0.0675
1000°C	0.040	0.025	0.045
1100°C	0.02	0.0125	0.0225
1200°C	0.00	0.0	0.0

Note that there is no need to consider lateral torsional buckling unless $\bar{\lambda}_{LT.\theta com} > 0.4$, and the correction factor of 1.2 simply allows for uncertainties.

- In the load resistance domain, buckling capacity at maximum temperature $\theta_{a.max}$ is

$$N_{b.fi.t.Rd} = Ak_{y.\theta.\max} f_y \left(\frac{\chi_{fi}}{1.2}\right) \frac{1}{\gamma_{M.fi}}$$

In the load resistance domain, buckling capacity at maximum temperature $\theta_{a.max}$ is reduced yield strength $= k_{y.q.max} f_y$ at $\theta_{a.max}$

- Reduction factor χ_{fi} for flexural buckling based on:
- Buckling curve (c)

TABLE 7.12 Critical Temperature $\theta_{a,cr}$ for Values of the Utilization Factor μ_o

μ_o	$\theta_{a,cr}$	μ_o	$\theta_{a,cr}$	μ_o	$\theta_{a,cr}$
0.22	711	0.42	612	0.62	549
0.24	698	0.44	605	0.64	543
0.26	685	0.46	598	0.66	537
0.28	674	0.48	591	0.68	531
0.30	664	0.50	585	0.70	526
0.32	654	0.52	578	0.72	520
0.34	645	0.54	572	0.74	514
0.36	636	0.56	566	0.76	508
0.38	628	0.58	560	0.78	502
0.40	620	0.60	554	0.80	496

- Effective lengths in fire
- Normalized slenderness is calculated as follows:

$$\bar{\lambda}_{\theta.\max} = \bar{\lambda}\sqrt{k_{y.\theta.\max}/k_{E.\theta.\max}}$$

7.7.4 Example: Strength Design for Steel Beams

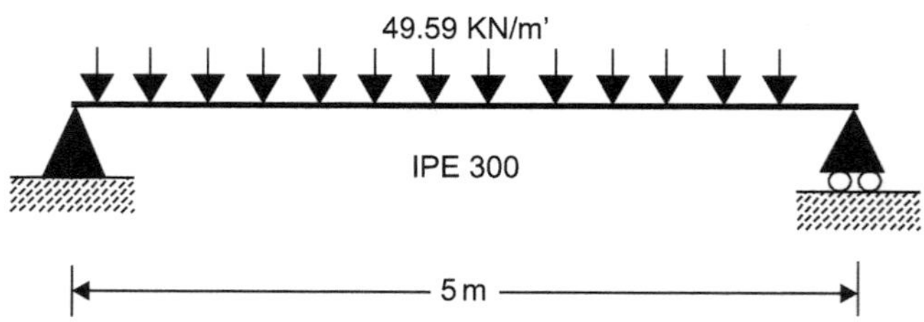

This a simply supported beam with section IPE 300; the span is 5 m under uniform load of 49.59 kN/m'. It is imperative to know the required temperature at which this beam will fail.

The bending moment on the member, which is a simple beam with uniform load, is calculated as follows.

$$M_d = 49.59 \times (5)^2/8 = 154.97\,\text{kNm}$$

Try IPE 300: $(300 \times 150 \times 42\ \text{kg/m})$

Section Classification:

$$\varepsilon = (235/f_y)^{0.5} = 0.92$$

Then check the compactness of the section:
$d/t_w = 248.6/7.1 = 37.5 < 72 \times 0.92$
$c/t_f = 7.0 < 10 \times 0.92$ ... The section is Class 1 (Compact section)

Moment Resistance:
There is a full restraint to top flange; so there is no effect of lateral-torsional buckling.
Resistance moment $M_{pl.Rd} = W_{pl.x} f_y/\gamma_{M.O} = 157$ kNm > 154.97

Shear Resistance:
Applied shear $V_d = 123.97$ kN
Shear area $A_v = 2567$ mm^2
Resistance $V_{pl.Rd} = 2567 \times 0.275/(1.732 \times [1.1]) = 370$ kN > 123.97

Steel Beam: Design Resistance at 20°C
Design loading in fire:

$$M_{f.d} = \eta_f M_d$$

Combination factor $\psi_{1.1} = 0.5$
$G_{k,1}/Q_k = 2.0$
Reduction factor $\eta_f = 0.46$

$$M_{f.d} = 0.46 \times 154.97 = 71.25\,\text{kNm}$$

Design resistance at 20°C, using fire safety factors:
For a Class 1 beam with uniform temperature distribution,
Resistance moment at temperature θ is $M_{fi.\ \theta.Rd} = k_{y.\ \theta}(\gamma_{M.1}/\gamma_{M.fi})M_{Rd}$
Strength reduction factor for 20°C: $k_{y.20} = 1.0$
$\gamma_{M.1} = [1.1]$ and $\gamma_{M.f} = [1.0]$
Resistance moment for strength at 20°C is $M_{Rd} = 157$ kNm
$M_{f.20.Rd} = 1.0 \times ([1.1]/[1.0]) \times 157 = 172.7$ kNm

$$M_{fi.t.Rd} = M_{fi.q.Rd}/k_1 k_2$$

$k_1 = [1.0]$ for beam exposed to fire from all sides
$k_2 = 1.0$ (note that k_2 is equal to 0.85 at the support for statically indeterminate beam)
$M_{f.t.Rd} = 172.7/([1.0] \times 1.0) = 172.7$ kNm

Critical Temperature of the Beam:
Degree of utilization $\mu_0 = 71.25/172.7 = 0.41$
Critical temperature of beam $\theta_{cr} = 616$°C

7.7.5 Steel Column: Strength Design

This example is a steel column with section HEB 180 under normal force equal to 991.8 kN and with a height of 3.5 m.

Design Loading: $N_d = 991.8$ kN

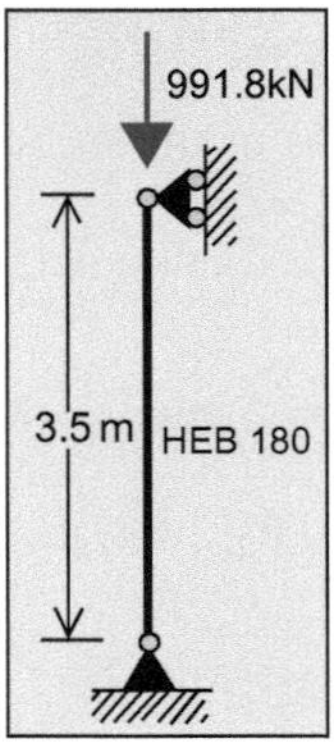

Try HEB 180: ($180 \times 180 \times 51.2$ kg/m, $i_{min} = 4.57$ mm, $A = 6530$ mm^2)
Section Classification: $\varepsilon = (235/f_y)^{0.5} = 0.92$
$d/t_w = 122/8.5 = 14.4 < 33 \times 0.92$
$c/t_f = 90/14 = 6.4 < 10 \times 0.92$ … Class 1

Compression Resistance:
Slenderness $\lambda = 3.5/0.046 = 76.6$
$\lambda_1 = 86.8$
Normalized slenderness $= \lambda/\lambda_1 = 0.88$
Reduction factor $\chi = 0.61$
$\beta_A = 1$ for Class 1 sections
Buckling resistance $N_{b.Rd} = \chi \beta_A A f_y / \gamma_{M.1}$
$= 0.61 \times 1 \times 6530 \times 0.275/1.1$
$= 997$ kN > 991.8
Steel column: design resistance at 20°C

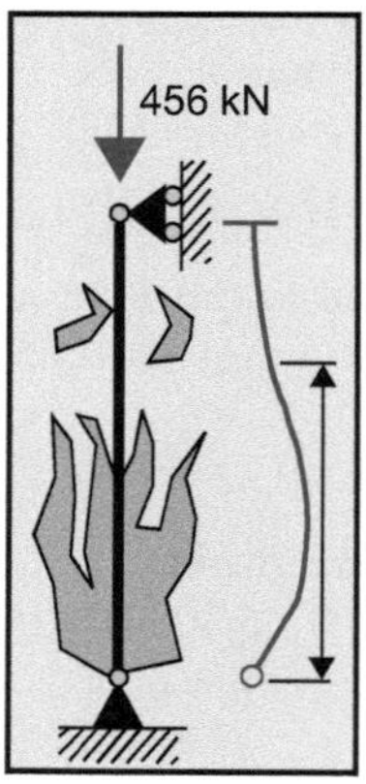

Design loading in fire: $N_{fi.d} = \eta_{fi} N_d$
Combination factor $\psi_{1.1} = 0.5$
$G_{k.1}/Q_k = 2.0$

Reduction factor $\eta_f = 0.46$
$N_{f.d} = 0.46 \times 991.8 = 456$ kN

Design Resistance at 20°C, using fire safety factors:
$N_{b.fi.t.Rd} = (\chi_{fi}/1.2)\, Ak_{y.q.max}\, (f_y/\gamma_{M.fi})$
Effective length factor, $k = 0.7$ (pinned base)
Slenderness λ, $kl/I = 0.7\ (3.5)/0.0457 = 53.6$
$\lambda_1 = 86.8$ as calculated above
Normalized slenderness $\bar{\lambda} = \lambda/\lambda_1 = 0.62$
$\lambda_{20} = 0.62\ (k_{y.20.max}/k_{E.20.max})$
For $\theta = 20°C$, there is no reduction on yield strength or the elastic modulus, as per Table 7.11.
$k_{y.20.max} = k_{E.20.max} = 1.0$
Reduction factor in fire, $\chi_f = 0.77$
$N_{b.f.t.Rd} = (0.77/1.2) \times 6530 \times 1 \times 0.275/1 = 1159.6$ kN

Critical Temperature of Column:
Degree of utilization $\mu_0 = 456/1159.6 = 0.39$
Critical temperature $\theta_{cr} = 622.4°C$

7.7.6 Case Study: Deck Fire

The platform in this case is a complex platform (consisting of many platforms), located in the Gulf of Suez in the Red Sea. The platform is used for drilling. The fire occurred in 2003 due to the sudden rupture of a gas riser (Figure 7.9). As a result of the rupture, the riser, which contained gas under pressure, caused a hydrocarbon fire jet that directly affected the bottom of the main deck (Figure 7.10). The assessment process was:

FIGURE 7.9 Fire on the main deck.

FIGURE 7.10 Fire-related effects on the main truss.

- An on-site inspection survey, in which different levels of damage were observed in various structural and piping components on the platform.
- Categorization of the damage level of different structural elements of the platform in accordance with API 579, Section 11.
- Measurement of the actual dimensions of the structural elements of the platform.
- Assessment of the material properties of different structural elements of the platform.
- Provision of the necessary information for structural modeling of the platform.

Figure 7.11 shows the fire-related deformation of the main and secondary beams. The part of the platform deck that was directly affected by the fire is illustrated in Figure 7.12. The main element of the platform deck that transfers the load to the piles is the truss system (see Figure 7.13). Unfortunately, in this case the truss system was also affected by the fire; the hatch part was exposed to temperatures greater than 732°C and other parts were exposed to temperatures from 427 to 732°C.

Figures 7.14 and 7.15 present the locations of the damage and zones highly affected by the fire on the platform deck for the main and secondary beams and the main truss system, respectively.

The isometric view in Figure 7.14 shows the deck beams that were directly affected by the fire. Site measurements were performed by taking a sample of structural elements of the platform not affected and affected by fire; the test included structural dimensions and hardness readings.

Estimation of the material properties is based on the hardness readings. All materials weaken with increasing temperature and steel is no exception. Strength loss for steel is generally accepted to begin at about 300°C and

FIGURE 7.11 Fire-related deformation of the main and secondary beams for the deck.

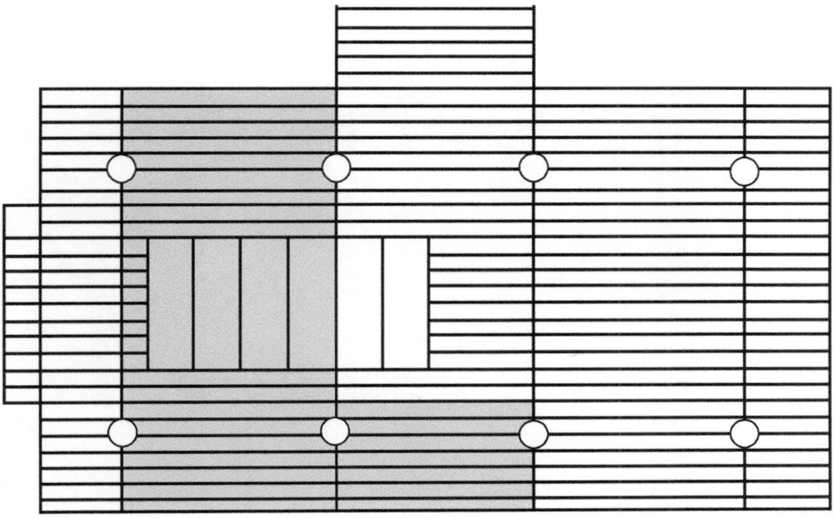

FIGURE 7.12 Deck layout showing the zone affected by fire.

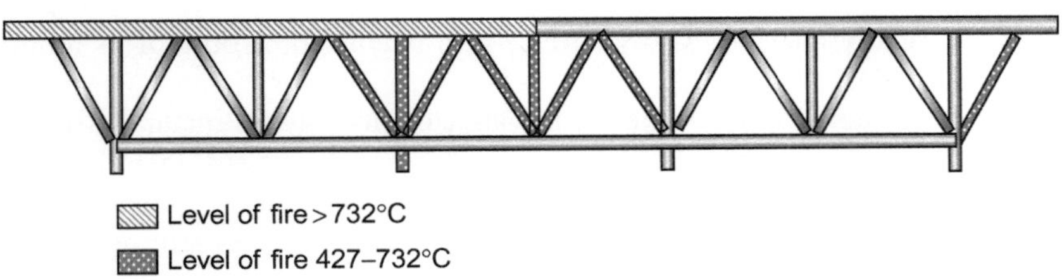

FIGURE 7.13 Fire classification effect on the main truss.

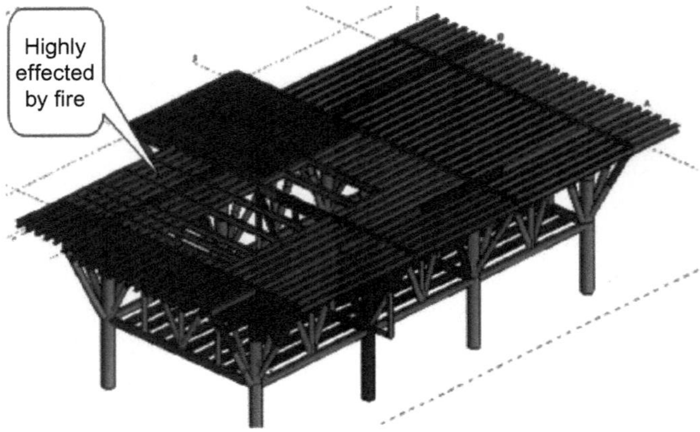

FIGURE 7.14 Deck categories based on the effects of fire.

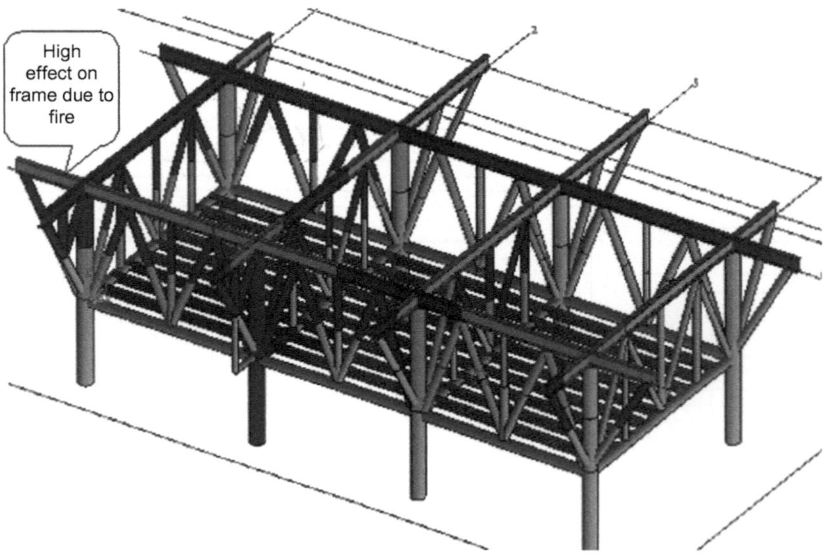

FIGURE 7.15 The main frame categories exposed to fire.

increases rapidly after 400°C; by 550°C, steel retains about 60% of its room-temperature yield strength.

An often quoted general rule for fire-affected hot-rolled structural steels is that, if the steel is straight and there are no obvious distortions, then the steel is fit for use. At 600°C, the yield strength of steel is equal to about 40% of its room-temperature value; it follows therefore that any steel still remaining straight after the fire and that had been carrying an appreciable load was probably not heated beyond 600°C, had not undergone any metallurgical changes and will probably be fit for re-use.

When calculating the yield strength by testing the steel, the following precaution, as specified by Brockenbrough (2003), should be considered. It should be recognized that the yield stress determined by standard ASTM methods and reported by mills and testing laboratories is somewhat greater than the *static* yield stress because of the dynamic effects of testing. Also, the test specimen's location may have an effect. These effects have already been accounted for in the nominal strength equations in the *specification*. However, when strength evaluation is done by load testing, these effects should be accounted for in test planning because yielding will tend to occur earlier than otherwise anticipated.

The static yield stress, F_{ys}, can be estimated from that determined by routine application of ASTM methods, F_y, by the following equation (Galambos, 1998):

$$F_{ys} = R(F_y - 4)$$

where F_{ys} = static yield stress, in ksi (MPa); F_y = reported yield stress, in ksi (MPa); $R = 1.00$ for tests taken from flange specimens; and $R = 0.95$ for tests taken from web specimens.

The R factor accounts for the effect of the coupon location on the reported yield stress. Prior to 1997, certified mill test reports for structural shapes were based on specimens removed from the web, in accordance with ASTM A6/A6M.

Subsequently, the specified coupon location was changed to the flange. During 1997–1998, there was a transition from web specimens to flange specimens as the new provisions of ASTM A6/A6M were adopted.

On the other hand, the material properties of the damaged part are estimated using a hardness tester in a fire-unaffected zone in order to obtain the original material ultimate strength and consequently the material grade. Table 7.13 provides the relation between the hardness reading using the Brinell method and the ultimate strength of the steel.

A beam with deformation should be replaced by a new beam. Difficulty arises on the main frame structure, which is very expensive to strengthen. Therefore, the hardness test is obtained, and the properties of the steel that have changed due to fire, such as the percentage of carbon, will affect the ductility of the materials of the structure. Because the offshore structure platform is always affected by wind and waves, the structure should be ductile enough to have drift without failure.

The structural analysis has revealed that the main deck is unsafe, the option is to reconstruct the deck. However, the decision requires consideration of the business as a whole, because the target is to deliver oil using this platform, and due to the age of the field, the best economic decision is to reduce the load on

TABLE 7.13 Relation between Brinell Number and Ultimate Strength

Brinell Reading	Ultimate Strength (MPa)	Brinell Reading	Ultimate Strength (MPa)	Brinell Reading	Ultimate Strength (MPa)
240	795	180	596	135	447
234	775	176	583	130	431
228	755	172	570	125	414
222	735	169	560	120	397
216	715	162	537	115	381
210	696	159	527	110	364
205	679	156	517	105	348
200	662	153	507	100	331
195	646	150	497	95	315
190	629	145	480	90	298
185	613	140	464	85	281

the deck and to prevent any work on the deck above. Obviously, this solution may not be the best for another platform from an economic and business perspective.

7.8 CASE STUDY: PLATFORM FAILURE

As already mentioned, most platforms worldwide are more than 30 years old. An integrity management system has been developed and implemented based on industrial practice. The role of integrity management is to perform assessment of existing structures based on API RP2A.

In general, API RP2A is tailored for platforms in the Gulf of Mexico. For fixed offshore structures elsewhere, the code needs to be modified and adapted for the platform location. The definition of risk for the structure is the probability of failure multiplied by the consequences of failure. Therefore, the risk is based on technical determination of the probability of failure and the business and economic aspects of the consequences.

In real life, this approach may need to be changed when dealing with the offshore structure maintenance plan, because, although the structural engineers should focus on the probability of failure, the management is focused on risk; therefore, you may have to ignore a remedial action for a structure with high probability of failure because it has a very low consequence.

Fatigue is the main factor governing a structure's lifespan. For example, an existing structure with a flooded member needs a comprehensive structural study of the platform considering fatigue. A remote operating vehicle (ROV) survey is usually used as part of API RP2A, but its requirements need to be changed according to the location and the environmental conditions.

This case study of platform failure and the reason for its failure clarify the pitfalls in the integrity management system philosophy and also in API RP2A for assessment of existing structures.

The platform shown in Figure 7.16 is a well-protector platform with 4 legs in a water depth of 39 m (130 feet). The platform was constructed in 1973 with an inverse K bracing system. A subsea ROV survey was performed 7 years before the failure. The survey showed 5 flooded members in the platform. The topside facilities' weight was 27 tons, with 6 risers and 4 conductors. The deck weight was 46 tons and the jacket weight was 73 tons. The platform was designed prior to the first edition of API RP2A; the design would have been done to engineering company specifications, possibly supplemented by owner recommendations.

There is no evidence that accidental loads, such as vessel impact, fatigue load and seismic load or the effects of corrosion or marine growth were accounted for in the design, because these were not design practice at the

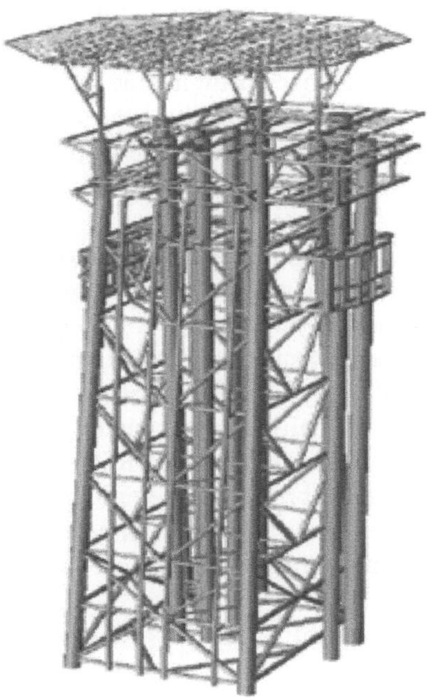

FIGURE 7.16 Platform configuration.

time. In addition, a pushover study using the SACS model was performed on this platform 5 years before failure and the reserve strength ratio was suitable.

Failure is usually caused by decreased strength and an increase in the affected load at the same time.

In this case, it was observed that the topsides were inclined 5° to 10° toward the south, indicating a significant horizontal force most likely applied well above the base of the jacket structure. The leg members above the pile stabbing points were cracked in a manner consistent with a horizontal shear-type action.

The subsea survey indicated multiple joint failures on the east and west faces, with bright steel on most visible failure surfaces and lack of corrosion and marine growth on most exposed pile surfaces where joint failure led to pull-out of the leg. Thus, these surfaces had only recently been exposed to seawater.

Failure was shown on the leg by shear cracks with rotation, which revealed that the leg had been under applied loads that caused shear stress with torsion effect. In addition, the ROV subsea inspection that was performed after failure showed a punching shear and detachment of brace member from the leg.

7.8.1 Strength Reduction

Four braces near the base of the jacket were reported to be flooded in previous subsea surveys, potentially indicating that some cyclic fatigue action had led to through-thickness cracking of the members, but it appears that the joint failure happened some time before complete structural failure, because the cracks had discolored surfaces and were covered by marine growth. It is considered unlikely that these flooded members led directly to the subsequent structural failure.

The tubular steel used offshore at the time would have had a high carbon equivalent value and consequently would be more brittle and less ductile than other offshore steel rolled sections.

Visual inspection of the welding was satisfactory, with no gross defects detected. No obvious weld fabrication problems were identified on this platform in particular or on any platform in the fleet in general.

In general, the K-braced system of a jacket with no joint-cans has little scope for redistributing load following initial joint failure.

Joint plasticity would normally be associated with structural overload but there was little evidence of this from subsea ROV inspections post-collapse. This may be due to the more brittle steel employed in the late 1960s/early 1970s when this platform was fabricated, and/or the relatively thin steel sections, because the member thicknesses are around 7–10 mm, and/or other factors discussed below.

In most cases, the cause of failure is a combination of many factors converging at the same time.

7.8.2 Environmental Load Effect

The catastrophic structural failure of the platform, an unmanned satellite platform, most likely was related to the peak of the storm preceding failure by two days, when onshore maximum wind gust speeds higher than 50 knots (25.7 m/s) were recorded at the nearest onshore location

The mode of failure was structural overload primarily caused by severe wave loading from the north and northwest (N-NW) impacting the platform over a prolonged period.

The failure characteristics were similar to the pushover collapse analyses performed 5 years before failure and another study for a similar platform 1 year before failure.

There was no evidence of vessel impact, a seismic event or another possible source of loading identified as a hazard to this platform, with the possible exception of an operated support vessel tying up to the boat landing to ride out a storm.

While the vessel tie-up option cannot be completely discounted, it is very unlikely because this platform is near a natural sheltered location and the bollards show no signs of damage.

The storm conditions recorded for 2 days before structural failure were typical of storms recorded around 3-4 times per year for the 7 years before failure. Based on the onshore gusts measured, the estimated maximum wave height and maximum 1-hour average wind speed during the storm are considered to be roughly wave height (H_{max}) equal to 6.5–7.5 m and wind speed (Ws) equal to 19.5 m/s, respectively.

A storm 2 months before failure was more severe and more typical of storms recorded around once every 2 years. Estimated maximum wave height and maximum 1-hour average wind speed during this storm were considered to be around $H_{max} = 6.7–8.7$ m and $Ws = 21$ m/s, respectively.

Other severe storms that happened 3, 4 and 7 years before failure had similar peak gusts and duration.

7.8.3 Structure Assessment

The platform's subsea jacket was considered to have been in a degraded condition prior to catastrophic collapse.

The collapse analysis of the platform was performed using the SACS model with no joint failure modeled and the estimated minimum reserve strength ratio (RSR), which is equal to the load at collapse divided by load under 100-year environmental conditions, was found to be equal to 1.25 at collapse. The pushover analysis revealed an RSR equal to 1.13 at first failure due to pile yield for example increasing environmental load 25% greater than maximum wave height with the associated wind speed. The wave contribution was found to be the most significant factor.

The collapse analysis for a similar platform was performed by USFOS pushover analysis, as per Soreide et al. (1986), and it was found that the minimum RSR was equal to 2.19 at collapse and 1.76 at first failure in the leg joints at −9.0 m elevation below the mean seal level, i.e., an environmental load from the northwest 119% greater than maximum wave height (H_{max}) combined with wind speed equal to 15.6 m/s. Note that 15% and 30% wall thickness loss reduce the RSR values to 2.02 and 1.55, respectively; however, in this analysis there was no corrosion loss considered for the platform.

The pushover analysis was performed without taking flooded members.

The ROV survey reported that there were four braces near the sea bed that were flooded, potentially indicating that some cyclic fatigue action had led to through-thickness cracking of the members. It is considered unlikely that these flooded members led directly to the subsequent structural failure, since ROV footage implied that two were intact after failure.

It is very important to remember that some joints appeared in the post-collapse subsea survey to have discolored surfaces and marine growth over cracks, which are good indications that the failure of the joints happened some time before platform structural failure.

The collapse analysis was performed regardless of the flooded member data because cracks in joints increase with time and could have caused a sudden failure with a high storm.

In-place analysis using the SACS model showed that 90% of the joints had a unity check (U_c) higher than 1.2. Also, some joints at elevations −11, −28 and −46 had U_c higher than 2.0.

In addition, the minimum pile capacity factor of safety in compression for storm condition was equal to 1.08, and the minimum factor of safety for tension was 1.00, which violates API RP2A safe conditions.

Collapse analysis is performed by taking pile-soil interaction, and in this case the RSR was equal to 1.25. An other collapse analysis was performed by using a dummy pile stub to define the RSR for the jacket, which in this case was equal to 1.65. Note that the joint flexibility option was included in the analysis.

From the dynamic point of view, the natural frequency is calculated through the eigenvalue problem, which takes the form:

$$K\phi = lM\phi \tag{7.26}$$

where K and M are the stiffness and mass matrices, respectively, and ϕ is the model shape factor of the structure. The software program calculated the lowest eigenvalue λ and the corresponding eigenvector. From the above, stiffness is the main factor in calculating dynamic analysis for the structure. The reason for the flooded members is very important to consider, because if there is a crack on the joint it will affect the global stiffness, so the natural frequency will increase, which will increase the dynamic amplification factor (DAF), which should be considered in the collapse analysis.

The DAF is calculated from:

$$DAF = \frac{1}{\sqrt{\left\{\left[1-\left(\frac{T_n}{T}\right)^2\right]^2+\left(2D\frac{T_n}{T}\right)^2\right\}}} \tag{7.27}$$

where T = the wave period in consideration, T_n = the natural period of the structure and D = damping ratio.

On the other hand, the reduction in stiffness under wave load will increase the vibration, which will affect fatigue and accelerate the opening of cracks, so it shortens the structure's lifespan.

The natural period of the structure was calculated and was found to be 2.78 s, which is less than 3 s, so there is no need to add DAF to the in-place analysis. Another pushover analysis was performed by USFOS software and in this study the natural period was 2.36, which added the DAF into consideration in the collapse analysis.

The effects of reduced stiffness, such as when the structure is losing stiffness, include an increase in the natural period and then a gradual increase in the DAF as well, as shown in Figure 7.17.

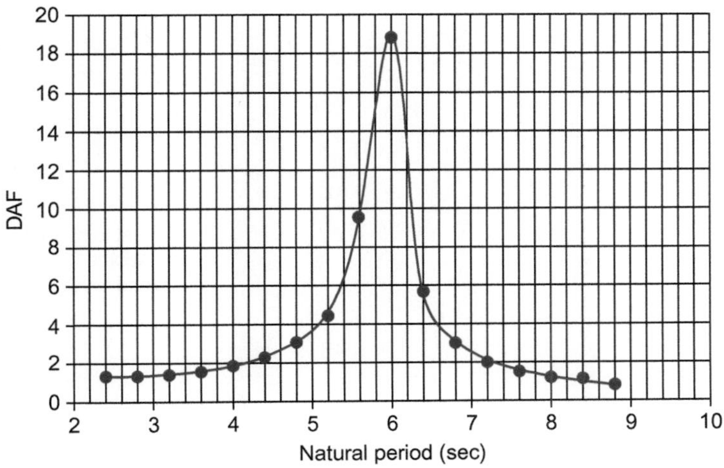

FIGURE 7.17 Relation between the natural period and the DAF.

The location of the first plastic joint (see Figure 7.18), after performing the pushover analysis. In this study, the RSR at first load failure was 1.76, 1.24 and 0.874 with 0%, 15% and 30% corrosion loss, respectively.

So the minimum RSR for an old structure with cut or flooded members should be considered the DAF that added value statically to the wave load on the structure.

All offshore platforms are under an integrity management system. In general, the integrity management system for offshore structures is based on risk-based inspection as clarified by AP SIM draft and API RP2A clause 17 about assessment of existing structures. The *risk* is a function of multiplying the probability of failure by its consequences. So, if the probability of failure increases, the risk may be still low, because the consequence may be very low. In this case, the platform's risk ranking was 50 among 240 offshore

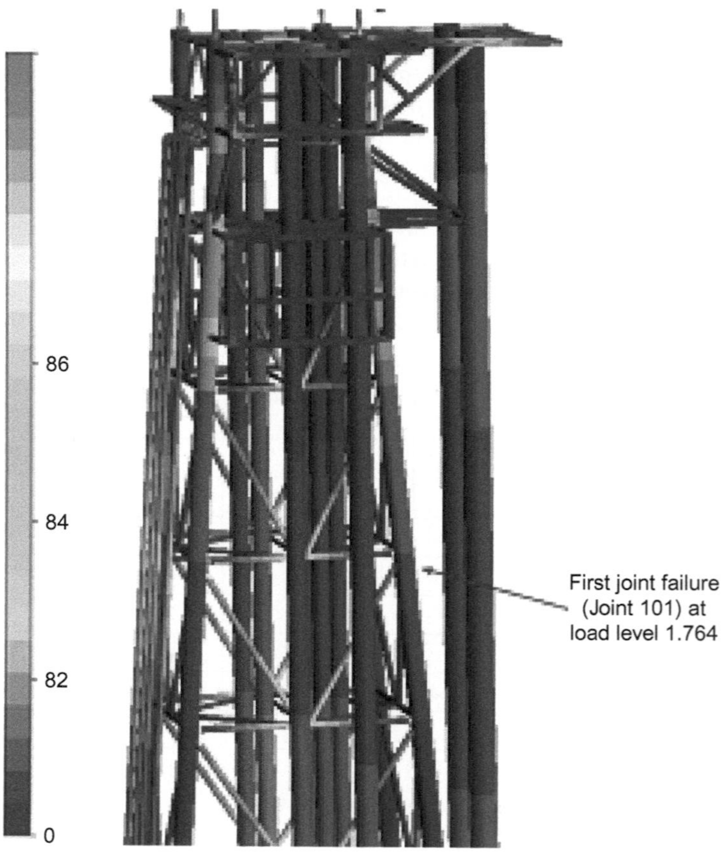

FIGURE 7.18 Location of first joint failure.

structures, which meant it was not at the highest risk. It indicates that risk assessment of the platform was a matter for managing the business and the goals of the owner organization and did not reflect the actual condition of the structure itself. Nevertheless, although risk may be low, failure can occur at any time. So, maintaining offshore structures from a structural point of view should focus on likelihood of failure.

In assessment of the platform, in-place analysis is very important for defining the condition of the structure. In most of our cases, the piles are out of code, as are the joints, as was true for this platform. It is important to concentrate the survey of connections to a short period with close visual inspection.

From an engineering point of view, if the probability of failure is high, the structure may fail at any time, regardless of its risk ranking.

In-place analysis is sufficient make a preliminary determination of which of the main members and joints could affect structural reliability (e.g., a cantilever with a very high unity check ratio). So the maintenance scope can be defined based on in-place analysis done by competent structural engineers familiar with assessment of mature structures.

Pushover analysis should present the exact condition of the platform; if there are any deviations, the reserve strength ratio (RSR) cannot identify exactly the condition of the platform. If there are any flooded members, fatigue assessment is critical, as is the dynamic analysis, because vibration of the structure is affected. The lesson is that the survey and collapse analysis should be very accurate and cover everything known or it might as well not be done, as there is no optimization in assessment of existing structures.

Furthermore, dynamic analysis with the fatigue study is more critical for defining the structure's lifespan than collapse analysis is. A checklist for pushover analysis is presented in Table 7.14.

For old platforms with cracked or cut members, the collapse analysis to calculate the RSR should consider the maximum amplification factor. Because the minimum RSR depends on the maximum wave height within a period of 100 years, it is reasonable for new platforms or platforms without joint cracks, but this analysis is useless and a waste of money if there are cracked joints.

7.9 ASSESSMENT OF PLATFORM

The criteria for assessment of the platform were under development at the time of preparation of this book and were only available as the draft of API SIM, which is the focus of the discussion here of the assessment of an existing structure.

The acceptance criteria will determine whether a structural assessment has confirmed the platform is fit for continued operation in its present condition.

The ultimate structural capacity of an offshore platform may be characterized either in terms of the platform reserve strength ratio (RSR) or in terms

TABLE 7.14 Checklist for Pushover Analysis

	Yes/No
General Checks	
Is scope of analysis clearly defined?	
Is content and standard of presentation reasonable?	
Are all units defined and consistent?	
Are all specifications, standards and codes listed?	
Does work satisfy the requirements of the design brief, design specs and/or other relevant documents?	
Have the correct conditions been included in the analysis (in-place, future, ship impact, fire, others)?	
Has the analysis been set up in a clear and logical file structure?	
Have the recommendations of the latest codes been adhered to?	
Do you have a project working folder and have you reported all relevant information on it?	
Checks for New or Converted Model	
Is the model oriented in the correct coordinate system?	
Have coordinates of key geometry points been checked?	
Have variables been defined in consistent units of weight, length and time?	
Have the beam element properties and orientation been checked?	
Did you modify the general sections properties?	
Have section, material, and other names (ID) been updated?	
Have basic and combined load cases been checked and/or updated?	
Did you check the member releases?	
Have the appropriate appurtenances been included?	
Did you define the joint cans and stubs?	
Have you defined or checked the brace offsets and gaps at joints?	
Have model sets been defined and checked?	
Has the model duplicated nodes?	
Have you checked that all section property shear factors are 1?	
Have correct C_d and C_m coefficients been used?	

TABLE 7.14 Checklist for Pushover Analysis—cont'd

	Yes/No
Checks for New or Converted Model	
Is water depth in accordance with latest air-gap measurements?	
Are wave heights in accordance with metocean data?	
Have the mud line and waterline been defined correctly?	
Have flooded members been defined correctly?	
Have member-specific Morison coefficients been defined correctly?	
Has marine growth thickness and mass been included?	
Has the associated current been defined correctly?	
Have anodes CD and CM been included? If not, increase 1.07 wave force in analysis.	
Are wave directions and headings consistent?	
Have the correct wave frequencies been defined?	
Have sea-state directions been checked?	
Have the correct wave amplitudes been defined?	
Has the wave steepness been checked?	
Have groups/sets been checked?	
Have base shear and OTM been checked/compared?	
Have you checked that buoyancy was not applied twice? (Storm case must not have buoyancy)	
Did you check the output file for possible warnings/errors?	
Have you reported all relevant information in the working folder?	
Checks for Modified Existing Model	
Have new or updated flooded members been defined correctly?	
Have the specific Morison, C_d and C_m coefficients and marine growth thickness and mass been defined correctly for new/updated members?	
Have anodes CD and CM been included? If not, increase 1.07 wave force in analysis.	
Have groups/sets been checked/updated/included?	
Did you check or compare the base shear and the overturning moment?	
Have you checked that buoyancy was not applied twice? (Storm case must not have buoyancy)	
Have you checked the output file for possible warnings/errors?	
Have you reported all relevant information in the working folder?	

(Continued)

TABLE 7.14 Checklist for Pushover Analysis—cont'd

	Yes/No
Model Conversion	
Foundation	
Have soil spring definitions been checked?	
Do you have the latest information about pile capacities?	
Have individual piles been limited in tension and compression according to latest information about pile capacities?	
Have group of piles been limited in tension and compression according to latest capacities?	
Have you reported all relevant information in the working folder?	
Geometry	
Have you correctly selected the node/element number?	
Have you defined the correct pile head node?	
Have the boundary conditions of piles, sleeves, etc., been defined correctly?	
Have the software basic load cases been checked?	
Have you checked that the genie sets are correct in the software?	
USFOS Analysis Checks	
Before Analysis	
Have you got the soil data for still and storm conditions?	
Have initial imperfections of 0.2% L for still and storm conditions been defined?	
Has MSL joint formulation been defined?	
Has hydrostatic pressure been included for member strength?	
Has fracture module been included in the analysis?	
Have appropriate load and material factors been applied?	
Have local buckling and lateral torsional buckling been considered? (That is, will local buckling occur before the full plastic capacity is achieved?)	
Has the software model been validated against results of known loading case?	

TABLE 7.14 Checklist for Pushover Analysis—cont'd

	Yes/No
USFOS Analysis Checks	
Before Analysis	
Have you checked the topside load combination center of gravity (CoG) via reaction loads?	
Has the wave OTM been checked via fixing the jacket at known elevation?	
Do pile reactions from output file add up and make sense against what is expected?	
Has the pile model been checked by plotting pile head axial force (with a linear jacket to ensure soil failure)?	
Have you considered *D/t* yield strength reduction in the header file (elastic and plastic local buckling)?	
During Analysis	
Has gravity (still water) pushover been completed for various and worst drilling rig locations with appropriate load factors (API and ISO)?	
Have soil data (factor = 0.80) for API still condition been used?	
Has 100-year storm (100-year wave + associated 1-year current) pushover for various/worst drilling rig locations been completed with appropriate load factors (API and ISO)?	
Have soil data (factor = 0.94) for API storm condition been used?	
Have you checked the serviceability limit?	
Has 10,000-year storm (10,000-year wave + associated current) pushover for various/worst drilling rig locations been completed with appropriate load factors (API and ISO)?	
Have soil data (factor = 1.0) for API and 1.15 F_y for ISO in the 10,000-year wave assessment been used?	
Have you checked API spreadsheet maximum member capacity predictions versus the software results?	
Have the output files been checked for additional failure information, such as buckling, joint failure and fracture, in order to understand the analysis behavior?	
Has a 100-increment solution been performed (without iterations) and checked with a 1000-increment solution (worst cases)?	

of the probability of increased environmental load due to wind, wave or current causing platform collapse. The RSR is defined as the ratio of the environmental load set causing collapse to the environmental load set with a 100-year probability of exceedance. Therefore, a platform with an RSR of 1.0 will have a deterministic probability of failure in the region of 1 in 100 (10^{-2}).

Note that design of an offshore platform to the 100-year event in accordance with API RP2A will deliver a structure with an ultimate capacity, or design capacity, in the range of approximately 1.8 to 2.5 times the design load. This design capacity represents the explicit factors of safety in the design code as well as the implicit margins associated with a variety factors, such as mean steel strength, conservatism in foundation capacity assumptions and presence of additional steel for temporary load conditions in case of construction, transportation and installation and the increased-capacity-associated system versus component failure of a 3-dimensional space frame structure.

Figure 7.19 shows a comparison of two geographical regions. The origin of the plot represents the design point for both regions. In both cases, the design capacity of the structures is assumed to be in the range of 1.8 to 2.5 times the 100-year event. This reserve above the 100-year design is the RSR.

The two curves shown in the figure represent the relationship between the load associated with the 100-year environmental event and the load from N-year events.

API SIM refers to these curves as *region hazard curves*. The figure shows that a typical platform structure in region 1 with a shallower hazard curve would

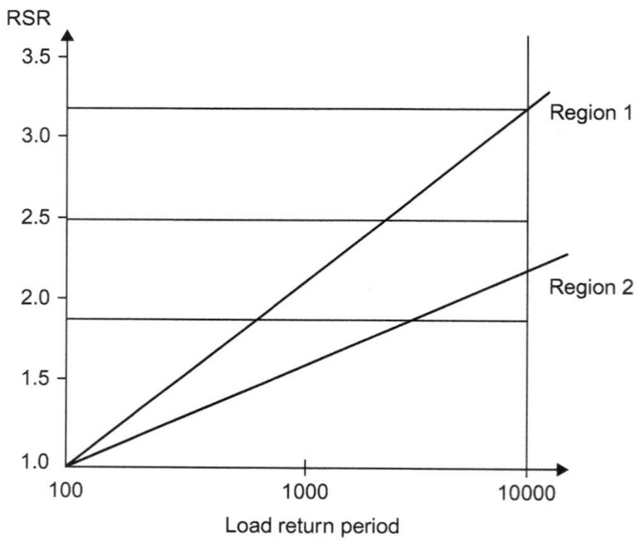

FIGURE 7.19 Comparison of regional hazard curves.

be expected to have sufficient reserve strength to survive a 10,000-year regional environmental event with an RSR of approximately 2.3 being required. On the other hand, a typical platform structure in region 2, with a steeper hazard curve, would not be expected to survive the 10,000-year environmental event, since it would require an RSR of approximately 3.4, which is greater than that explicit in API RP2A.

This example illustrates that, where regions have different hazard curves, the same RSR will be associated with significantly different probabilities of failure or structural reliability. Therefore, in order to meet the objective of consistent minimum reliability against environmental overload for offshore platforms around the world, it is apparent that acceptance criteria need to be region-specific, because they should take account of the slope of the regional environmental hazard curve.

The probability of structural failure of offshore platforms that exist in the region in question should be understood very well. This is achieved by carrying out a structural reliability analysis of the platforms selected as representative of the fleet. Statistical sensitivity studies may be required to establish whether certain characteristics are significant.

The probability of failure is estimated using reliability analysis to account for uncertainties in the derivation of both the loading and the response.

7.9.1 Nonlinear Structural Analysis in Ultimate Strength Design

Conventional structural analysis practice relies on an idealized linear-elastic model to determine forces in the structure members. The components' adequacy is then determined by comparing the applied element forces with parametric code-check capacity equations that are based on checking every element's failure data.

In ultimate strength analysis, nonlinearities associated with the plasticity and large deformations of components under extreme load are included explicitly in the element modeling. The analysis tracks the interaction between components as member end restraints are modified and forces are redistributed in response to local stiffness changes. The sequence of nonlinear events leading to a global collapse mechanism and the associated system capacity are determined.

Thus, while the typical linear design process checks for the adequacy of each individual component, nonlinear ultimate strength analysis models the performance of the system as a whole.

Elastic structural analysis and nonlinear structural analysis are shown in Figures 7.20 and 7.21.

In general, nonlinear analysis has four basic techniques:

- General-purpose nonlinear beam column models
- Plastic hinge beam column models

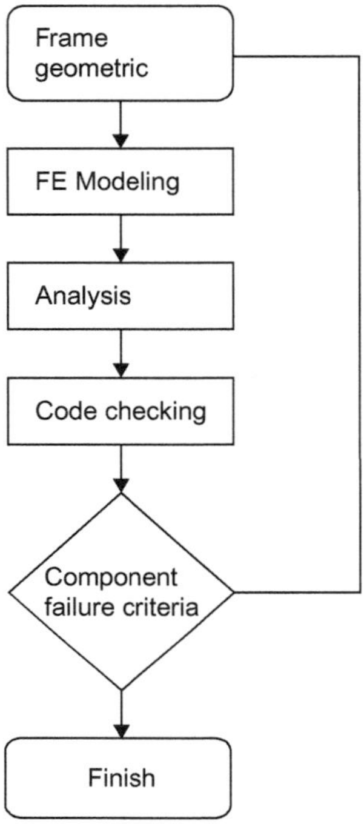

FIGURE 7.20 Procedure for conventional analysis.

- Phenomenological models
- Shell finite element models.

Nonlinear Beam Column Models

Beam elements don't generally incorporate geometric nonlinearity at the element level and multiple beam elements along a member are required to accurately model buckling responses.

The equation of equilibrium is usually evaluated in the deformed condition, which is the largest displacement.

Stiffness is integrated numerically from the stress distribution at points across the section due to the combined action of axial forces and moments.

For detailed stress-strain material modeling, formulations, including strain hardening, may be available but they may require definition in actual elemnt performance rather than engineering format. The structure element may also be modeled by shell elements or with a combination of shell elements, as for joints that were damaged or have a crack and representation of members by beam element.

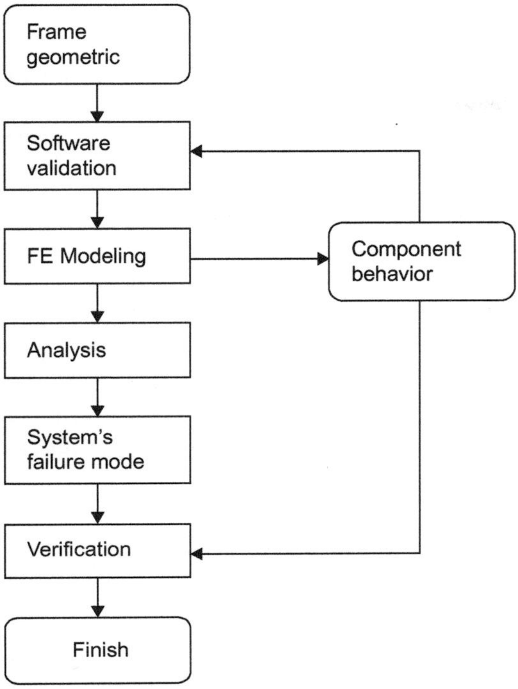

FIGURE 7.21 Nonlinear ultimate strength analysis.

Certain techniques have been developed specifically for ultimate strength analysis of frame structures. The elements have been derived to model beam column behavior, and each column member is usually represented by one single element member in the model that takes the buckling effect into consideration.

Plasticity may be presented by modeling the propagation of yield through the section and along the element sections.

Depending on the structure element, hinges will form at the ends or in the midpoint of the member, based on the whole structure system.

Elastic plastic behavior analysis or gradual strain hardening can be expressed as the energy dissipated in the element. Specification can be adjusted to account for imperfection, rather than mean component strength provided by the material test data.

Phenomenological Models

In phenomenological models, component responses are prescribed with force and deformation relationships, which can be empirically related to element geometry or determined through analysis. A single element represents the member behavior.

In some software, the type of failure must be defined and anticipated for each element of the structural component and loading mode before starting the analysis, and the element type and its nonlinear characteristics must be defined accordingly. Element types include:

- Elastic members
- Buckling members
- Beam members
- Frame members
- Joints.

Material nonlinearity and initial imperfections are taken into consideration in this method and the material data will be obtained from direct test results. The yield of the section may be determined from the full section under the yield stress.

Shell Finite Element Models

Any member or joint-can component may also be modeled by shell elements, or it can be modeled with a combination of shell elements and beam elements.

Steel usually yields before it fractures because the uni-axial yield stress is significantly less than the stress required for a cleavage fracture. However, in three-dimensional stress fields, the individual stress components can be much greater than the uni-axial yield stress. Without yield occurring in the vicinity of a crack tip, a simple elastic calculation of the stress distribution combined with the von Mises equation predicts that the stresses may reach 2.5 times the uni-axial yield stress before yield occurs.

The fracture mechanism is usually complex, as it is much more dependent on the largest stress and is less affected by the tri-axial stress pattern. Therefore, in the tri-axial stress field, the applied stress may exceed the yield stress, so that fracture becomes the failure mechanism. The main factor in the fracture is the plane strain condition due to effect of the Poisson ratio.

Brittle fracture is more likely to occur in cold conditions and at high rates of loading (e.g., under impact). The toughness grade of material specified for any application should be carefully selected to account for these factors as well as the likely stress levels, crack growth through fatigue, inspection procedures and consequences of failure.

Modeling the Element

The sliding action of piles within legs should be modeled with the approximate constraint conditions, which allow unrestrained differential axial displacement and rotations but couple the lateral displacements of piles and legs.

Grouted piles can be modeled as composite leg-pile members. On the other hand, leg and pile members can be modeled as separate elements with full coupling between end degrees of freedom.

For ungrouted piles, the lateral displacements for pile and leg are usually coupled at each horizontal elevation, but some differential movement can occur between elevations.

Conductors make a limited contribution to the strength of the steel frame system. In most cases, they can be modeled as pure load-attracting members; as long as their load contributions are correctly captured, the conductors can be omitted from the strength calculations. However, for structures with limited foundation resistance, conductors can contribute significantly to the foundation stiffness and collapse strength of the structure. In that case, the conductor should also be modeled and analyzed as a structural element and should be included in the integrated structure soil model. The conductor guide framing at the mud line may then be highly loaded and may need a more detailed inclusion in the structural model as primary framework.

Conductor Connectivity

Conductor guides should be modeled in sufficient detail to transfer the required loads into the primary framework. Local overstresses in conductor guides should not cause concern if they occur in areas where the model has been purposely simplified and the surrounding primary members have sufficient capacity.

If conductors are included as structural members, the sliding action of conductors with the conductor guide frames should be represented in the structural model taking into consideration that the differential rotation is unrestrained.

In the case of a curved conductor, the connection between curved conductors and their guide frames due to the imposed conductor curvature and friction may require specific consideration.

7.9.2 Structural Modeling

Structures must be modeled in sufficient detail to ensure that the nonlinear analysis program adequately captures the relevant global and local failure modes and load redistribution.

The models for component strength, such as member compressive strength and joint strength, are semi-empirical. They have a theoretical basis but are formulated to conform to experimental data. In general, all theoretical formulations need some calibration in order to represent the behavior of "real" structural components with sufficient accuracy.

Moreover, it should be possible for an engineer to select specific failure criteria and have the analysis tools calculate the structure's strength based on those criteria. In such a case, the requirement for the analysis tool is not to present theoretically correct solutions for the structure, but to present a consistent strength estimate based on the engineer's specifications.

This implies that the structure analysis software should be able to represent different failure criteria for different structural elements as specified by different codes of practice.

Instead of modeling the structure out of purely geometric considerations, the modeling must consider the analysis tool that will be used for the nonlinear analysis, and the mathematical formulation that is embedded within the program. Some nonlinear analysis programs have built-in features for calibrating component failure modes to specific criteria. For other programs, the engineer must give special consideration during the modeling of the structure in order to make the program represent the required failure modes or limiting criteria. Which consideration to take depends on the component as member, joint or foundation and on the mathematical formulation.

If there is any doubt about the FE formulation, a simple model should be subjected to a well-defined load and deformation path. This will allow the results to be judged and calibrated against engineering practice.

This discussion gives a set of modeling recommendations to help make the nonlinear analysis tool produce reliable results that conform to recognized failure criteria and design formulations.

Frame Modeling

In frame modeling, the space frame model should describe the three-dimensional geometry of the platform.

The model for ultimate strength assessment can usually be significantly simpler than models required for design and fatigue analysis. The primary framework essential in maintaining overall integrity of the structure for the in-place condition must be included in the structural model. Secondary structures and members generating dead loads and/or environmental loading need only be represented in sufficient detail to introduce the relevant loads on the primary structure.

The analytical models should consist primarily of beam elements. The structural members of the framework may be modeled using one or more beam elements for each span between the nodes of the model of the framework.

Primary Framework

The primary framework of the structure comprises those members that provide the stiffness and strength of the structures; these are usually the legs, the piles, the vertical diagonal members and the main plan frame bracing members.

Secondary Framework

The secondary framework consists of members that don't contribute to the global stiffness and strength of the framework. In general, their structural contribution may be neglected and they do not need to be included in the model as structural members. Boat landing, fenders, lowest deck and walkways are examples of secondary members. Secondary framework should be modeled in sufficient detail to transfer the required loads into the primary framework.

Some local overstresses in secondary framework should be accepted if they occur in areas where the model has been purposely simplified and where the adjacent primary members show sufficient capacity. However, they should be subject to separate justification in each case.

The following secondary framework should be included in the model:

- Members or joints that are essential for transfer of reaction loads from conductors and appurtenances, etc., to the main structural elements.
- Members or joints that are highly loaded by local wave action. A separate assessment may be done on the local behavior. The global assessment may be performed with a simplified model if it is demonstrated that the load can be carried and transferred by the secondary framework.
- Secondary members associated with launch framing, mud mats, conductor support during transportation and others should be included in the model if they provide significant support to primary members and thus contribute to the system capacity.

When neglecting the structural contribution of secondary members, their load-attracting properties—that is, loading due to self-weight or hydrodynamic loading—should still be accounted for and included in the appropriate loading condition.

Conductors and other appurtenances, such as launch cradles, mud mat, J-tubes, risers, skirt pile guides and others, should be included in the model if they contribute significantly to the overall strength of the structure or foundation. Otherwise, they may be disregarded as structural elements.

Dented Beams and Cracked Joints

Denting of members usually occurs when an object is dropped on the bracing, in which case the strength of the member will be reduced by the following formula in the case of normal strength and bending strength:

$$\left(\frac{N_U}{N_P}\right) = \exp\left(-0.08\frac{D_d}{t}\right)$$

$$\left(\frac{M_U}{M_P}\right) = \exp\left(-0.06\frac{D_d}{t}\right)$$

where D_d is the denting depth and t is the wall thickness.

The damaged member should be modeled in sufficient detail to assess the impact of the damage on the global behavior of the structure. A lower bound on the remaining structure's strength can usually be obtained by removing the affected member(s) from the model. A less conservative strength estimate is obtained by modeling the damage in the nonlinear analysis. Some nonlinear analysis software has special formulations for modeling dented or distorted members. Alternatively, the damage can be modeled explicitly by shell elements or a reduced cross-sectional area can be specified in the damage zone.

The same philosophy can be applied in the case of cracked joints to eliminate the member from the model. The less conservative approach is to reduce the strength of the affected joint by some fraction. Note that the presence of a crack will also limit the ductility and the cyclic capacity of the cracked joint. In the structure analysis the cracked joint will be simulated near to its strength capacity. In some cases, for special joints, one can use the finite element method to define the joint capacity. A crack will incur a reduction factor, F_{JR}:

$$F_{JR} = \left(1 - \frac{A_C}{A}\right)\left(\frac{1}{Q_\beta}\right)^{m_q}$$

where A_C is the cracked area and A is the weld length multiplied by the thickness, T. Q_β is known as the geometrical modifier, usually used in design codes to account for the increasing capacity of uncracked tubular joints at high β (where β is the brace diameter/chord diameter) and the maximum strength occurs at β values equal to 0.6:

$$Q_\beta = 1 \quad \text{for} \quad \beta \leq 0.6$$
$$Q_\beta = 0.3/\beta(1 - 0.833\beta) \quad \text{for} \quad \beta > 0.6$$

m_q is the power allocated to Q_β and depends on the approach used to estimate the capacity of the uncracked joint. In case of tubular joints containing part-thickness flaws, $m_q = 0$.

For tubular joints containing through-thickness flaws, validated correction factors giving lower-bound estimates of the collapse load are at present limited to joints with b ratios less than 0.8 and the following configurations:

- K joints with a through-thickness crack at the crown subjected to balanced axial loading
- Axially loaded T and DT joints with a through-thickness crack at the saddle.

7.9.3 Determining the Probability of Structural Failure

It is necessary to understand the probability of structural failure for the range of offshore platforms that exist in the region in question. This is achieved by carrying out a structural reliability analysis of the platforms selected as representative of the fleet. As mentioned already, statistical sensitivity studies may be required to establish whether certain characteristics are significant or not.

The probability of failure is estimated using reliability analysis to account for uncertainties in the derivation of both the loading and the response. The analysis procedure is outlined in the flow chart in Figure 7.22.

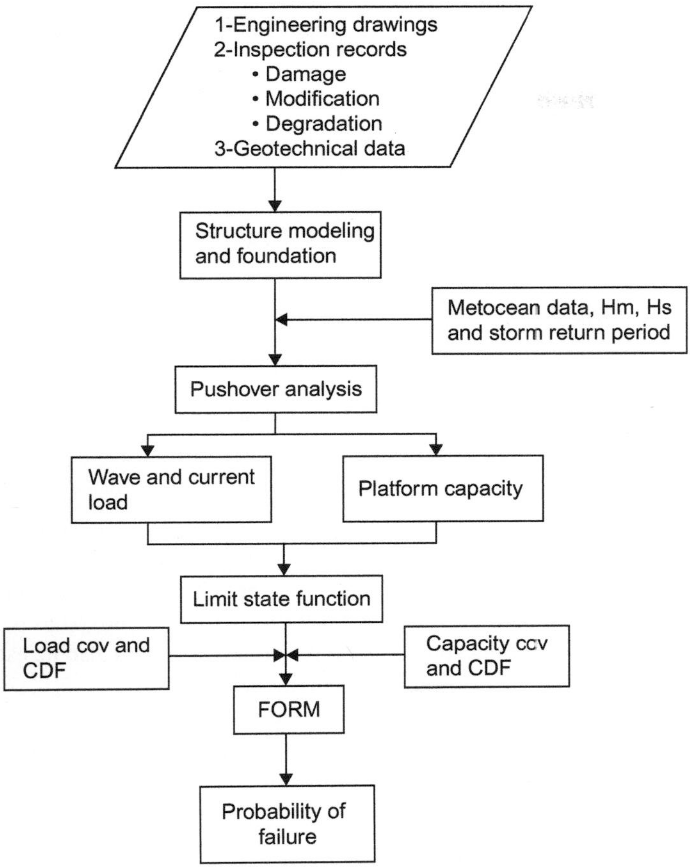

FIGURE 7.22 Procedure for calculation of probability of failure. CDF = cumulative distribution function; cov = coefficient of variation; FORM = first-order reliability method.

7.9.4 Offshore Structure Acceptance Criteria

In order to establish the assessment acceptance criteria, it is necessary to establish the relationship between the probability of failure of the representative structures and their reserve strength, defined in terms of RSR.

In the reliability analysis procedure described above, the structural capacity of the platform, R, is assumed to be a random variable with a log-normal distribution. The mean value of R is estimated from an ultimate strength (or pushover) analysis, as described. By arbitrarily moving the mean structural capacity to represent a range around the calculated capacity, it is possible to estimate the probability of failure of a range of designs of the same structural type. This allows the relationship between RSR and probability of failure to be

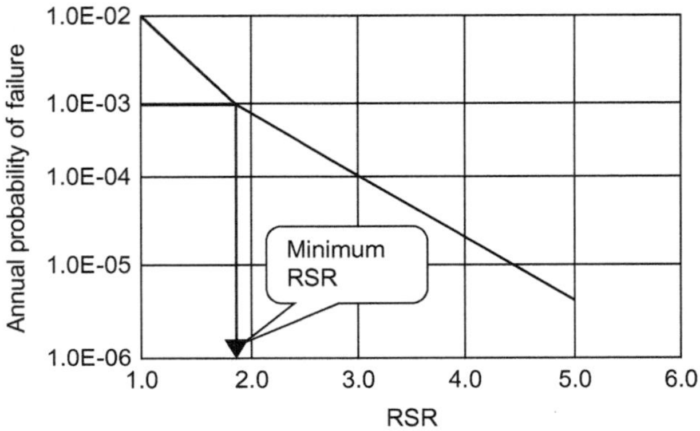

FIGURE 7.23 Relation between RSR and the probability of failure.

approximated and plotted. It should be noted that the approach is an approximation and assumes that the load on the structure remains constant. In fact, it may be argued that increasing RSR would perhaps also increase load; assuming larger members and therefore increased fluid drag loading—and conversely reducing the RSR—may reduce the load. If this effect were judged to be significant for the structures in question, the effect on the load could also be estimated and included by adjustment of the estimated mean load.

Figure 7.23 illustrates the relation between RSR and probability of failure. As shown in the figure, if the probability of failure is defined to be equal to 1×10^{-3} the corresponding minimum RSR can be obtained.

7.9.5 Reliability Analysis

For platforms that experience storm waves from a dominant approach direction, probability of failure may be calculated for the dominant direction. However, it may be necessary to calculate the failure probability for up to eight wave-approach directions (four orthogonal directions and four diagonal directions) for platforms that do not have a dominant wave-approach direction. In this case, failure in each direction is treated as a failure element, and the multiple-direction failure forms a series system for reliability calculations.

For each wave direction, a system reliability approach is also required for estimation of the annual probability of platform failure. During the system reliability analysis, a platform is modeled in terms of two basic subsystems:

- The jacket, including foundation
- The topsides.

The platform fails if either the jacket subsystem or the deck subsystem fails. The platform system reliability analysis models the two correlated subsystems as a system in series. In most cases, the failure of the topsides subsystem does not occur in advance of failure of the jacket/foundation subsystem, even with large wave force effect on the topsides under extreme storm conditions. The failure of the topsides subsystem is not considered acceptable criteria against environmental overload.

Limit State Function

The limit state function for estimating the failure probability may be defined as

$$g(X) = R - L \tag{7.28}$$

where R is the platform capacity in terms of maximum lateral load that the platform can withstand before system failure or collapse. L is the total environmental load, which consists of wave and current load, W_v and wind load, W_l, i.e.,

$$g(X) = R - W_v - W_l \tag{7.29}$$

The limit state function $g(X)$ defines a failure criterion that is a function of all random variables X. Failure occurs when the load L exceeds the capacity R or when $g(X) < 0$.

First-Order Reliability Method (FORM)

With the establishment of a limit state function, the conventional formula for computing the probability of failure is:

$$P_f = \int_0^{\infty} (1 - F_s(x)) f_R(x).dx \tag{7.30}$$

where F_s is the cumulative distribution function (CDF) for the total environmental load variable S, and f_R is the probability density function (PDF) for the capacity variable R.

The probability of failure can be calculated using the Rackwitz-Fiessler FORM method. In this method, the probability of failure is estimated using first-order approximation to the limit state at the design point. The Rackwitz-Fiessler FORM method includes the following steps:

1. Transform non-normal-distribution random variables that are used to define the limit state function to equivalent normal-distribution variables. The CDFs and the PDFs of the actual variables and the equivalent normal variables are equal at the iteration point on the failure surface.
2. Iteratively search the most probable point (MPP) of failure in the limit state surface based on Newton-type recursive formula.
3. Compute the reliability index β.

4. Determine the probability of failure using the equation $P_f = \Phi(-\beta)$, (2.3-2), where Φ is the CDF of a standard normal variable.

7.9.6 Software Requirement

The fundamental requirement for the software is that it should adequately represent the relevant failure modes for the basic components in framed offshore structures:

- Members
- Joints
- Foundation
- Loading.

The software should have clear documentation of which facilities are available and how they should be applied. The software should include specification of any limits of validity for special features (e.g., *D/t* limits for a particular local buckling formulation, β-range for a joint capacity formulation, etc.).

From the quality assurance point of view, it is essential that the software can document compliance with theoretical solutions and test results for single components, substructures and structural systems.

If there is any doubt about the formulation, a simple model should be subjected to a well-defined load and deformation path. This will allow the results to be judged and calibrated against engineering practice.

Simple program input reduces the possibilities for modeling errors and errors in result interpretation. The input should be given in familiar engineering terms. Unfamiliar or specialized input parameters increase the possibilities for input errors.

The software should include pre-processing tools and default parameters to reduce the need for detailed information from the user. This is especially relevant for specialized information outside the main engineering focus, such as parameters concerning numerical integration, mathematical stability or detail parameters for special program features.

Program default parameters should be listed with a description of what they imply and what any variation may represent.

The primary and essential validation of nonlinear analysis results comes from understanding the development of the global collapse mechanism.

The software should present the analysis results efficiently, so that the structural behavior is easily understood by the engineer and is readily conveyed to others. Extensive use of computer graphic capabilities is recommended.

Critical members should be identified, along with documentation of their strength and buckling load.

The software should contain a self-checking mechanism so that clear indications are given if, at any stage in the analysis, the results violate basic assumptions of the theory.

The treatment of different failure modes will vary from formulation to formulation. Different failure modes may typically be treated at one of the following levels:

1. As specialized features implemented in the program (e.g., local buckling criteria implemented in the program, including dent growth and modification of post-buckling load shedding)
2. As modeling guidelines (e.g., describing how the program's input parameters should be modified to capture the appropriate reduction in axial capacity and the accelerated load shedding in the post-collapse range)
3. As a provision for separate, manual checking after the analysis is completed.

Furthermore, program modules separate from the structural analysis module are often used to calculate soil parameters and environmental loading.

The interface between the modules should then be well defined and clearly documented, to prevent user errors or misunderstandings during transfer of data.

The following is a list of general modeling requirements for nonlinear analysis of framed offshore structures.

Material properties

- Yielding/yield hinges
- Strain hardening
- Strain rate effects

Section properties

- First fiber yield
- Gradual plastification of cross-section
- Fully plastic capacity
- Interaction between axial force and bending capacity
- Strain hardening

Member properties

General

- Elastic
- Compression (crushing) failure
- Yield (tension) failure
- Stability failure
- Post-collapse behavior

Behavior modes

- Beam bending
- Column buckling
- Residual stresses/initial imperfections
- Member ductility

- Local buckling
- Hydrostatic pressure

Special formulations

- Dented members
- Cracked members
- Grouted members
- Cyclic degradation

Tubular joint properties

Formulae

- API
- HSE
- User-defined
- Mean
- Characteristic

Behavior modes

- Elastic flexibility
- Ultimate capacity
- Nonlinear deformation

Special formulations

- Grouted joints
- Ring-stiffened joints
- Cracked joints
- Ground joints
- Cyclic degradation

Foundation properties

- Behavior modes
- Lateral soil failure
- Axial failure
- Monotonic behavior
- Fully degraded behavior

General finite element modeling

General

- Joint eccentricities
- Linear dependencies
- Shim elements
- Locked-in forces

- Linear springs
- Nonlinear springs
- Pinned supports
- Fixed supports
- Spring supports
- Prescribed displacement
- Prescribed acceleration
- Load modeling

Loads

- Load combinations
- Concentrated nodal loads
- Linearly distributed member loads
- Thermal loading
- Environmental loading
- Self-weight calculated from density and section properties

Wave kinematics

- Stokes fifth-order
- Airy
- Wheeler
- Stream function
- Current loading
- Buoyancy loads
- Marine growth

Loading algorithms

- Initial loads (self-weight and buoyancy)
- Proportional loads
- Nonproportional loading
- Wave height incrementation
- Wave-in deck forces
- Cyclic storm loading

As presented by van de Graaf et al. (1994), the relation between the probability of failure and the RSR for offshore structures in the Gulf of Mexico, southern North Sea (SNS) and northern North Sea (NNS) is presented in Figure 7.24.

Based on El-Reedy and Ahmed (2002), the reliability analysis of the tubular joints of offshore structures under axial tension, compression, in-plan bending and out-of-plan bending at yield is discussed considering the capacity of the tubular joint.

The probabilistic analysis is performed using Monte Carlo simulation technique and is based on the API RP2A-LRFD model. The analysis includes the variability of the steel yield strength and dimensions of the brace and chord.

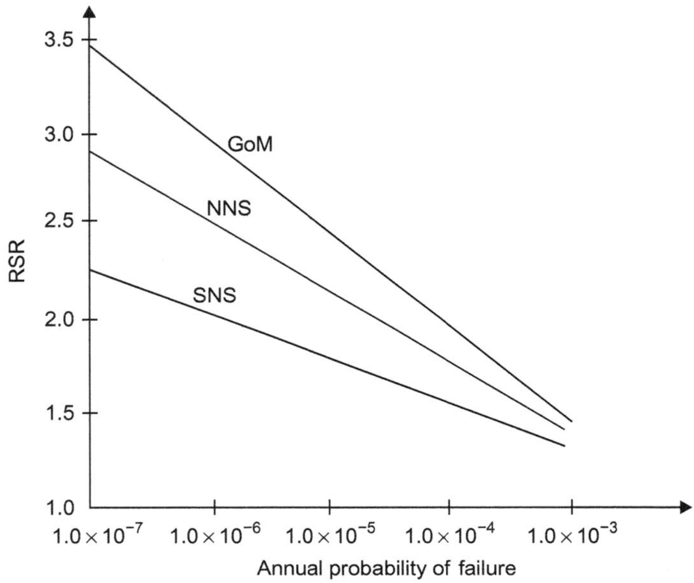

FIGURE 7.24 Relationship between RSR and annual probability of failure in different locations.

The ratio between the tubular joint capacity mean value and its nominal value is presented, and, in any case, the tubular joint capacity follows a gamma distribution. The parametric study is performed taking into consideration a different chord to branch ratio and different values of steel yield strength.

This study is compared with an experimental test done to tubular joints for offshore structures and the variation of the model calculation is presented.

Evaluating the strength variability of tubular joints is an essential requirement in developing probability-based design criteria in assessing the safety of existing design.

The capacity of the tubular joint in all cases of loading and different values of β is well presented by a gamma distribution, with an average bias factor 1.15 and coefficient of variation (cov) of 0.28.

The value of the chord thickness has a major effect on the variation in the capacity of the tubular joint. So, thickness of the tubular member must be under rigid quality control. The bias factor has a slightly smaller effect on the values of β.

7.10 CASE STUDY: PLATFORM DECOMMISSIONING

The platform in this case study is shown in sections, due to its failure. It was necessary to demolish this platform, because visual survey showed that the current condition of the platform would not allow for safe operations for long-term topsides intervention, due to instability of the platform. The recommendation was to prepare the structure for a controlled toppling. In addition, it was determined that diver operations were not possible due to significant overhead and dropped object hazards.

Once the platform is resting on the bottom, the wells can be secured using methodology similar to that applied before to numerous demolished platforms and wells projects. The main principle of the toppling process is to disconnect the risers near the sea floor and to create hinges in the conductors and north legs, thus creating the potential for the platform to fall in a general southerly direction. All subsea operations would be conducted by ROVs. Tugboats would then be utilized to pull the platform over for final toppling.

The dynamic nature of this type of work means that the work plan may change due to weather, safety or operational restrictions to ensure continued work-flow efficiencies.

Also, the platform could fail with continued regular weather forces that affect the structure, and it is possible that collapse could occur after any structural cutting. So it is important to prepare for this potential situation by proposed tooling packages that can be used for toppling, post-toppling debris removal, and any intermediate situation.

A toppling analysis was developed through basic overturning and structure analysis calculations. The analysis revealed that the bracing had to be cut in defined locations, as shown in Figure 7.25. Analysis showed that the platform would require around 85 to 120 tons of force for toppling after proposed cuts to

FIGURE 7.25 Locations for cutting the bracing.

the riser and reduction of the conductor and leg capacity by making a notch to convert their behavior to hinges from a structural point of view.

The platform will be rigged at two points to ensure that the deck structure and jacket collapse together. A barge will be used to transfer personnel and rigging to the platform. The barge will be used to handle the long rigging and to make the connections to the buoy.

In this type of project, a crane vessel must be on standby on site, as per the site safety procedure and safe boarding plan, to help shelter the platform from weather and for use in transporting personnel from the platform in case of any issues with the barge crane.

All platform boarding and pre-rigging will be completed after ROV survey of the jacket below waterline and in a period of limited wind or waves.

All slings and shackles used in the toppling operation will be used at a safety factor of 2. This allows the slings to be smaller and greatly improves the ease of handling, which is essential for maintaining safe working conditions on the platform.

In most cases, to obtain better overhead access to the well-head deck below, it is necessary to cut and remove the member; in this case, the southern fence on the helideck.

A structural survey should be performed to identify the integrity of the jacket and conductors before any cutting operations are started. This survey will include review of previous structural data regarding existing failures and will be used to track any additional failures that occur during the cutting portion of the toppling.

All survey information will be used to develop the three-dimensional model, which will be continuously updated in the field to be matched with any new events throughout the project until the platform is lying on the seabed after toppling.

If, during the cutting process, significant structural movement is noticed, a brief survey should be conducted to review any additional changes to the global structure.

It is very important to give the ROVs safe access to the conductor bay for notching or cutting operations. In this case, the first important step is to inspect all the other members before cutting to ensure they are functionally carrying load during the cutting and dropping of the brace and the horizontal member.

On the south side of the platform there are two risers that will be notched on the south face. Great care will be taken with these risers because they may be providing some restraining force to the platform.

On the other hand, there are four risers on the north face of the platform that will be cut.

In addition, when rigged up, the vertical bracing on the north face should be cut prior to final cutting of the north legs.

After cutting the risers, the next step is to cut the legs. The expected final cuts will be the complete severing of the north legs. This operation will require running two 36-inch diamond wire saws simultaneously. The notching and cutting machines are shown in Figures 7.26 and 7.27, respectively.

Operations will proceed by connection of the tugboats to the toppling rigging on the buoys and standby for direction. The diamond wire saw will be

FIGURE 7.26 Notching machine.

FIGURE 7.27 Cutting machine.

set up on deck, the hydraulics attached and tested and the diamond wire saw will be lowered to the sea floor by crane and then unhooked by the ROV. The ROV will maneuver the saws into place (just above the mud line) and clamp the saws onto the leg utilizing surface hydraulics. The ROV will monitor cutting from a safe distance.

For the north side jacket leg cuts, the tugs will tie onto the platform before cutting operations begin. The first tug will pull directly to the south, while the second tug will pull to the southeast. During this time, tugs will be directed based on the reaction of the global structure, to ensure that once the platform gets momentum it is guided to continue falling.

During the toppling work on any platform requiring decommissioning, there is a potential for pollution from pipeline/risers (if they have not been recently pigged), trapped annular and other well-bore fluids, deck storage tanks and the like. Pollution prevention/control entails necessary response vessels, containment boom, sorbent boom and skimmer/separator equipment. All equipment should be ready to be deployed.

After the platform is toppled, it will be necessary to survey and to depict the new orientation and altitude of the platform on the seabed. An ROV visual survey and single side band sonar data will be collected to create a revised three-dimensional model of the platform resting on the seabed. Once the platform has been surveyed in its new orientation, any additional structural debris removal required to clear the direct well bay area should be completed.

It is anticipated that the bell will be kept approximately −100 feet FSW. The structure will need to be surveyed for any high points that are above this depth and these items will need to be removed as well.

Toppling analysis should determine the pulling force required to topple the platform after the cutting and notching take place on the legs. Then the location of each tug that will be pulling against the single rigging point can be defined in the analysis. In addition, the analysis should provide the required towline distances and compare them with the fall path to enable smooth toppling without problems.

Figure 7.28 shows the rigging arrangement that will be used for each tugboat.

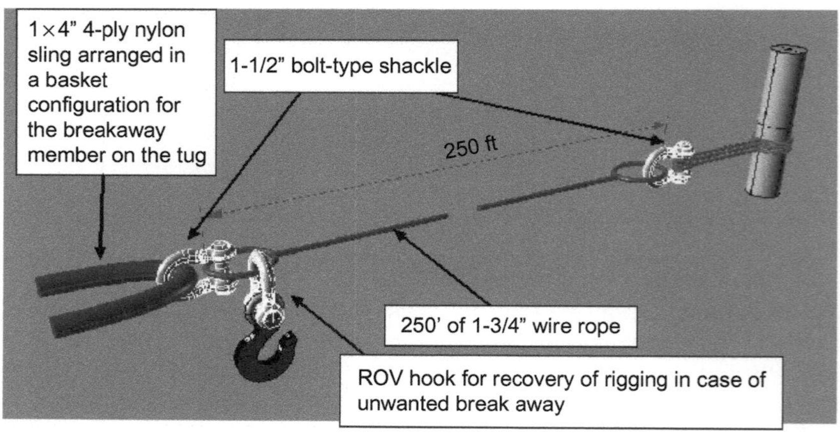

FIGURE 7.28 Rigging arrangement.

7.11 SCOUR PROBLEM

There are several methods for scour prevention and repair. Angus and Moore (1982) detail the methods used in the southern North Sea. Gravel grout bags and sand bags have been used effectively to fill in the scoured area and raise the seabed profile up to the original level. Car tires have also been used successfully. They are tied together to suit the geometry of the local scour holes, are dropped to the seabed and then are moved into position by the divers. Plastic seaweed is another useful method. The seaweed reduces the water velocity and thus mediates deposition of sediments. The seaweed system usually comprises continuous lines of overlapping buoyant polypropylene fronds that, when activated, create a viscous drag barrier that significantly reduces current velocity around the piles. The frond lines are secured to a polyester webbing mesh base that is itself secured to the seabed by anchors pre-attached to the mesh base by polyester webbing lines.

The action of reducing current velocity immediately prevents seabed sediment in the immediate area of the fronds from being transported (i.e., scoured out) and causes sediment transported across the fronded area to fall into, and collect within, the fronds.

At the design stage, jacket structures can be made less sensitive to scour by providing stronger piles and jacket legs. According to API, this is done by considering scour as occurring by about 1.5 times the pile diameter (and this value can be 2 times the pile diameter in special situations, depending on the soil type and previous experience).

7.12 OFFSHORE PLATFORM REPAIR

7.12.1 Deck Repair

The topsides incorporate robustness through consideration of the effects of all hazards and their probabilities of occurrence, to ensure that consequent damage is not disproportionate to the cause. Damage from an event with a reasonable likelihood of occurrence should not lead to complete loss of integrity of the structure. The structural integrity in the damaged state should be sufficient to allow a process system closedown and/or a safe evacuation. Framing patterns that provide load paths are preferred.

The maintenance plan should include the risk-based inspection technique. The first step of the inspection is visual inspection, which should be performed periodically to monitor for degradation of the structure. Figures 7.29 and 7.30 present examples of what can be found during inspection (such as corrosion of the main steel for the stairs and the main supports on the helideck truss support). Repair of the deck structure is usually easy to perform: the stair can be replaced by a new one and the helideck can also usually be replaced by a new one and then repaired and transferred to another platform if there is a big fleet. If there is no fleet, the helideck will be repaired.

FIGURE 7.29 Severe corrosion damage to the stair.

FIGURE 7.30 Severe corrosion of helideck main truss.

7.12.2 Load Reduction

Marine Growth Removal

Marine growth accumulates on the legs and bracing with time. As a result, the diameter of the members affected by waves will increase, so the lateral load due to waves will increase as well with time and can become critical if the marine growth thickness increases more than the predicted marine growth thickness in design. Therefore, removing marine growth can enhance structural capacity. Removal of marine growth can be performed by different techniques. A new technique for marine growth removal uses a mechanism that is fixed on the leg and bracing, as shown in Figure 7.31.

This technique generates substantial savings in cleaning and prevention costs (up to 80% reduction) for existing structures, reduces fabrication and installation costs for new structures and offers both removal and prevention capabilities. The mechanism can be easily installed by riggers or abseilers from above the water surface and by divers or ROV underwater. It enables instant visual inspection of the substructure without prior cleaning and it is harmless to the environment. It is effective in reducing the weight of marine growth for decommissioning of structures and eliminates safety hazards encountered by divers using high-pressure water jetting, which is the traditional method, especially in the splash zone environment.

The fouling system is a painting technique that prevents the accumulation of marine growth on the members and maintains existing structures free of regrowth after cleaning of previous growth.

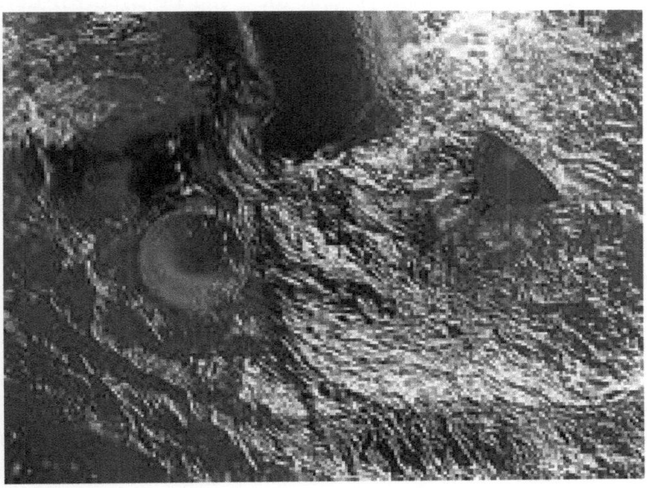

FIGURE 7.31 Working mechanism for marine growth removal.

Vibration Monitoring

Structural vibration monitoring offers a low-cost method of assessing structural integrity. The cost is low because monitoring is done exclusively above water at deck level by a small crew using lightweight portable equipment.

Platform natural frequencies and mode shapes are typically measured using sensitive accelerometers. These are mounted horizontally and detect the small sway movements of the platform. These movements occur primarily at the wave period but the platform's natural frequencies are usually clearly identifiable. By measuring movements at different locations on the deck, it is possible to distinguish between sway and torsional natural frequencies.

Similarly, by measuring movements at different elevations on the jacket, it is theoretically possible to distinguish between the first sway natural frequency and higher-order sway natural frequencies, although in practice higher-order natural frequencies tend not to be strongly excited.

The effect of damage of a single member on the overall natural frequencies will depend on the level of member redundancy of the jacket, and also on the contribution the specific member makes to the dynamic stiffness of the platform at that particular natural frequency.

The offshore platform natural frequencies depend, of course, on deck mass as well as jacket stiffness. Other factors may have a secondary effect, including variations in the effective mass of entrained water and nonlinearity of foundation stiffness. Further, the mathematical calculation of natural frequencies is a statistical process and each estimate of natural frequency has an associated error. Natural frequency data from continuously monitored platforms can be used to show day-by-day variations.

The above discussion indicates that natural frequencies are adequately stable and so sensitive to damage that they can be used for detection of changes in stiffness of the order of $\pm 3\%$ ($\pm 1\%$ change in natural frequency). Damage of a bracing can be detected on platforms with low-redundancy member configurations, whereas several or many member failures may occur on higher-redundancy structures before a change is detected. For example, a loss of a diagonal on a K-braced structure results in a frequency change of 9.5% to 11.5%, whereas a similar loss on an X-braced structure results in a change of only 1% to 2%. The former is detectable and indicative of a significant loss of overall stiffness.

Topside accelerometers, shown in Figure 7.32, are cabled back to a central data-collection station using conventional methods. A significant change in recent years has been the development of cableless underwater sensor packages. These systems are battery powered, and data are transmitted by hydro-acoustic telemetry.

Nowadays, vibration monitoring is widely used because it is so much less expensive than follow-up and because monitoring the performance of the natural structure allows the required action to be taken in a reasonable time.

FIGURE 7.32 Vibration monitoring attached to the deck beam.

7.12.3 Jacket Repair

There are many methods for strengthening and repairing an offshore structure platform. The main element of repair is the clamp.

Figure 7.33 presents the traditional procedure for repairing severe corrosion of a leg. The repair depends on removing the corroded part of the leg, fixing the half cylinder around the leg, then placing the other half, which will connect by bolts, and finally reinstalling the bracing by a clamp to the repaired leg (as shown in the figure).

The method of repair when a buckling bracing member requires strengthening is shown in Figure 7.34. As shown in the figure, the buckling member is connected from the midpoint to the leg by two clamps on the leg and on the bracing.

Corrosion of a horizontal member is shown in Figure 7.35. The main principle of strengthening this member is adding a new horizontal member that will be connected to the corroded member by clamps.

Clamps can also be used to strengthen all the jacket faces, as shown in Figure 7.36.

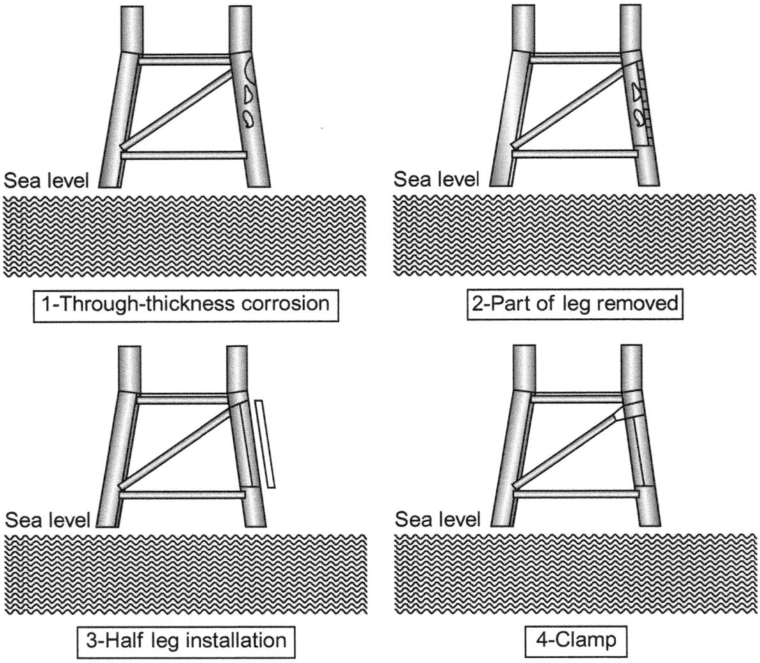

FIGURE 7.33 Repair procedure.

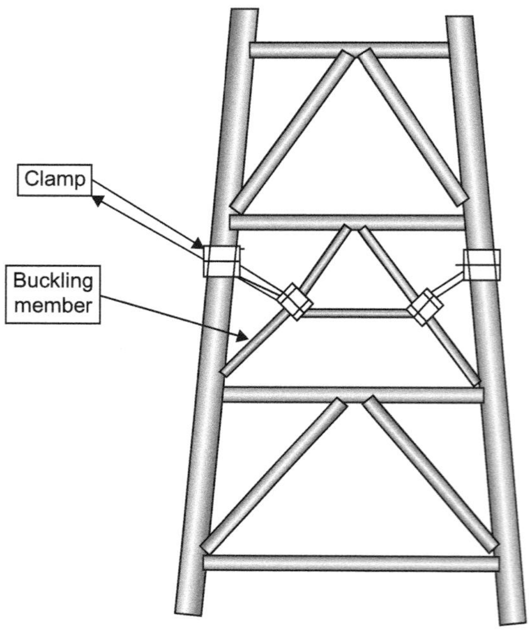

FIGURE 7.34 Bracing repair.

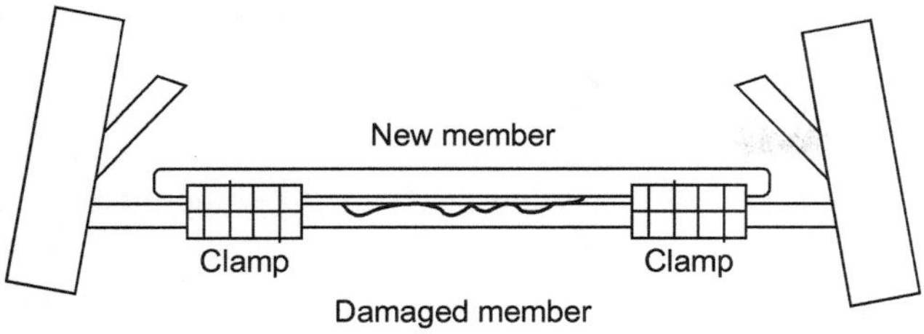

FIGURE 7.35 Corroded member repair.

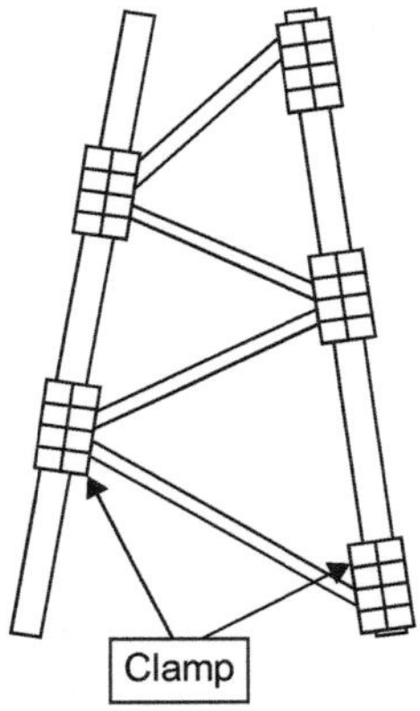

FIGURE 7.36 Repair using clamps.

7.12.4 Dry Welding

Welding is often regarded as the best modification and repair technique for strengthening and no doubt would be used even more often if it were not for the operational difficulties in its execution. Several welding techniques and welding processes can be considered:

- Dry welding topside
- Dry welding at or below the sea surface at one atmosphere using cofferdam or pressure-resisting chambers. All normal welding processes can be used, but gas tungsten arc welding (GTAW), shielded metal arc welding (SMAW) and, to a lesser extent, flux cored arc welding (FCAW) are the main methods used in practice
- Hyperbaric welding using habitats. The main processes used are GTAW and SMAW, although FCAW and gas metal arc welding (GMAW) are sometimes employed.

Repairs by both cofferdam and hyperbaric habitat welding techniques have proven track records. Since 1970, particularly in the North Sea, hyperbaric welding has been used as an underwater technique.

Most of the time required for an underwater welding repair is accounted for by the preparatory work, rather than the welding process, and the preparation must be planned with considerable care. Around the more complex node geometries, the time required for assembling and sealing the welding chamber can equal the time needed for the welding operation. A compromise in the execution is to divide the chamber into as few pieces as possible, thereby minimizing the amount of underwater assembly work (which increases cost), and minimizing the safety risk to the divers by reducing the size of the chamber components, especially where such work is to be carried out in tidal or splash-zone conditions. Furthermore, it should not be assumed that the actual geometry of the structure is precisely as designed, and it is important to perform a survey of the location before building the welding chamber. It will be necessary to clean portions of the structure to ensure that an effective seal can be achieved, allowing the chamber to be dewatered. The welding chamber must be made sufficiently large to enable the welders to have effective access to the weld site.

Dry Welding Topside

For repair of the topside structure above the water level, dry welding is routine. The area where dry welding is undertaken is normally designated a temporary hazardous area and all routine safety procedures must be followed. Hot works are a consideration when welding topside. Platform shut-in may be required, depending on the location of the welding relative to the well bay and equipment/piping.

Dry welding topside is the most widely used form of welding. The only limitation is the requirement for hot works.

Dry Welding at or below the Sea Surface

Since a large body of welding technology exists relating to normal atmospheric pressure, a logical approach to underwater welding repair is to duplicate surface welding conditions by providing a one-atmosphere environment at the repair site, but this strategy is limited to shallow water depths. Two methods are available that can achieve this:

- Cofferdam: This essentially is a watertight structure that surrounds the repair location and is open to the atmosphere. The structure can be open-topped or it can have a closed top with an access shaft to the surface.
- Pressure-resistant chamber: The work site is surrounded by a chamber constructed as a pressure vessel, capable of withstanding the water pressure at the depth of the repair location. Once the chamber is in place and sealed to the structure, it is dewatered and the pressure can then be reduced to one atmosphere. The repair crew can transfer to the welding chamber in a one-atmosphere environment, within a diving bell, to perform the repair.

Dry welding is also possible at pressure below the sea surface using a hyperbaric habitat or chamber. The chamber is filled with gas equal to the hydrostatic head at the weld depth. The following factors govern the selection of habitat:

- The extent of welding required
- The complexity of repair site geometry
- Depth of repair
- Welding process and ancillary equipment
- Environmental conditions.

Given that conditions within the cofferdam or welding chamber duplicate those on the surface, any normal welding process could be used. In practice, GTAW and SMAW predominate, with minor usage of FCAW.

The primary limitation for atmospheric welding below the sea surface is depth. Differential pressure at depth precludes the use of cofferdams or pressure-resistant chambers due to size and related cost.

In summary, dry welding at or below the sea surface guarantees a good-quality weld. Water depth and potential costs are limitations.

Hyperbaric Welding

Hyperbaric welding is the most widely used dry weld repair technique for primary structures and pipelines. The repair site is again enclosed within a working habitat, which is dewatered by filling the habitat with gas. Since the gas and water will be at equal pressure at a point close to the bottom of the chamber, the maximum differential pressure will be at the top of the chamber, and obviously depends on the height of the chamber. This differential pressure [normally a few tenths of a bar (10's of kPa)] is easily resisted by lightweight habitats and simple flexible seals, making deployment and sealing of the work chamber operationally feasible. The depth for hyperbaric welding is only limited by access.

A typical hyperbaric welding operation requires the following equipment:

- Purpose-built cofferdam or pressure-resistant chamber
- Diving support
- Environmental control equipment
- Pre- and post-weld heating equipment
- Welding equipment (often GTAW and one other)
- Weld inspection equipment
- Equipment to remove marine growth and grit blast
- Temporary holding clamps to take the weight of additional members and maintain root gaps, as needed
- Crane capacity.

The problem with hyperbaric techniques is that the environmental pressure at which the weld is carried out is essentially that of the work site. These elevated pressures affect the gas, slag and metal reactions for all welding processes, and the high-density gas enhances the rate of heat loss from the weld. Hyperbaric welding research is mainly concerned with ensuring that, for any specific environmental pressure and composition, welding parameters can be specified that will ensure the production of welded joints with properties acceptable to the project specification and the supervisor who is responsible for the structure that will use this type of weld. Given that the welding process has to be specially optimized for hyperbaric conditions, the number of techniques used has been limited. The great majority of hyperbaric welding is carried out using GTAW and SMAW techniques, with small amounts of FCAW and GMAW.

A variety of habitats have been used, depending on such factors as the extent of welding required, the complexity of the repair-site geometry, depth of repair, welding process and ancillary equipment and environmental conditions. Generally, designs of dry hyperbaric habitat fall into one of the following four groups:

- Lightweight steel habitats: These have stiffened plate construction and are fabricated in two or more sections to allow their placement around jacket members. They may have an open grate floor with an access hole, or they may be fitted with a closed floor and access shaft. The latter is used in shallow depths where the shaft acts as a surge tube, thereby reducing the volume and pressure changes in the habitat that otherwise could affect diver physiology.
- Inflatable flexible habitats: Where the differential pressures are expected to be low, flexible habitats of sufficient strength are practicable and have been used. The skin of the inflatable habitat takes a shape dictated by the skin membrane stresses and the depth-dependent differential pressure, and it is the same shape that would be obtained onshore by turning the habitat upside down and filling it with water.
- Mini habitats: These habitats are small, with just enough room for the arms and sometimes the head of the welder/diver. In essence, mini habitats protect only the welding head and a small area around the weld.
- Portable dry spot habitats: These protect only the welding head and a small area around the weld. A clear plastic box, fitted with sponge or flexible rubber seals, moves with the head. These devices have not undergone as much development as either large habitat welding or wet welding.

In general, the hyperbaric welding conditions require that the welding be specially optimized for the elevated pressure and for the fact that the high-density gas enhances the rate of heat loss from the weld.

A hyperbaric welding operation requires the following equipment and tools:

- Purpose-built habitat
- Saturation diving support
- Environmental control equipment

- Pre- and post-weld heating equipment
- Welding equipment
- Weld inspection equipment
- Equipment to remove marine growth and grit blast
- Temporary holding clamps to take the weight of additional members and to maintain root gaps, as needed
- Crane capacity.

7.12.5 Example: Platform Underwater Repair

The platform is a four-legged jacket platform operating in 21 m of water. The jacket structure is horizontally braced at four levels with K-diagonal braces. The platform was inspected immediately after Hurricane Andrew. The inspection discovered that two midpoint joints, located at EL (−)12 m between the four legs, were damaged.

Divers performed the wet welding repair work to replace the two damaged joints.

The repair procedure was:

- New joints were designed (see Figure 7.37)
- The proper welding technique was determined by collection of specimens from jacket structure and measuring the carbon equivalent
- Wet welding specimens were tested
- The two damaged joints were removed

FIGURE 7.37 New replacement part.

- Grit blasting of the remaining structural members was done in preparation for welding
- Final dimensions were measured to ensure that the new joints fit
- New joints were installed.

All welds were inspected by magnetic particle inspection (MPI) equipment.

Other repair options considered included mechanical clamps and dry hyperbaric welding. It was estimated that wet welding saved 40% and 60% of the costs of dry hyperbaric welding and mechanical clamps, respectively.

7.12.6 Example: Platform "Shear Pups" Repair

The platform consists of two 8-legged jackets (in 8 ft of water) that are connected above water by a common deck structure and by horizontal braces.

In a large storm, the platform sustained several buckled members as well as small cracks at several above- and below-water joints. Subsea survey found small cracks in the bracing member that were identified by MPI. The damaged members were repaired using traditional slip-sleeve replacement members. The cracked joints were repaired using "shear pups," as shown in Figure 7.38, to strengthen the joints. The shear pups consisted of 1/3 circumference pipe pieces about 24 inches long that piggy-back on the brace at the joint in order to provide an additional path for loads from the brace into the joint. The shear pups are wet welded into place. The shear pups were considered a better repair than mechanical clamps due to the complexity of the joint configuration, which made it difficult to clamp. In addition, the cracks were small at these locations, were identifiable only by MPI, or were only a few inches or less.

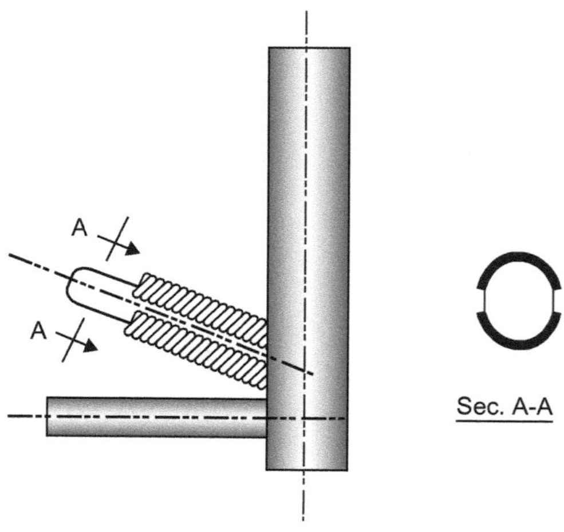

FIGURE 7.38 Shear pup.

Repair procedures were:

- As with any wet weld, a material sample of the jacket was taken to ensure that wet welding could be performed
- The joint was cleaned for fitting the shear pup and wet welding
- Multiple 600 mm shear pups were cut and fabricated on deck, including approximate fitting of ends at the joint connection
- Ends of the shear pup opposite the joint were constructed by transition to a narrow section in order to eliminate the potential for fatigue cracking
- Diver fit the shear pup into place
- Final shear pup was installed above water
- Diver wet welded shear pup into place and inspected welds.

7.12.7 Case Study: Underwater Repair for Platform Structure

In this case, the platform is a 4-legged, fixed steel jacket platform operating in 43 m of water. The jacket is a vertical diagonal braced structure.

Underwater inspection revealed a buckled vertical diagonal. The buckled area is located approximately in the center of the member. Replacement of the member and wet welding work for the buckled vertical diagonal member were preformed.

The repair procedure was:

- Remove vertical brace, leaving a 2-foot stub at the top and a 6-inch stub at the bottom
- Install a single, telescoping, vertical diagonal member over the stubs
- Wet weld the sliding sleeve, as shown in Figure 7.39.

7.12.8 Case Study: Platform Underwater Repair

The platform in this case is a fixed steel jacket platform operating in 7.8 m of water. The jacket structure is braced with vertical diagonal bracing members.

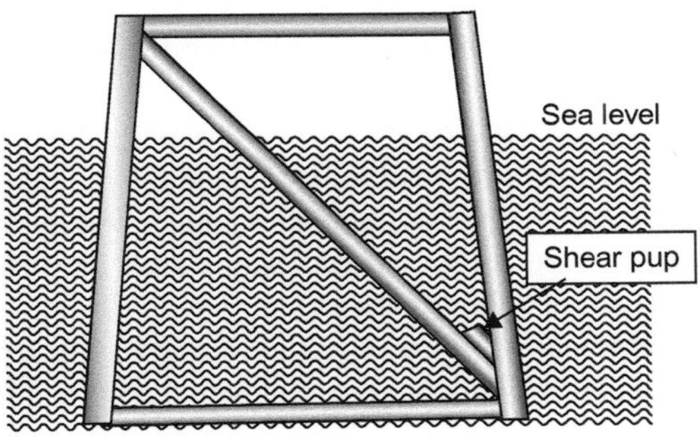

FIGURE 7.39 Sketch of proposed wet weld repairs.

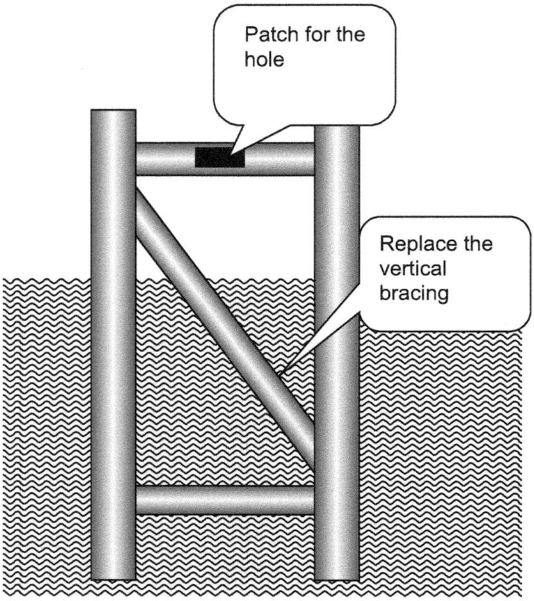

FIGURE 7.40 Repairs of the members.

An inspection revealed a 10¾-inch diameter vertical member, with a 30-inch crack located at midpoint around the circumference, as shown in Figure 7.40. An additional 6-inch diameter hole was located on the horizontal bracing member.

Contractors performed the wet welding replacement work on the vertical diagonal member and the dry welding work on the corrosion hole patch.

7.12.9 Clamps

Repair clamps are normally constructed from low-carbon steel and have a reinforcing sleeve that may include brace attachments. For ease of installation, the clamps can be split or hinged. Whether used for repair of a member or a node, a clamp will typically take on the same general appearance as the part of the structure it is to reinforce. Typical clamps weigh between 0.5 and 10 tons, although clamps weighing 50 tons are known, and they may require grouting, bolting or some other structural bonding technique.

Clamps may be used to repair a member by installing an additional brace, and they are normally used for a connection to a riser or caisson if there has been loss of support to such an appurtenance.

However, the principal use of steel repair clamps is in repairing primary structural joints. Four clamp types are traditionally used in offshore platform repair. Their classification is based upon the installation and fixing method, rather than terms of usage.

- The stressed mechanical (friction) clamp uses long stud-bolts to produce a friction grip on the repaired structural elements.
- The unstressed grouted repair clamp relies upon grout shear strength to affect load transference, working in a similar way to a grouted pile sleeve connection.
- The stressed grouted clamp is a hybrid of the above two clamp types, with load being transferred partly by grout bond but mainly by friction. This type is popular for its ability to tolerate dimensional variation while achieving load transfer in a reasonable sleeve length.
- The stressed elastomer-lined clamp is similar to a mechanical clamp except for the elastomer, which is usually a Neoprene liner that lies between the member and the clamp. Such clamps are often selected for repairs to caissons and other secondary structural elements.

The basis of any clamp design must be the establishment of the forces in the structural elements under operating and storm loading conditions. Normally, the forces are evaluated for the undamaged condition and then the clamp is designed so that load transfer is accomplished within the body of the repair clamp.

The technology for clamp installation is highly developed and a number of specialist companies are now able to provide the necessary equipment and materials on a contract basis.

Stressed Mechanical (Friction) Clamps

A stressed mechanical clamp is a steel-to-steel friction clamp that is connected by long tension stud-bolts. The clamp comprises two or more segments of closely fitting stiffened saddle plates, stressed directly onto a tubular section by means of long stud-bolts. The strength of the mechanical connection is obtained from steel-to-steel friction generated by compressive forces normal to the tubular/clamp saddle interface, applied by the external bolt loads.

Stressed mechanical clamps are generally used for strengthening and repair of damaged members or for connecting new members. A high degree of tolerance is required for close contact between the clamp and the tubular member. Note that these clamps are unsuitable for repair of tubular joints. Furthermore, this type of clamp requires full contact between the clamp and member surface interface, so the clamp offers minimal translation or angular tolerances. Thus, extremely accurate offshore surveys are needed for its proper construction and the clamp requires very tight tolerances in fabrication. The connection between clamp and tubular member is susceptible to crevice corrosion; therefore, the clamp needs periodical inspections to confirm bolt tension.

The major advantage of a mechanical clamp is that large forces can be transferred through friction over a short clamp length, limited only by the hoop resistance of the member.

The time required for installation of stressed mechanical clamps varies depending on the complexity of the clamp (for instance, number of clamp segments), space limitations and water depth at the repair site. Typically, installation times are 2–6 days, depending on size and complexity.

Over the past 20 years there have been numerous applications of stressed mechanical clamps worldwide. Their use has principally been to repair or to strengthen jacket components damaged by boat impact, fabrication flaws, fatigue cracking and corrosion.

Stressed mechanical clamps are one of the quickest clamps to deploy and have good transfer capacity. They require a detailed survey and good fabrication tolerances. They are ideal for clamping on intact tubular members to strengthen, replace or add members.

For diver installation of a stressed mechanical clamp, the following items of equipment/support are required:

- Diving spread and divers
- Crane for lifting and placing in position
- Rigging for installation
- Underwater cutting and grinding equipment, if obstructions have to be removed
- Stud-bolt tensioning equipment
- Monitoring equipment, such as video camera
- Equipment to remove marine growth and grit blast.

Unstressed Grouted Clamp Connections

The grouted sleeve clamp uses short bolts. An unstressed grouted clamp or sleeve connection comprises sleeves placed around a tubular member or joint with the annular space filled with grout. The sleeves may be split, as in the case of a clamp, or continuous, as in a sleeve connection. For split sleeves, short bolts are provided and are tightened prior to injection of grout into the annulus.

The bond at the grout/steel interface provides the only means of transfer of load between the tubular member and the clamp. To increase the capacity of the clamp, its length needs to be increased. Shear keys, usually provided in the form of weld beads, can increase the clamp capacity, but it should be kept in mind that the underwater welding option is very expensive.

Unstressed grouted clamps and connections offer a versatile means for strengthening or repairing tubular joints and members since they require less accurate offshore survey. Both angular and translation tolerances can be readily accommodated by the annulus. However, the loading regime and the availability of space are dominant in deciding the suitability of an unstressed grouted connection or clamp.

The installation time for this type of clamp varies depending on the complexity of the clamp and the number of pieces in which the clamp is installed. The number of bolts that have to be tensioned and the amount of grout will also influence the installation time. As an example, two X joints strengthened using unstressed grouted split-sleeve connections were reported to require a total dive time of 50 hours: 10 hours each for survey and cleaning, 20 hours for clamp installation and bolting, 5 hours for grouting and 5 hours for inspection.

It is important to note that curing time should be included to ensure that the unstressed grouted clamp/sleeve connection is not subjected to loading before the grout has gained sufficient strength. In some instances, temporary clamps may be necessary.

Several applications of unstressed grouted clamps and sleeve connections are evident. Pile or sleeve connections for numerous jackets make use of this technique. For repair/strengthening, unstressed grouted clamps/sleeve connections have often been used to overcome fatigue cracks and members damaged by boat impact.

The advantages of this technique are its reasonable transfer capacity, its good tolerance for fit-up and its ideal potential for clamping on joints and members. It is particularly good for strengthening dented members.

The disadvantages of the technique are that, without the use of shear keys, the required connection length may be unacceptably long and that the grout seal, if improperly fitted, often results in leakage of grout, resulting in loss of friction.

The equipment required for diver installation of an unstressed grouted connection or clamp is:

- Diving spread and divers
- Crane for lifting and placing in position
- Rigging for installation
- Underwater cutting and grinding equipment, if obstructions have to be removed
- Bolt torque/tension equipment for spilt-sleeve clamp
- Grouting spread
- Monitoring equipment (e.g., video/camera)
- Equipment to remove marine growth and grit blast.

Stressed Grouted Clamps

A stressed grouted clamp is formed when two or more segments of strengthened saddle plates are stressed by means of long stud-bolts onto a tubular member after grout has been injected and allowed to cure in the annular space between the clamp and the tubular member.

This type of clamp is a hybrid between a stressed mechanical clamp and an unstressed grouted clamp. The strength of the clamp is obtained from a combination of "plane pipe" bond and grout/steel friction developed because of the compressive force applied normal to the grout/tubular surface interface by the stud-bolt tension.

Stressed grouted clamps offer the benefits of stressed mechanical clamps of high strength-to-length ratio, and the benefits of unstressed grouted clamps in the ability to absorb significant tolerances.

As an example, eight stressed grouted clamps were fully installed on a platform in the Gulf of Mexico in a total of 18 days. About one-third of this time was used in removal of obstructions, while cleaning, clamp installation, grouting and stud-bolt tensioning took an equal amount of time. It is important to recognize, however, that installation time is very dependent on the complexity of the clamp, access at the repair site, water depth at the repair site and environmental conditions that may severely limit dive time and greatly influence weather downtime.

Stressed grouted clamps can be viewed as representing the strength advantages of stressed mechanical clamps with the tolerance advantages of unstressed grouted clamps; it is therefore not surprising to note that stressed grouted clamps are the most popular form of clamp used today.

The disadvantages of stressed grouted clamps include the grout seals, which, if improperly fitted, often result in leakage of grout and loss of friction. The advantages of these clamps are good transfer capacity, good tolerance for fit-up and their ideal potential for clamping on joints and members. They are particularly good for repair of joints.

The required equipment and tools are the same as for the unstressed grouted clamps.

Stressed Elastomer-Lined Clamp

Elastomer-lined clamps have not been used for primary structural repairs, because of concerns that the flexibility of the liner may reduce the efficiency of the repaired system. Therefore, this type of repair has been limited to secondary components where stiffness is not critical to effectiveness. Typically, elastomer-lined clamps are used to seal holes in caissons and for stub connections to appurtenances.

Installation time for stressed elastomer-lined clamps varies depending on the complexity of the clamp (for instance, number of clamp segments, space limitations and water depth at the repair site). In general, the times are similar to those for stressed mechanical clamps.

Stressed elastomer-lined clamps are very similar to stressed mechanical clamps, except that an elastomer lining is bonded to the inside faces of the clamp saddle plates. The lining is a solid polychloroprene (Neoprene) sheet.

The strength of the clamp is derived from external bolt loads, which impart compressive force normal to the interface of the liner and the tubular member. This type of clamp is not recommended for transfer of forces between structural components. It is recommended for the attachment of components, such as guides for appurtenances.

The elastomer lining offers a degree of angular and translation tolerance, eliminating the need for very accurate offshore surveys, as are required for stressed mechanical clamps.

The relatively low stiffness of the liner gives rise to significant stud-bolt load fluctuations due to elastomer relaxation. Fatigue of stud-bolts therefore requires careful consideration. Large-scale tests have demonstrated this type of clamp does not transfer structural forces. Furthermore, the typical coefficient of friction between liner and steel is considerably lower than previously assumed values. These clamps require periodic inspection to confirm bolt tension.

In summary, a stressed elastomer-lined clamp is one of the quickest clamps to deploy and requires a reasonably detailed survey. It has poor axial and bending load transfer capacity. It is ideal for clamping on intact tubular members or for adding members and appurtenance supports.

The required equipment for installation is the same as for the stressed grouted clamp.

7.12.10 Example: Drilling Platform Stabilization after Hurricane Lili

A drilling platform located in 71 m of water was severely damaged when Hurricane Lili passed through the Gulf of Mexico in 2002. In order to stabilize the platform during well plug, in 2003 additional piles were installed to strengthen the damaged platform.

The leeward pile was severed at 6.3 m below the base of jacket leg. The pile severance was located at the mud line due to the lean of the platform. In addition, the remaining three piles were assumed to be damaged at or below the mud line due to platform rotation.

Divers performed member cleaning, clamp installation and grouting operations. Once the grout had sufficiently cured within the clamp annulus, the stud-bolts were tensioned to the required tensile loads. Heavy lifting and pile installation were carried out.

Note that several trials using different repair methods were considered, such as guy wires to piles and attached to the undamaged nearest platform. Analyses indicated that templates were required in addition to severed pile reinstatement to provide sufficient torsional resistance against potential hurricane wave loads. The strengthening scheme consisted of two templates, a stressed grouted clamp and five piles.

The repair procedure for the platform was:

- Clean exposed pile
- Cut holes in base of severed pile and place support clamp
- Install sleeve clamp around exposed pile using support clamp
- Close clamp and hand-tighten several bolts
- Check seals for leakage and place grout
- Once grout has reached required cube strength, stress the 32 2-inch stud-bolts to required tensile load
- Lower piles through sleeves to self-penetration depth
- Hammer piles to the required 200 feet of penetration
- Grout pile sleeves.

7.12.11 Grouting

Using grout to fill structural members is a cheap and effective solution to several repair and strengthening problems. It is most beneficial in these conditions:

- Compressively loaded dent-damaged elements, where the grout prevents any further deformation of the tubular section. In such cases, the grout filling need only be extended to the immediate region of the damage.
- Improving the strength and fatigue performance of a tubular joint by grouting the chord in the location of the entire tubular joint.

There is no benefit accrued in the following situations:

- Tensile loads in the element (unless there is a problem with hydrostatic/tension interaction collapse).
- Unchanging compressive or bending loads in the element (because a repair can carry only a load that is applied after the grout has set).
- Partially filled compressive elements (because there is no guaranteed load-transfer mechanism between the grout and the steel).

The grouting work should be done by a competent company that can deliver a grouting design mix that reduces the heat of hydration effects, grout shrinkage and grout port locations and that ensures complete element filling if it is deemed essential.

In general, grout-filled elements and joints will be stiffer than their pure steel counterparts and as such they may attract more load in a statically indeterminate frame. In seismic zones, the additional mass associated with grout filling may need to be considered in the structural analysis. On any structure, the additional weight imposed by grout may constitute a significant load, particularly if the grouted joint or element lies in a horizontal plane, as is the case with conductor bracing.

Based on typical offshore installation times, grouting should be achievable with 2–3 days of offshore work.

Joint Grouting

Joint grouting is generally performed by filling a chord with grout in the region of a tubular joint. Grouted joints have the chord member fully filled with a cementitious grout material. Double-skin joints are those in which the chord member contains a pile and a grouted annulus.

It is relatively easy to fill a chord member over its full length if the member is anything other than a jacket leg, as this avoids the necessity of cutting windows in the member to insert seals to localize the grout plug. The grout can be placed through small-diameter inlets and outlets, which can be drilled and tapped into the tubular wall.

The geometric shape of the annulus that should be filled with grout between the leg and pile is shown in Figure 7.41.

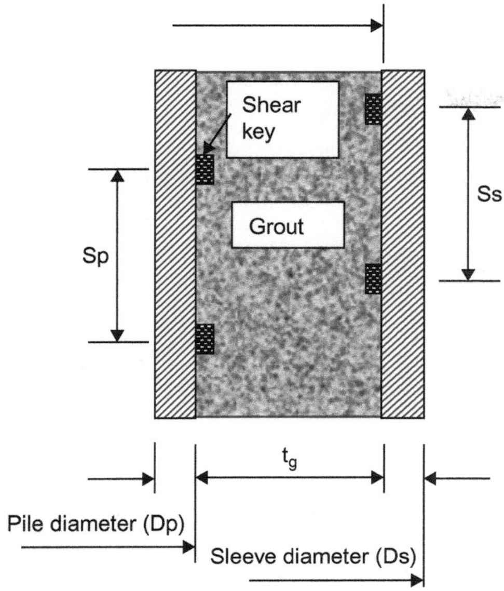

FIGURE 7.41 Grouting the annulus between piles and legs.

Filling tubular chord members with a cementitious material will increase their strength, improve ductility, and increase the radial stiffness of the member due to grout. The grout restricts local chord wall deformations, reducing deformation-induced bending stresses and associated stress concentration factors, so it will improve fatigue life.

Therefore, grout filling of tubular chord elements is used to improve the static strength of the joint and to extend the fatigue life of the connections made at the joint. The repair method has the advantage of introducing no additional wave and current loads on the platform, but local dead loads can be increased.

Only grout-filled joints have been considered. It is therefore assumed that the material is cementitious and not reinforced. If the joint is simply pumped full of grout, then a simple grout-filled joint is created. In some instances there will be a concentric pile within the joint, as is the case with many leg sections, and the resulting construction is then termed double-skinned. The grout is assumed to completely fill the available annulus in the joint.

Ring stiffening is sometimes used at the joint to increase the resistance capacity of the chord wall to applied member forces.

In most cases, grouting the joint takes 3–4 days, because time is needed to install grout bag seals, to allow seals to set and cure, and then to grout the plug.

Void formation is a potential problem at ring-stiffened joints and at joints with an expanded can diameter.

In summary, joint grouting increases the strength and fatigue performance of a tubular joint. It doesn't increase hydrodynamic load but it does add mass to the structure. If a large volume of grout is used, excessive heat generated while setting will need addressing. Consideration of mix design and heat loss will be required.

The equipment required for a typical grout-filling operation is:

- Diving spread
- Drilling and tapping tools for grout ports
- Underwater cutting and grinding equipment if temporary seals are to be placed
- Grouting spread
- Monitoring equipment, such as video or still camera
- Marine growth removal equipment.

Grout Filling of Members

Grout filling of a member increases both its cross-sectional strength and its overall stability. This process provides a relatively easy way of strengthening tubular members, particularly compression members with or without bending. The grout filling the compression member increases the axial load capacity and prevents local buckling.

As a direct effect of grout filling, the additional mass of the member, the additional earthquake load due to increased inertia and the extra stiffness of the member and tubular joints along its length need to be considered in design.

This technique is most beneficial to tubes with low L/D ratio and high *D/t* ratios.

It is important that the grout completely fill up the tubular member, as small voids close to the tubular inner wall may reduce the load-carrying capacity of the strengthened member significantly. Void formation is also a potential problem for all ring-stiffened tubular members.

If there is no damage to the member but an increase in axial load-carrying capacity is sought, then a few preliminary calculations will indicate if there is likely to be sufficient increase in member capacity. If the member is short and has a low *kl/r* ratio, then the axial load capacity in compression will be increased if the element is completely filled and load transference can be made from the node-capping plate to the grout body. If the element has a high D/T ratio, then there may be an improvement in load-carrying capacity due to the prevention of local instability.

A large volume of grout will be used, so the grout design mix should be done professionally to prevent generation of excess heat during setting.

On the other hand, grout filling has little benefit for tension members. In addition, full grout filling must be achieved and the increase in mass due to the filling of a member with grout may result in overstress of the member under seismic and/or in-place conditions.

In summary, grout filling of members increases the strength of a member but does add mass to the structure.

Grout filling of members requires the same equipment as specified above for joint grouting.

Grouting of Piles

Platform loads may be transferred to steel piles by grouting the annulus between the jacket leg (or sleeve) and the pile. The load is transferred to the pile from the structure across the grout. Experimental work indicates that the mechanism of load transfer is a combination of bond and confinement friction between the grout and the steel surfaces and the bearing of the grout against mechanical aids, such as shear keys.

Centralizers should be used to maintain a uniform annulus or space between the pile and the surrounding structure. A minimum annulus width of 1 1/2 inches (38 mm) should be provided where grout is the only means of load transfer. Adequate clearance between pile and sleeve should be provided, taking into account the shear keys' outer dimension.

Note that packers should be used as necessary to confine the grout. Proper means for the introduction of grout into the annulus should be provided, to minimize the possibility of dilution of the grout or formation of voids in the grout.

The use of wipers or other means of minimizing mud intrusion into the spaces to be occupied by piles should be considered at sites having soft mud bottoms.

Computation of Allowable Axial Force

Until now there has been little research into the transfer of axial force from the pile to the leg through the grouting. The transfer of axial force from the pile to the leg is accomplished by the bond strength between the pile and the leg through the grouting. The allowable axial load transfer should be calculated as the smaller value from pile or leg of the force calculated by a multiplication of the contact area between the grout and steel surfaces and the allowable bond strength due to axial load f_{ba}.

The value of the allowable bond strength due to axial load, f_{ba}, should be taken as 0.138 MPa for API loading conditions 1 and 2, and 26.7 psi (0.184 MPa) for loading conditions 3 and 4, where the API loading conditions are:

1. $E_o + D_L + L_{max}$
2. $E_o + D_L + L_{min}$
3. $E_D + D_L + L_{max}$
4. $E_D + D_L + L_{min}$

where E_o is the operating environmental condition, D_L is the dead load, L_{max} and L_{min} are the maximum and minimum live load, respectively, and in both cases the live load is appropriate to the normal operating condition of the platform. E_D is the design environmental condition in extreme environmental conditions.

In case of higher values of axial load, shear keys are required at the interface between steel and grout, and the value of the nominal allowable bond strength for axial load transfer f_{ba}, for loading conditions 1 and 2 should be taken as:

$$f_{ba} = 0.138 + 0.5 f_{cu} \frac{h}{s} \tag{7.31}$$

For loading conditions 3 and 4, f_{ba} should be taken as:

$$f_{ba} = 0.184 + 0.67 f_{cu} \frac{h}{s} \tag{7.32}$$

where f_{cu} = unconfined grout compressive strength (in MPa), h = shear key outer dimension (in mm) and s = shear key spacing (in mm).

Shear keys designed according to Equations (7.31) and (7.32) should be detailed in accordance with the following requirements:

1. Shear keys may be circular hoops at spacing s or a continuous helix with a pitch of S_s.
2. Shear keys should be one of the types indicated in Figure 7.42.
3. For driven piles, shear keys on the pile should be applied to sufficient length to ensure that, after driving, the length of the pile in contact with the grout has the required number of shear keys.
4. Each shear key cross-section and weld should be designed to transmit the part of the connection capacity that is attributable to the shear key for loading conditions 1 and 2.

The shear key and weld should be designed at basic allowable steel and weld stresses to transmit an average force equal to the shear key bearing area multiplied by 1.7 f_{cu}, except for a distance of 2 pile diameters from the top and the bottom end of the connections, where 2.5 f_{cu} should be used.

The following limitations should be observed when designing a connection.

- 17.25 MPa $\leq f_{cu} \leq$ 110 MPa
- Sleeve geometry: $D_s/t_s \leq 80$
- Pile geometry: $D_p/t_p \leq 40$

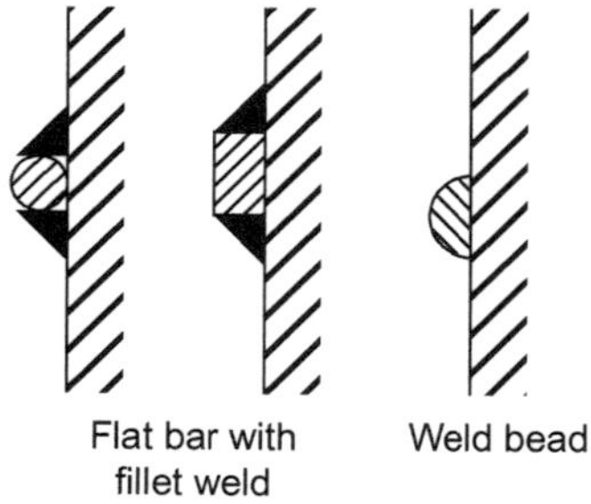

FIGURE 7.42 Recommended API shear key detail.

- Grout annulus geometry: $7 \leq D_g/t_g \leq 45$
- Shear key spacing ratio: $2.5 \leq D_p/s \leq 8$
- Shear key ratio: $h/s \leq 0.10$
- Shear key shape factor: $1.5 \leq w/h \leq 3$
- $f_{cu}\,(h/s) \leq 5.5$ MPa.

7.12.12 Composite Technology

The definition of composite technology is use of a combination of two or more materials. Composite materials have been extensively used onshore; however, the use of composites for offshore structural applications has been somewhat limited. Composites have been used widely for pipeline and piping on topsides and composites are seeing greater use for tertiary steel structures, such as handrails and grating.

Two main groups of composites have been used in offshore applications, namely:

- Reinforced epoxy grout—A rebar-reinforced epoxy grout typically used for the repair of conductors.
- FRP (fiber-reinforced polymer)—A laminate structure that contains an arrangement of unidirectional fibers or woven fiber fabrics embedded within a thin layer of light polymer matrix material. The fibers, typically composed of carbon or glass, provide the strength and stiffness. The matrix commonly consists of polyester, epoxy or nylon, which binds and protects the fibers from damage and transfers the stresses between fibers. The matrix also provides a bond and thus an interface for the transfer of forces between the parent structure and the composite component. FRP can be molded or pultruded to form many shapes for topside applications.

To date, the use of composites, particularly for offshore applications, has been limited largely by lack of design guidance, designers' lack of familiarity with them, combustibility concerns, lack of performance data and fabrication scale. These limitations have been addressed, with specialist contractors offering off-the-shelf solutions and proof-of-performance data. A range of resins are available for different applications, including low levels of smoke generation and toxicity in fires.

The application of composites differs between repair of topsides and subsea applications. Composite technology to date has shown little success in the use of composites subsea when the laminates are formed in situ. Preformed components can be used, but they are typically limited to nonstructural applications. Composites are used topside as:

- Repair wraps for tubular members
- Wraps for deck legs for corrosion protection
- Grating

- Stair treads and ladders
- Handrails
- Mud mats
- Fire and blast walls
- Access platforms.

Many of the composite applications topside can be utilized as part of strengthening and repair or for new construction. A number of notable new-build offshore facilities have made extensive use of composites for a range of applications. As much as 10% of their topside weight is represented by composite components.

Reinforced Epoxy Grout

A rebar-reinforced epoxy grout is typically used for the repair of conductors. Reinforced epoxy grout provides a method of repair that requires little preparation onshore prior to implementation of the repair offshore. The repair technique can readily conform to the structure's geometry without the requirement for a detailed survey. The main advantages of using epoxy grout are:

- Transportation to site and installation are very easy
- Can be implemented without the requirement for hot works
- Minimal lift equipment is required, particularly if rebar cage is assembled in situ
- Significantly improved corrosion resistance compared to steel
- Prevents further deterioration, particularly corrosion of conductors
- Limited requirement for design, with minimal fabrication.

Reinforced epoxy grout has been mainly limited to axially loaded members, such as conductors under compression loading, and mainly has been used for above-water applications. In addition, it is easily transported and utilized for repair to conductors.

In general, the following equipment and support may be needed:

- Grouting spread
- Composite materials
- Other components, depending on repair type, such as temporary outer jacket to provide formwork
- Installation aids.

FRP Composites

Linear elastic response of FRP to axial stress is one of the main attractions of its use in structural engineering. The response of FRP to axial compression depends on the relative proportion in volume of fibers, the properties of the fiber and resin and the interface bond strength. FRP composite compression failure occurs when the fibers exhibit the extreme, often sudden and dramatic, lateral or sideways deflection called fiber buckling.

The base fiber matrix is available in many forms, from random mixed fibers to woven fiber blankets consisting of either glass fibers or carbon fibers. The woven matrix provides the necessary bidirectional fiber strength, depending on the strength requirements and application.

FRP's response to transverse tensile stress is very much dependent on the properties of the fiber and matrix, the interaction between the fiber and matrix and the strength of the fiber-matrix interface. Generally, however, tensile strength in the direction perpendicular to the fiber direction is very poor, so the greatest tensile strength is in the direction of the fibers. The shear strength of FRP is difficult to quantify. To avoid that, some products on the market have the fiber in two directions, while other products have fiber in two diagonal directions so that it can carry the load in the four directions. This type of FRP composite will be used for repair on site, but prefabricated FRP used as grating usually has fiber in one direction.

Besides FRP's high strength, its most relevant features include excellent durability and corrosion resistance. Furthermore, the high strength-to-weight ratio is of significant benefit; a member composed of FRP can support larger live loads, since its dead weight does not contribute significantly to the loads that it must bear. Other features include ease of installation, versatility, electromagnetic neutrality, excellent fatigue behavior, fire resistance and possible maintenance-free use. So, FRP has increased reliability due to good corrosion resistance and structural superiority because of weight savings, higher stiffness and ability to better tailor the structure to the load.

The composite repair method takes less time offshore and involves light material handling. The method often proves to be more cost-effective than other repair solutions.

FRP has the following limitations, which can be easily overcome through design and application and search for advanced products on the market:

- Brittle behavior
- Susceptibility to deformation under long-term loads
- UV degradation and photodegradation (from exposure to light)
- Temperature and moisture effects.

Limitations are also related to the contractor's previous experience. However, the number and experience of contractors are increasing and are paralleled by continuous research and development aimed at enhancing the properties of FRP.

7.12.13 Example: Using FRP

Davy and Bessemer are two shallow-water monopod platform structures located in the southern North Sea that commenced production in 1995. These two platforms used FRP. The fiberglass components consisted of 10% by weight of composites, including the following:

- Office and equipment modules
- Diesel- and water-storage tanks
- Pultruded glass/phenolic gratings for floors
- Ladders
- Walkways
- Handrails
- Enclosures and heat-protection for walls.

Glass-reinforced epoxy was also widely used for the pipe work and tubular members.

The main benefit of FRP was the reduction in weight, which was important for these lightweight structures because it reduced the cost, reduced the topside weight and eliminated the cost of regular maintenance.

7.12.14 Case Study: Conductor Composite Repair

In this case, an 8-legged jacket drilling platform, installed in 1964, was operating in 75 m of water in the Gulf of Mexico. Its original conductors were heavily corroded. The 5-year conductor maintenance plan identified the required conductor restoration work.

Cost analysis revealed that composite repair would be the most cost-effective repair method. The process included:

- Underwater and above-water repair-site inspection
- Planning the staging area for equipment and material
- Surface preparation, including removal of excess scale, removal of grout that was not structurally sound and grit blasting to near-white metal
- Installation of shear lug, rebar cage, and translucent FRP jacket outside rebar cage
- Pumping of epoxy grout into FRP jacket from bottom up
- Installation of wear pads and conductor centralizers at guide bell.

7.12.15 Fiberglass Access Decks

For the past decade, use of fiberglass for grating and handrails has been growing in offshore structures. Fiberglass is used for both new platforms and for replacing existing ones.

Fiberglass access decks and stair towers are lighter than the steel equivalent and are easier to install.

Fiberglass structural decks can be designed to meet any load requirements. The decks can be custom built and installed offshore with no welding and or heavy lift equipment required.

Fiberglass access decks consist of pultruded structural sections that are bolted together. Fiberglass grating and handrails complete the full deck.

The benefits of fiberglass access decks include the fact that they can be installed without hot work and there is no long-term maintenance or painting. They can be readily transported to the site and can be handled using lighter lifting equipment than is needed for steel decks. The main advantage of fiberglass access decks is that they do not corrode, eliminating maintenance and costly repairs or replacement.

Fiberglass access decks, primarily well-access decks, have been used throughout the world on projects ranging from minimal structures to deep-water platforms.

Fiberglass grating products and resins are available to cover a range of requirements, including:

- Strength
- Chemical resistance
- Impact resistance
- Fire resistance.

A number of attachment options are available, including bolting, welding, clips and friction welding. The choice of attachment depends on the location (wave zone or non-wave zone) and whether or not hot works are possible.

FIGURE 7.43 Deck from FRP.

Fiberglass grating systems have been proven to perform even after being subjected to hurricane wave forces. Grating is typically designed to last for the life of the structure, without replacement.

Fiberglass grating does not corrode, which is the primary cause for replacement of steel grating and plating.

Fiberglass access stairs and treads are lighter than the steel equivalent and easier to install.

Fiberglass stairs can be designed to meet any access requirements.

Fiberglass stairs consist of pultruded structural sections with fiberglass stair treads. Fiberglass handrails complete the full stair. Fiberglass stairs can be replacement items for steel stairs, and they can also be used for new design or new access requirements. Fiberglass stairs are installed using lighter lifting equipment than is needed for the steel equivalent.

Fiberglass treads have been proven to perform even after being subjected to hurricane wave forces. Stairs and treads are typically designed to last for the life of the structure and do not require replacement.

Fiberglass stairs and treads, shown in Figure 7.44, do not corrode, which is the primary cause for the replacement of steel stairs.

Fiberglass stairs and treads have been used throughout the world on offshore platforms.

FIGURE 7.44 Non-wave-zone fiberglass stairs.

FIGURE 7.45 Handrail made from glass fiber-reinforced polymer (GFRP).

Fiberglass handrails, shown in Figure 7.45, can be used for the jacket walkways, boat landings and deck perimeters.

Handrail sockets are typically fabricated from stainless steel and are welded to the structure. If hot work is not permitted, attachment of the handrails can be modified for installation; for example, bolted attachments are available.

Fiberglass handrail systems have been proven to perform even after being subjected to hurricane wave forces. Handrails are typically designed to last for the life of the structure, without replacement.

Fiberglass handrails are not subject to corrosion, which is the primary reason for replacement of steel handrails.

7.12.16 Fiberglass Mud Mats

Fiberglass mud mats provide an economical alternative to traditional steel systems. Fiberglass mud mats have high potential flexural strength (as high as 427 N/mm^2) and high flexural stiffness (EI as high as 64,121 N/mm^2), and they can span two to three times more than unstiffened steel plate mud-mat skins. As many as one-half to two-thirds of the beams required to support a steel mud-mat skin can be eliminated when fiberglass mud mats are used.

In addition, fiberglass mud mats, shown in Figure 7.46, reduce steel fabrication tonnage and eliminate the need for the mud-mat cathodic protection that steel mud mats require. The service life of any mud-mat system is very short, and after the piles have been driven and welded to the jacket structure, mud mats provide no further service; however, a steel mud-mat system continues to draw from the platform's cathodic protection system.

FIGURE 7.46 Fiberglass mud-mat system.

A fiberglass mud-mat system is significantly lighter than steel, ranging from 44 to 58 kg/m^2 in air and 49 to 58 kg/m^2 submerged.

Mud mats constructed of fiberglass have the following advantages:

- High flexural strength
- High flexural stiffness
- Reduced number of support beams
- Reduction in fabrication costs
- Reduction in anodes
- Lighter weight
- Easy installation—no bolting.

Several companies have used fiberglass mud mats on approximately 20 jackets in the Gulf of Mexico since the late 1990s.

7.12.17 Case Study: Repair of the Flare Jacket

The flare in this case was constructed in 1970. A survey showed cracks on the bracing due to corrosion (Figure 7.47). The flare jacket leg-brace clamps were designed in 1984 but were not fitted until 1994. They are ungrouted, loose and heavily corroded. The vertical diagonal braces in all three elevations in vertical intervals between the upper bay are totally corroded, with extensive holes through most of perimeter. The horizontal braces in all three elevations are totally corroded, with extensive holes through most of the perimeter. Visual inspection revealed that the roller bearings at deck level supporting the flare bridge are highly corroded and

FIGURE 7.47 Corroded bracing.

are frozen. In addition, the dead load deflection of the last span of the flare bridge has caused a rotation of the roller shoe beam, with the result that the flare bridge load is carried only on the inner one or two rollers and thermal expansion/contraction is causing excessive loading of the jacket structure.

Prior to the start of diving operations, the current condition of the submerged structure was assessed by ROV. The underwater inspection made a general inspection of the structure for potential hazards to divers, with a particular focus on the first bay up and vertical diagonal braces and horizontal plan bracing members.

Face brace members on elevations, as shown in Figure 7.48, should be generally inspected. If cyclic lateral movements of any braces or leg clamps are detected in a running sea, remedial measures should be taken, up to and including removal of members that could present a hazard and that would not further reduce the strength and stability of the structure.

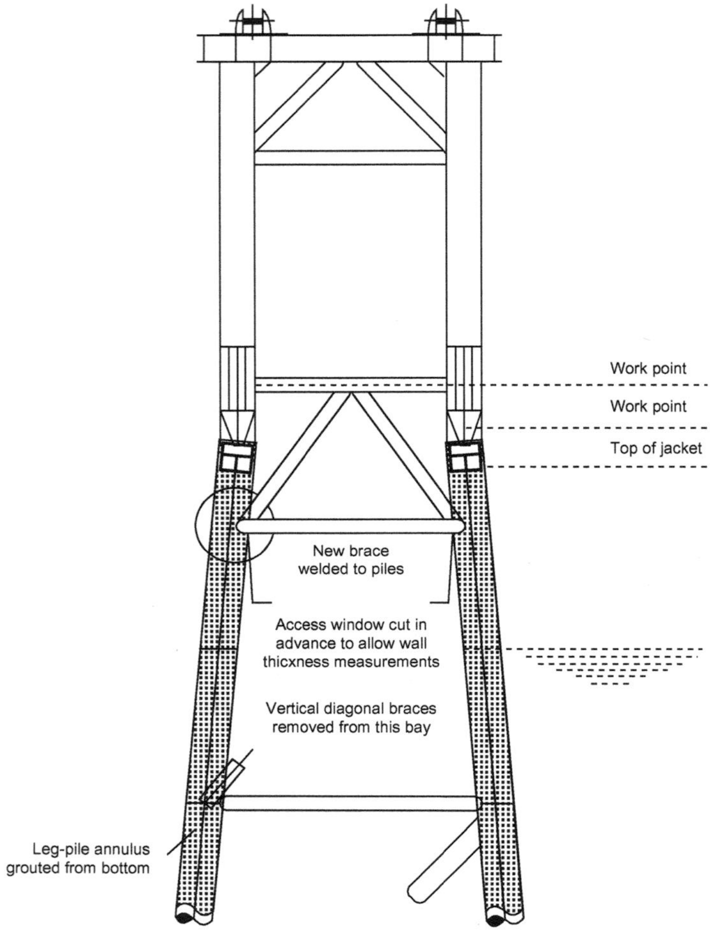

FIGURE 7.48 Flare jacket configuration and location of grouting.

Human access to the underside roller-bearing deck was required for attaching beam clamps to support the weight of the new brace members during installation. Scaffolding to enable safe access to these locations from the flare bridge and to provide a working platform was constructed prior to the start of repair operations.

Repairs were carried out in the following sequence:

1. Pile wall-thickness survey
2. Refurbishment of roller bearings
3. Grouting of leg/pile annuli
4. Cutting of legs or crown weld shims
5. New horizontal braces for the two levels.

7.12.18 Case Study: Repair of Bearing Support

A bearing support for a bridge on which there is severe corrosion is shown in Figure 7.49.

Figure 7.50 presents a sketch for the original design for the roller support over the flare support. It is shown that it allows horizontal movement of the bridge

(a)

(b)

FIGURE 7.49 **(a)** Corrosion of bearing, **(b)** Corrosion of the bearing stud.

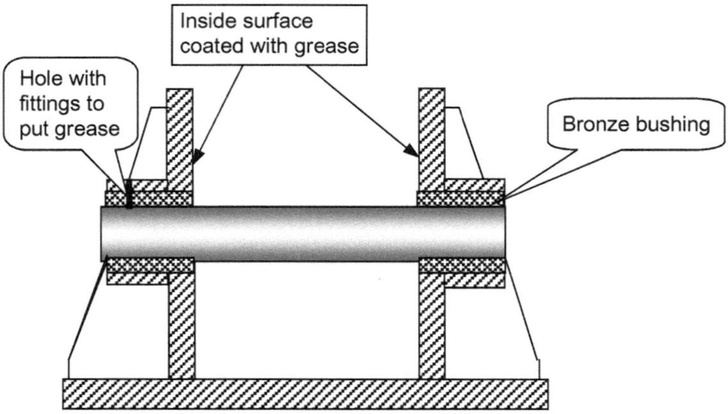

FIGURE 7.50 Sketch for roller support on the bridge.

with thermal expansion due to the grease of the internal surface and the presence of bronze to reduce the friction. Therefore, in the mature structure, corrosion will increase the coefficient of friction, which may cause higher stresses on the bridge by preventing movement (the coefficient of friction may be greater than 10% due to corrosion and depletion of the bronze bushing).

The procedure for refurbishing the roller bearings is critical because it is based on removal of the old, corroded rollers and cleaning up the existing steelwork and bronze bushings.

1. The first step is to install a temporary lateral bracing to support flare-bridge lower chord during refurbishment of rollers. Then install hydraulic jacks to lift and support the flare bridge as required and lift the flare bridge by jack-up for 40 mm to unload rollers. This distance is the maximum allowable, obtained from structural analysis, that will not place overstress on the bridge.
2. The repair procedure starts with removal of all corrosion on the outer parts using tools like a needle-gun or grit-blast to reduce the diameters of the protruding parts so that they can pass through the bushings without causing damage. Cut the roller and then remove it by hammering the remaining roller portions through the bronze bushings from outside and removing them.
3. After cleaning up and removing corrosion and rust scale from surfaces, apply paint to the bearing support steelwork and also inspect, clean and grease the bronze bushings. The bearings are believed to be intact. If there is wear or corrosion, then they should be cleaned up as far as possible and the surfaces of the bushings polished without increasing the hole diameter by more than 2 mm. Finally, install the new rollers, grease the rollers and apply paint to roller ends.

BIBLIOGRAPHY

American Institute of Steel Construction (AISC), 1969. Specification for the Design, Fabrication and Erection of Structural Steel for Buildings, Seventh ed. American Institute of Steel Construction, New York.

American Institute of Steel Construction (AISC), 1978. Specification for the Design, Fabrication and Erection of Structural Steel for Buildings, Eighth ed. American Institute of Steel Construction, Chicago.

American Institute of Steel Construction (AISC)-ASD, 1989. Specification for Structural Steel Buildings—Allowable Stress Design and Plastic Design, Ninth ed. American Institute of Steel Construction, Chicago.

American Welding Society, 1972. AWS D1.1, Structural Welding Code, First ed. American Welding Society, USA.

API RP2A, 1969. Recommended Practice for Planning, Designing and Constructing Fixed Offshore Platforms, First ed. American Petroleum Institute, Washington DC.

API RP2A, 1971. Recommended Practice for Planning, Designing and Constructing Fixed Offshore Platforms, Second ed. American Petroleum Institute, Washington DC.

API RP2A, 1972. Recommended Practice for Planning, Designing and Constructing Fixed Offshore Platforms, Third ed. American Petroleum Institute, Washington DC.

API RP2A, 1972. Recommended Practice for Planning, Designing and Constructing Fixed Offshore Platforms, Fourth ed. American Petroleum Institute, Washington DC.

API RP2A, 1974. Recommended Practice for Planning, Designing and Constructing Fixed Offshore Platforms, Fifth ed. American Petroleum Institute, Washington DC.

API RP2A, 1975. Recommended Practice for Planning, Designing and Constructing Fixed Offshore Platforms, Sixth ed. American Petroleum Institute, Washington DC.

API RP2A, 1976. Recommended Practice for Planning, Designing and Constructing Fixed Offshore Platforms, Seventh ed. American Petroleum Institute, Washington DC.

API RP2A, 1977. Recommended Practice for Planning, Designing and Constructing Fixed Offshore Platforms, Eighth ed. American Petroleum Institute, Washington DC.

API RP2A, 1977. Recommended Practice for Planning, Designing and Constructing Fixed Offshore Platforms, Ninth ed. American Petroleum Institute, Washington DC.

API RP2A, 1979. Recommended Practice for Planning, Designing and Constructing Fixed Offshore Platforms, Tenth ed. American Petroleum Institute, Washington DC.

API RP2A, 1980. Recommended Practice for Planning, Designing and Constructing Fixed Offshore Platforms, Eleventh ed. American Petroleum Institute, Washington DC.

API RP2A, 1981. Recommended Practice for Planning, Designing and Constructing Fixed Offshore Platforms, Twelfth ed. American Petroleum Institute, Washington DC.

API RP2A, 1982. Recommended Practice for Planning, Designing and Constructing Fixed Offshore Platforms, Thirteenth ed. American Petroleum Institute, Washington DC.

API RP2A, 1984. Recommended Practice for Planning, Designing and Constructing Fixed Offshore Platforms, Fourteenth ed. American Petroleum Institute, Washington DC.

API RP2A, 1984. Recommended Practice for Planning, Designing and Constructing Fixed Offshore Platforms, Fifteenth ed. American Petroleum Institute, Washington DC.

API RP2A, 1986. Recommended Practice for Planning, Designing and Constructing Fixed Offshore Platforms, Sixteenth ed. American Petroleum Institute, Washington DC.

API RP2A, 1987. Recommended Practice for Planning, Designing and Constructing Fixed Offshore Platforms, Seventeenth ed. American Petroleum Institute, Washington DC.

API RP2A, 1989. Recommended Practice for Planning, Designing and Constructing Fixed Offshore Platforms, Eighteenth ed. American Petroleum Institute, Washington DC.

API RP2A, 1991. Recommended Practice for Planning, Designing and Constructing Fixed Offshore Platforms, Nineteenth ed. American Petroleum Institute, Washington DC.

API RP2A, 1993. Recommended Practice for Planning, Designing and Constructing Fixed Offshore Platforms, Twentieth ed. American Petroleum Institute, Washington DC.

API RP2A-LRFD, 1989. Recommended Practice for Planning, Designing and Constructing Fixed Offshore Platforms—Load and Resistance Factor Design, Draft ed. American Petroleum Institute, Washington DC.

API RP2A-LRFD, 1993. Recommended Practice for Planning, Designing and Constructing Fixed Offshore Platforms—Load and Resistance Factor Design, First ed. American Petroleum Institute, Washington DC.

API RP2A-WSD, 1969. Recommended Practice for Planning, Designing, and Constructing Fixed Offshore Platforms, First ed. American Petroleum Institute, Washington DC.

API RP2A-WSD, 2000. Recommended Practice for Planning, Designing, and Constructing Fixed Offshore Platforms, Twentieth ed, Supplement 1, December, 2000. American Petroleum Institute, Washington DC.

API RP2A-WSD, 2007. Recommended Practice for Planning, Designing, and Constructing Fixed Offshore Platforms, Twentieth ed, Supplement 3, October, 2007. American Petroleum Institute, Washington DC.

Angus, N.M., Moore, R.L., 1982. Scour repair methods in the southern North Sea. In: Proceedings of the Fourteenth Annual Offshore Technology Conference, Houston, TX. OTC paper 4410.

Barltrop, N.D.P., Adams, J.A., 1991. Dynamics of Fixed Marine Structures, Butterworth and Heinemann, USA.

Barltrop, N.D.P., Mitchell, G.M., Atkins, J.B., 1990. Fluid Loading on Fixed Offshore Structures, OTH 90 322. HMSO, London.

Brockenbrough, R.L., 2003. AISC Rehabilitation and Retrofit Guide: A Reference for Historic Shapes and Specifications. American Institute of Steel Construction, Inc, Chicago.

Dean, R.G., 1965. Stream function representation of non-linear ocean waves. J. Geophys. Res. 70, 4561–4572.

El-Reedy, M.A., Ahmed, M.A., 2002. Reliability-Based Tubular Joints. In: Procedings of 8th International Conference of Structural Safety and Reliability, ICOSSAR 01, USA.

Forristall, G.Z., 1978. On the statistical distribution of wave heights in a storm. J. Geophys. Res. 83, 2353–2358.

Galambos, T.V., 1998. Guide to Stability Design Criteria for Metal Structures. John Wiley, USA.

Health and Safety Executive (HSE), 1997. Offshore Technology Report, OTO 97 040, UK.

Health and Safety Executive (HSE), 1990. Offshore Installations: Guidance on Design, Construction and Certification, London.

Health and Safety Executive (HSE), 1986. Offshore Installations: Guidance on Design, Construction and Certification, London.

Health and Safety Executive (HSE), 1984. Offshore Installations: Guidance on Design, Construction and Certification, Third ed. London.

Health and Safety Executive (HSE), 1999. Offshore Technology Report, Assessment of the Historical Development of Fixed Offshore Structure Design codes, OTO, UK.

Le Mehaute, B., 1976. An Introduction to Hydrodynamics and Water Waves. Springer-Verlag, Dusseldorf.

Norske Veritas, 1999. Ultiguide: Best Practice Guidelines for Use of Non-Linear Analysis Methods in Documentation of Ultimate Limit States for Jacket type Offshore Structures. Høvik Det Norske Veritas, Oslo, Norway.

OTO report index, 1999. http://www.hse.gov.uk/research/otohtm/1999/index.htm

Soreide, T.H., Amdahl, J., Granli, T., Astrup, O.C., 1986. Collapse analysis of framed offshore structures. OTC 5302, Houston, TX.

YingzhiYin, 2003. Analysis of Catenary Action in Steel Beams under Fire Conditions, Manchester Centre for Civil and Construction Engineering University of Manchester & UMIST Manchester, http://www.steelinfire.org.uk/downloads/YYJ_stiff_2003.pdf, access date 2011.

Tromans, P.S., Anaturk, A.R., Hagemeijer, P., 1991. A new model for the kinematics of large ocean waves—Application as a design wave. In: Proceedings of the First ISOPE Conference, vol. 3. Edinburgh, pp. M-71.

Van de Graaf, J.W., Tromans, P.S., Shell Research, B.V., Efthymiou, M., 1994. The reliability of offshore structures and its dependence on design code and environment, Offshore Technology Conference, OTC 7382, Houston, TX.

Chapter 8

Risk-Based Inspection Technique

8.1 INTRODUCTION

This chapter presents risk assessment methodology based on current industrial practice. In addition, methods of subsea inspection and methods for implementing inspection programs are illustrated at every inspection level.

The implementation of a structure integrity management (SIM) system is presented in Figure 8.1. The first step is to collect and evaluate all available data about all the offshore structures in the fleet. Structures should include supports, such as flare, and bridges connecting platforms if they exist. In general, all available data should be on hand, and, if there are missing data that are critical, a special survey or study can be performed to obtain the data.

The data should include:

- Year of design
- Year of construction
- Water depth
- Calculation report
- Construction drawings
- As-built drawings
- Management of change records, additional risers, conductors, equipment or deck extensions or other additional loads or changes of configuration
- Number of risers and conductors
- Pile depth and driving records
- Existing metocean data
- Last inspection findings
- Soil report
- Any previous structural study performed
- Records of any accidents or fires or other events affecting the structure.

The next step is to evaluate the structure's integrity and to obtain its fitness for service. Also in this phase, the data can be used to define the structure's risk assessment ranking. Based on the risk rankings for all offshore structures in the fleet, the overall inspection philosophy, strategy and plan for all the structures can be defined.

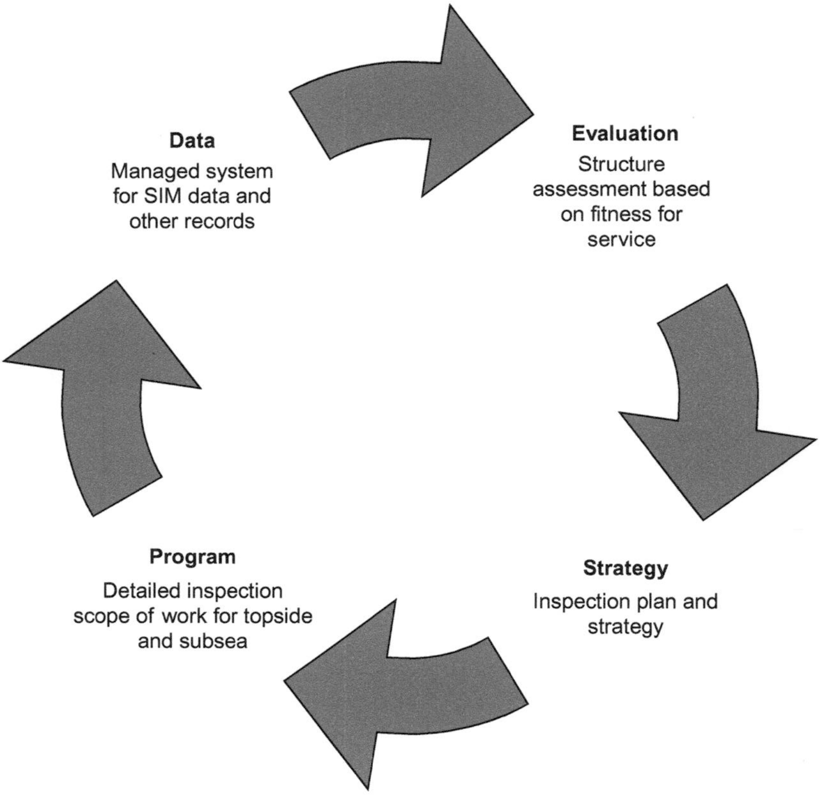

FIGURE 8.1 SIM main steps.

After the plan has been determined, the scope of the inspection will be implemented and the inspection program will be executed. Finally, when the data are updated, the cycle begins again, as shown in Figure 8.1.

8.2 SIM METHODOLOGY

The structure integrity management (SIM) system needs to be a strong system for maintaining a reliable structure during its lifespan. The SIM system requires procedures, resources, budget and a timeline.

The steps of SIM (summarized in Figure 8.1) start with collection of data about all offshore fixed platforms in the fleet of the owner company and input of the data into the computer. The second step is to evaluate the data and to identify each structure's reliability and its fitness for service. The next step is to develop an inspection and repair plan designed for needed minor or major repairs. Then a comprehensive inspection plan and detailed scope of work are developed for the ROV crew. After the ROV inspection has been performed, data are again input into the system, the structure is re-evaluated and so on until all the structures in the fleet are under control and their condition is clear. In future decision making, management can define the required annual budget for maintaining the SIM cycle without stopping.

The core of the structural evaluation and ranking is to perform qualitative risk assessment by identifying the likelihood of failure and the impact of that failure. For example, the likelihood of failure can be the result of the strength and the affected loads as per the following factors:

- The date of construction
- The structure's configuration
- The number of legs
- Marine growth
- Date of the last inspection.

The consequence of failure can affected by:

- Manned or unmanned platform
- Distance (near or far away) from the shoreline.

Factors that indicate platform-strength deterioration are:

- Splash-zone corrosion
- Cathodic protection (CP) survey
- Damaged or missing members
- Inspection history
- Cutting member
- Others.

Factors that indicate extreme platform loads are:

- Boat landings
- Marine growth
- Scour
- Additional conductors or risers
- Design loading
- Others.

8.3 QUALITATIVE RISK ASSESSMENT FOR FLEET STRUCTURES

The philosophy adopted for developing the offshore structure jacket ranking methodology is risk-based. Risk in this context is the combination of the likelihood of failure and the consequence of failure for the platform (Risk = Probability of failure × Consequence). The likelihood of failure is defined as the probability of failure or the failure rate.

The consequence of failure is the combination of losses that could be incurred as a result of a failure.

Classification factors that do not require any analytical assessment are presented that enable the likelihoods and consequences of failure to be quantitatively evaluated or qualitatively categorized.

The likelihood of structural collapse is a function of two primary factors, the platform strength or capacity, and the extreme loads that affect the offshore structure. The likelihood categorization system identifies the platform characteristics or factors that affect the platform strength and loads. For example, factors that decrease the strength of the platform and its capacity to carry the applied load will increase the likelihood of platform failure. Factors that contribute to increasing the applied load on the platform or the offshore structure in general will also increase the probability of failure.

The overall consequence of failure is the sum of three main components: environmental losses, business losses and injury/safety-related losses. The effect of each of the three consequences is converted to a dollar value (in USD) and the effects are summed to give the overall consequence. While the resulting dollar value does not represent the total amount of money that could be lost due to a failure, the monetary concept allows the three components to be combined.

After each platform is categorized in terms of likelihood and consequence of failure, the categories are entered into the risk matrix to establish the overall relative risk ranking. Note that the risk matrix can be a 4x4 or 5x5 matrix. The relative risk rankings are high (red), moderate (yellow), low (green) and insignificant (white).

In parallel with the risk assessment, the platforms have also been grouped into "families" that represent platforms with similar configurations and similar duties. Hence, in addition to identifying and ranking the high-risk platforms across the overall fleet of offshore structures, high-risk platforms can also be identified within each family. The two rankings make possible rational decisions about where to focus and how to distribute the resources of the rehabilitation project.

8.3.1 Likelihood (Probability) Factors

This section presents the factors that influence the likelihood of jacket structural failure (the likelihood is equal to the probability of failure in quantitative risk assessment). The factors fall into two groups, those that relate to the strength of the platform, listed in Table 8.1, and those that relate to the loads applied to the platform, listed in Table 8.2.

The factors should be defined separately for different locations, based on the expertise of the owner companies and regional data about the factors that can affect the structures. In addition, as more detailed inspections are undertaken,

TABLE 8.1 All Strength Factors Related to Platform Likelihood of Failure

Item	Factor	Influence of the Factor	Weight	Score
1	Design practice	To consider the modification in the new revision of codes and specifications.	6	0–60
2	Number of legs and bracing type	To consider the effect of the number of legs and the bracing system on the structure's probability of failure.	10	10–100
3	Pile system	To consider the effect that the pile system has on the likelihood of failure and to account for the uncertainty results in the case of potential damage to the pile or if there are no recorded data on the pile driving.	10	0–120
4	Risers and conductors	To account for the number of conductors and risers that affect the hydrodynamic loading and how the structure's carrying hazard materials affects its probability of failure.	7	0–70
5	Boat landings	To account for non-designed hydrodynamic loads from boats.	5	0–50
6	Grouted piles	To recognize the reduced likelihood of failure if piles are grouted.	3	0–30
7	Damaged, missing and cut members	To account for the effect of damaged, removed or cut members on the platform's likelihood of failure.	21	0–210
8	Splash zone	To consider the platform's probability of failure due to corroded members in the splash zone.	8	0–80

(Continued)

TABLE 8.1 All Strength Factors Related to Platform Likelihood of Failure—cont'd

Item	Factor	Influence of the Factor	Weight	Score
9	Flooded members	To account for the effect that flooded members have on the likelihood of failure.	8	0–80
10	CP surveys and anode depletion	To consider the condition of the CP system or the lack of CP system.	6	0–60
11	Inspection history	To account for the inspection history on the platform.	6	0–60
12	Remaining thickness	To account for remaining wall as a percentage of nominal wall thickness or number of members marked as corroded. Penalizes platforms where inspections have detected actual member wall corrosion.	8 × BL	0–80
Strength only—subtotal				10–1000

TABLE 8.2 All Load Factors Related to Platform Likelihood of Failure

Item	Factor	Effect Factor	Weight	Score
13	Design loading	To account for the effect that design loading has on the likelihood of failure. This differs from the design practice factor, which accounts for design details of the structure, such as joints and framing schemes.	8	0–60
14	Marine growth	To consider the increase of platform likelihood of failure by increased marine growth.	5	0–50
15	Scour	To account for the effect that scour has on the likelihood of failure.	6	0–60
16	Topside weight change	To account for the likelihood of failure of the platform due to changes in deck weight due to deck extensions, additional equipment or change of use.	10	0–100
17	Additional risers, caissons and conductors	To account for the effect that additional risers, caissons and conductors, over and above those that the structure was originally designed for, have on the likelihood of failure of the platform.	10	0–100
18	Wave-in deck	To consider the increase of platform probability of failure as a result of storm-wave crests' hitting the lower platform deck.	25	0–250
19	Earthquake	To account for whether design considered earthquake loads. Penalizes platforms that have not been designed for earthquakes in areas where earthquakes are likely.	8	0–80
Load only—subtotal				0–700
Total score				10–1700

the number of factors can be increased. The values in Table 8.2 are just guidelines to illustrate this approach. The condition of operation, environment, the back history and the data you have collected should all be examined. For example, if for a certain structure, no member thicknesses have been measured to check for corrosion, member thickness is not included. However, once thickness measurements become available, this factor should be introduced. Examples of factors are:

Strength factors:

- Design practice
- Number of legs and bracing type
- Foundation
- Risers and conductors
- Boat landings
- Grouted pile
- Damaged, missing and cut members
- Splash zone
- Flooded members
- Cathodic protection survey and anode depletion
- Inspection history
- Remaining wall thickness

Load factors:

- Design loading
- Marine growth
- Scour
- Topside weight change
- Additional risers, conductors and caissons
- Wave-in deck
- Earthquake load.

Interactions

Most of the above factors are not truly independent. Some of the factors have strong interactions. For example, the bracing type and the number of legs affect how much a damaged member decreases the strength of the platform and how much it increases the likelihood of failure.

According to De Franco et al. (1999), the complex interactions between different platform characteristics might make it impossible to develop a risk-based inspection system that doesn't depend on a great deal of structural details. The team evaluated the interaction of each pair of factors, and the resulting relations between the likelihood factors and their interactions are presented in the matrix in Table 8.3. The H, M and L mean high, moderate and low interactions, respectively.

TABLE 8.3 Matrix for Correlation between Failure Likelihood Factors

	Year of Design	Bracing System	Location	Number of Legs	Grouted Piles	Sensitivity to Corrosion	Splash Zone	Damaged & Flooded Members	Deck Elevation	Scour	Inspection History	Remaining Wall Thickness	Topside Weight Change
Year of design		L	**H**	M	**H**	M	L	H	H	M	L	L	M
Bracing system			H	H	M	**H**	**L**	**H**	H	H	M	H	L
Location				**H**	**L**	H	**L**	**H**	L	H	H	M	L
Number of legs					M	**H**	L	H	L	M	L	L	H
Grouted piles						L	L	M	L	L	L	L	L
Sensitivity to corrosion							H	H	L	L	H	H	M
Splash zone								M	L	L	H	H	M
Damaged & flooded members									L	L	H	M	L

(Continued)

TABLE 8.3 Matrix for Correlation between Failure Likelihood Factors—cont'd

	Year of Design	Bracing System	Location	Number of Legs	Grouted Piles	Sensitivity to Corrosion	Splash Zone	Damaged & Flooded Members	Deck Elevation	Scour	Inspection History	Remaining Wall Thickness	Topside Weight Change
Deck elevation										L	L	L	L
Scour											H	L	L
Inspection history												H	L
Remaining wall thickness													L
Topside weight change													

The investigators arrived at the matrix by identifying the following:

- If the parameters were related to each other, that is, the first factor correlated to the second parameter or not. For example, the design year of the platform implies that certain factors of safety and certain detailing practices were followed in the platform's design, so, in this case, the year of the platform design correlates with the design practice in that year so that it will be correlated with the design load, as described in Chapter 7.
- The weight of the correlation between two factors. For example, the number of legs and the bracing system together strongly affect redundancy and possibility of damage to the platform. In contrast, the effect of marine growth on the loads is considered small.

For various reasons, not all strong interactions were covered. In some cases, when the interaction was reconsidered during construction, it was decided that it was either too complex to describe adequately or the interaction was not as strong as originally believed. In other cases, such as in the case of corrosion, the effect of locality, which is captured in the calculation, was determined to be more important than the interaction with the type of structural system, and as a result it was omitted. Furthermore, two other parameters were already bound to the bracing systems and leg factors; tying on a third would make this factor overpower most of the other factors that affect likelihood.

In the case of the interaction between location and grouted piles, where the initial assessment ranked it as low, a factor was developed for a reason other than the physical interaction of piles with the local load environment. Instead, the factor for grouted piles included location to reflect design, detailing and construction practices that would lessen the impact of retrofitting a platform by grouting its piles.

Likelihood Calculation for Strength

The definition of the parameters mainly depends on the characteristics of the fleet of fixed platforms to be maintained, as some factors exist in some locations but don't exist in other locations. The discussion of the factors here is just an example of the effect of each factor, which should be tailored to each location and region world-wide. This is done by collecting the data for the whole structure and then a comprehensive study is required to evaluate the role of each factor. For example, one owner may have an old platform constructed in the year 1995, while another owner may have a platform from 1960, so the weight and the score will be different. The numbers in Tables 8.1 and 8.2 are just a guideline and cannot be mentioned in any standard. The normal procedure for qualitative risk assessment is that a team, consisting of the owner, the operator and the engineering group with expertise, hold a meeting and use brainstorming or other techniques to define the interaction between the factors and their weight.

Design Practice

Design practice accounts for platform design attributes, such as member sizing, joints, pile design and others. The intent is to quantify the inherent strength in a platform depending upon the point in time in which it was designed. The year it was designed was selected because it is most reflective of the strength of the platform. The year of design should be different from the date of installation; in some cases, the platform may have been designed several years prior to being installed.

The indicated dates generally match up with the continued development of the API RP2A recommended practice for design of offshore platforms in U.S. waters. The first version of API RP2A, issued in 1969, was the first good guideline for platform design. The year 1970 is used here (conservatively) to reflect the fact that, although the first edition was issued in 1969, the document did not reflect common practice until about a year later. The next major update to RP2A was the inclusion of joint design practices, in approximately 1975. In 1979, API RP2A was updated to the point that the API Section 17 Committee on assessment of existing platforms deemed that the guideline had reached maturity and that platforms designed to the guideline since that time are of good quality.

Table 8.4 presents an example for a score between 0 and 10 for the year of design of the platform, a factor that estimates the level of confidence that can be placed on it due to the state of design practices prevalent at that time. A platform designed later than 1979 scores 0, to indicate that the platform is reliable due to the good and improved design practices. Platforms designed before 1970 score 10 because they tend to have the highest probabilities of failure. Note that the overall weight of the total score is 5.

TABLE 8.4 Design Practice Factor

Year of Design		Score
≥	≤	
	1970	10
1971	1974	7
1975	1979	4
1979		0
Unknown		9

Sometimes the year of design isn't known for old platforms, but the drawings for them indicate the construction or installation year, so the design year is taken to be 2 years prior to the year of installation.

If a platform design has been repeated as a standard, the year of design should be taken as the first time of design or the installation year for the first platform the design was applied to.

The "year designed" input parameter is in the general category of the platform data. As mentioned, this parameter is used for estimating the reliability of the platform based on design practices. This factor is independent of other factors because it does not relate to any of the others. However, the "year designed" input parameter is also used in the "year-designed and location" factor and the "grouted piles" factor because the input has a high interaction with the loading and resistance aspects in these factors.

Number of Legs and Bracing Configuration

It is well known that the number of legs and the framing scheme contribute to the strength of a platform and, more importantly, to its ability to sustain damage and still "survive" extreme loading events. (The relation between the structure's configuration and its strength is discussed in Chapter 7.) For example, a 3-legged platform with several damaged K braces has a significant decrease in capacity compared to the same number of damaged members on an 8-legged platform that is X-braced. The number of legs provides for an increased number of major load paths, while the X-brace scheme provides redundant framing at each major joint. Note that the "preferred" platform framing scheme per API for earthquake regions (where redundancy and ductility are important) is an 8-legged (or more) structure that is X-braced.

In terms of score, an X-braced 8-legged (or more) structure was assigned the highest value and the K-braced 3-legged (or less) structure was assigned the lowest value. Values in between were based upon subjective ranking, based upon experience, where available. For example, according to Puskar et al. (1994), Hurricane Andrew in 1992 caused failure in some 8-legged K-braced platforms, but numerous 4-legged X-braced platforms survived without any failures. Hence the score for 8-legged K-braced structures is slightly higher than the score for 4-legged X-braced structures.

This criterion is required to account for the effect that the number of legs and the bracing system have on the likelihood of failure. The number of legs and the bracing system should be identified. Table 8.5 presents the criteria and score based on the number of legs and the bracing system.

If the platform has different types of bracing schemes in different directions (e.g., K in the transverse direction and diagonal in the longitudinal direction), then the framing scheme that results in the highest score is used (K in this example).

TABLE 8.5 Bracing Configuration and Number of Legs Factor

Bracing System	Number of Legs			
	3	4	6	8
K and Diamond	10	8	6	5
Diagonal	7	6	4	3
X	5	4	2	1
Unknown	10	7	5	4

Note that an increase in the number of legs increases the redundancy of the platform. Of the bracing systems, X bracing is the most efficient, while K bracing is the least efficient. The overall weight in the total score is 10.

The inputs for this factor are available in Table 8.5. There is a high interaction between the number of legs and the bracing system because both these inputs give strong indication of the damage tolerance of the platform. Because failure in a platform initiates at the brace, the bracing pattern dictates the damage sensitivity of the platform.

As mentioned elsewhere, this factor interacts significantly with damaged members and flooded members. The significance of any damage affecting the likelihood of failure is dependent on the damage tolerance of the platform.

Pile System

This factor accounts for the effect that the pile system has on the likelihood of failure. Uncertainty results from both the lack of records relating to the design and installation of the piles and to potential damage to the piles from soil disturbance caused by the feet of jack-up rigs. For this factor, you should answer the following three questions.

1. Does pile design exist (Y/N)?
2. Are there actual pile penetration records (Y/N)?
3. If answers to the above are Y, does actual penetration meet target requirement (Y/N)?

From the above answers, select the appropriate score in Table 8.6. The overall weight of the total score is 10.

Note that the accuracy factor of the foundation design itself has not been included in Table 8.5.

TABLE 8.6 Pile Strength Factor

Criteria	Answers to Input Questions	Score <1975	Score ≥1975
Existing pile penetration achieves target	Y, Y, Y	2	0
Existing pile penetration known but not target	N, Y, (N/A)	3	2
Target pile penetration known but not existing pile penetration	Y, N, (N/A)	7	3
Existing pile penetration does not meet target	Y, Y, N	8	7
Target and existing pile penetration unknown	N, N, (N/A)	10	10

Disturbance from Jack-up Feet	Factor
No seabed disturbance	1
Depression >0.3 m <5 m from pile	1.4
No data	1.4

Score = Score from table × factor (max score equals 14)

Risers and Conductors

Previous factors have been associated directly with the strength and redundancy of the platform and how the particular issue affects the platform's ability to withstand extreme events, such as waves and earthquakes. However, appurtenances like risers and conductors also contribute to the platform's likelihood of failure and need to be included in the quantitative risk calculation, since inspection of these items is also typically included in an underwater inspection.

Caissons for fire-control water and other important activities are considered a riser. Note that appurtenances like boat landings, walkways and barge bumpers have been excluded since they generally play a much smaller role in platform risk than risers and conductors do.

Risers and conductors that contain hydrocarbon gas or oil can result in an explosion or fire that can damage the platform or cause it to fail. The score increases with the number of risers or conductors, since there is an increased likelihood of damage as the number of them increases.

In terms of weight, risers were assigned a higher value than conductors: 1 riser was assumed to equal 5 conductors. This bias to risers is necessary because conductors are less likely to fail than risers are, for the following reasons:

- They are often located on the interior of the platform and therefore are less vulnerable to boat impact or other external damage.
- They have multiple layers of pipe and drill string, which decrease the probability of a complete breach.
- They typically have subsurface safety valves that shut off quickly in the event of a breach.

The 20% value was based upon testing of the weighted score for several platforms in the Amoco database and determining how much of an effect there was on the rank of the platform if the number of risers and/or conductors was increased or decreased.

This criterion accounts for the effect that risers and conductors have on the likelihood of failure. Risers and conductors attract hydrodynamic loading, which increases the global loading on the structure, increasing its likelihood of failure. Also, damage to risers may cause loss of containment, which could result in a major incident, significantly increasing the likelihood of failure.

The input data should include the number of risers, the number of conductors, whether the riser or conductor carries oil or gas and, in the case of gas, whether it is high- or low-pressure gas. Key caissons, such as firewater caissons, should be counted as "oil" risers.

Table 8.7 illustrates an example of a score between 0 to 10 for both risers and conductors depending upon their contents. These scores are then multiplied by a weighted factor (1 for risers and 0.2 for conductors) and the sum of the two

TABLE 8.7 Risers and Conductors

Number	Type		
	Risers and Caissons		Conductors
	Oil and Water	Gas	
0	0	0	0
<3	0	1	0
4–7	1	3	1
8–14	2	6	2
14–20	3	8	3
>20	4	10	3

Score = Riser score × 1.0 + Conductor score × 0.2

TABLE 8.8 Additional Boat Landings

Number of Boat Landings	Score
0	0
1	5
2	8
≥3	10
No data	10

scores gives the total score. The total score can vary from 0 to a maximum of 13.6. The overall weight of the total score is 5.

Risers and conductors do not have any interdependence with others, as the appurtenances do not directly contribute to the redundancy or strength of a platform. On the other hand, riser failure increases the overall risk in a platform.

Damages to appurtenances are usually detected through different levels of inspection. The causes of damages are mainly internal or external corrosion, external impact, vibration or natural hazards.

Boat Landings

Both excessive marine growth and additional boat landings, for which the jacket has not been designed, increase the global loading on the structure, increasing the likelihood of failure. Corrosion and uncontrolled modifications may result in boat-landing failure, creating a dropped-object hazard to the jacket. The score for the effect of boat landings is presented in Table 8.8.

The number of boat landings should be identified, because the score for this factor depends on this number and ranges from 0 to 10. The overall weight of the total score is 5.

Grouted Piles

Grouting is usually used to fill the annulus between the leg and pile and increases the effective wall thickness of the leg, since it is then be a combined section of leg, grout and pile together, which increases the capacity of the joints of braces that frame into the leg. In many areas, particularly older Gulf of Mexico platforms, the weak links in the platform design are these joints. Therefore, grouting of the annulus between the leg and pile of some platforms is a reasonable measure for reducing the likelihood of failure.

The 1975 threshold date is meant to capture the impact of API RP2A factors at that time that resulted in better joint design, and hence a lower score, since after 1975 the joints may be well designed and grouting will not provide much

TABLE 8.9 Grouted Piles

Non-Grouted Piles	Year Designed	
	<1975	≥1975
	10	4

Score = tabulated score × weight

improvement. Similarly, the North Sea factor was included since platforms in this region tend to have better joint design and therefore grouting will have less of an impact.

It is worth mentioning that grouted piles will increase the structure's strength, as discussed in Chapter 7, so they will reduce the likelihood of failure.

If the piles are grouted, they are ignored in scoring an increased the likelihood of failure.

As presented in Table 8.9, the score for non-grouted piles is dependent on the year of design of the platform. From a practical point of view, this score can be reduced by around 60% if the platform is located in the North Sea, as joint design there tends to have been much better than in other areas of the world, and therefore grouting would have a much lower effect, as discussed in Chapter 7.

As evident from Table 8.9 for this factor, grouted piles will interact with the year designed and location. A weight factor has been added to reduce the effect considerably in the North Sea, where platform joints have historically been better designed.

Damaged, Missing and Cut Members

In the initial risk-based underwater inspection (RBUI) calibration work done in 1997, the change in the failure probability was estimated primarily using results from the Joint Industry Projects and Puskart et al. (1994) relating the RSR to the failure probability for a number of locations world-wide. The calculation of the likelihood requires collecting numerous data about the platform's configuration and its current condition to assign risk. A key component is the number of damaged members, as they can have a significant impact on the strength of the platform.

The AIM Phase III project, undertaken in 1988, evaluated a platform with numerous postulated damaged conditions, including several scenarios of damaged members, ranging from 1 to 3 or more. The damaged members were located at different regions of the jacket, including the waterline, mid-depth and near the seabed. The results indicated that overall platform capacity was reduced little with one damaged member, but that numerous damaged members, for example more than 5 members, caused noticeable reduction in platform capacity. This led to the ranges shown in Table 8.9.

As discussed by Puskar et al. (1994) and Gebara et al. (1998), K-braced platforms are more sensitive to damage than X-braced platforms. Based upon this background, it was decided that the K-braced platforms would be ten times more sensitive to damaged members than X-braced platforms.

The probability of failure is affected by the number of damaged, missing or cut members, and damaged members are defined as members that remain in place but have dents, holes, cracks, out-of-straightness or other defects that reduce their strength.

The number of missing, cut or damaged members should be known from the underwater survey or can be estimated from previous experience in limited cases. If there is no available underwater survey information, the score depends on platform type, as well-head protectors are particularly vulnerable to having members cut out.

Note that if the platforms have been only partly surveyed, the number of damaged, missing or cut members should be extrapolated from the available survey data.

Tables 8.10 and 8.11 illustrate scores for damaged and missing members. The platform's damage tolerance is accounted for by multiplying the tabulated scores by the "leg-bracing" score and dividing by 10. The resulting score is limited to a maximum of 10. The overall weight of the total score is 20.

The number of damaged members is an inspection result. Consequently, this factor is taken as one of the strongest indicators of the current condition of the platform. Inspection-related inputs are given a stronger overall weight in the scoring, because the goal of the system is to examine the effect of different inspection policies on risk.

The number of damaged members also interacts with the type of framing system in any given platform. Three-legged, K-braced platforms have few

TABLE 8.10 Damaged Members

Number of Damaged Members	Score
0	0
1–4	2
5–9	3
More than or equal to 10	5
No data for well-head, tender or drilling platforms	5

TABLE 8.11 Missing or Cut Members

Number of Missing Members	Score
0	0
1–3	6
4–9	8
More than or equal to 10	10
No data for well-head, tender or drilling platforms	10

Total score (limited to 10) = (tabulated score for damaged members + tabulated score for missing members) × leg-bracing score/10

redundant load paths and are, therefore, not very damage-tolerant, while 8-legged, X-braced platforms are not as affected by missing or damaged members.

The total weight given to damaged members is 10.5. This value was obtained by testing the database, which showed that this level of weight was required to increase the risk of platforms with significant damage: for example, if the platform has four damaged members, the platform's rank should be changed to a number that qualifies this platform to have more frequent underwater inspections. Therefore, these scores are just a guide and definition of reasonable scores and score weights for the fleet of platforms in the same location should be done by trial and error.

Splash-Zone Corrosion and Damage

In some regions of the world, such as the Red Sea, splash-zone corrosion is a major hazard. Survey data for splash-zone members should be collected.

Where no survey information is available, the score depends on topside weight growth. Tables 8.12 and 8.13 are a guide for scoring this factor. The overall weight of the total score is 10.

This factor is essential when the pile-leg annulus is not grouted and the jacket therefore hangs from the crown weld, so that the highest utilized leg members are in the splash zone.

Flooded Members

Flooded member detection is a common approach to inspection of offshore platforms. A flooded member provides an indication of the current condition of the member and possible existence of damage or defect, since most platform

TABLE 8.12 Splash-Zone Corrosion and Damage

Extent of Corrosion	Score
None and light corrosion	0
Moderate corrosion (approx 50% coverage)	3
Heavy corrosion	10
No corrosion data	10 or go to Table 8.13

TABLE 8.13 Corrosion Condition against Topside Loads Where No Data Exist on Extent of Corrosion

	Original Topside Weight	
	Known	Estimated
No weight growth or known/believed weight growth ≤10%	3	7
Known/believed weight growth >10% or no data	8	10

members are designed to be buoyant. Basically, damage like a crack or a hole has allowed water to enter the flooded member. In some cases, the cause of the flooding cannot be found, and this is the condition considered here. The basic assumption is that, although the specific damage is unknown, it is still significant enough that it will affect member strength, and if enough members are flooded, it will affect platform strength.

In terms of the scores assigned to the number of flooded members, the logic previously described for damaged members, when applied to flooded members, will be different and have little impact on the platform strength. A larger number of flooded members has a greater impact. As with damaged members, the number of legs and the type of framing scheme also have an impact on the score.

The number of flooded members does not include members known to be "damaged." For damaged members, the damaged member factor overrides the flooded member likelihood of failure score. Only flooded members that have an unknown cause of flooding are counted.

Table 8.14 is very similar to the damaged members table. It illustrates a score between 0 and 10, where a score of 0 corresponds to no flooded members and hence no increase in the likelihood of failure, and a score of 10 corresponds to 10 or more flooded members and a significantly increased likelihood of

TABLE 8.14 Flooded Member Effect

Number of Flooded Members		Score
≥	≤	
	0	0
1	3	3
4	9	8
10		10
No data available		10 (or estimated)

Total score = tabulated score × leg-bracing score/10

failure. The platform's damage tolerance is accounted for by multiplying the tabulated scores by the "leg-bracing" score and dividing by 10. The resulting scores range from 0 to 10. The overall weight of the total score is 6.

If underwater survey data relate to a percentage of the members, the number of flooded members is to be factored up to reflect the total number of members. For example, if a survey of 25% of the members detected one flooded member, then a total of 4 flooded members should be assumed for the structure.

The number of flooded members is an inspection result and gives a strong indication of the current condition of the platform. As the objective of the system is to examine the effect of different inspection policies on risk, inspection-related inputs are given a stronger overall weight in the scoring.

The number of flooded members also interacts with the type of framing system in any given platform. Three-legged, K-braced platforms have few redundant load paths and are, therefore, not very damage-tolerant, while 8-legged, X-braced platforms are not as affected by missing or flooded members.

CP Surveys and Anode Depletion

An ineffective CP system will result in corrosion, which will increase the likelihood of failure.

Anode survey data by "grade" should be defined. Note that, if an overall grade for the platform has not been established, then the platform grade is calculated by averaging the anode survey results of the worst 30% of the structure's anodes. Table 8.15 illustrates a score between 0 and 10, where a higher score signifies greater risk. The overall weight of the total score is 6.

Inspection History

Inspection history is twofold: it includes the time period and the level of survey, because longer time periods between inspections increase the probability of

TABLE 8.15 Cathodic Protection

Criteria	Score
Grade 1: <10% consumed	0
Grade 2: <50% consumed	3
Grade 3: >50% consumed	7
Grade 4: 100% consumed	10
No data (assume worst)	10

platform failure, as there is a high probability of unpredicted local failure during the gap between inspections. On the other hand, because underwater survey is a more detailed inspection, so it is less likely that an undetected local failure or deficiency in strength or an increased load will occur. A well-inspected platform is defined as a platform that has been checked for flooded members and for which several joints have been cleaned/inspected; it should be able to go longer without an inspection update than a platform that was inspected with only a swim-by.

When a platform is inspected, the data obtained are used to recompute the risk and the platform is re-ranked. In this way, if two platforms have the same likelihood of failure prior to the inspection of one, the inspected platform's likelihood drops, as does its overall risk. The platform with lowered risk is then a lower inspection priority while the uninspected platform becomes a higher inspection priority.

The quality of the inspection also affects the ranking. Two platforms with the same risk prior to being inspected—one with a Level II inspection and the other with a Level III—will have the same risk immediately after the inspection. While not strictly true, it is assumed that both levels of inspection will be able to determine the current condition of the platform with the same level of confidence. However, the Level III or close visual inspection is assumed to be able to detect impending problems with more confidence than the Level II. Four years after the inspection, the score of the platform with the Level II inspection increases to 8 while the score for the Level III platform increases to 5, indicating the system considers that the predictive capabilities of the Level III inspection are significantly better than those of the Level II inspection.

Level I inspections, defined in API RP2A as above-water walk-around inspections, have been excluded because they should be performed every year for all platforms.

As shown in Table 8.16, the input data will be for the year of last Level II, III or IV inspection. The overall weight for the inspection history factor in the total score is 5.

TABLE 8.16 Inspection History

Years Since Inspection	Level of Last Inspection		
	Level II	Level III	Level IV
1	0	0	0
2	2	0	0
3	4	2	1
4	8	5	3
5	9	7	5
6	9	8	6
7	10	9	8
8	10	10	9
9	10	10	10

For each inspection level, calculate tabulated scores, *SII* for Level II, *SIII* for Level III and *SIV* for Level IV. Score = min (*SII*, *SIII*, *SIV*)

For each inspection Level II through III, apply the following formula:

$$\text{The years since inspection} = \min(9, \text{this year} - \text{inspection year})$$

The last inspection factor does not interact with any other factors, as it is totally independent of the design or the condition of the platform; it is a matter of operation and the maintenance plan, which depend mainly on the budget and the existing robust plan. This factor highlights the necessity of different levels of inspection in order to eliminate or lower the perceived risk factors of a platform.

Remaining Wall

Similar to damaged and flooded members, the number of corroded members has a major effect. However, in this factor, the *amount* of corrosion is also important (for damaged or flooded members, the amount of damage is not considered).

In terms of remaining wall thickness, it was felt that at least a 20% decrease is required to affect strength. Corrosion is typically localized on the member and this has been accounted for in the approximation, noting that the score would be higher if the entire member were corroded. Corrosion above 50% was considered extreme and was assigned the highest score.

The extent of corrosion refers to the number of members affected. The quantitative calculation for this factor follows the same logic as for damaged and flooded members. Note that, in this case, the score is inverted compared to

TABLE 8.17 Remaining Wall Thickness

Remaining Wall (%)		Score
>	≤	
	50	10
50	80	8
80	99	3
100		0
Number of Members		Weight
≥	≤	
	4	0.2
5	9	0.4
10		1

Total score = tabulated score × weight

damaged and flooded members, since the remaining wall is the key factor. Remaining wall is expressed as a percentage of the nominal thickness.

Table 8.17 presents a score between 0 and 10, where a score of 0 corresponds to a wall retaining 100% thickness and a score of 10 corresponds to a wall that has thinned to less than or equal to 50%. This score is then multiplied by another factor that is determined by the number of members thinned. The resulting scores range from 0 (no risk of failure) to 10 (increased risk of failure). The overall weight of this factor in the total score is 7.5.

The remaining wall and the extent are both inspection results. Consequently, the corrosion factor is taken as a strong indicator of the current condition of the platform.

However, for this factor, the effect on likelihood of failure has been kept independent of bracing pattern and number of legs, because a member with a corroded wall would still be considered to be contributing to the structural stiffness and thus would carry load with a reduced capacity. When the remaining wall turns into a through-thickness hole, a crack or a dent, it would be re-categorized as a damaged member, for which separate qualitative calculations exist, as described above.

Likelihood Calculation for Load

Design Loading

Previous studies concluded that the first generation of platforms in the Gulf of Mexico were designed for a hydrodynamic loading that is 55% of today's design load. The metocean data were subsequently revised and the design loading became more severe as the years progressed.

As per the discussion in Chapter 7, in the North Sea, the earlier platforms of the late 1960s and early 1970s were designed for loading that is 70% of today's design load. However, in the years between 1980 and 1986, the metocean data produced design loads that were 45% more severe than the present loads. This explains the separate tables for the North Sea.

At other locations, the Gulf of Mexico trend is used.

The design loading factor differs from the design practice factor, which accounts for the level of confidence in the design of structural components such as joints and members.

Note that the "year designed" can generally be taken as 2 years prior to the year of installation.

The scores for this factor vary from 0 to 10 for different ranges of year designed. Low scores correspond to higher values of design loads and thus low likelihood of failure, while high scores indicate lower values of design loads and increased likelihood of failure. This factor has two tables, one for the North Sea and the other for elsewhere (combined into one table in Table 8.18). However, the scoring principle is the same for both. The scores vary from 1 to 10 for different ranges of year designed. The overall weight in the total score is 5.

Where the design is a repeat of a standard design, the year designed should be taken as the year designed for the first of the standard platforms. Where the year designed for the first of the standard platforms is not known, a date 2 years

TABLE 8.18 Year of Design

	Year Designed		Score
North Sea	≥	≤	
		1972	10
	1973	1980	6
	1981	1986	1
	1987		5
Another Location	**Year Designed**		**Score**
	≥	≤	
		1970	10
	1971	1974	7
	1975	1978	4
	1979		1

prior to the year of installation may be taken, although this may not be conservative.

It is evident from Table 8.18 that the year designed parameter interacts with location to distinguish the North Sea from the rest. The scores are based on the design hydrodynamic loading adopted from the metocean data as per practice in the year of design.

Marine Growth

Marine growth is a very simple factor but can cause many problems, as discussed in Chapter 3. The wave force that affects the load on the structure is a function of member diameter, so marine growth will increase member diameter and increase the wave force correspondingly. This factor uses actual measured marine growth (mm)/design marine growth (mm).

Table 8.19 illustrates scores for this factor between 0 and 10, where a higher score signifies greater risk. If the measured marine growth exceeds the design value, it increases the environmental load on the platform and therefore increases the likelihood of failure for the platform. The table allots a score of 10 to platforms whose measured marine growth exceeds the design value by 50% or more.

If there are no data available, the score may be reduced to 5 if underwater surveys indicate that the levels of marine growth are low, that is <50 mm thick.

This parameter does not have any interdependence with other factors or parameters in the likelihood category. The measured marine growth is an inspection item. It is divided by the design marine growth (design data sheet) to obtain the ratio in terms of percentage.

Although marine growth is an inspection-related input, the parameter that causes failure is the environmental loading. Excessive marine growth, especially when it far exceeds the design growth, results in higher loading on the platform than it was originally designed for.

TABLE 8.19 Marine Growth Factor

Measured/Allowable Marine Growth in Design		Score
>	≤	
	90%	0
90%	135%	3
135%	150%	8
150%		10
No data	10	

Note that there is not a 1:1 correspondence in the percent increase in marine growth and the increase in global base shear on the structure. This is because the computed ratio is the percent increase in marine growth only, with the marine growth typically only being a small portion of the total diameter of a tubular member (the member diameter is by far the largest contributor). This is why the marine growth ratio has to exceed a large value (150%) before there is any significant effect.

Scour

Although scour this is an inspection-related input, the parameter to cause failure is the resultant stress.

This factor uses measured scour depth (ft or m) divided by the design scour depth (ft or m). Table 8.20 presents a score between 0 and 10, where a higher score signifies greater risk. If the measured scour exceeds the design value, it increases the stresses on the pile and lower portion of the jacket and therefore increases the likelihood of failure for the platform. This table allots a score of 10 to platforms whose measured scour exceeds the design value by 150% or more.

Although scour is a concern, it does not typically result in a significant increase in stresses on the piles and lower portion of the jacket (although the stresses are higher than if there were no scour). Therefore it has been assigned a low overall weight of 2.

A "No data available" score may be reduced to 3 if underwater surveys undertaken across the assets indicate that typical scour levels are low, that is, <1 m.

This factor does not have any interdependence with other factors or parameters in the likelihood category. The measured scour is divided by the design scour depth (design data sheet) to obtain the ratio in terms of percentage.

TABLE 8.20 Scour Factor

Scour Measured/Design		Score
≥	≤	
	90%	0
90%	125%	3
125%	150%	7
150%		10

Topside Weight Change

This factor accounts for the likelihood of failure of the platform due to changes in deck weight due to deck extensions, additional static or rotating equipment or change of use. Defining deck weight change is not easy for platforms from the past (e.g., over 40 years old) because, without computerized archives of drawings and reports, it is very difficult to make the calculations. The main sources of data are the original drawings, if they still exist, the actual condition of the platform, and collection of information from people who worked on the platform before.

There are four possible scenarios: you have no data at all; you have minor data but you cannot know if the weight has increased or decreased from the original; you have data and estimate that the growth in weight has been less than 10%; or you estimate the weight growth has been more than 10%. The score for this factor is shown in Table 8.21.

Additional Risers, Caissons and Conductors

This factor is intended to account for the effect that additional risers, caissons and conductors (over and above those that the structure was originally designed for) have on the likelihood of failure of the platform.

The score depends on both the number of risers, caissons and conductors that the structure was designed for and the additional number subsequently installed (Table 8.22). Note that the overall weight of the total score is 5.

If there is uncertainty concerning the basis of the original design, a conservative estimate of the number of additional risers, caissons and conductors is used.

If you have the diameter of the original riser and conductor design and the diameters of the actual existing riser and conductors, this factor can be calculated by obtaining the ratio of the sum of the diameters of the existing risers and conductors to the sum of the diameters for the original design of conductors and risers.

TABLE 8.21 Topside Weight Change

Criteria	Score
Not known if weight has increased or decreased	2
Estimated weight growth ≤10%	5
Estimated weight growth >10%	8
No data	10

TABLE 8.22 Additional Risers or Conductors

Number of Additional Risers, Caissons and Conductors	Number of Risers, Caissons and Conductors Designed for		
	1–10	11–20	>20
0	0	0	0
2	2	1	0
4	4	3	1
6	6	4	2
8	8	5	2
>8	10	7	3
Not known	5	3	2

Wave-in-Deck

The wave-in-deck factor is intended to account for the likelihood of failure of the platform due to storm wave crests impinging on and inundating the platform's lowest deck. This might occur if the platform has improper design or if there has been a change in metocean data so that the lowest deck elevation fails to satisfy the minimum air gap requirement. This would mean that, in the case of an extreme event, the maximum wave could potentially hit the deck levels and cause damage. This affects the reliability if the structure in terms of both structural integrity and human safety. Therefore, it is essential that this situation be reported to the operating and engineering group.

Since this problem seldom occurs, if the review study reveals that there is a problem, it should be dealt with at once to take a decision.

If the wave-in-deck factor does not apply, it will be neglected, but if it does apply, it will have the higher score, which is 10. The overall weight of the total score is 20.

Earthquake Load

Amoco's platforms in Trinidad lie in an earthquake zone. Although recent platforms installed in the area have been designed for earthquakes, a majority (if not all) of the older platforms did not account for earthquakes in their original design. This inadequacy in design increases the likelihood of failure for those platforms when compared with platforms that have been designed for earthquakes or are located in non-earthquake zones.

ISO Zone 2 correlates to an effective ground acceleration of 0.2 g (ISO zones have ground accelerations that are 10% of the zone value: i.e., Zone 1 = 0.1 g, Zone

TABLE 8.23 Seismic Load

ISO Seismic Zone		Score
≥	≤	
0	2	0
3	3	4
4	4	7
5	5	10

2 = 0.2 g, etc.). Typical offshore platforms that are designed for wave only can normally withstand ground accelerations of 0.15 to 0.2, provided that they were initially designed for a reasonable wave height and have good joint detailing. Ground accelerations higher than this start causing problems, with an ISO Zone 4 or larger requiring specific earthquake design features for ductility (special kl/r values, extra-heavy cans, redundant framing schemes, etc.).

Obviously, it is important to define the earthquake zone. Table 8.23 presents scores between 0 and 10 for seismic load. Higher scores are used for platforms existing in higher earthquake zones to penalize them for design inadequacy. In an earthquake zone, seismic load is a major design consideration; therefore, the overall weight in the total score is 8.

This factor does not have any interdependence with other factors or parameters in the likelihood category. It is formulated with the earthquake zone and earthquake design (yes or no) parameters from the design category of the data sheets.

Likelihood Categories

Most factors provide a score between 0 and 10. Factors like bracing-legs and year-load location have values ranging from 1 to 10, and scores for appurtenances range from 0 to 12. A value of 0 indicates that the factor has the least effect on increasing the probability of failure. A value of 10 indicates that the factor has a maximum detrimental effect on the likelihood of failure.

The weights indicate how strongly the effect assessed by each factor affects the overall likelihood of failure of the platform. The total score is calculated by the formula $\Sigma i\, w_i s_i$, where w_i is the weight of the *i* factor and s_i is the score. This score can mathematically range from 15 to 2255, but for practical purposes scores range from about 40 to 350.

The total weighted scores are converted to a likelihood category according to the ranges in Table 8.24. These ranges have been developed from the ranges presented in the DNV paper.

After scores for all the platforms in the fleet have been obtained, the cumulative percentage can be calculated and, using statistical data, the

TABLE 8.24 Likelihood Category

Category	Range
1	≤S1
2	S1–S2
3	S2–S3
4	S3–S4
5	>S4

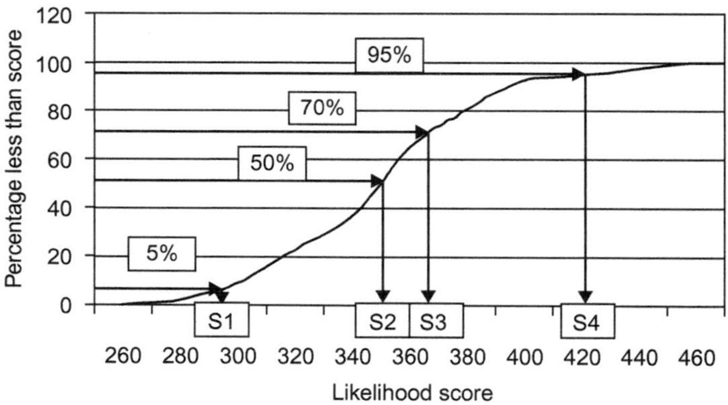

FIGURE 8.2 Cumulative density function (CDF) for likelihood score strength.

cumulative density function can be obtained, as shown in Figure 8.2. Because the category limits identify the best and worst 5% of the platforms, it is assumed that 5% of the platforms have a high likelihood of failure and 5% of the platforms have an insignificant likelihood of failure. The intermediate category limits are intended to represent the 50th and 70th percentiles. This distribution of the category limits assigns two category ranges for platforms that have better-than-average likelihood scores and three ranges for platforms that have worse-than-average scores. Consequently, it will be necessary to review and modify the ranges once data from the jacket evaluation and scoring become available, to ensure that the intent is being realized. The values of the categories S1, S2, S3 and S4 will be obtained from the cumulative density function (CDF).

8.3.2 Consequence Factors

In general, the consequences of platform failure are the sum of three main factors, where the components account for:

- Environmental losses (C_E)
- Business losses (C_B)
- Injuries and safety-related losses (C_S).

Each of the above consequences is converted to a monetary value in USD, and the three factors are then summed to give the overall consequence.

While the summation of the monetary values obtained does not represent the total amount of money that could be lost due to a failure, the concept of a monetary value for consequences has been adopted so that the effects of safety, environmental, and business losses can be measurable and combined.

Whenever two consequences have an equal abstract monetary value, they should represent two events that have an equivalent detrimental effect on the offshore platform fleet.

While each type of consequence is measured in the common unit of USD, different factors determine the value of each type of loss. When possible, the consequence calculations are quantitative and are related to actual economic values, such as the price of oil and gas and the cost of offshore construction. For other consequences, the values are subjective and may not be actual economic values.

The factors affecting each type of consequence and the factors for calculating their monetary value are described next. Every category will be calculated separately.

The total cost will be calculated as follow:

$$C_T = C_E + C_B + C_S \tag{8.1}$$

Environmental Losses

Environmental losses are calculated based on the amount of liquid that was spilled, which is calculated using the daily production or the amount of liquid released if it can be computed. There is a fixed cost, which includes the mobilization of personnel and equipment to perform the clean-up, in addition to the regulatory cost. The variable cost will depend on the volume of the oil spill. It is worth mentioning that the fixed and variable costs vary with global location, to reflect the regional differences in clean-up and regulatory costs (see Table 8.25). For example, strong wave action in some regions will quickly break up, dissipate and volatize spilled oil, reducing the overall environmental impact of an oil release.

The cost is adjusted by multiplying the calculated cost by a factor to account for the distance offshore. The distance from shore is important because both the clean-up cost and the damaging effect of an oil spill increase if oil is washed up

TABLE 8.25 World-wide Variation in Default Values for Consequence Costs

Variable	Location					Units	Description
	G	N	C	M	E		
i	10%	10%	10%	10%	10%	%	Discount rate
P_{oil}	$120.00	$130.00	$110.00	$130.00	$110.00	$/bbl	Oil revenue per bbl
P_{gas}	$1.50	$3.00	$1.50	$1.50	$1.50	$/mscf	Gas
C_{ex}	$1 MM	$1 MM	$1 MM	$1 MM	$1 MM	$	Personnel marginal exposure cost
n	200	545	180	180	400	Days	Default deferred production period
FC	$500 M	$750 M	$400 M	$350 M	$500 M	USD	Fixed environmental cost in open sea
VC	$500	$500	$500	$500	$500	$/bbl	Variable environmental cost in open sea
d	50	65	25	25	50	Miles	Default distance offshore
d_m	100	100	100	100	100	Miles	Distance offshore to open sea
—	$5 M	$5 M	$5 M	$5 M	$5 M	$/ton	Default replacement value
M	$150 M	$150 M	$150 M	$150 M	$150 M	$	Default daily production value
RV	250	250	250	250	250	Bbl	Default spill volume
N_{crew}	20	20	20	20	20	Persons	Default crew size: service unknown
N_{crew}	30	30	30	30	30	Persons	Default crew size: production platform
N_{crew}	20	20	20	20	20	Persons	Default crew size: drilling platform

TABLE 8.25 World-wide Variation in Default Values for Consequence Costs—cont'd

Variable	Location					Units	Description
	G	N	C	M	E		
N_{crew}	50	180	50	50	50	Persons	Default crew size: quarters platform
N_{crew}	10%	10%	10%	10%	10%	%	Percent of crew exposed if evacuated
N_{crew}	2	2	2	2	2	Persons	Default crew size: unmanned platform
—	20%	20%	20%	20%	20%	%	The percent increase in the expected safety losses in gas production platforms.

G = Gulf of Mexico, N = North Sea, C = Canada, M = Malaysia and E = elsewhere.

on shore. Furthermore, an oil spill can affect coral reefs or can destroy a tourist area, as in Egypt, where the oil fields are near the tourist areas.

Estimated environmental loss calculations are presented in the following equation.

$$C_E = f(d) \times \{F_C + V_C \times \min(D_P, R)\} \tag{8.2}$$

Where

$$f(d) = 1 \quad d > d_m \tag{8.3}$$

$$f(d) = 1 + \left(\frac{d_s - d}{d_s}\right)^2 \quad d \le d_m \tag{8.4}$$

and where:

C_E is the environmental loss, in USD; V_C is the variable cost, in USD/bbl; F_C is fixed environmental cost, in USD; R is minimum released oil volume, in bbl; D_P is daily production, in barrel of liquid, bpl; d is distance offshore, in kms; and d_s is the maximum significant distance offshore, which is the distance from the shoreline in km.

The calculations are followed only when the environmental loss, C_E, has not been determined for a given platform. The estimate is based on the assumption that the environmental losses are proportional to daily production. The fixed cost, F_C, and the marginal (variable) cost, V_C, depend on the location, in order to account for the variations world-wide in the costs associated with clean-up, regulations, and other factors.

Equation (8.2) also accounts for the increased environmental effects of near-shore spills, where a spill is more likely to threaten beaches and other ecosystems. In these locations, clean-up and regulation costs are likely to be higher than in the open sea, where wave action and weather can dissipate and volatize a large portion of a spill.

The value of the significant distance should be defined in each region separately for each geographic location

Note that, for input variables that are unknown, the variables can revert to default values.

It is worth mentioning that the environmental factor alone made the oil spill that happened in the Gulf of Mexico in 2010 a disaster for BP: the cost may reach 60 billion USD according to new analysis from Moody's, the rating agency. In addition, there is also an effect on the reputation to the company. The reputation factor should be considered another consequence of platform failure.

Business Losses

To assess the business consequences of a platform failure, two terms are considered. First, there is the replacement cost of constructing a new platform.

Second, there is the impact of deferred production, which reflects the cost of not being able to produce for the time it takes to replace the platform.

For water and gas injection platforms and jackets supporting vent lines and flares, this value should reflect the loss of production, should the structure fail.

Equation (8.1) estimates the total business loss component of the failure consequences. The two main factors that influence the business loss cost are:

- Replacement cost (e.g., the cost of constructing a new platform).
- Deferred production due to complete or partial platform structural failure (that is, the cost of stopping production and shutting off the well until replacement of the platform or major repair). In this calculation, the oil and gas reserve is modeled as an investment present value (PV). The value of the deferred production is equal to the difference between the present value and the value of the reserve, discounted for the time it takes to return it to production after a failure at a given interest rate.

The owner company should define the estimated cost of the platform (e.g., it can estimate the cost as \$5,500 per ton of platform) if there is no given replacement value for the platform.

The deferred production losses are a function of the platform's production rate. The production rate can be expressed as either the value of a month's production or the volume of oil and gas produced in a month. When the volume is given, the value of the production is estimated using default oil and gas prices, which vary with location, to reflect the actual value of the oil and gas in any given region of the world.

The value of all the oil in the reserve is simplistically modeled as the net present value of an annuity that makes monthly payments equal to the monthly production. If the reserve lifetime is unknown, it is assumed to be infinite and the value of the reserve is the "capitalized cost" of the reserve. This is equal to monthly production in dollars divided by the monthly interest rate (M/I). This is the value of the reserve if production is never interrupted. But if the platform fails and production must be deferred, the value of the reserve decreases because monthly payments on the annuity are suspended for the period it takes to replace the platform.

The deferred production loss is the difference between the value of the uninterrupted production and the present value of the interrupted production. The day production resumes, the value of the reserve is equal to the value of the reserve based on the uninterrupted assumption. This is converted to a net present value by discounting the value of the reserve for the interruption period. By default, this is calculated at an interest rate of 1% per month.

In determining the business loss for collapse of the structure, if C_{DP} is not given, then determine its default value.

Calculate the net monthy production from the following equation:

$$P_M = (D_{Poil} \times P_{oil} + D_{Pgas} \times P_{gas}) \times (30 \text{ days}) \tag{8.5}$$

where D_P = daily production in bpd or MSCFD, in USD; P_{oil} is oil price/bbl, in USD/bbl; P_{gas} is gas price/mscf, in USD/MSCF; P_m is the net monthly production.

The present value will be as shown in Equation (8.6), where r_m is the monthly return on investment, equal approximately to (r/12), which will be obtained by defining r, the annual rate of return on investment.

$$PV = P_M / r_m \tag{8.6}$$

$$C_{DP} = P_m \sum_{k=1}^{n} \left(\frac{1}{1+r} \right)^k \tag{8.7}$$

where n is calculated from N/30, where N is the total number of downtime days.

Note that the above equation assumes the volume of oil is fixed for the period of shutdown, where as in most cases there is a decline in production, depending on the reservoir characteristics.

$$C_B = C_{DP} + C_R \tag{8.8}$$

where C_R is replacement cost, in USD; C_{DP} is deferred production loss, in USD; C_B is business loss, in USD. The value of deferred production is based on the total present value of production. If this is not given, PV is estimated from the monthly production in dollars by assuming an infinite reserve life.

The values for deferred production, replacement and total business loss can be input into the system from a variety of sources. The values here are default values for these losses when no additional information is available. The defaults for net oil price, expected return on investment, etc., can be assigned from region to region to reflect the differences in each home market.

The loss from failure of the structure is the sum of the replacement cost and deferred production. If the replacement cost is unknown, it can be estimated using the approximate weight of the structure.

Safety Consequences

To assess the safety consequences of a platform failure, two terms are used. The calculation is simply equal to the product of the crew size and a marginal safety cost per crew member. Because there is an increased hazard potential when the platform handles gas, the product is increased by 20% for gas-producing and gas-compression platforms. The consequence value is recommended to increase by around 50% when the platform handles H_2S. If the platform is bridge-linked to another platform, the consequence value is halved.

The default safety loss is proportional to the average number of people on a platform. The estimated safety loss is the product of a mean per capita safety loss and the average platform crew. As in other consequences, default crew sizes can vary from location to location, so the resulting consequences reflect the relative differences in hazard exposure costs world-wide.

This calculated loss is adjusted downward when the platform is evacuated during severe storms. The expected consequence is increased if the platform handles gas production, because gas production implies that fire could result from an underwater structural failure. In the case of unmanned platforms, a minimum safety loss is assumed.

Regional factors, such as workplace safety regulations, fines, or other legal costs, are accounted for by using regionally specific default mean per-capita safety losses.

The safety-related losses, C_S, are proportional to the number of people exposed, the location and the marginal safety loss per person as a result of failure

$$C_S = C_{ex} \times N \times G \tag{8.9}$$

where N is the number of crew on a platform.

The penalty that increases expected losses by 20% for gas platforms is

$$G = 1 \quad \text{if } DP_{gas} = 0$$
$$G = 1.2 \text{ if } DP_{gas} \neq 0$$

Input data depend on location, platform service and the type of production. Default values for each of the inputs can be specified for different world-wide locations to reflect local conditions and expectations.

Consequence Categories

The estimated loss in dollars is converted to consequence categories. In order to determine the ranges for the categorization of your fleet of platforms, it will be necessary to calculate the total loss for all the platforms, and then plot the cumulative distribution of the estimated total losses on probability paper. The results are used to estimate the 5%, 50%, 75% and 95% fractiles. The four categories C1, C2, C3 and C4 are used as the bounds to define the consequence categories, as shown in Table 8.26.

The environmental, business and safety losses are summed to calculate an overall loss in terms of dollars or the appropriate currency depending on the country. This overall loss is converted to a consequence category according to the ranges presented in Table 8.26.

As with the likelihood of failure, the consequence category limits are intended to identify the best and worst 5% of the platforms. The intermediate category limits are intended to represent the 50th and 70th percentiles. This distribution of the category limits assigns two category ranges for platforms that have better-than-average likelihood scores and three ranges for platforms that have worse-than-average scores.

TABLE 8.26 Relationship between Estimated Consequence and Consequence Category

Category	Percentage of the Platform Fleet	Consequence (US $M)
A	<5%	<C1
B	5–50%	C1 to C2
C	50–75%	C2 to C3
D	75–95%	C3 to C4
E	>95%	>C4

8.3.3 Overall Risk Ranking

After each platform has been categorized in terms of likelihood (1 to 5) and consequence (A to E), the categories are entered into the risk matrix in Figure 8.3 to establish the overall relative risk ranking. The overall rankings are high (red), moderate (yellow), low (green) and insignificant (white).

The highest-risk items fall into Category E5 and the lowest-risk items fall into category A1. The different colors show that the relative risks are grouped into high, moderate, low and insignificant risk levels.

Where guidance in terms of ranking within each of the four risk categories may be useful, a preliminary assessment can be made by multiplying the total likelihood score by the total loss in abstract dollars. This approach is far from rigorous, so the resulting listing should be treated as indicative only.

Figure 8.3 shows the qualitative risk matrix proposed by the API Committee for Refinery Equipment and based on Bea et al. (1988) for the assessment of refinery and petrochemical plants. This same five-by-five matrix has been adopted for the RBUI system. While categorizing refinery equipment using this matrix differs from categorizing platform structures, the philosophy of the ranking is the same.

Puskar et al. (1994) delivered other risk matrices that have been used for offshore risk assessment. Many of these systems are based on two-by-two or three-by-three classification schemes. The five-by-five system shown in Figure 8.3 was selected because it offers enough range in likelihood and consequence that it can distinguish the differences between platforms. Larger matrices are possible; however, it becomes more difficult to categorize with confidence any given platform as the number of likelihood and consequence categories increases.

The categorization scheme summarized by the API risk matrix in Figure 8.3 appears to be biased toward high-consequence events. High-consequence and moderate-likelihood equipment (E3) is classified as a high risk, while high-likelihood, moderate-consequence equipment (C5) is classified as a moderate risk. Because the API group that developed this matrix felt that avoiding

Consequence

Likelihood		A	B	C	D	E
	1	Y	Y	Y	R	R
	2	G	G	Y	Y	R
	3	W	W	G	Y	R
	4	W	W	G	G	Y
	5	W	W	G	G	Y

FIGURE 8.3 Risk matrix proposed by API for the categorization of qualitative risk.

very-high-consequence failures was a priority, the bias was intentional. It was included as a precaution against uncertainties in categorizing the likelihood of high-consequence equipment, but since neither the likelihood nor consequence categories correspond exactly to an exact quantitative value, this bias may not be as significant as it appears in the diagram.

Note that API RP2-SIM recommends the matrix in Table 8.27 to present the risk level of the platform.

The structure should be classified into one of three categories:

- Red = high risk
- Yellow = medium risk
- Green = low risk.

An other way of managing overall risk is to rank all the platforms as shown in Table 8.27 and then to divide the platforms into three categories with equal numbers of platforms in the red, yellow and green categories. For example, if you have a 150 platforms, each category will contain 50 platforms and the priority for inspection will be based on a platform's color category. The qualitative risk assessment method is very easy to use to manage a large number of offshore structure platforms.

It is very important to highlight that, as engineers, we focus on the likelihood of failure; however, the company management usually focuses on the business, so the consequence should be in the risk equation. Therefore, if you implement the risk categories based on the consequence and the likelihood of failure, the maintenance plan will be business-oriented and the company may spend no money on a platform with a high probability of failure but very low consequence cost, even if you should predict that the platform will fail. But if the maintenance and inspection plan is intended to prevent any failures, the platforms should be ranked by probability of failure only.

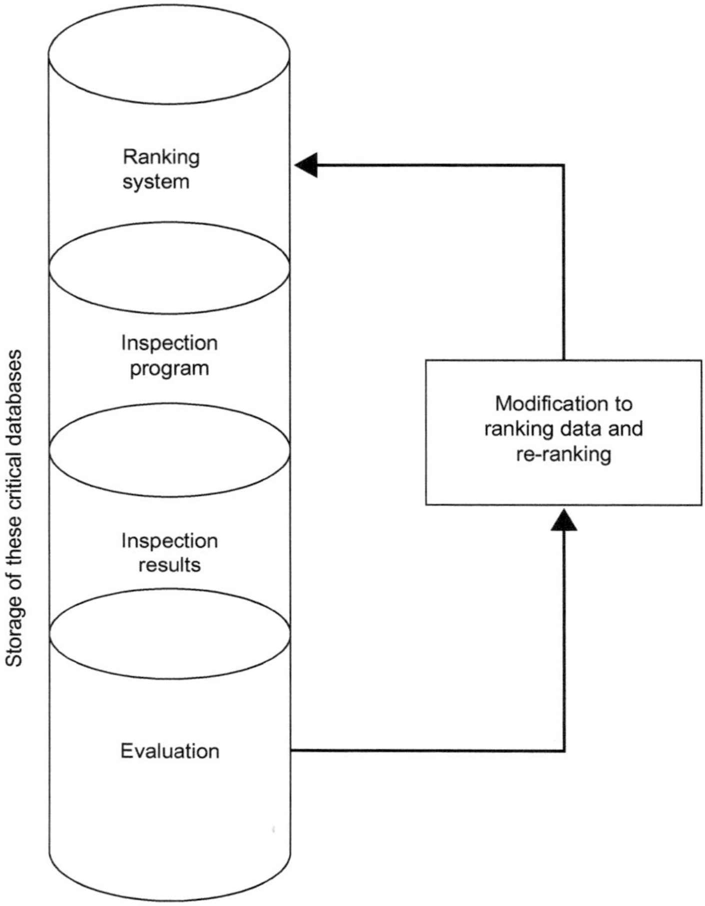

FIGURE 8.4 Risk ranking procedure.

TABLE 8.27 Risk Matrix Based on API SIM

		Likelihood of Failure		
		Low	**Medium**	**High**
Consequence of Failure	High	Risk level 2	Risk level 1	Risk level 1
	Medium	Risk level 3	Risk level 2	Risk level 1
	Low	Risk level 3	Risk level 3	Risk level 2

TABLE 8.28 Data for Platforms Constructed 1980–1990

Platform	Design Year	Total Score
P49	1985	410
P50	1985	470
P51	1984	465
P52	1986	473
P53	1985	399
P54	1985	394
P55	1985	394
P56	1985	323
P57	1985	405
P58	1985	416
P59	1985	391
P60	1984	418
P61	1987	423
P62	1981	483
P63	1984	348
P64	1986	463
P65	1983	439
P66	1984	458
P67	1984	463
P68	1987	420
P69	1985	503
P70	1985	333
P71	1982	463
P72	1983	473
P73	1989	407
P74	1989	449
P75	1989	444
P76	1984	458
P77	1989	421

(*Continued*)

TABLE 8.28 Data for Platforms Constructed 1980–1990—cont'd

Platform	Design Year	Total Score
P78	1982	503
P79	1985	463
P80	1985	645
P81	1981	458
P82	1981	662

8.4 UNDERWATER INSPECTION PLAN

After defining the risk ranking for all fixed offshore platforms in the fleet, the next step is to define the future inspection plan for the topsides and for the subsea structure. The inspection program focuses on underwater inspection because it is more costly: it requires a vessel and divers or an ROV.

8.4.1 Underwater Inspection (According to API SIM 2005)

Periodic underwater inspection should be carried out at intervals consistent with the SIM strategy adopted by the owner or the operator. Inspection intervals are provided based on platform consequence of failure; they are default intervals in lieu of intervals developed based on risk of failure. Periodic underwater inspection intervals may also be influenced by operational considerations, such as riser importance and condition, seabed movement, CP system condition, etc.

The purpose of routine underwater inspection is to provide the information necessary to evaluate the condition of the platform and appurtenances.

In the absence of a risk-based SIM strategy, default inspection intervals, shown in Table 8.29, based on the consequence of platform failure, should be used. These default intervals are based on historic industry practice that has resulted in satisfactory in-service performance of the platform, cathodic protection system and appurtenances.

Note that the timing of the first underwater periodic inspection is determined from the date the baseline inspection was completed.

If the owner or operator who is responsible for SIM has adopted a risk-based SIM strategy, the inspection intervals shown in Table 8.29, which are based on the risk of platform failure, should be used.

The risk-based inspection interval should not exceed 5 years for high-consequence platforms where the consequence category is driven by the presence of pipeline risers at the platform.

TABLE 8.29 Default Inspection Intervals

Consequence Category	Maximum Inspection Interval
High	3 years
Medium	5 years
Low	5 years

TABLE 8.30 Risk-based Inspection Interval Ranges

Risk Category	Inspection Interval Ranges
High	3–5 years
Medium	6–10 years
Low	11 years or more

API SIM defines the requirement for survey level as shown in Table 8.31. The general visual survey in level II, as mentioned in Table 8.31, is the basis for initiation of a level III survey. The scour survey in level II will be done if the seafloor is loose sand, if previous experience suggests the need for scour survey or if the seafloor is suspected to be unstable. As presented in Table 8.31 for the category survey, the visual corrosion survey in level III in the medium-consequence category is not required.

Based on the number of platforms that need inspection, we can estimate a budget cost so that management can define the required budget every year for the inspection plan.

8.4.2 Baseline Underwater Inspection

A baseline underwater inspection is required to define the as-installed condition of the platform. The minimum scope of work should consist of the following, unless data are available from the installation survey:

- A general visual survey of the platform from the mud line to the top of jacket, including all the members and joints and also including coating integrity through the splash zone.
- Below-water boat-landing and barge-bumper integrity.
- Anode count and verification of their presence and integrity, in addition to measurement of the CP reading.
- Appurtenance survey.

TABLE 8.31 API SIM Definitions of Requirements for Survey, by Level

Subsea Survey Level	Consequence Categorization		
	Low	Medium	High
Level II			
General visual survey	•	•	•
Damage survey	•	•	•
Debris survey	•	•	•
Marine growth survey	•	•	•
Scour survey	•	•	•
Anode survey	•	•	•
Cathodic potential	•	•	•
Riser/J-tubes/caissons	•	•	•
Level III			
Visual corrosion		•	•
Flooded member detection		•	•
Weld/joint close Visual		•	•
Level IV			
Weld/joint nondestructive test NDT		•	•
Wall thickness			•

- Measurement of the mean water surface elevation, as installed, with appropriate correction for tide and sea-state conditions.
- Tilt and platform orientation.
- Riser and tube soil contact.
- Scour survey (seabed profile).

8.4.3 Routine Underwater Inspection Scope of Work

Routine underwater platform inspections are required to detect and properly measure and record any platform defects, deterioration or anomalies that affect the structural integrity and performance. Platform deterioration may include excessive corrosion of welds and members, weld/joint damage (including overload and fatigue damage) and mechanical damage in the form of dents, holes, bows and gouges. Platform anomalies may include a non-operating or ineffective corrosion protection system, scour, seafloor instability, hazardous or detrimental debris and excessive marine growth.

Consistency, accuracy and completeness of inspection records are important since these data form an integral part of the structural integrity management system.

The underwater inspection program may include one or more of the following surveys:

- General visual survey
- Damage survey
- Debris survey
- Marine growth survey
- Scour survey
- Anode survey
- CP surveys
- Visual corrosion survey
- Appurtenance inspection
- Flooded member detection survey
- Weld/joint close visual inspection
- Weld/joint NDT (nondestructive testing)
- Wall thickness.

8.4.4 Inspection Plan Based on ISO 9000

The inspection plan is defined after determination of the exposure level, as shown in Table 8.32.

It is important to highlight that the timing of the first periodic level I inspection should be determined from the date the platform installation was completed. However, the timing of the first periodic level II and level III inspections should be determined from the date of the baseline inspection.

ISO identifies levels of survey as presented in Table 8.33.

A level I inspection consists of a below-water verification of the performance of the CP system (for example, a drop cell), and of an above-water visual survey. This inspection includes a general examination of all structural members in the splash zone and above water, concentrating on the condition of the more critical areas, such as topside legs, girders, trusses, etc. If above-water damage is detected,

TABLE 8.32 Determination of Exposure Level

Life-safety Category	Consequence Category		
	C1 (High Consequence)	C2 (Medium Consequence)	C3 (Low Consequence)
S1 Manned non-evacuated	L1	L1	L1
S2 Manned evacuated	L1	L2	L2
S3 Unmanned	L1	L2	L3

TABLE 8.33 Maximum Inspection Intervals for Default Periodic Inspection Program

Exposure Level	Level I Inspection	Level II Inspection	Level III Inspection	Level IV Inspection
L1	Annual	3 years	5 years	Determined from level III inspection results
L2	Annual	5 years	10 years	Determined from level III inspection results
L3	Annual	5 years	Not required	Not required

NDT is used when visual inspection cannot fully determine the extent of damage. If the level I inspection indicates that underwater damage is possible, a level II inspection should be conducted as soon as conditions permit.

A level II periodic inspection consists of a general underwater visual inspection. The inspection includes the measurement of cathodic potentials of pre-selected critical areas. Detection of significant structural damage during a level II inspection initiates a level III inspection. The level III inspection, if required, should be conducted as soon as conditions permit. The level II should also detect the following:

- accidental or environmental overloading
- scour, sea floor instability
- damage detectable in a visual swim-round survey
- design or construction deficiencies
- presence of debris
- excessive marine growth
- measurement of cathodic potentials of pre-selected critical areas.

A level III periodic inspection consists of an underwater close visual inspection (CVI) of pre-selected areas and/or areas of known or suspected damage. Such areas should be sufficiently cleaned of marine growth to permit thorough inspection. Pre-selection of areas should be based on results of an engineering evaluation of areas particularly susceptible to structural damage and to areas where repeated inspections are desirable in order to monitor integrity over time. Flooded member detection (FMD) can provide an acceptable alternative to CVI of pre-selected areas. Engineering judgment should be used to determine optimal use of FMD and/or CVI of joints. CVI of pre-selected areas for corrosion monitoring is included as part of the level III inspection.

Detection of significant structural damage during a level III inspection initiates a level IV inspection where visual inspection alone cannot determine the extent of damage. The level IV inspection, if required, should be conducted as soon as conditions permit.

FMD is used to determine the integrity of the tubular bracing members of steel jacket structures. The method relies on the detection of through-wall cracks causing the leakage of water into the internal, air-filled space and the successful detection of that water. The detection accomplished by the use of gamma radiation or ultrasonic techniques, both of which are applied externally to a tubular member, usually by means of a remotely operated vehicle (ROV) in the case of gamma FMD, or by diver, or potentially an ROV, in the case of ultrasonic FMD.

A level IV inspection also includes underwater NDT of areas pre-selected from results of a level III inspection or from known or suspected damage. A level IV inspection includes detailed inspection and measurement of damaged areas.

A level III and/or level IV inspection of fatigue-sensitive joints and/or areas susceptible to cracking can be necessary to determine if damage has occurred. Monitoring fatigue-sensitive joints, and/or reported crack-like indications, may be an acceptable alternative to analytical verification.

An example of a survey scope of work is presented in Table 8.34, and it can be considered a guideline in preparing the inspection scope of work.

TABLE 8.34 Inspection Survey Scope of Work Overview

Survey	Generic Scope of Work	Recommendation
General Visual	Underwater general visual inspection of full structure with particular reference to members, joints and appurtenance connections, to detect the presence of excessive corrosion, accidental or environmental overloading, fatigue damage, design or construction deficiency.	All damage and all other anomalies found to be reported in accordance with owner or operating company specification. On plan (1st level above waterline), confirm or add location of jacket legs, conductor guides, conductors, J-tubes, casings, risers and boat landings.
Marine Growth	Measurements of compressed marine growth at different locations that should be defined based on the water depth and the location of the platform.	Present schematic for marine growth measurement locations. Visually check for areas of existing or missing marine growth. Document and report, immediately, any locations of marine growth to the company representative.
Scour	Visually inspect the sea floor up to 7.5 m (25 ft) out for possible erosion and scouring or instability. Check if the piles are exposed and whether the mud mat and bottom braces are on the bottom. Take readings of the bottom depths.	If the pile is exposed, note if there is any corrosion or damage and check for mud mats. Note any bottom irregularities and bottom depth.

(*Continued*)

TABLE 8.34 Inspection Survey Scope of Work Overview—cont'd

Survey	Generic Scope of Work	Recommendation
Cathodic Protection	Readings to be taken on legs on outboard truss chord, legs at the mean water level and halfway between each horizontal framing plan readings.	See schematic for CP reading locations. Measure all potentials in mV by Ag/AgCl half-cell.
Anodes	Visual examination to locate, count, grade and confirm the physical condition as loose, missing or damaged of all original or retrofit anodes. Potential readings and detailed dimensional survey to be taken on two anodes.	Grade 4 anodes should be reported as anomalous in accordance with the owner or operator specification. Anodes identified as loose, damaged or missing should be recorded as anomalous.
Debris	Visual inspection to record the quantity, type, size and location of debris hanging on or in the jacket, in contact with the bottom elevation, on the sea floor up to 10.0 m (30 ft) out or in contact with any risers, pipelines or conductors.	Report the amount of debris inside, on, and around the platform. Note whether the debris is metallic, hazardous, obstructing or items that could have caused damage. Include a map of debris on the seabed. Remove, subject to confirmation from the company.
Risers	Visually inspect the risers. Take CP readings between riser clamps. Record riser coating type and depth of termination. Inspect the pipelines to the point of entry into the sea floor or 10 m (30 ft) from platform, whichever is closer.	Document and report immediately to the company representative, any soil disturbance or pipeline distortion within 10 m (30 ft) of the platform or any hydrocarbon leaks from any source. If pipeline is suspended, note distance to point of contact and distance away from the shore.
Caissons	Visual inspection of casings, casing clamps and intakes. Take CP readings on clamps and casing mid-sections. General visual survey of the conductors, with particular attention given to examination for conductor movement.	Note total number of conductor slots, number of conductors driven, and if any are undrilled. Intakes should be cleaned of marine growth if necessary.
Flooded Members	Flooded member detection (FMD) for all the members or specific members in the platform.	Provide a schematic for FMD locations. If flooding is detected, the company representative may request an additional survey to establish the cause.

8.4.5 Inspection and Repair Strategy

The inspection and repair strategy should be defined based on the probability of failure and the economics of inspection and minor and major repair. El-Reedy (2003) presents the optimal inspection and repair strategy.

Inspection alone does not improve reliability unless it is accompanied by corrective action in the event of a discovered defect. Some policies and strategies used in a wide range of offshore structures include:

- Monitoring (e.g., until the crack depth reaches a certain proportion of the material thickness), then repair
- Immediate repair upon detection of indications
- Repair at a fixed time (e.g., 1 year) after detection of indications
- Repair as new (i.e., repair welding).

Generally, it is assumed that inspections are performed at constant time intervals, as shown in Figure 8.5, since the inspection authorities often prefer constant inspection intervals to facilitate planning. Inspection intervals are chosen so that the expected cost of inspection, repair and failure is minimized.

Most offshore structure inspection techniques are visual or use NDT. The ability to detect damage is dependent on the quality of the inspection technique being used. A higher-quality inspection method will provide a more dependable assessment of damage. No repair will be made unless the damage is detected.

Inspection does not affect the probability of failure of the structure. After an inspection, a decision must be made regarding repair if damage is found. The repair decision will depend on the inspection quality. With advanced inspection methods, repair work can be effective, since even a small defect can be detected and repaired.

Higher-quality inspection may lead to higher-quality repair, which could bring the reliability of the structure closer to its original condition (although the reliability of the structure decreases with age). After inspection and repair, the structure's capacity should be the same as the design condition, as shown in Figure 8.6.

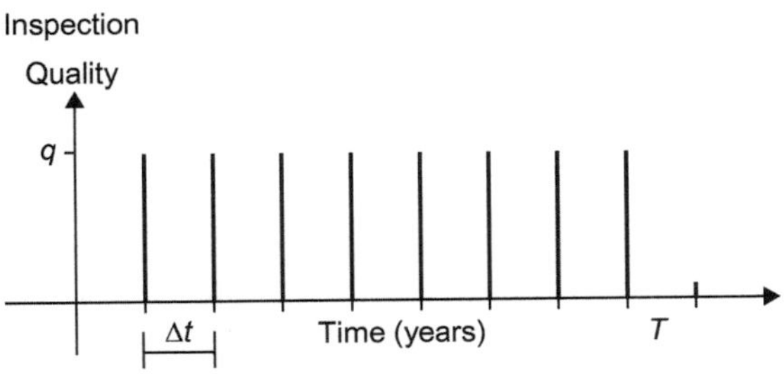

FIGURE 8.5 Inspection strategy.

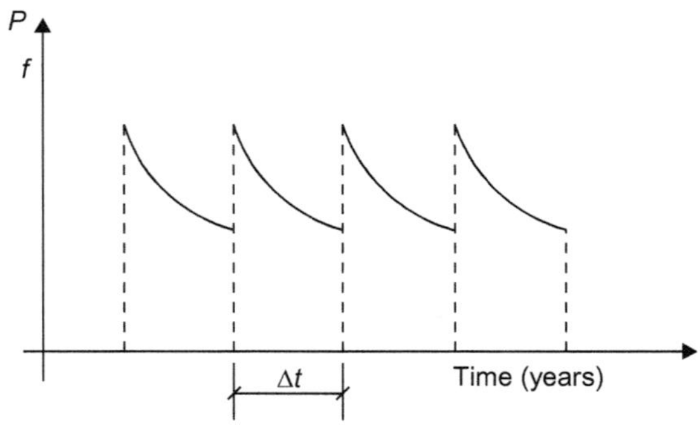

FIGURE 8.6 Offshore structure performance.

Expected Total Cost

The first step is to determine the service life of the structure. Assume service life is 75 years and routine maintenance is scheduled once every 2 years. It starts at $t = 2$ years and continues until $t = 74$ years. Consequently, preventive maintenance work will be performed 37 times during the life of the structure. Therefore, the lifetime routine maintenance cost becomes

$$C_{FM} = \mathrm{C}_{\mathrm{m}2} + \mathrm{C}_{\mathrm{m}4} + \mathrm{C}_{\mathrm{m}6} + \ldots + \mathrm{C}_{\mathrm{m}74} \tag{8.10}$$

If the total expected cost in its lifespan (T) is based on the present value, according to Turan (1990), the expected lifetime preventive maintenance cost becomes

$$C_{IR} = C_{IR2}\frac{1}{(1+r)^2} + C_{IR4}\frac{1}{(1+r)^4} + \ldots\ldots.. + C_{IR74}\frac{1}{(1+r)^{74}} \tag{8.11}$$

where $r =$ net discount rate of money.

In general, for a strategy involvng m lifetime inspections, the total expected inspection cost is

$$C_{ins} = \sum_{i=1}^{m} C_{ins} + C_R\frac{1}{(1+r)^{T1}} \tag{8.12}$$

where $C_{ins} =$ inspection cost based on inspection method, $C_R =$ repair cost and $r =$ net discount rate.

Finally, the expected total cost, C_{ET}, is the sum of its components, including the initial cost of the structure, the expected cost of routine maintenance, the expected cost of preventive maintenance, which includes the cost of the inspection and the repair for and maintenance, and the expected cost of failure.

Accordingly, C_{ET} can be expressed as

$$C_{ET} = C_T + (C_{ins} + C_{rep})(1 - P_f) + C_f \bullet P_f \tag{8.13}$$

The objective remains to develop a strategy that minimizes C_{ET} while keeping the lifetime reliability of the structure above a minimum allowable value.

Optimization Strategy

To implement an optimum lifetime strategy, the following problem must be solved:

$$\text{Minimize } C_{ET} \text{ subject to } P_{f\,life} \leq P_{\max}$$

where P_{max} = maximum acceptable lifetime failure probability. Alternatively, considering the reliability index,

$$\beta = \phi^{-1}(1 - P)$$

where ϕ is the standard normal distribution function, and the optimum lifetime strategy is defined as the solution of the following mathematical problem.

$$\text{Minimize } C_{ET} \text{ subject to } \beta_{life} \geq \beta_{\min}$$

The optimal inspection strategy with regard to costs is determined by formulating an optimization problem.

The objective function (C_{ET}) in this formulation is defined as including the periodic inspection and minor joint repair costs, as well as the failure cost, which includes the cost of major joint repair. The inspection periodic time (Δt) is the optimization variable, which is constrained by the minimum index β specified by the code and the maximum periodic time.

The optimization problem may be mathematically written as: find the Δt that minimizes the objective function

$$C_{ET}(\Delta t) = (C_{IR})(1 - P_f(\Delta t))\left(\frac{(1+r)^T - 1}{\left((1+r)^{\Delta T} - 1\right)(1+r)^T}\right) + C_f P_f(\Delta t)\left(\frac{(1+r)^T - 1}{\left((1+r)^{\Delta T} - 1\right)^T (1+r)^T}\right) \tag{8.14}$$

subject to $\beta(\mathrm{t}) \geq \beta\mathrm{min}$, $\Delta t \leq \mathrm{T}$, where C_{IR} is the periodic inspection and minor repair cost per inspection, C_f is the major repair cost, r is the real interest rate and β min is the minimum acceptable reliability index. C_{IR} and C_f are assumed to be constant with time. Although the failure cost is minimized (as a part of the total cost), it is often necessary to put a constraint on the reliability index to fulfill code requirements.

8.4.6 Flooded Member Inspection

The two principal means of flooded member detection (FMD) are ROV-deployed gamma FMD and diver-deployed ultrasonic FMD.

Gamma FMD requires a gamma source and a detector to be fixed to either end of a yoke that is attached to an ROV. The ROV pilot then positions the yoke over the member and notes the reading count from the detector. This is compared with an expected count that is calculated taking account of the absorption characteristics of water between the yoke and the member, the member steel thickness and the air within the member. If the member is flooded, the count is lower than expected due to the absorption of the gamma radiation by the water and the member is flagged as flooded.

Ultrasonic FMD employs an ultrasound probe that is fixed to the outside of the member after suitable cleaning to provide good acoustic contact. (Figure 8.7 provides a schematic.)

Water is a better transmitter of ultrasound than air is, and a flooded member is detected by observing echoes that are received back at the probe and viewed on a screen. For an air-filled member, the echo will be that of the back wall. For a water-filled member, the ultrasound is transmitted through the water and a second echo will be generated at the interface between the water and the back face of the tube. This back-face echo is an indication that the member is flooded. Figure 8.8 shows a typical UT trace for a flooded member.

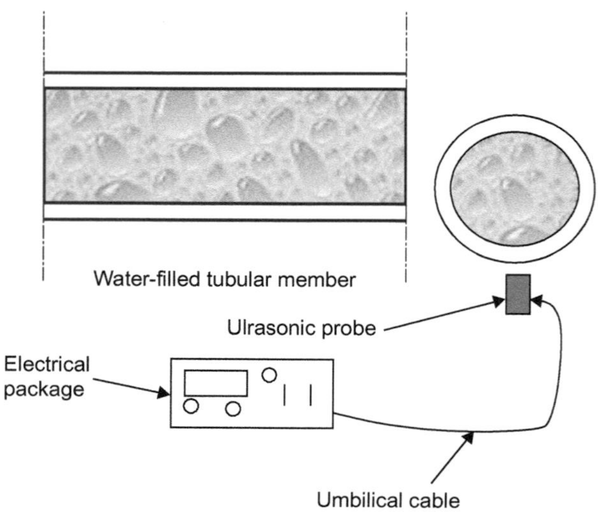

FIGURE 8.7 Typical arrangement of UT FMD equipment.

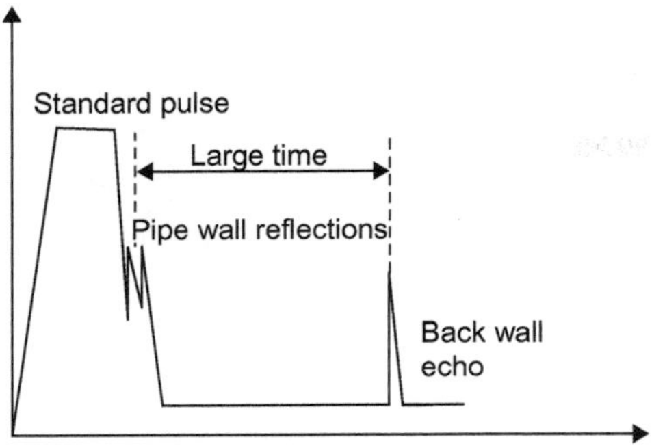

FIGURE 8.8 Typical UT signal for a flooded member.

For gamma FMD, the maximum distance between source and detector determines the maximum diameter of member that can be inspected. The distance between the two is limited not only by the size of the yoke, which is dependent on the payload capability of the host vehicle, but also by the strength of the radioactive element relative to the thickness of the steel walls of the members requiring inspection. The size of a gamma source is normally varied according to member diameter and wall thickness. In general, as large a source as possible is used to give meaningful results but without saturating the detector so that a drop cannot be observed.

The host vehicle should be of sufficient size to allow access and to carry the FMD equipment, and it should be able to operate in the environmental conditions present.

The range of member diameters typically surveyed by gamma FMD methods is from 12 inches (0.3 m) to 2.6 m. FMD is currently performed on members with a wall thickness in the range of 8 mm to 63.5 mm.

Orientation of the yokes can be adjusted to any position for taking readings on inclined members.

Figure 8.9 shows that the positions in which readings may be taken on the member can also be varied. Typically, readings are taken at the "6 o'clock–12 o'clock" or the "3 o'clock–9 o'clock" positions. The yoke can be adjusted to make the necessary readings possible.

The majority of flooded or cracked members occur as a result of unforeseen circumstances, such as the presence of a gross defect caused by poor fabrication, poor design practice, or accidental damage. Poor fabrication or design errors manifest themselves in the form of unexpected fatigue cracking, where joints crack at a time well before their design lives. Because the occurrence of this damage is not expected, the inspection strategy should enable detection of unforeseen damage.

Reference to design fatigue lives is not a suitable means of justifying the use of FMD. The traditional S-N fatigue design process does not legislate for large defects or poor design, assuming welded joints to be "nominally perfect."

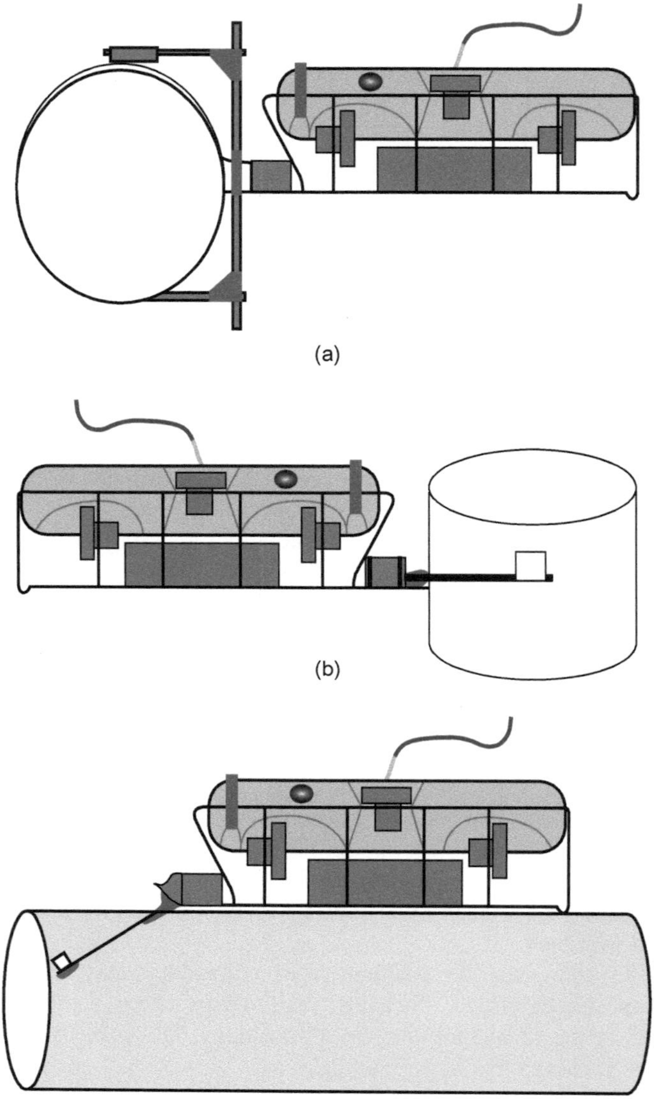

(a)

(b)

(c)

FIGURE 8.9 (a) Horizontal member standard survey mode for FMD. (b) Vertical member survey mode for FMD. (c) Horizontal member survey mode for FMD.

Operationally, it should be assumed that a large defect or design error could potentially exist in every member and could lead to through-thickness failure of the member and subsequent flooding, and could potentially develop to a size that could lead to final severance.

The most appropriate means of assessing an individual member for the effect of a crack of a size allowing flooding is through a fracture mechanics assessment. A fracture mechanics assessment will determine, for the size of defect, whether failure by fracture or collapse will occur under the prescribed conditions.

It should first be determined that the member can sustain through-thickness damage of a size sufficient to cause leakage without failing completely under the anticipated loading conditions. If this cannot be demonstrated, then FMD is not a suitable method for ensuring that the member's integrity is maintained. If it is determined that through-wall damage is possible but that complete severance is not, then it is necessary to demonstrate that FMD can detect the flooded condition before final severance or some other degree of damage is detected over a particular inspection interval. Given the extreme variability associated with all factors influencing fracture and collapse performance, a probabilistic fracture mechanics assessment should be undertaken to establish the probability of failure of an individual weld and, ultimately, the probability of failure of a complete member.

An alternative method for predicting failure frequencies is the use of historical data. There are many studies of a fracture mechanics approach to calculation of the probability of a member's leaking, and the probability of complete severance. The studies consider member failure rates obtained from a historical review. It should also be noted that other analytical approaches, many of them based on fracture mechanics, may be employed to determine the probability of failure of a member.

Final Inspection Reporting

The inspection contractor's final report is submitted upon completion of all inspection activities.

The format for the final report is:

- Title page
- Report Title/Author/Company/Date, etc.
- Contents
- Summary
- Contractor's name
- Dates of survey
- Inspections completed and inspections omitted and reasons for omission
- Anomalies found
- Conclusions
- Outstanding actions
- Inspection results by area
- Written summary
- Inspection datasheets
- Drawings
- Anomaly sheets (including NDT reports where available)
- Photographs/Video stills
- Operations analysis

- Summary of time breakdown (chart)
- Summary of work breakdown (chart)
- Daily reports
- Inspection team details
- Copies of up-to-date CVs
- Film/video logs
- Summary of all film/video logs.

Table 8.35 shows an example annual summary sheet for each fixed platform; the summary sheet should be updated to reflect the existing conditions of the offshore fleet.

8.5 ANODE RETROFIT MAINTENANCE PROGRAM

As discussed above, the inspection plan includes survey of the condition of the anodes, which provide the cathodic protection to the jacket to prevent corrosion.

An anode survey during underwater inspection will focus on the condition of the anodes (i.e., whether they are depleted or not; Table 8.36) as well as on a reading of the potential of the CP system (Table 8.37), in order to determine if the jacket is protected or not.

The anode condition and the CP reading will have different criticality rankings. For example, if the anode is in a bad condition and the reading is very low, the jacket is not protected against corrosion. On the other hand, if the CP reading is good but the anode is in bad condition, after a short time the CP reading will be low. The findings about these two factors can define a maintenance plan for anode retrofit of the fleet of platforms. Of course, anode retrofit is expensive because it requires divers and a vessel, which cost a lot more than the anode material itself. So the maintenance plan should be clear, based on the priority of anode condition and budget.

Ultimately, the integrity engineer should have a table like Table 8.38, defining the estimated cost and the status of the anodes for the platforms.

8.6 ASSESSMENT PROCESS

8.6.1 Collecting Data

Collecting data is the first step in the assessment process. The required information includes the above-water structure's deck loads, which are obtained from review of the weight report or comparison of the original facilities' locations and weights with the current condition. Facilities on the deck may have been changed due to operational requirements. The most common change on the topside is adding a deck extension or new separators or new compressors, but sometimes the original facilities have been removed.

In addition, jacket structure information will be available from the underwater inspection.

TABLE 8.35 Annual Summary Sheet for Each Fixed Platform

General Information	Construction Year	Water Depth	Type of Platform	Conductors	Risers	Adding Riser or Conductor	Last ROV Inspection	Other
Summary								
Topside condition survey								
Subsea inspection								
Structure analysis								
Weight control								
Document control								
Prepared by: Checked by: Approved by:				Date				

TABLE 8.36 Anode Condition

Grade of Anode	Anode Depletion (%)
Grade 4	100 (red cell)
Grade 3	>50 (yellow cell)
Grade 2	<50 (yellow cell)
Grade 1	<10 (green cell)

TABLE 8.37 Cathodic Protection (CP) Readings

CP Readings (mV)	Protection Condition
<850	Unprotected (red cell)
850–900	Marginal protection (yellow cell)
900–950	Satisfactory (yellow cell)
950–1000	Good (yellow cell)
>1000	Very good (green cell)

Soil data and metocean data should be available as well. So the required information includes:

- Deck extensions
- New conductors or risers
- Additional facilities with their weights and locations
- Soil data
- Metocean data.

8.6.2 Structure Assessment

Structure assessment is discussed in detail in Chapter 7, but, in general, structure assessment uses different methods according to the structure's condition and the available data.

Simple Methods

Simple methods for assessment are used instead of the more complex and time-consuming platform-specific analyses. Simple methods are typically used for a platform in a certain class where prior studies are available or when previous analyses are available. Guidance for simple methods is provided here. If

TABLE 8.38 Maintenance Plan for Anode Retrofit

Structure Name	WD (ft)	RA Rank	Last Inspection	Inspection Reading (mV)	Reading Condition	Anode Condition	Anode Installation Date	Budget Cost	Color Code
PL1	123	2	2000	691–708	Unprotected	Grade 4			Red
PL2	123	5	2000	684–690	Unprotected	Grade 4			Red
PL3	123	6	2001	686–699	Unprotected	Grade 3–Grade 4			Red
PL4	113	11	2000	811–907	Unprotected/marginal	Grade 3–Grade 4			Red
PL5	280	16	2002	871–906	Marginal	Grade 4			Red
PL6	127	20	2001	721–739	Unprotected	Grade 4			Red
PL7	155	23	2003	1008–1014	Very good	Grade 3			Green
PL8	130	26	2003	995–1012	Very good	Grade 2–Grade 4			Yellow
PL9	134	27	2001	1004–1020	Very good	Grade 2–Grade 3			Green
PL10	122	32	2001	850–938	Marginal–satisfactory	Grade 4			Yellow

(Continued)

TABLE 8.38 Maintenance Plan for Anode Retrofit—cont'd

Structure Name	WD (ft)	RA Rank	Last Inspection	Inspection Reading (mV)	Reading Condition	Anode Condition	Anode Installation Date	Budget Cost	Color Code
PL11	130	39	2003	954–980	Good	Grade 4			Yellow
PL12	250	40	2003	941–962	Good–Very good	Grade 3–Grade 4			Green
PL13	123	44	2001	716–719	Unprotected	Grade 4			Red
PL14	123	45	2001	714–725	Unprotected	Grade 4			Red
PL15	123	46	2001	707–719	Unprotected	Grade 3–Grade 4			Red
PL16	132	47	2003	919–924	Satisfactory	Grade 2–Grade 4			Yellow
PL17	150	104	2002	664–675	Unprotected	Grade 4			Red

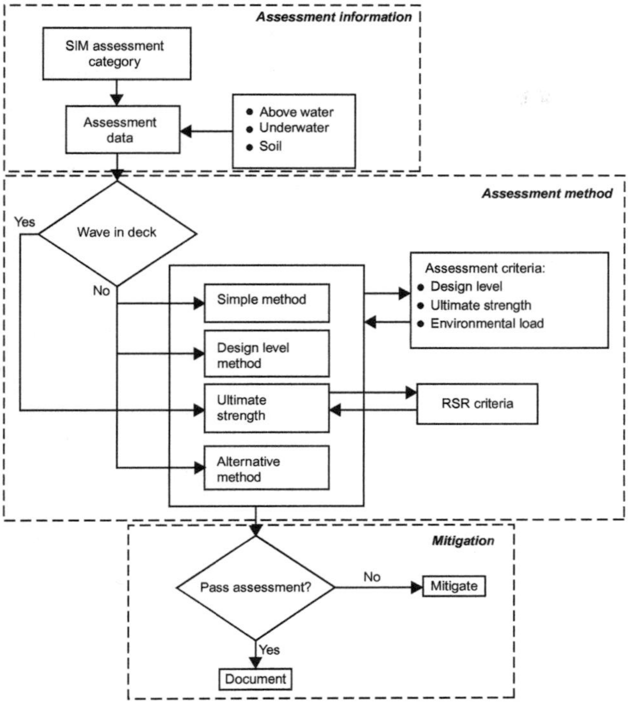

FIGURE 8.10 Assessment process.

there is any concern that a simple method does not meet the requirements discussed here, then a more detailed assessment approach should be used.

There are three basic types of simple methods:

The **design-level method** and **ultimate strength method** use simple procedures provided that such procedures have been validated. Several investigators have developed simplified procedures for the evaluation of the adequacy of existing platforms. Successful use of these procedures requires intimate knowledge of the assumptions upon which they were based.

The **results from previous analyses** for the platform are used provided that the analysis is representative of the platform's condition at the time of the assessment trigger.

Comparison with a similar platform uses assessment results available from a similar platform.

Design-Level Method (DLM)

DLM involves linear analysis using the standard API RP2A approach for new platform design by checking the platform on a component basis. The same factors of safety used for new design are used for DLM. The key difference for assessment is the use of special assessment criteria. The platform must be shown to perform linearly at loads equal to or greater than the assessment criteria.

DLM is typically the first level of direct analysis of the platform. DLM is a simpler and more conservative check than the ultimate strength method, which is more complex and less conservative. It is generally more efficient to begin with DLM because it is usually simpler to implement.

Ultimate Strength Method (USM)

USM typically involves the use of nonlinear analysis to determine the maximum environmental loading that the platform can sustain without collapse. USM is performed using special assessment criteria. A DLM with all safety factors and sources of conservatism removed is also permitted, as this provides a conservative estimate of ultimate strength. In this case, ultimate strength acceptance criteria must be used. There may also be an existing computer model of the platform that was used for design of upgrades or other modifications that can be readily updated for platform assessment.

USM is typically used when a platform does not pass DLM, since it reduces conservatism. A platform that does not pass DLM may pass USM.

USM is always required if the initiating trigger is wave-in-deck loading, because this type of stepwise increase in environmental loading is difficult to capture properly with DLM.

Nonlinear analysis is intended to demonstrate that a platform has adequate strength and stability to withstand the ultimate strength criteria, with local overstress and damage allowed, but without collapse. At this level of analysis, stresses have exceeded linear levels and modeling of overstressed members, joints, and foundations must recognize ultimate capacity as well as post-buckling behavior, rather than the linear load limit.

The ultimate strength of a platform is often determined using a nonlinear pushover analysis, which applies an increasing lateral load to the platform until the platform collapses. The lateral pushover load should be representative of the metocean loads acting on the platform at the instance of collapse. The pushover load profile should consider all of the metocean issues for the assessment, including wind, wave, current, marine growth, etc.

There are several structural analysis programs that contain semi-automated approaches to performing a pushover analysis. These programs should always be used and the results should be interpreted by competent personnel.

The following loading guidelines should be considered for USM.

Gravity loading includes the actual loads on the platform as well as future planned or temporary loads (for example, the drilling rig).

Environmental loading should consider the actual configuration of the structure at the time of the assessment, such as the actual number of conductors or risers, the drill rig and metocean data. Future planned or temporary loads should also be considered.

If there is a wave-in-deck loading trigger, then the procedures provided should be followed. Alternative wave-in-deck loading methods can be used as long as they are justifiable. Even though there may not be a wave-in-deck loading trigger, wave loads may act on other deck areas, such as sump and spider decks, and the wave loading should be determined in the appropriate manner.

In most cases, dynamic effects should be considered for platforms in water depths greater than 400 ft. Dynamic effects should also be considered for damaged platforms that may sustain higher motions in the damaged condition than in the intact condition. This can occur for any water depth.

The global structural model should be three-dimensional. Special attention should be given to defensible representation of the actual stiffness of damaged or corroded members and joints. The following guidelines should be considered for the USM.

Damage Modeling

The ultimate strength of undamaged members, joints, and piles can be established using the formulas from API RP2A with all safety factors taken as equal to 1.0. Nonlinear interactions may also be utilized where justified. The ultimate strength of joints may also be determined using a mean "formula or equation" versus the lower-bound formulas for joints design. Alternatively, the ultimate strength of damaged or repaired elements of the structure may be evaluated using a rational, defensible engineering approach, including special procedures developed for that purpose.

Actual Yield Stress

Actual yield stress or expected mean yield stresses may be used instead of nominal yield stresses. However, mean yield strength should not be used when it is greater than actual laboratory test or mill certificate strength. Increased strength due to strain hardening may also be acknowledged if the section is sufficiently compact, but not rate effects beyond the normal (fast) mill tension tests.

Effective Length Factors

Studies and tests have indicated that effective length (K) factors are substantially lower for elements of a frame subjected to overload than those specified in API RP2A 3.3.1d. Lower values may be used if it can be demonstrated that they are both applicable and substantiated.

Soil Strength

For USM, it is usually appropriate to use best-estimate soil properties as opposed to conservative interpretations. This is particularly true for dynamic analyses, where it is not always clear what constitutes a conservative interpretation.

Pile foundations should be modeled in sufficient detail to adequately simulate their response. It could be possible to simplify the foundation model to assess the structural response of the platform. However, such a model should realistically reflect the shear and moment coupling at the pile head. Furthermore, it should allow for the nonlinear behavior of both the soil and pile. Lastly, a simplified model should accommodate the development of collapse within the foundation for cases where this is the weak link of the platform system.

Alternative Assessment Methods

Alternative assessment methods involve the use of techniques other than a direct structural evaluation to assess an existing platform. There are two basic types of alternative methods.

Historical Performance

The platform must have survived, with no or little damage, environmental loading that is as severe as, or more severe than, that required for USM.

Explicit Probabilities of Survival

This is a platform assessment using explicit probabilities of survival of the platform for the appropriate assessment criteria.

Acceptance Criteria

Two types of acceptance criteria are provided. The first is specific environmental loading criteria, such as wave height, current, etc., that the platform should be shown to withstand without collapse. The second is based upon the reserve strength ratio (RSR), which is a measure of the platform loading relative to loads caused by 100-year environmental conditions used for new platform design.

Environmental lateral loading is computed using API RP2A criteria for new design.

The required minimum RSR is based upon the platform's exposure category and the version of API RP2A used for design. Platforms designed prior to API RP2A should be considered pre-nineteenth edition. The required minimum RSR for the Gulf of Mexico and other U.S. locations is shown in Table 8.39.

8.7 MITIGATION AND RISK REDUCTION

Structures that do not meet the assessment requirements using any of the methods discussed in this section will need mitigation actions. Mitigation should be considered at all stages of assessment and may be used in lieu of more complex assessment.

Mitigation is defined as modifications or operational procedures that reduce loads, increase capacities or reduce exposure.

TABLE 8.39 Acceptable Minimum Reserve Strength Ratios

		Reserve Strength Ratio (RSR)	
Offshore Location	Assessment Category	API RP2A 19th Edition and Earlier	API RP2A 20th Edition and Later
Gulf of Mexico	A-1	1.2*	1.6 (draft)
	A-2	0.8	1.2 (draft)
	A-3	0.6**	1.0 (draft)
Other U.S. offshore areas	A-1	1.6	2.0

**This RSR is applicable only for continued use of the platform for its present purpose. Not applicable for change-of-use conditions.*
***Not to be used for water depth greater than 400 ft.*

Mitigation includes such measures as demanding, either during a forecast event or completely, a hydrocarbon inventory reduction that reduces the severity and consequence of platform failure. Mitigations like repairs should be designed to meet the requirements of this section, so that they do not reduce the overall strength of the platform.

Owners or operators who wish to continue to operate structures that have been assessed in accordance with API RP2A Section 6 and do not meet the fitness-for-purpose acceptance criteria appropriate for their consequence of failure category are required to reduce the risk of operating the platform. Either or both of the following can reduce the operating risk:

- Mitigate the consequences of structural failure.
- Reduce the probability of structural failure.

Competent assessment engineering should determine the need for, and appropriate selection of, risk-reduction options.

8.7.1 Consequence Mitigation

If mitigation is an action taken to reduce the consequences of failure of the platform, and if the consequence of failure is related to life safety, then mitigation measures have to reduce the risk to life safety sufficiently to lower the consequence category. For example, implementing operational procedures to evacuate a manned platform, such as in the Gulf of Mexico during an extreme event, reduces the consequence of failure during that event from high to medium. If the consequence of failure is related to environmental impact or some other consequence consideration, then other mitigation measures will be required and may include one or more of the following.

- Shut down the platform either completely or during extreme events if it is possible from an operational point of view
- Operate subsurface safety valves that are manufactured and tested in accordance with applicable API specifications
- Remove or reduce hydrocarbon storage or inventory volume
- Remove or reroute major oil pipelines
- Remove or reroute large-volume gas flow lines
- Plug and abandon nonproducing wells
- Operate pipeline shutting systems to reduce the potential for hydrocarbon release.

8.7.2 Reduction of the Probability of Platform Failure

Reduction is defined as an action taken to reduce the likelihood of failure of the platform. In general, two methods exist to reduce the likelihood of failure of the platform during an extreme event:

- Load reduction reduces the load to be resisted by the structure during an extreme event.
- Strengthening increases the global, or system, strength of the structure.

Competent assessment engineering should determine the need for, and appropriate selection of, either load reduction or strengthening options or both.

Load Reduction

Gravity and Hydrodynamic Loading

During operation of the platform, the actual topside loading may be significantly lower than the loads assumed for the design of the platform. For example, operational procedures can be implemented to reduce and control topside loads by removal of unnecessary equipment or structures, by applying effective weight-control management procedures with defined weight limits, by use of lightweight drilling rigs or operations without a rig or by using cantilever jack-up drilling operations.

The major effect of load reductions is to reduce leg and pile stresses and pile reactions. Reduced mass generally has a beneficial effect on platform dynamics, although not necessarily for earthquake response, but in most instances this effect will generally be small. On platforms with pile tips founded in sand layers, tensile pile capacities may need to be checked. One potential beneficial interaction associated with weight reduction is a possible associated reduction of wind area.

Nonessential components, depending on their function and mode of operation (for example, a main crude oil pump not currently in use) can be removed to lighten the weight. Since conductors may contribute to the capacity of the platform foundation, it is often most efficient to remove only the upper section in the wave zone—this should be confirmed during the assessment process.

Removal, or relocation, of equipment on lower deck elevations will significantly reduce loads on the platform in the event of wave inundation of this deck.

In some instances, structural members in the jacket may be removed where it can be shown that the removal results in an increase in the overall system reliability. Examples of structural members that can be removed include launch truss members and redundant structural members. In an older platform fleet, change of the mode of operation with time usually has resulted in redundant risers, and their removal can reduce load significantly. However, removal of these risers requires performing a comprehensive study, because removing risers is not easy and will be expensive, although it can be the best solution to maintain the structural integrity.

Another solution is to decrease the hydrodynamic load by using marine growth removal, as described in Chapter 7.

In some cases, painting the surface of legs and bracing with a coating prevents marine growth from accumulating.

Other actions include marine growth removal on a regular basis by means of diver-held water-jetting equipment. A new trend is to put a piece of plastic around the steel member so that the plastic rotates with wave movement and provides a clean surface all the time. Inspections of actual growth levels in comparison to design values may sometimes show design values to be incorrect, sometimes positively and sometimes negatively. Marine growth control also has the added benefit of virtual mass reduction on platforms subjected to dynamic excitation by waves or earthquakes.

Raising the Deck

If, after comparing the platform's elevation with the recent metocean data it is found that the wave crest is expected to hit the deck, raising the deck out of the wave crest will significantly reduce the structure's global loading. In raising the deck, the effects of increased unbraced deck leg lengths must be evaluated. Due to the high cost and operational impact of raising the deck, cost-benefit analysis should be done on a case-by-case basis.

An alternative to raising the deck is to remove or to relocate equipment and nonessential structures from the lower deck elevations; this results in lower hydrodynamic forces and will reduce equipment damage from direct wave loads.

The other suitable solution is to install deck grating, instead of plating; this can be beneficial in reducing vertical loads on the underside of the deck by allowing encroaching water and trapped air to dissipate more easily.

In some locations, field subsidence has caused a general settling of the sea floor. Mitigation alternatives for this case often rely on reservoir pressure techniques, such as water or gas injection. This approach does not recover lost height, but it can be used to slow future subsidence.

Strengthening

There are many strengthening and repair techniques available, as is illustrated in Chapter 7. Platform assessment will determine whether platform strengthening or repair is required to meet the assessment acceptance criteria. If strengthening and repair are to be considered, the assessment model should be used to develop strengthening options. Strengthening and repair of existing platforms require specialist competence to provide reliable and economical solutions that can be efficiently and safely installed.

The need for accurate inspection data is emphasized; usually, a special inspection with a detailed dimensional survey is required to ensure that no problems are caused by a lack of data during installation.

Some strengthening and repair techniques are discussed in Chapter 7, such as use of external bracing, which is the traditional method of strengthening a fixed offshore structure. It is important to recognize that, in all cases, accurate fabrication and clear and efficient installation procedures are as important as the design of the strengthening scheme. Specific design guidance for the techniques discussed is not provided herein but is available in specialist references.

One of the cost-effective methods of increasing the global capacity of the structure is to grout the annulus between the jacket legs and piles. The grout, pile and legs will work as a composite section. The effect can be especially pronounced on jackets that have skirt piles, because the increased leg stiffness will tend to take the load from the skirt piles and move it to the jacket main piles. The grout, in effect, mobilizes the pile cross-section and forces the jacket leg and pile to act compositely against joint ovalization, thereby increasing joint capacity for both compression and tension loads.

The grout causes the pile and jacket leg to act compositely, which equalizes the stresses within the two members.

The grouting of the annulus between the jacket legs and piles has the added benefit of locally strengthening the jacket joints for bracing loads.

Two issues that should be considered before grouting main piles are the impact on the platform in case of decommissioning and the increase in dynamic mass.

Member Flooding

Intentional flooding of structural members that are subjected to combined structural and hydrostatic loading can be used to increase the load-carrying capacity of the member. The impact of the increased gravity loads and dynamic mass should be considered in the assessment process, as well as possible decommissioning implications.

8.8 OCCURRENCE OF MEMBER FAILURES WITH TIME

The detection of damage is almost entirely a function of when a weld or member is inspected; damage could exist for many years until discovered by the next programmed inspection. Fatigue lives should not be inferred from inspection results unless the inspection intervals are very short. The occurrence of accidental damage is easier to identify because an accidental event will be noted and in most cases investigated.

A review of the data in the context of when the damage was discovered is informative and demonstrates trends in failure rates.

A study of platforms installed between 1966 and 1977 in the southern sector of the North Sea suggests that the platforms may have been operated for 40 years before damage occurred. However, a major inspection campaign undertaken in the 1980s discovered damage and it is unknown when the members actually cracked.

Structures that were installed from 1971 to 1980 in deep water in central and northern sectors of the North Sea continue to sustain through-thickness damage. A proposed explanation for this is that, while fabrication defects resulted in cracking in the first few years of life that was discovered, assumptions and errors made during design yielded structural defects and stresses that are resulting in ongoing fatigue damage. Structures from this period were designed without the detailed knowledge available at present, particularly analysis software and fatigue design factors.

Structures installed from 1981 to 1985 suggest that an unforeseen problem, such as a fabrication defect, will be exposed at an early stage of the structure's operational life, before it settles into a period when little or no damage occurs unexpectedly. However, with improved analytical techniques and detailing, long-term fatigue performance has improved dramatically and has resulted in no damage being noted after initial fabrication-defect problems have shaken out.

No gross damage has been noted in structures installed between 1986 and 1995; the explanation appears to be good fabrication and design practices and the use of NDT. However, newer structures have seen less service life than older structures.

BIBLIOGRAPHY

API, 1996. Base Resource Document—Risk-Based Inspection—Preliminary Draft, API 581. American Petroleum Institute, Washington DC.

API RP2A-SIM, 2006. Structural Integrity Management of Fixed Offshore Structures. American Petroleum Institute, Washington DC.

API RP2A-WSD, 2000. Recommended Practice for Planning, Designing, and Constructing Fixed Offshore Platforms, Twentieth ed. American Petroleum Institute, Washington DC. Supplement 1.

Bea, R.G., Puskar, F.J., Smith, C., Spencer, J.S., 1988. Development of AIM (Assessment, Inspection, Maintenance) Programs for Fixed and Mobile Platforms. In: Offshore Technology Conference, OTC Paper 5703 Houston, TX.

DeFranco, S., O'Connor, P., Andy, Tallin, Roy, R., Puskar, F., 1999. Development of a Risk Based Underwater Inspection (RBUI) Process for Prioritizing Inspections of Large Numbers of Platforms. In: Offshore Technology Conference, OTC10846, Houston, TX.

Det Norske Veritas (USA) Inc. (DNV), 2002. Risk Based Underwater Inspection—Phase II, for Amoco, Houston. Document Number 221-8823, 8 February 2002.

El-Reedy, M.A., 2003. Life-Cycle Cost Design of Deteriorating Offshore Structures. In: Offshore Mediterranian Conference Proceeding, OMC2003, Ravenna, Italy.

Gebara, J.M., Westlake, H., DeFranco, S., O'Connor, P., 1998. Influence of framing configuration on the robustness of offshore structures. Offshore Technology Conference, OTC Paper 8736, Houston TX.

http://www.telegraph.co.uk/finance/newsbysector/energy/oilandgas/8462057/BPs-Gulf-of-Mexico-oil-spill-bill-could-hit-60bn-Moodys-warns.html, 2011.

PMB Engineering, Inc., 1988. AIM—Assessment, Inspection and Maintenance, Phase III Final Report. San Francisco, CA.

Puskar, F.J., Aggarwal, R.K., Cornell, C.A., Moses, F., Petrauskas, C., 1994. A comparison of Analytically Predicted Platform Damage to Actual Damage During Hurricane Andrew. In: Offshore Technology Conference, OTC Paper 7473, Houston, TX.

Stahl, B., 1986. Reliability engineering and risk analysis. MCClelland, B., Reifel, M.D. (Eds.), Planning and Designing of Fixed Offshore Platforms, Van Nostrand Reinhold, New York.

Stahl, B., De Franco, S. Internal Presentation Material Developed by Amoco, Amoco WE&C, Houston, TX.

Tallin, A.G., Puskar, F.J, Matos, S., 1998. Risk-Based Underwater Inspection, Report for Amoco WE&C, DNV Materials Engineering Consulting Report 232-8507.

Gonen, T., 1990. Engineering Economy for Engineering Managers. John Wiley and Sons, Hoboken, NJ.

Index

Page numbers in *italics* indicate figures and tables

A

D

G

H

M

N

O

P

Q

R

S

Kolibri 360 GmbH
Ferdinand-Jühlke-Straße 7
99095 Erfurt, Deutschland
produktsicherheit@kolibri360.de